AF705220

High-Gain Observers in Nonlinear Feedback Control

Advances in Design and Control

SIAM's Advances in Design and Control series consists of texts and monographs dealing with all areas of design and control and their applications. Topics of interest include shape optimization, multidisciplinary design, trajectory optimization, feedback, and optimal control. The series focuses on the mathematical and computational aspects of engineering design and control that are usable in a wide variety of scientific and engineering disciplines.

Editor-in-Chief

Ralph C. Smith, North Carolina State University

Editorial Board

Siva Banda, Air Force Research Laboratory
Stephen L. Campbell, North Carolina State University
Michel C. Delfour, University of Montreal
Fariba Fahroo, Air Force Office of Scientific Research
J. William Helton, University of California, San Diego
Arthur J. Krener, University of California, Davis
Kirsten Morris, University of Waterloo
John Singler, Missouri University of Science and Technology
Stefan Volkwein, Universität Konstanz

Series Volumes

Khalil, Hassan K., *High-Gain Observers in Nonlinear Feedback Control*
Bauso, Dario, *Game Theory with Engineering Applications*
Corless, M., King, C., Shorten, R., and Wirth, F., *AIMD Dynamics and Distributed Resource Allocation*
Walker, Shawn W., *The Shapes of Things: A Practical Guide to Differential Geometry and the Shape Derivative*
Michiels, Wim and Niculescu, Silviu-Iulian, *Stability, Control, and Computation for Time-Delay Systems: An Eigenvalue-Based Approach, Second Edition*
Narang-Siddarth, Anshu and Valasek, John, *Nonlinear Time Scale Systems in Standard and Nonstandard Forms: Analysis and Control*
Bekiaris-Liberis, Nikolaos and Krstic, Miroslav, *Nonlinear Control Under Nonconstant Delays*
Osmolovskii, Nikolai P. and Maurer, Helmut, *Applications to Regular and Bang-Bang Control: Second-Order Necessary and Sufficient Optimality Conditions in Calculus of Variations and Optimal Control*
Biegler, Lorenz T., Campbell, Stephen L., and Mehrmann, Volker, eds., *Control and Optimization with Differential-Algebraic Constraints*
Delfour, M. C. and Zolésio, J.-P., *Shapes and Geometries: Metrics, Analysis, Differential Calculus, and Optimization, Second Edition*
Hovakimyan, Naira and Cao, Chengyu, *$\mathcal{L}_1$ Adaptive Control Theory: Guaranteed Robustness with Fast Adaptation*
Speyer, Jason L. and Jacobson, David H., *Primer on Optimal Control Theory*
Betts, John T., *Practical Methods for Optimal Control and Estimation Using Nonlinear Programming, Second Edition*
Shima, Tal and Rasmussen, Steven, eds., *UAV Cooperative Decision and Control: Challenges and Practical Approaches*
Speyer, Jason L. and Chung, Walter H., *Stochastic Processes, Estimation, and Control*
Krstic, Miroslav and Smyshlyaev, Andrey, *Boundary Control of PDEs: A Course on Backstepping Designs*
Ito, Kazufumi and Kunisch, Karl, *Lagrange Multiplier Approach to Variational Problems and Applications*
Xue, Dingyü, Chen, YangQuan, and Atherton, Derek P., *Linear Feedback Control: Analysis and Design with MATLAB*
Hanson, Floyd B., *Applied Stochastic Processes and Control for Jump-Diffusions: Modeling, Analysis, and Computation*
Michiels, Wim and Niculescu, Silviu-Iulian, *Stability and Stabilization of Time-Delay Systems: An Eigenvalue-Based Approach*
Ioannou, Petros and Fidan, Barış, *Adaptive Control Tutorial*
Bhaya, Amit and Kaszkurewicz, Eugenius, *Control Perspectives on Numerical Algorithms and Matrix Problems*
Robinett III, Rush D., Wilson, David G., Eisler, G. Richard, and Hurtado, John E., *Applied Dynamic Programming for Optimization of Dynamical Systems*
Huang, J., *Nonlinear Output Regulation: Theory and Applications*
Haslinger, J. and Mäkinen, R. A. E., *Introduction to Shape Optimization: Theory, Approximation, and Computation*
Antoulas, Athanasios C., *Approximation of Large-Scale Dynamical Systems*
Gunzburger, Max D., *Perspectives in Flow Control and Optimization*
Delfour, M. C. and Zolésio, J.-P., *Shapes and Geometries: Analysis, Differential Calculus, and Optimization*
Betts, John T., *Practical Methods for Optimal Control Using Nonlinear Programming*
El Ghaoui, Laurent and Niculescu, Silviu-Iulian, eds., *Advances in Linear Matrix Inequality Methods in Control*
Helton, J. William and James, Matthew R., *Extending H^∞ Control to Nonlinear Systems: Control of Nonlinear Systems to Achieve Performance Objectives*

High-Gain Observers in Nonlinear Feedback Control

Hassan K. Khalil
Michigan State University
East Lansing, Michigan

Society for Industrial and Applied Mathematics
Philadelphia

Copyright © 2017 by the Society for Industrial and Applied Mathematics

10 9 8 7 6 5 4 3 2 1

All rights reserved. Printed in the United States of America. No part of this book may be reproduced, stored, or transmitted in any manner without the written permission of the publisher. For information, write to the Society for Industrial and Applied Mathematics, 3600 Market Street, 6th Floor, Philadelphia, PA 19104-2688 USA.

Trademarked names may be used in this book without the inclusion of a trademark symbol. These names are used in an editorial context only; no infringement of trademark is intended.

MATLAB is a registered trademark of The MathWorks, Inc. For MATLAB product information, please contact The MathWorks, Inc., 3 Apple Hill Drive, Natick, MA 01760-2098 USA, 508-647-7000, Fax: 508-647-7001, *info@mathworks.com*, *www.mathworks.com*.

Simulink is a registered trademark of The MathWorks, Inc.

Publisher	David Marshall
Acquisitions Editor	Elizabeth Greenspan
Developmental Editor	Gina Rinelli Harris
Managing Editor	Kelly Thomas
Production Editor	Louis R. Primus
Copy Editor	Bruce Owens
Production Manager	Donna Witzleben
Production Coordinator	Cally Shrader
Compositor	Lumina Datamatics
Graphic Designer	Lois Sellers

Library of Congress Cataloging-in-Publication Data

Names: Khalil, Hassan K., 1950-
Title: High-gain observers in nonlinear feedback control / Hassan K. Khalil, Michigan State University, East Lansing, Michigan.
Description: Philadelphia : Society for Industrial and Applied Mathematics, 2017. | Series: Advances in design and control ; 31 | Includes bibliographical references and index.
Identifiers: LCCN 2017012787 (print) | LCCN 2017016390 (ebook) | ISBN 9781611974867 (ebook) | ISBN 9781611974850 (print)
Subjects: LCSH: Observers (Control theory) | Feedback control systems. | Stochastic control theory.
Classification: LCC QA402.3 (ebook) | LCC QA402.3 .K4287 2017 (print) | DDC 629.8/314–dc23
LC record available at https://lccn.loc.gov/2017012787

SIAM is a registered trademark.

Contents

Preface

This book is the result of over twenty-five years of research on high-gain observers in nonlinear feedback control. Most of the results of the book were generated by my research team at Michigan State University. I am indebted to my students and collaborators who worked with me on these problems. I am also indebted to many colleagues whose work on high-gain observers motivated many of our results. I am grateful to Michigan State University for an environment that allowed me to write the book and to the National Science Foundation for supporting my research.

The book was typeset using LaTeX. Computations were done using MATLAB and Simulink. The figures were generated using MATLAB or the graphics tool of LaTeX.

Hassan Khalil

Chapter 1

Introduction

1.1 ▪ Brief History

The use of high-gain observers in feedback control appeared first in the context of linear feedback as a tool for robust observer design. In their celebrated work on loop transfer recovery [36], Doyle and Stein used high-gain observers to recover, with observers, frequency-domain loop properties achieved by state feedback. The investigation of high-gain observers in the context of robust linear control continued in the 1980s, as seen in the work of Petersen and Hollot [116] on H_∞ control. The use of high-gain observers in nonlinear feedback control started to appear in the late 1980s in the work of Saberi [86, 129], Tornambe [151], and Khalil [37]. Two key papers, published in 1992, represent the beginning of two schools of research on high-gain observers. The work by Gauthier, Hammouri, and Othman [50] started a line of work that is exemplified by [21, 25, 35, 49, 51, 57, 154]. This line of research covered a wide class of nonlinear systems and obtained global results under global growth conditions. The work by Esfandiari and Khalil [39] brought attention to the peaking phenomenon as an important feature of high-gain observers. While this phenomenon was observed earlier in the literature [109, 117], the paper [39] showed that the interaction of peaking with nonlinearities could induce finite escape time. In particular, it showed that, in the lack of global growth conditions, high-gain observers could destabilize the closed-loop system as the observer gain is driven sufficiently high. It proposed a seemingly simple solution for the problem. It suggested that the control should be designed as a globally bounded function of the state estimates so that it saturates during the peaking period. Because the observer is much faster than the closed-loop dynamics under state feedback, the peaking period is very short relative to the time scale of the plant variables, which remain very close to their initial values. Teel and Praly [149, 150] built on the ideas of [39] and earlier work by Tornambe [152] to prove the first nonlinear separation principle and develop a set of tools for semiglobal stabilization of nonlinear systems. Their work drew attention to [39], and soon afterward many leading nonlinear control researchers started using high-gain observers; cf. [5, 8, 26, 54, 62, 65, 67, 71, 72, 89, 94, 98, 99, 102, 103, 118, 119, 123, 128, 133, 138, 139, 141, 158]. These papers have studied a wide range of nonlinear control problems, including stabilization, regulation, tracking, and adaptive control. They also explored the use of time-varying high-gain observers. Khalil and his coworkers continued to investigate high-gain observers in nonlinear feedback control for about twenty-five years converging a wide range of problems; cf. [1, 2, 4, 32, 46, 73, 74, 76, 83, 104, 113, 134, 135, 136, 140, 153]. Atassi and Khalil [18] proved a separation principle that adds a new dimension to the result of Teel and Praly [149]; namely, the combination of fast observer with control saturation enables the output feedback controller to recover the trajectories of the state feedback controller as the observer gain is made sufficiently high.

1.2 ▪ Motivating Examples

We use a few examples to motivate the design of high-gain observers and discuss their main features. Example 1.1 illustrates the robustness of the observer and its peaking phenomenon. Example 1.2 shows that the interaction of peaking and nonlinearities can lead to a finite escape time. It shows also how saturating the controller overcomes the destabilizing effect of peaking. An important feature of high-gain observers in feedback control is illustrated in this example, where it is shown that the output feedback controller recovers the performance of the state feedback controller as the observer's dynamics become sufficiently fast. This performance recovery property is due to the combined effect of a fast observer and a saturated control. Example 1.3 shows that this property does not hold by designing a fast observer without saturated control. Example 1.4 introduces the reduced-order high-gain observer and compares it with the full-order one.

Example 1.1. Consider the two-dimensional system

$$\dot{x}_1 = x_2, \qquad \dot{x}_2 = \phi(x,u), \qquad y = x_1,$$

where $x = \text{col}(x_1, x_2)$, ϕ is locally Lipschitz, and $x(t)$ and $u(t)$ are bounded for all $t \geq 0$. To estimate x, we use the observer

$$\dot{\hat{x}}_1 = \hat{x}_2 + h_1(y - \hat{x}_1), \qquad \dot{\hat{x}}_2 = \phi_0(\hat{x}, u) + h_2(y - \hat{x}_1),$$

where $\phi_0(x,u)$ is a nominal model $\phi(x,u)$. We can take $\phi_0 = 0$, which simplifies the observer to a linear one. Whatever the choice of ϕ_0 is, we assume that

$$|\phi_0(z,u) - \phi(x,u)| \leq L\|x - z\| + M$$

for some nonnegative constants L and M, for all (x,z,u) in the domain of interest.[1] In the special case when $\phi_0 = \phi$ and ϕ is Lipschitz in x uniformly in u, the foregoing inequality holds with $M = 0$. The estimation error $\tilde{x} = x - \hat{x}$ satisfies the equation

$$\dot{\tilde{x}} = A_o\tilde{x} + B\delta(x, \tilde{x}, u), \qquad \text{where} \qquad A_o = \begin{bmatrix} -h_1 & 1 \\ -h_2 & 0 \end{bmatrix}, \quad B = \begin{bmatrix} 0 \\ 1 \end{bmatrix},$$

and $\delta(x, \tilde{x}, u) = \phi(x,u) - \phi_0(\hat{x}, u)$. We view this equation as a perturbation of the linear system $\dot{\tilde{x}} = A_o\tilde{x}$. In the absence of δ, asymptotic error convergence is achieved by designing $H = \text{col}(h_1, h_2)$ such that A_o is Hurwitz. In the presence of δ, we need to design H with the additional goal of rejecting the effect of δ on $\tilde{x}$. This is ideally achieved, for any δ, if the transfer function from δ to $\tilde{x}$,

$$G_o(s) = \frac{1}{s^2 + h_1 s + h_2}\begin{bmatrix} 1 \\ s + h_1 \end{bmatrix},$$

[1] Throughout the book, $\|x\| = \sqrt{x^T x}$.

is identically zero. While this is not possible, we can make $\sup_{\omega\in R}\|G_o(j\omega)\|$ arbitrarily small by choosing $h_2 \gg h_1 \gg 1$. In particular, taking

$$h_1 = \frac{\alpha_1}{\varepsilon}, \qquad h_2 = \frac{\alpha_2}{\varepsilon^2}$$

for some positive constants α_1, α_2, and ε, with $\varepsilon \ll 1$, it can be shown that

$$G_o(s) = \frac{\varepsilon}{(\varepsilon s)^2 + \alpha_1 \varepsilon s + \alpha_2}\begin{bmatrix} \varepsilon \\ \varepsilon s + \alpha_1 \end{bmatrix}.$$

Hence, $\lim_{\varepsilon\to 0} G_o(s) = 0$. This disturbance rejection property of the high-gain observer can be seen in the time domain by scaling the estimation error. Let

$$\eta_1 = \frac{\tilde{x}_1}{\varepsilon}, \qquad \eta_2 = \tilde{x}_2. \tag{1.1}$$

Then

$$\varepsilon\dot{\eta} = F\eta + \varepsilon B\delta, \qquad \text{where} \quad F = \begin{bmatrix} -\alpha_1 & 1 \\ -\alpha_2 & 0 \end{bmatrix}. \tag{1.2}$$

The matrix F is Hurwitz because α_1 and α_2 are positive. The matrices A_o and F/ε are related by the similarity transformation (1.1). Therefore, the eigenvalues of A_o are $1/\varepsilon$ times the eigenvalues of F. From (1.2) and the change of variables (1.1), we can make some observations about the behavior of the estimation error. Using the bound $|\delta| \le L\|\tilde{x}\| + M \le L\|\eta\| + M$ and the Lyapunov function $V = \eta^T P\eta$, where P is the solution of $PF + F^T P = -I$, we obtain

$$\varepsilon\dot{V} = -\eta^T\eta + 2\varepsilon\eta^T PB\delta \le -\|\eta\|^2 + 2\varepsilon L\|PB\|\,\|\eta\|^2 + 2\varepsilon M\|PB\|\,\|\eta\|.$$

For $\varepsilon L\|PB\| \le \frac{1}{4}$,

$$\varepsilon\dot{V} \le -\tfrac{1}{2}\|\eta\|^2 + 2\varepsilon M\|PB\|\,\|\eta\|.$$

Therefore (see [80, Theorem 4.5]), $\|\eta\|$, and consequently $\|\tilde{x}\|$, is ultimately bounded by εcM for some $c > 0$, and

$$\|\eta(t)\| \le \max\left\{be^{-at/\varepsilon}\|\eta(0)\|,\ \varepsilon cM\right\} \qquad \forall\, t \ge 0$$

for some positive constants a and b. This inequality and the scaling (1.1) show that

$$|\tilde{x}_1| \le \max\left\{be^{-at/\varepsilon}\|x(0)\|,\ \varepsilon^2 cM\right\}, \qquad |\tilde{x}_2| \le \max\left\{\frac{b}{\varepsilon}e^{-at/\varepsilon}\|x(0)\|,\ \varepsilon cM\right\}.$$

Hence, $\|\tilde{x}(t)\|$ approaches the ultimate bound exponentially fast, and the smaller ε, the faster the rate of decay, which shows that for sufficiently small ε the estimation error $\tilde{x}$ will be much faster than x. The ultimate bound can be made arbitrarily small by choosing ε small enough. If $M = 0$, which is the case when $\phi_0 = \phi$, then $\tilde{x}(t)$ converges to zero as t tends to infinity. Notice, however, that whenever $x_1(0) \ne \hat{x}_1(0)$, $\eta_1(0) = O(1/\varepsilon)$.[2] Consequently, the solution of (1.2) will contain a term of the form $(1/\varepsilon)e^{-at/\varepsilon}$ for some $a > 0$. While this exponential mode decays rapidly for small ε, it exhibits an impulsive-like behavior where the transient response peaks to $O(1/\varepsilon)$ values before it decays rapidly toward zero. In fact, the function $(a/\varepsilon)e^{-at/\varepsilon}$

[2] A function $f(\varepsilon)$ is $O(\varepsilon^n)$ if $|f(\varepsilon)| \le k\varepsilon^n$ for some $k > 0$, where n can be positive or negative.

approaches an impulse function as ε tends to zero. This behavior is known as the *peaking phenomenon*. It has a serious impact when the observer is used in feedback control, as we shall see in the next example. We use numerical simulation to illustrate the foregoing observations. Consider the system

$$\dot{x}_1 = x_2, \qquad \dot{x}_2 = -x_1 - 2x_2 + a x_1^2 x_2 + b \sin 2t, \qquad y = x_1,$$

with $a = 0.25$ and $b = 0.2$. It can be shown that for all $x(0) \in \Omega = \{1.5x_1^2 + x_1x_2 + 0.5x_2^2 \leq \sqrt{2}\}$, $x(t)$ is bounded [80, Example 11.1]. We use the high-gain observer

$$\dot{\hat{x}}_1 = \hat{x}_2 + \frac{2}{\varepsilon}(y - \hat{x}_1), \qquad \dot{\hat{x}}_2 = -\hat{x}_1 - 2\hat{x}_2 + \hat{a}\hat{x}_1^2\hat{x}_2 + \hat{b}\sin 2t + \frac{1}{\varepsilon^2}(y - \hat{x}_1),$$

with two different choices of the pair $(\hat{a}, \hat{b})$. When $a = 0.25$ and $b = 0.2$ are known, we take $\hat{a} = 0.25$ and $\hat{b} = 0.2$. This is a case with no model uncertainty and $\phi_0 = \phi$. The other case is when the coefficients a and b are unknown. In this case we take $\hat{a} = \hat{b} = 0$. Figure 1.1 shows simulation results for both cases. Figures 1.1(a) and (b) show the estimation errors $\tilde{x}_1$ and $\tilde{x}_2$ in the no-uncertainty case for different values of ε. The estimation error $\tilde{x}_2$ illustrates the peaking phenomenon. We note that the peaking phenomenon is not present in $\tilde{x}_1$. While peaking is induced by $\tilde{x}_1(0)$, it does not appear in $\tilde{x}_1$ because $\tilde{x}_1 = \varepsilon\eta_1$. Figures 1.1(c) and (d) show $\tilde{x}_2$ for the uncertain model case when $\hat{a} = \hat{b} = 0$. Comparison of Figures 1.1(b) and (c) shows that the presence of uncertainty has very little effect on the performance of the observer when ε is sufficiently small. Figure 1.1(d) demonstrates the fact that the ultimate bound on $\tilde{x}_2$ is $O(\varepsilon)$. $\triangle$

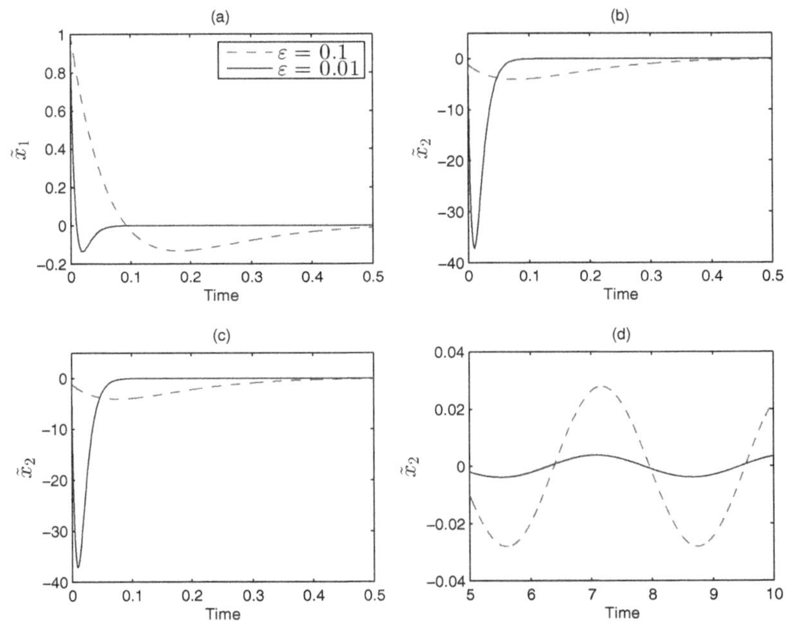

Figure 1.1. *Simulation of Example* 1.1. *Figures* (a) *and* (b) *show the estimation errors* $\tilde{x}_1$ *and* $\tilde{x}_2$ *in the case* $\hat{a} = a$ *and* $\hat{b} = b$. *Figures* (c) *and* (d) *show the transient and steady-state behavior of* $\tilde{x}_2$ *in the case* $\hat{a} = \hat{b} = 0$.

Example 1.2. Consider the two-dimensional system

$$\dot{x}_1 = x_2, \qquad \dot{x}_2 = \phi(x,u), \qquad y = x_1,$$

where $x = \text{col}(x_1, x_2)$. Suppose $u = \gamma(x)$ is a locally Lipschitz state feedback controller that stabilizes the origin of the closed-loop system

$$\dot{x}_1 = x_2, \qquad \dot{x}_2 = \phi(x, \gamma(x)).$$

To implement this control with output feedback, we use the high-gain observer

$$\dot{\hat{x}}_1 = \hat{x}_2 + (\alpha_1/\varepsilon)(y - \hat{x}_1), \qquad \dot{\hat{x}}_2 = \phi_0(\hat{x}, u) + (\alpha_2/\varepsilon^2)(y - \hat{x}_1),$$

where ϕ_0 is a nominal model of ϕ, and α_1, α_2, and ε are positive constants with $\varepsilon \ll 1$. We saw in the previous example that if

$$|\phi_0(z,u) - \phi(x,u)| \le L\,\|x - z\| + M$$

over the domain of interest, then for sufficiently small ε, the estimation errors $\tilde{x}_1 = x_1 - \hat{x}_1$ and $\tilde{x}_2 = x_2 - \hat{x}_2$ satisfy the inequalities

$$|\tilde{x}_1| \le \max\left\{ b e^{-at/\varepsilon} \|\tilde{x}(0)\|,\ \varepsilon^2 c M \right\}, \qquad |\tilde{x}_2| \le \left\{ \frac{b}{\varepsilon} e^{-at/\varepsilon} \|\tilde{x}(0)\|,\ \varepsilon c M \right\}$$

for some positive constants a, b, c. These inequalities show that reducing ε diminishes the effect of model uncertainty and makes $\tilde{x}$ much faster than x. The bound on $\tilde{x}_2$ demonstrates the peaking phenomenon; namely, $\tilde{x}_2$ might peak to $O(1/\varepsilon)$ values before it decays rapidly toward zero. The peaking phenomenon might destabilize the closed-loop system. This fact is illustrated by simulating the system

$$\dot{x}_1 = x_2, \qquad \dot{x}_2 = x_2^3 + u, \qquad y = x_1,$$

which can be globally stabilized by the state feedback controller

$$u = -x_2^3 - x_1 - x_2.$$

The output feedback controller is taken as

$$u = -\hat{x}_2^3 - \hat{x}_1 - \hat{x}_2, \qquad \dot{\hat{x}}_1 = \hat{x}_2 + (2/\varepsilon)(y - \hat{x}_1), \qquad \dot{\hat{x}}_2 = (1/\varepsilon^2)(y - \hat{x}_1),$$

where we take $\phi_0 = 0$. Figure 1.2 shows the performance of the closed-loop system under state and output feedback. Output feedback is simulated for three different values of ε. The initial conditions are $x_1(0) = 0.1$, $x_2(0) = \hat{x}_1(0) = \hat{x}_2(0) = 0$. Peaking is induced by $[x_1(0) - \hat{x}_1(0)]/\varepsilon = 0.1/\varepsilon$ when ε is sufficiently small. Figure 1.2 shows a counterintuitive behavior as ε decreases. Since decreasing ε causes the estimation error to decay faster toward zero, one would expect the response under output feedback to approach the response under state feedback as ε decreases. Figure 1.2 shows the opposite behavior, where the response under output feedback deviates from the response under state feedback as ε decreases. This is the impact of the peaking phenomenon. The same figure shows the control u on a much shorter time interval to exhibit peaking. Figure 1.3 shows that as we decrease ε to 0.004, the system has a finite escape time shortly after $t = 0.07$.

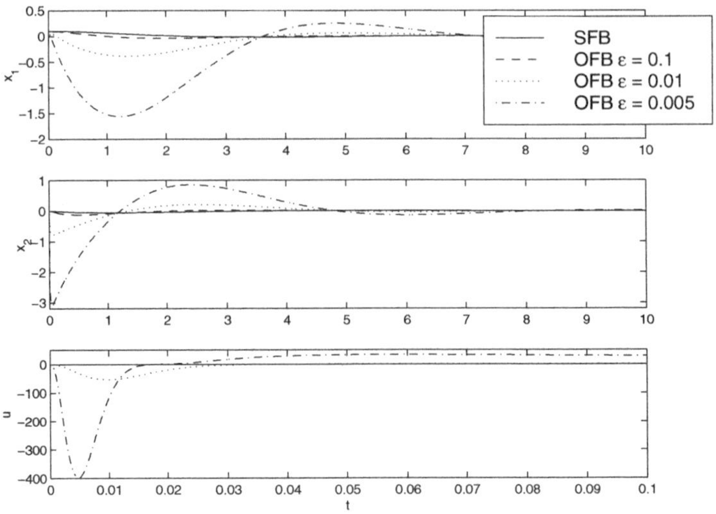

Figure 1.2. *Simulation of Example* 1.2. *Performance under state (SFB) and output (OFB) feedback. Reprinted with permission of Pearson Education, Inc., New York, New York* [80].

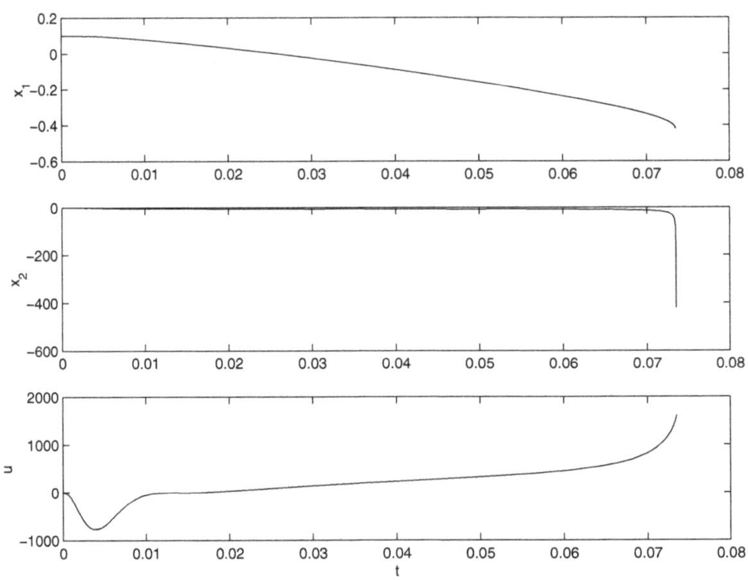

Figure 1.3. *Simulation of Example* 1.2. *Instability induced by peaking at* $\varepsilon = 0.004$. *Reprinted with permission of Pearson Education, Inc., New York, New York* [80].

Fortunately, we can isolate the observer peaking from the plant by saturating the control outside a compact set of interest. Writing the closed-loop system under state feedback as $\dot{x} = Ax$ and solving the Lyapunov equation $PA + A^T P = -I$, we have

$$A = \begin{bmatrix} 0 & 1 \\ -1 & -1 \end{bmatrix} \quad \text{and} \quad P = \begin{bmatrix} 1.5 & 0.5 \\ 0.5 & 1 \end{bmatrix}.$$

Then $V(x) = x^T P x$ is a Lyapunov function for $\dot{x} = Ax$ and $\dot{V}(x) = -x^T x$. Suppose all initial conditions of interest belong to the set $\Omega = \{V(x) \leq 0.3\}$. Because Ω is positively invariant, $x(t) \in \Omega$ for all $t \geq 0$. By maximizing $|x_1 + x_2|$ and $|x_2|$ over

Ω, it can be shown that for all $x \in \Omega$, $|x_1 + x_2| \leq 0.6$ and $|x_2| \leq 0.6$. Hence, $|u| \leq |x_2|^3 + |x_1 + x_2| \leq 0.816$. Saturating u at ± 1 results in the globally bounded state feedback control

$$u = \text{sat}(-x_2^3 - x_1 - x_2),$$

where $\text{sat}(y) = \min\{|y|, 1\}\,\text{sign}(y)$. For all $x(0) \in \Omega$, the saturated control produces the same trajectories as the unsaturated one because for $x \in \Omega$, $|u| < 1$ and the saturation is not active. In output feedback, the state x in the foregoing saturated control is replaced by its estimate $\hat{x}$, namely,

$$u = \text{sat}(-\hat{x}_2^3 - \hat{x}_1 - \hat{x}_2).$$

During the peaking period, the control saturates. Figure 1.4 shows the performance of the closed-loop system under saturated state and output feedback. The control u is shown on a shorter time interval that exhibits control saturation during peaking. The peaking period decreases with ε. The states x_1 and x_2 exhibit the intuitive behavior we expected earlier; namely, the response under output feedback approaches the response under state feedback as ε decreases. Note that we decrease ε to 0.001 beyond the value 0.004 where instability was detected in the unsaturated case. Not only does the system remain stable, but the response under output feedback is almost indistinguishable from the response under state feedback. What is also interesting is that the region of attraction under output feedback approaches the region of attraction under saturated state feedback as ε tends to zero. This is shown in Figures 1.5 and 1.6. The first figure shows the phase portrait of the closed-loop system under $u = \text{sat}(-x_2^3 - x_1 - x_2)$. It has a bounded region of attraction enclosed by a limit cycle. The second figure shows the intersection of the boundary of the region of attraction under $u = \text{sat}(-\hat{x}_2^3 - \hat{x}_1 - \hat{x}_2)$ with the x_1–x_2 plane. It approaches the limit cycle as ε tends to zero.

The behavior we see in Figures 1.4 and 1.6 will be realized with any globally bounded stabilizing function $\gamma(x)$. During the peaking period, the control $\gamma(\hat{x})$ saturates. Since the peaking period shrinks to zero as ε tends to zero, for sufficiently small ε the peaking period becomes so small that the state of the plant x remains close

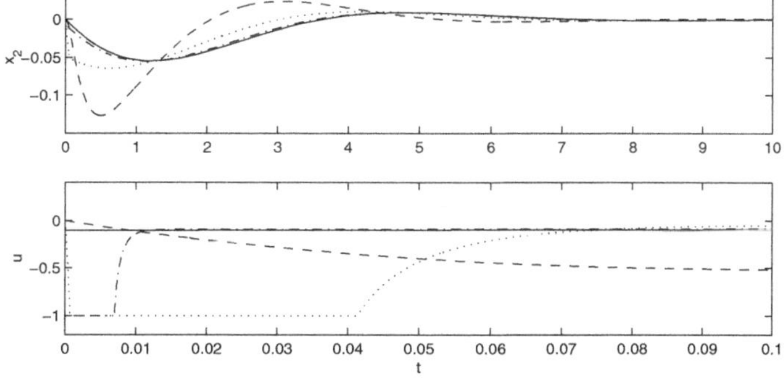

Figure 1.4. *Simulation of Example* 1.2. *Performance under state (SFB) and output (OFB) feedback with saturation. Reprinted with permission of Pearson Education, Inc., New York, New York* [80].

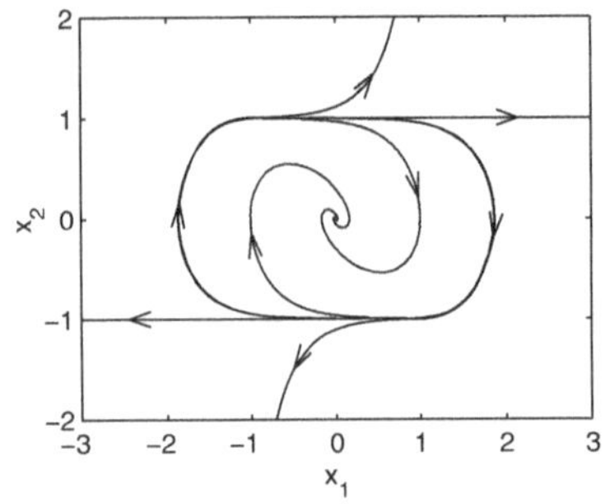

Figure 1.5. *Simulation of Example 1.2. Phase portrait of the closed-loop system under* $u = \text{sat}(-x_2^3 - x_1 - x_2)$. *Reprinted with permission of Pearson Education, Inc., New York, New York* [80].

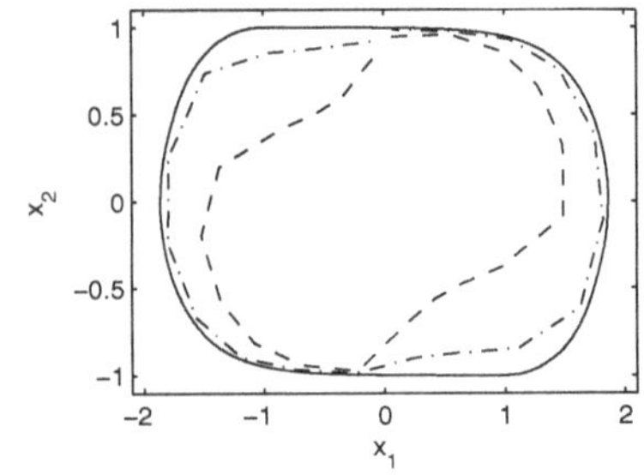

Figure 1.6. *Simulation of Example 1.2. Output feedback with* $\varepsilon = 0.08$ *(dashed) and* $\varepsilon = 0.01$ *(dash-dot). Reprinted with permission of Pearson Education, Inc., New York, New York* [80].

to its initial value. After the peaking period, the estimation error becomes $O(\varepsilon)$ and the feedback control $\gamma(\hat{x})$ becomes close to $\gamma(x)$. Consequently, the trajectories of the closed-loop system under output feedback asymptotically approach its trajectories under state feedback as ε tends to zero. This leads to recovery of the performance achieved under state feedback. The global boundedness of $\gamma(x)$ can be always achieved by saturating the state feedback control, or the state estimates, outside a compact set of interest. $\triangle$

The foregoing example shows that the design of the output feedback controller is based on a separation procedure, whereby the state feedback controller is designed as if the whole state was available for feedback, followed by an observer design that is independent of the state feedback design. By choosing ε small enough, the output feedback controller recovers the stability and performance properties of the state feedback controller. This is the essence of the separation principle that is presented in Section 3.1. The separation principle is known in the context of linear systems where the closed-loop eigenvalues under an observer-based controller are the union of the eigenvalues under state feedback and the observer eigenvalues; hence stabilization under output feedback can be achieved by solving separate eigenvalue placement problems for the state feedback and the observer. Over the last two decades there have been several results that present forms of the separation principle for classes of nonlinear systems. It is important to emphasize that the separation principle in the case of high-gain observers has a unique feature that does not exist in other separation-principle results, linear systems included, and that is the recovery of state trajectories by making the observer sufficiently fast. This feature has significant practical implications because it allows the designer to design the state feedback controller to meet transient response specification and/or constraints on the state or control variables. Then, by saturating the state estimate $\hat{x}$ and/or the control u outside compact sets of interest to make $\gamma(\hat{x})$ and $\phi_0(\hat{x}, u)$ globally bounded in $\hat{x}$, the designer can proceed to tune the parameter ε by decreasing it monotonically to bring the trajectories under output feedback close enough to the ones under state feedback. This feature is achieved not only by making the observer fast but also by combining this property with the global boundedness of γ and ϕ_0. We illustrate this point in the example to follow.

Example 1.3. Consider the linear system

$$\dot{x}_1 = x_2, \qquad \dot{x}_2 = u, \qquad y = x_1,$$

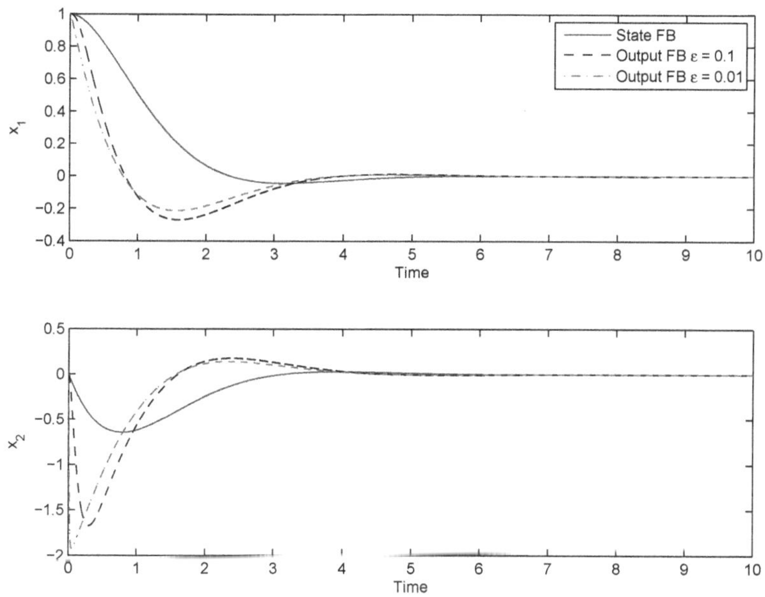

Figure 1.7. *Simulation of Example 1.3. The state trajectories under state and output feedback for linear example with unsaturated control. Reprinted with permission from John Wiley and Sons, Ltd* [84].

which is a special case of the system of Example 1.1 with $\phi = u$. A linear state feedback controller that assigns the eigenvalues at $-1\pm j$ is given by $u = -2x_1 - 2x_2$. A high-gain observer is taken as

$$\dot{\hat{x}}_1 = \hat{x}_2 + (3/\varepsilon)(y - \hat{x}_1), \quad \dot{\hat{x}}_2 = u + (2/\varepsilon^2)(y - \hat{x}_1).$$

It assigns the observer eigenvalues at $-1/\varepsilon$ and $-2/\varepsilon$. The observer-based controller assigns the closed-loop eigenvalues at $-1\pm j, -1/\varepsilon$, and $-2/\varepsilon$. The closed-loop system under output feedback is asymptotically stable for all $\varepsilon > 0$. As we decrease ε, we make the observer dynamics faster than the state dynamics. Will the trajectories of the system under output feedback approach those under state feedback as ε approaches zero? The answer is shown in Figure 1.7, where the state x is shown under state feedback and under output feedback for $\varepsilon = 0.1$ and 0.01. The initial conditions of the simulation are $x_1(0) = 1$ and $x_2(0) = \hat{x}_1(0) = \hat{x}_2(0) = 0$. Contrary to what intuition may suggest, we see that the trajectories under output feedback do not approach the ones under state feedback as ε decreases. In Figure 1.8 we repeat the same simulation when the control is saturated at ± 4, that is, $u = 4\,\text{sat}((-2\hat{x}_1 - 2\hat{x}_2)/4)$. The saturation level 4 is chosen such that $4 > \max_\Omega |-2x_1 - 2x_2|$, where $\Omega = \{1.25x_1^2 + 0.5x_1x_2 + 0.375x_2^2 \le 1.4\}$ is an estimate of the region of attraction under state feedback control that includes the initial state $(1,0)$ in its interior. This choice of the saturation level saturates the control outside Ω. Figure 1.8 shows a reversal of the trend we saw in Figure 1.7. Now the trajectories under output feedback approach those under state feedback as ε decreases. This is a manifestation of the performance recovery property of high-gain observers when equipped with a globally bounded control. Figure 1.9 shows the control signal u with and without saturation during the peaking period for $\varepsilon = 0.01$. It demonstrates the role of saturation in suppressing the peaking phenomenon. △

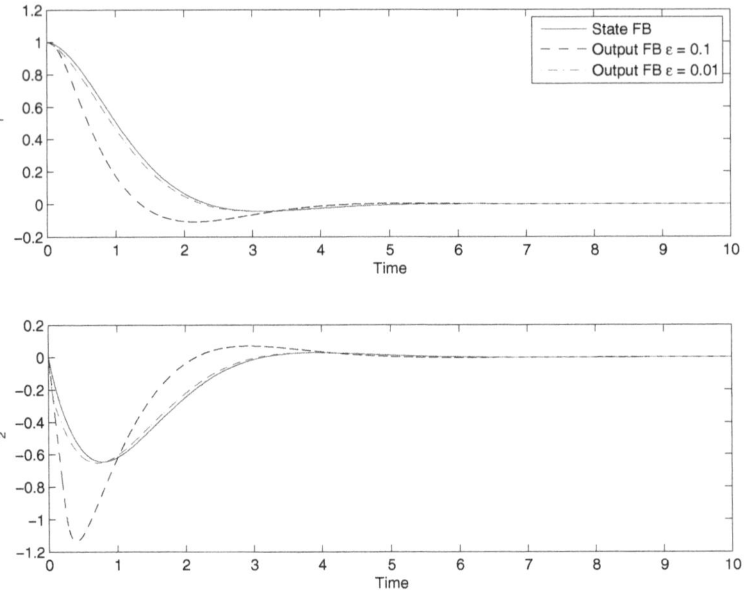

Figure 1.8. *Simulation of Example* 1.3. *The state trajectories under state and output feedback for linear example with saturated control. Reprinted with permission from John Wiley and Sons, Ltd* [84].

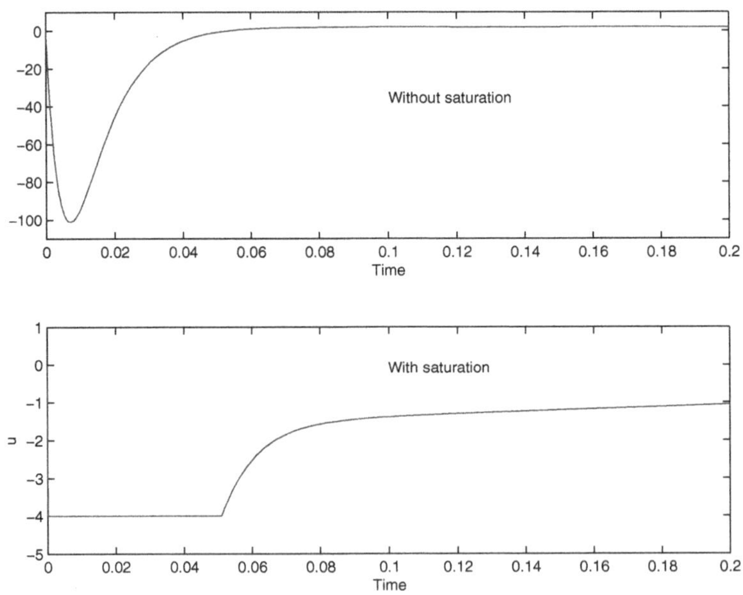

Figure 1.9. *Simulation of Example* 1.3. *The control signal for the linear example with and without control saturation when* $\varepsilon = 0.01$. *Reprinted with permission from John Wiley and Sons, Ltd* [84].

The full-order observer of Example 1.1 estimates both states x_1 and x_2. Since $y = x_1$ is measured, we can design a reduced-order observer of the form

$$\dot{z} = -h(z + hy) + \phi_o(\hat{x}, u), \qquad \hat{x}_2 = z + hy,$$

to estimate $\hat{x}_2$ only. When the observer gain h is chosen as $h = \alpha/\varepsilon$ for some positive constants α and ε with $\varepsilon \ll 1$, it becomes a high-gain observer that has features similar to the full-order high-gain observer. In particular, the estimation error decays rapidly while exhibiting the peaking phenomenon. The example to follow compares the full- and reduced-order observers.

Example 1.4. Consider the van der Pol oscillator

$$\dot{x}_1 = x_2, \qquad \dot{x}_2 = -x_1 + (1 - x_1^2)x_2,$$

and suppose we want to estimate x_2 from a measurement of x_1. We design a full-order high-gain observer of dimension two and a reduced-order high-gain observer of dimension one. In both cases we do not include a model of the function $x_1 + (1 - x_1)^2 x_2$ so that the observers would be linear. The full-order observer is taken as

$$\dot{\hat{x}}_1 = \hat{x}_2 + (2/\varepsilon)(y - \hat{x}_1), \qquad \dot{\hat{x}}_2 = (1/\varepsilon^2)(y - \hat{x}_1),$$

and its transfer function from y to $\hat{x}_2$ is

$$G_2(s) = \frac{s}{(\varepsilon s + 1)^2}.$$

The reduced-order observer is taken as

$$\dot{z}_1 = -(1/\varepsilon)[z_1 + (1/\varepsilon)y], \qquad \hat{x}_2 = z_1 + (1/\varepsilon)y,$$

and its transfer function from y to $\hat{x}_2$ is

$$G_1(s) = \frac{s}{\varepsilon s + 1}.$$

Comparison of the two transfer functions shows that the magnitude of the frequency response of $G_2(s)$ rolls off beyond the cutoff frequency of $1/\varepsilon$, while that of $G_1(s)$ remains constant beyond the cutoff frequency. If the measurement of x_1 is corrupted by noise, that is, $y = x_1 + v$, the full-order observer will have better attenuation of high-frequency noise. For example, if $v(t) = N \sin \omega t$ and $\omega = k/\varepsilon$ with $k \gg 1$, the magnitude of the steady-state component of $\hat{x}_2$ due to noise will be $(kN/\varepsilon)/(1+k^2) \approx N/(\varepsilon k)$ for the full-order observer and $(kN/\varepsilon)/\sqrt{(1+k^2)} \approx N/\varepsilon$ for the reduced-order observer. The high-frequency noise is attenuated much better by the full-order observer. The same observation can be seen from simulation. Figure 1.10 shows the estimation error $\tilde{x}_2 = x_2 - \hat{x}_2$ for the two observers. The simulation is run with $\varepsilon = 0.001$ and initial conditions $x_1(0) = x_2(0) = 1$, $\hat{x}_1(0) = \hat{x}_2(0) = z_1(0) = 0$. The measurement noise is a uniformly distributed random signal with values between ± 0.0001, generated by the Simulink icon "Uniform Random Number" with sample time 0.00001. The estimation error is plotted over the time period $[0, 0.01]$ to show the peaking behavior of $\tilde{x}_2$ and over the period $[9.9, 10]$ to show the steady-state behavior where the effect of noise is significant. In both cases $\tilde{x}_2$ peaks to $O(1/\varepsilon)$ values, but peaking is larger in the reduced-order observer due to the fact that $\hat{x}_2$ has a component proportional to y/ε. What is significant in this simulation is the effect of measurement noise. Comparison of Figures 1.1(b) and (d) shows that the full-order observer attenuates the effect of measurement noise better by an order of magnitude. Moreover, it filters out the high-frequency content of the noise. $\triangle$

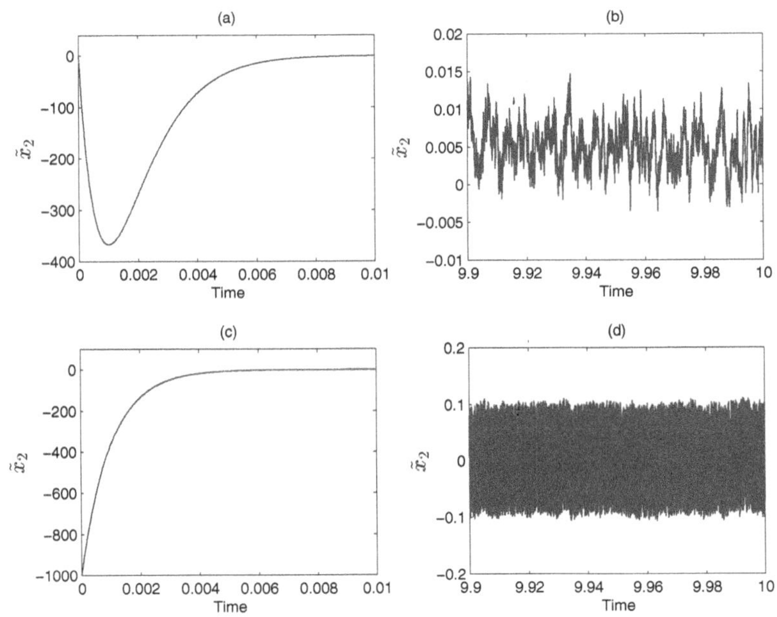

Figure 1.10. *Simulation of Example* 1.4. *Figures* (a) *and* (b) *are for the full-order observer. Figures* (c) *and* (d) *are for the reduced-order observer. All figures show the estimation error* $\tilde{x}_2 = x_2 - \hat{x}_2$.

1.3 ▪ Challenges

1.3.1 ▪ High-Dimensional Observers

A high-gain observer for the system

$$\begin{aligned}\dot{x}_i &= x_{i+1} \quad \text{for } 1 \le i \le \rho - 1,\\ \dot{x}_\rho &= \phi(x, u),\\ y &= x_1\end{aligned}$$

is given by

$$\begin{aligned}\dot{\hat{x}}_i &= \hat{x}_{i+1} + \frac{\alpha_i}{\varepsilon^i}(y - \hat{x}_1) \quad \text{for } 1 \le i \le \rho - 1,\\ \dot{\hat{x}}_\rho &= \phi_0(\hat{x}, u) + \frac{\alpha_\rho}{\varepsilon^\rho}(y - \hat{x}_1),\end{aligned}$$

where ϕ_0 is a nominal model of ϕ, ε is a sufficiently small positive constant, and α_1 to α_ρ are chosen such that the polynomial

$$s^\rho + \alpha_1 s^{\rho-1} + \cdots + \alpha_{\rho-1} s + \alpha_\rho$$

is Hurwitz; that is, its roots have negative real parts. This observer faces a numerical challenge if its dimension, ρ, is high. The observer gains are proportional to powers of $1/\varepsilon$, with $1/\varepsilon^\rho$ as the highest power. During the transient period, the internal states of the observer could peak to large values, which are proportional to powers of $1/\varepsilon$, with $1/\varepsilon^{\rho-1}$ as the highest power. These features pose a challenge in the numerical implementation of the observer when ρ is high because in digital implementation both parameters and signals have to be represented within the finite word length of the digital system. It is worthwhile to note that saturating the state estimates or the control signal before applying the control to the plant does not prevent peaking in the observer's internal variables. To address this numerical challenge, modified high-gain observers are presented in Chapter 7.

1.3.2 ▪ Measurement Noise

The most serious challenge to the application of high-gain observers is measurement noise. This is not surprising because the observer estimates the derivatives of the output. When the output is corrupted by measurement noise, it is expected that noise will have a serious effect on the accuracy of the estimates. The following example explores the effect of measurement noise on a second-order observer.

Example 1.5. Reconsider the system of Example 1.1 and suppose the measurement y is corrupted by bounded noise v; that is,

$$\dot{x}_1 = x_2, \qquad \dot{x}_2 = \phi(x,u), \qquad y = x_1 + v,$$

where $|v(t)| \le N$. Equation (1.2) and the inequality satisfied by $\dot{V}$ change to

$$\varepsilon\dot{\eta} = F\eta + \varepsilon B\delta - \frac{1}{\varepsilon}Ev, \qquad \text{where} \qquad E = \begin{bmatrix} \alpha_1 \\ \alpha_2 \end{bmatrix},$$

and

$$\varepsilon\dot{V} \le -\tfrac{1}{2}\|\eta\|^2 + 2\varepsilon M\|PB\|\,\|\eta\| + \frac{2N}{\varepsilon}\|PE\|\,\|\eta\|.$$

Therefore, the ultimate bound on $\|\tilde{x}\|$ takes the form

$$\|\tilde{x}\| \le c_1 M\varepsilon + \frac{c_2 N}{\varepsilon} \tag{1.3}$$

for some positive constants c_1 and c_2. This inequality shows a trade-off between model uncertainty and measurement noise. An illustration of the ultimate bound in Figure 1.11 shows that decreasing ε reduces the ultimate bound until we reach the value $\varepsilon_1 = \sqrt{c_2N/(c_1M)}$. Reducing ε beyond this point increases the ultimate bound.

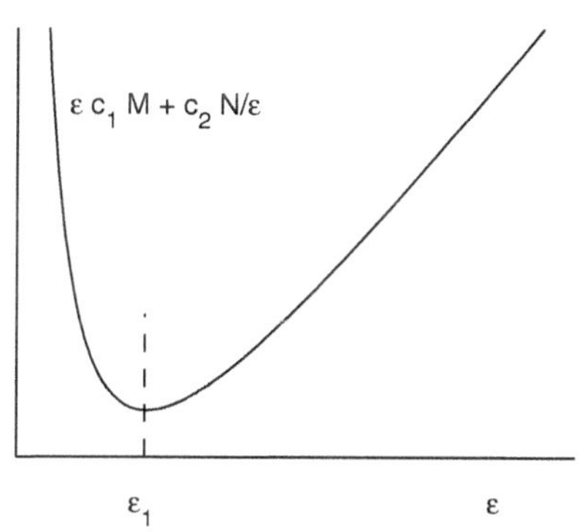

Figure 1.11. *Illustration of the ultimate bound* (1.3).

Thus, the presence of measurement noise puts a lower bound on ε. Another trade-off exists between the speed of convergence of the observer and the ultimate bound on the estimation error. For the separation properties we saw in Example 1.2, we need to choose ε sufficiently small to make the observer dynamics sufficiently faster than the dynamics of the closed-loop system under state feedback. A lower bound on ε limits the speed of convergence. For the high-gain observer to be effective, the ratio N/M should be relatively small so that ε can be chosen to attenuate the effect of uncertainty and make the observer sufficiently fast. Even if there was no model uncertainty, that is, $M = 0$, we still need N to be relatively small so that we can design the observer to be sufficiently fast without bringing the ultimate bound on the estimation error to unacceptable levels. It is worthwhile to note that the bound (1.3) does not take into consideration the low-pass filtering characteristics of the observer at frequencies higher than $1/\varepsilon$, as we saw in Example 1.4. △

Ideas to reduce the effect of measurement noise are presented in Chapter 8.

1.4 ▪ Overview of the Book

Chapter 2 starts by describing the class of nonlinear systems for which high-gain observers are designed. The observer design is then presented in terms of a small positive parameter ε, which can be thought of as the observer's time constant or the reciprocal of the observer's bandwidth when the observer is linear. The observer gains are proportional to negative powers of ε. Theorem 2.1 gives an upper bound on the estimation error, which illustrates three key features of the observer: the peaking phenomenon, the fast decay of the error, and the small ultimate bound on the error. The main step in the observer design is a pole placement problem where the eigenvalues of a matrix are assigned in the left-half plane. This step can be carried out by solving Lyapunov or Riccati equations, as shown in Section 2.3. An interesting observation about the Lyapunov-equation design is that it assigns all the eigenvalues of the observer at $-1/\varepsilon$. For a system with a chain of ρ integrators, the dimension of the observer is ρ. Since the first state of the chain is the output, it is possible to design an observer of dimension $\rho-1$. This reduced-order observer is described in Section 2.4. The chapter concludes by presenting the observer design for a special class of multi-output systems, which covers important physical problems where the measured variables are positions and the estimated variables are velocities, accelerations, etc. References are given for more general classes of multi-output systems.

Chapter 3 deals with stabilization and tracking problems. Theorem 3.1 is the separation principle for the stabilization of time-invariant systems. Its proof is a must-read for anyone who wants to understand how high-gain observers work in feedback control. Elements of that proof are repeated in several proofs throughout the book. While the separation principle is presented only for time-invariant systems, extensions to time-varying systems are referenced. For minimum-phase systems, robust stabilization and tracking problems are presented in Sections 3.2 and 3.3, respectively. In both cases the design of output feedback control follows a separation approach where state feedback control is design first; then its performance is recovered by a high-gain observer. A key idea of Section 3.2 is the use of a high-gain observer to reduce a relative degree-higher-than-one system to a relative-degree-one system, for which the control design is straightforward.

Chapter 4 considers adaptive control of nonlinear systems modeled globally by an nth-order differential equation. The dynamics of the system are extended by adding integrators at the input side, which results in a state model in which the state variables are either states of the added integrators or derivatives of the output. By using a high-gain observer to estimate the derivatives of the output, the output feedback controller can recover the performance of the state feedback controller. A Lyapunov-based state feedback adaptive control is designed in Section 4.2, and its output feedback implementation is given in Section 4.3. Convergence of the tracking error to zero is shown without additional conditions, but convergence of the parameter errors to zero is shown under a persistence of excitation condition. After illustrating the performance of the adaptive controller and the persistence of excitation condition by examples in Section 4.4, robustness to bounded disturbance is shown in Section 4.5. It is shown that the ultimate bound on the tracking error can be made arbitrarily small by adding a robust control component. The robustness properties of the adaptive controller allows its application to nonlinear systems where a nonlinear function is approximated by a function approximator, such as neural network, with a small bounded residual error; this is the subject of Section 4.6.

The nonlinear regulation problem is the subject of Chapter 5. The controller includes a servocompensator that implements an internal model of the exogenous (reference and disturbance) signals. The exogenous signals are generated by a linear exosystem that has simple eigenvalues on the imaginary axis. The problem is treated first for the special case of constant exogenous signals in Sections 5.1 and 5.2; then the more general cases is treated in the rest of the chapter. In addition to the usual tool of using a high-gain observer to recover the performance of state feedback control, the chapter has a number of results that are unique to the regulation problem. First, it shows that for nonlinear systems the internal model is not simply a model of the exosystem; it has to include harmonics of the sinusoidal signals, which are induced by nonlinearities. Second, it deals with impact the servocompensator has on the transient response of the system by introducing the conditional servocompensator, which ensures that the transient response of the system under output feedback can be made arbitrarily close to the transient response of a state feedback sliding mode controller that has no servocompensator. Third, the chapter deals with the case when the internal model is uncertain or unknown. The effect of uncertainty is studied in Section 5.5, and adaptation is used in Section 5.6 to estimate the unknown model.

Chapter 6 presents the extended high-gain observer. It shows two uses of the extended observer, one as a disturbance estimator and the other as a soft sensor of the internal dynamics. As a disturbance estimator, it is shown in Section 6.1 how to implement feedback linearization by estimating the uncertain terms and compensating for them. This is shown for single-input–single-output systems, then for multi-input-multi-output systems. It is also shown how to combine the extended high-gain observer with dynamic inversion to deal with nonaffine control or uncertain control coefficient. Most of the high-gain observer results in Chapters 3 to 5 apply only to minimum-phase systems. This is mainly because the high-gain observer does not estimate the states of the internal dynamics. In Section 6.3 it is shown that the extended high-gain observer can provide information about the internal dynamics by sensing a term in the external dynamics that can be viewed as an output for the internal dynamics. This is used for stabilization of a nonminimum phase system or for designing an observer that estimates the full state of the system. In the later case, two observers are used: the extended high-gain observer is used for the external dynamics, and an extended Kalman filter is used for the internal dynamics.

Chapters 7 and 8 address the challenges described in Section 1.3. The challenge with a ρ-dimensional observer when ρ is high is that the observer gain is proportional to $1/\varepsilon^{\rho}$ and the peak of its internal states during the transient period is the order of $1/\varepsilon^{\rho-1}$. Chapter 7 addresses this challenge by introducing observers where the gain and the peaking signal are limited to the order of $1/\varepsilon$. Such observers are constructed as cascade connections of first- or second-order observers where the gains are of the order of $1/\varepsilon$. Peaking is limited by inserting saturation functions in between the cascaded observers. Two such observers are designed. The first observer is a simple cascade connection of one second-order observer and $\rho-2$ first-order observers. This observer is robust to model uncertainty, and its performance in feedback control is comparable to the standard high-gain observer, but the steady-state estimation errors are orders of magnitude higher than the standard observer's errors. Moreover, the estimation error does not converge to zero under conditions where the standard observer's error converges to zero. The second observer is a cascade connection of $\rho-1$ second-order observers with feedback injection. This observer has the same steady-state and convergence properties of the standard observer.

Chapter 8 starts by characterizing the effect of measurement noise on the estimation error. The general result for bounded noise shows that the ultimate bound on the estimation error is of the order of $N/\varepsilon^{\rho-1}$, where N is the bound on the noise and ρ is the dimension of the observer. While the rest of the chapter is concerned with this case, Section 8.1 shows less conservative bounds when the frequency of the noise is much lower or much higher than $1/\varepsilon$. The effect of noise on feedback control is studied in Section 8.2, where it is shown that a result similar to the separation principle of Section 3.1 can be shown if the amplitude of the noise is restricted. Even in this case, the presence of noise puts a lower bound on ε. In the tracking problem of Section 8.3 it is shown that the component of the tracking error due to measurement noise does not depend on a negative power of ε. Its ultimate bound is of the order of N. In Section 8.4 two techniques are discussed for reducing the effect of measurement noise. In the first technique, the high-frequency content of the noise is filtered out before feeding the measurement into the observer. The second technique uses a nonlinear gain to adjust ε so that a smaller ε is used during the transient period to achieve fast convergence and a larger one is used at steady state to reduce the effect of noise.

Digital implementation of high-gain observers is discussed in Chapter 9. The observer is discretized with a sampling period proportional to ε. The nonlinear observer is discretized using the Forward Difference method, while other discretization methods can be used with linear observers. Digital control with zero-order hold is shown to have properties similar to continuous-time controllers when ε and the sampling period are sufficiently small. Finally, a multirate digital control scheme is presented. This scheme is useful for computationally demanding controllers where the control sampling period cannot be reduced beyond a certain value, which might not be small enough to implement the fast high-gain observer. The multirate scheme allows the observer to run with a sampling period shorter than the controller's sampling period.

Chapter 2

High-Gain Observers

2.1 ▪ Class of Nonlinear Systems

Consider a class of single-output nonlinear systems represented by

$$\dot{w} = f_0(w,x,u), \tag{2.1}$$
$$\dot{x}_i = x_{i+1} + \psi_i(x_1,\dots,x_i,u) \qquad \text{for } 1 \le i \le \rho-1, \tag{2.2}$$
$$\dot{x}_\rho = \phi(w,x,u), \tag{2.3}$$
$$y = x_1, \tag{2.4}$$

where $w \in R^\ell$ and $x = \text{col}(x_1,x_2,\dots,x_\rho) \in R^\rho$ form the state vector, $u \in R^m$ is the input, and $y \in R$ is the measured output. Equations (2.1)–(2.4) cover two important classes of nonlinear systems. The normal form of a single-input–single-output nonlinear system having relative degree ρ is a special case of (2.1)–(2.4) where $\psi_i = 0$ for $1 \le i \le \rho-1$, f_0 is independent of u and $\phi(w,x,u) = \phi_1(w,x) + \phi_2(w,x)u$.[3] A nonlinear single-input–single-output system of the form

$$\dot{\chi} = f(\chi) + g(\chi)u, \qquad y = h(\chi), \tag{2.5}$$

has relative degree ρ in an open set $\mathscr{R}$ if, for all $\chi \in \mathscr{R}$,

$$L_g L_f^{i-1} h(\chi) = 0 \quad \text{for } i = 1,2,\dots,\rho-1; \qquad L_g L_f^{\rho-1} h(\chi) \neq 0,$$

where $L_f h(\chi) = \frac{\partial h}{\partial \chi} f$ is the Lie derivative of h with respect to the vector field f. Under appropriate smoothness conditions on f, g, and h, there is a change of variables that transforms a relative degree ρ system into the normal formal, at least locally. Under some stronger conditions, the change of variables will hold globally [64, Chapter 9].

When w is absent, equations (2.2)–(2.4) reduce to

$$\dot{x}_i = x_{i+1} + \psi_i(x_1,\dots,x_i,u) \qquad \text{for } 1 \le i \le \rho-1, \tag{2.6}$$
$$\dot{x}_\rho = \phi(x,u), \tag{2.7}$$
$$y = x_1. \tag{2.8}$$

[3]See [64, Chapter 4] or [80, Chapter 8].

This system is observable, uniformly in u, in the sense of differential observability [49, Chapter 2]. It is not hard to verify that, if the functions ψ_i are sufficiently smooth, the state x can be uniquely expressed as a function $y, \dot{y}, \ldots, y^{(\rho-1)}, u, \dot{u}, \ldots, u^{(\rho-2)}$. A system whose state can be expressed as a function of its output, input, and their derivatives up to order $\rho-1$ is said to be differentially observable of order ρ. As a special case of (2.6)–(2.8), consider a ρ-dimensional, nonlinear, single-input–single-output system of the form (2.5) with sufficiently smooth functions f, g, and h. Suppose the system is observable uniformly in u in the sense that on any infinitesimally small time interval $[0,T]$ and for any measurable bounded input $u(t)$ defined on $[0,T]$, the initial state $x(0)$ is uniquely determined from the trajectories of the output $y(t)$ and the input $u(t)$ over $[0,T]$. Then the map

$$\Phi(\chi) = \text{col}\Big(h(\chi),\ L_f h(\chi), \ldots, L_f^{\rho-1} h(\chi)\Big)$$

is a local diffeomorphism, and the change of variables $x = \Phi(\chi)$ transforms the system (2.5), at least locally, into the form[4]

$$\dot{x}_i = x_{i+1} + g_i(x_1,\ldots,x_i)u \qquad \text{for } 1 \le i \le \rho-1, \tag{2.9}$$

$$\dot{x}_\rho = \phi_1(x) + \phi_2(x)u, \tag{2.10}$$

$$y = x_1. \tag{2.11}$$

2.2 ▪ Observer Design

In the motivating examples of Section 1.2 we saw that high-gain observers have two features. First, the estimation error decays rapidly toward small (order $O(\varepsilon)$) values within a time interval that decreases toward zero as ε decreases. Second, the observer is robust with respect to uncertain nonlinear functions. The task of this section is to design a high-gain observer for a nonlinear system of the form (2.1)–(2.4), that is,

$$\dot{w} = f_0(w,x,u), \tag{2.12}$$

$$\dot{x}_i = x_{i+1} + \psi_i(x_1,\ldots,x_i,u) \qquad \text{for } 1 \le i \le \rho-1, \tag{2.13}$$

$$\dot{x}_\rho = \phi(w,x,u), \tag{2.14}$$

$$y = x_1, \tag{2.15}$$

where f_0, $\psi_1, \ldots, \psi_{\rho-1}$, and ϕ are locally Lipschitz in their arguments, and $w(t)$, $x(t)$, and $u(t)$ are bounded for all $t \ge 0$.[5] In particular, let $w(t) \in W \subset R^\ell$, $x(t) \in X \subset R^\rho$, and $u(t) \in U \subset R^m$ for all $t \ge 0$, for some compact sets W, X, and U. The fast decay of the estimation error can be achieved only when we estimate x. Therefore, the high-gain observer is a partial-state observer that estimates only x, not the full state (w,x). The robustness feature of the observer can be achieved only with respect to the nonlinearity ϕ; therefore, we assume that ψ_1 to $\psi_{\rho-1}$ are known. Furthermore, we require that for any compact set $S \subset R^\rho$, there are positive constants L_1 to $L_{\rho-1}$ such that the functions ψ_1 to $\psi_{\rho-1}$ satisfy the Lipschitz conditions

$$|\psi_i(x_1,\ldots,x_i,u) - \psi_i(z_1,\ldots,z_i,u)| \le L_i \sum_{k=1}^{i} |x_k - z_k| \tag{2.16}$$

[4]See [49, Theorem 4.1 of Chapter 3] or [50, Theorem 2].

[5]This assumption is needed because the system is operated in open loop. In later chapters when the observer is studied within a feedback loop, the boundedness of the state under feedback will be established in the analysis.

for all $x, z \in S$, and $u \in U$. The observer is taken as

$$\dot{\hat{x}}_i = \hat{x}_{i+1} + \psi_i^s(\hat{x}_1, \dots, \hat{x}_i, u) + \frac{\alpha_i}{\varepsilon^i}(y - \hat{x}_1) \qquad \text{for } 1 \le i \le \rho - 1, \tag{2.17}$$

$$\dot{\hat{x}}_\rho = \phi_0(\hat{x}, u) + \frac{\alpha_\rho}{\varepsilon^\rho}(y - \hat{x}_1), \tag{2.18}$$

where ϕ_0 is a nominal model of ϕ, ε is a sufficiently small positive constant, and α_1 to α_ρ are chosen such that the polynomial

$$s^\rho + \alpha_1 s^{\rho-1} + \cdots + \alpha_{\rho-1} s + \alpha_\rho \tag{2.19}$$

is Hurwitz.[6] The functions ψ_1^s to $\psi_{\rho-1}^s$ are locally Lipschitz and satisfy the conditions

$$\psi_i^s(x_1, \dots, x_i, u) = \psi_i(x_1, \dots, x_i, u) \qquad \forall\, x \in X \tag{2.20}$$

and

$$|\psi_i^s(x_1, \dots, x_i, u) - \psi_i^s(z_1, \dots, z_i, u)| \le L_i \sum_{k=1}^{i} |x_k - z_k| \tag{2.21}$$

for all $x, z \in R^\rho$ and all $u \in U$. The difference between (2.16) and (2.21) is that (2.16) holds for x and z in a compact set, while (2.21) holds for all x and z. If the Lipschitz condition (2.16) holds globally, we can take $\psi_i^s = \psi_i$. Otherwise, we define ψ_i^s by saturating the x_1 to x_i arguments of ψ_i outside the compact set X. Choose $M_i > \max_{x \in X} |x_i|$ and define ψ_i^s by

$$\psi_i^s(x_1, \dots, x_i, u) = \psi_i\left(M_1 \operatorname{sat}\left(\frac{x_1}{M_1}\right), \dots, M_i \operatorname{sat}\left(\frac{x_i}{M_i}\right), u\right), \tag{2.22}$$

where sat$(\cdot)$ is the standard saturation function defined by

$$\operatorname{sat}(y) = \min\{|y|, 1\} \operatorname{sign}(y) = \begin{cases} y & \text{if } |y| \le 1, \\ -1 & \text{if } y \le -1, \\ 1 & \text{if } y \ge 1. \end{cases}$$

The function ψ_i^s satisfies (2.21) because

$$\begin{aligned} |\psi_i^s(x_1, \dots, x_i, u) - \psi_i^s(z_1, \dots, z_i, u)| &\le L_i \sum_{k=1}^{i} \left| M_k \operatorname{sat}\left(\frac{x_x}{M_k}\right) - M_k \operatorname{sat}\left(\frac{z_k}{M_k}\right) \right| \\ &\le L_i \sum_{k=1}^{i} |x_k - z_k|, \end{aligned}$$

where the first inequality follows from (2.16) and the second inequality holds because the saturation function is globally Lipschitz with Lipschitz constant equal to one.

We assume that ϕ_0 is locally Lipschitz in its arguments and

$$|\phi(w, x, u) - \phi_0(z, u)| \le L\, \|x - z\| + M \tag{2.23}$$

for all $w \in W$, $x \in X$, $z \in R^\rho$, and $u \in U$. Because

$$\phi(w, x, u) - \phi_0(z, u) = \phi(w, x, u) - \phi_0(x, u) + \phi_0(x, u) - \phi_0(z, u)$$

[6]A polynomial is Hurwitz when all its roots have negative real parts.

and $\phi_0(x,u)$ can be chosen to be globally Lipschitz in x by saturating its x-argument outside the compact set X, (2.23) requires the modeling error $\phi(w,x,u)-\phi_0(x,u)$ to be bounded. We can choose $\phi_0=0$, which would be a natural choice if no information were available on ϕ. In this case, (2.23) holds with $L=0$. On the other hand, if ϕ is known and either it is not function of w or w is measured, we can take $\phi_0=\phi$ with the x-argument of ϕ_0 saturated outside X. In this case, (2.23) holds with $M=0$.

Theorem 2.1. *Under the stated assumptions, there is $\varepsilon^*\in(0,1]$ such that for $0<\varepsilon\le\varepsilon^*$, the estimation error $\tilde{x}_i=x_i-\hat{x}_i$, for $1\le i\le\rho$, of the high-gain observer* (2.17)–(2.18) *satisfies the bound*

$$|\tilde{x}_i|\le\max\left\{\frac{b}{\varepsilon^{i-1}}\|\tilde{x}(0)\|\,e^{-at/\varepsilon},\ \varepsilon^{\rho+1-i}cM\right\} \tag{2.24}$$

for some positive constants a, b, c. ◇

Proof: Define the scaled estimation errors

$$\eta_1=\frac{x_1-\hat{x}_1}{\varepsilon^{\rho-1}},\ \eta_2=\frac{x_2-\hat{x}_2}{\varepsilon^{\rho-2}},\ \dots,\ \eta_{\rho-1}=\frac{x_{\rho-1}-\hat{x}_{\rho-1}}{\varepsilon},\ \eta_\rho=x_\rho-\hat{x}_\rho. \tag{2.25}$$

It can be shown that $\eta=\text{col}(\eta_1,\eta_2,\dots,\eta_\rho)$ satisfies the equation

$$\varepsilon\dot{\eta}=F\eta+\varepsilon\delta(w,x,\tilde{x},u), \tag{2.26}$$

where

$$F=\begin{bmatrix}-\alpha_1 & 1 & 0 & \cdots & 0\\ -\alpha_2 & 0 & 1 & \cdots & 0\\ \vdots & & \ddots & & \vdots\\ -\alpha_{\rho-1} & & & 0 & 1\\ -\alpha_\rho & 0 & \cdots & \cdots & 0\end{bmatrix}, \tag{2.27}$$

$$\delta=\text{col}(\delta_1,\delta_2,\cdots,\delta_\rho),$$

$$\delta_i=\frac{1}{\varepsilon^{\rho-i}}[\psi_i(x_1,\dots,x_i,u)-\psi_i^s(\hat{x}_1,\dots,\hat{x}_i,u)]\quad\text{for}\quad 1\le i\le\rho-1,$$

$$\delta_\rho=\phi(w,x,u)-\phi_0(\hat{x},u).$$

The matrix F is Hurwitz by design because its characteristic polynomial is (2.19). Using (2.20) and (2.21) we see that δ_1 to $\delta_{\rho-1}$ satisfy

$$\begin{aligned}|\delta_i|&=\frac{1}{\varepsilon^{\rho-i}}\,|\psi_i^s(x_1,\dots,x_i,u)-\psi_i^s(\hat{x}_1,\dots,\hat{x}_i,u)|\\ &\le\frac{L_i}{\varepsilon^{\rho-i}}\sum_{k=1}^{i}|x_k-\hat{x}_k|=\frac{L_i}{\varepsilon^{\rho-i}}\sum_{k=1}^{i}\varepsilon^{\rho-k}|\eta_k|=L_i\sum_{k=1}^{i}\varepsilon^{i-k}|\eta_k|.\end{aligned}$$

The preceding inequality and (2.23) show that

$$\|\delta\|\le L_\delta\|\eta\|+M, \tag{2.28}$$

where L_δ is independent of ε for all $\varepsilon\le 1$. Let $V=\eta^TP\eta$, where $P=P^T>0$ is the solution of the Lyapunov equation $PF+F^TP=-I$. Then

$$\varepsilon\dot{V}=-\eta^T\eta+2\varepsilon\eta^TP\delta.$$

Using (2.28) we obtain

$$\varepsilon\dot{V} \le -||\eta||^2 + 2\varepsilon||P||L_\delta||\eta||^2 + 2\varepsilon||P||M||\eta||.$$

For $\varepsilon||P||L_\delta \le \frac{1}{4}$,

$$\varepsilon\dot{V} \le -\tfrac{1}{2}||\eta||^2 + 2\varepsilon||P||M||\eta||.$$

Hence,

$$\varepsilon\dot{V} \le -\frac{1}{4}||\eta||^2 \quad \forall\ ||\eta|| \ge 8\varepsilon M||P||.$$

We conclude that [80, Theorem 4.5]

$$||\eta(t)|| \le \max\left\{be^{-at/\varepsilon}||\eta(0)||, \varepsilon cM\right\}$$

for some positive constants a, b, c. From (2.25) we see that $||\eta(0)|| \le ||\tilde{x}(0)||/\varepsilon^{\rho-1}$ and $|\tilde{x}_i| = \varepsilon^{\rho-i}|\eta_i|$, which yield (2.24). □

The two terms on the right-hand side of (2.24) show bounds on the estimation error due to two sources. The term $(b/\varepsilon^{i-1})\, e^{-at/\varepsilon}$ is due to the initial estimation error $\tilde{x}(0) = x(0) - \hat{x}(0)$. It exhibits the peaking phenomenon of the high-gain observer. It can be seen from (2.25) that peaking will not happen if the initial estimation error satisfies $x_i(0) - \hat{x}_i(0) = O(\varepsilon^{\rho-i})$ for $i = 1, \dots, \rho$. This term shows also that the estimation error decays rapidly to $O(\varepsilon)$ values. In particular, given any positive constant K, it can be seen that

$$\frac{b}{\varepsilon^{\rho-1}} e^{-at/\varepsilon} \le K\varepsilon \quad \forall\ t \ge T(\varepsilon) \stackrel{\text{def}}{=} \frac{\varepsilon}{a} \ln\left(\frac{b}{K\varepsilon^\rho}\right). \tag{2.29}$$

Using l'Hôpital's rule, it can be seen that $\lim_{\varepsilon\to 0} T(\varepsilon) = 0$. The second term $\varepsilon^{\rho+1-i}cM$ is due to the error in modeling the function ϕ. This error will not exist if the observer is implemented with $\phi = \phi_0$. Thus, $\lim_{t\to\infty} \tilde{x}(t) = 0$ when $\phi = \phi_0$.

2.3 ▪ Lyapunov and Riccati Equation Designs

In the observer (2.17)–(2.18), the observer gain can be written as

$$H = \frac{1}{\varepsilon^\rho} D(\varepsilon) H_o, \tag{2.30}$$

where $D(\varepsilon) = \text{diag}[\varepsilon^{\rho-1}, \dots, 1]$ and $H_o = \text{col}(\alpha_1, \alpha_2, \dots, \alpha_\rho)$ assigns the eigenvalues of the matrix F of (2.27) at desired locations in the open left-half plane. The matrix F can be written as $F = A - H_o C$, where

$$A = \begin{bmatrix} 0 & 1 & 0 & \cdots & 0 \\ 0 & 0 & 1 & \cdots & 0 \\ \vdots & & \ddots & & \vdots \\ 0 & & & 0 & 1 \\ 0 & 0 & \cdots & \cdots & 0 \end{bmatrix}_{\rho\times\rho} \quad \text{and} \quad C = \begin{pmatrix} 1 & 0 & \cdots & \cdots & 0 \end{pmatrix}_{1\times\rho}.$$

In this section we present two special designs of H_o, which are obtained by solving Lyapunov and Riccati equations. The Lyapunov equation–based design comes from early work on high-gain observers [50], while the Riccati equation–based design has its roots in loop transfer recovery techniques of linear systems.[7]

[7] See [127] and the reference therein.

2.3.1 ▪ Lyapunov Equation Design

Let P be the positive definite symmetric solution of the Lyapunov equation

$$P(A+\tfrac{1}{2}I)+(A+\tfrac{1}{2}I)^T P - C^T C = 0 \tag{2.31}$$

and take

$$H_o = P^{-1}C^T. \tag{2.32}$$

The existence and uniqueness of P is guaranteed by the fact that $-(A+\frac{1}{2}I)$ is Hurwitz and (A, C) is observable.[8] The fact that $A - H_o C = A - P^{-1}C^T C$ is Hurwitz can be seen from the Lyapunov equation

$$P(A-P^{-1}C^T C)+(A-P^{-1}C^T C)^T P + C^T C + P = 0, \tag{2.33}$$

which is obtained from (2.31) by adding and subtracting $C^T C$. Equation (2.33) shows that $A - P^{-1}C^T C$ is Hurwitz because $C^T C + P$ is positive definite. What is more interesting is the fact that all the eigenvalues of $A - P^{-1}C^T C$ are assigned at -1. For example, for $\rho = 2$, 3, and 4, the matrix $A - P^{-1}C^T C$ is given by

$$\begin{bmatrix} -2 & 1 \\ -1 & 0 \end{bmatrix}, \quad \begin{bmatrix} -3 & 1 & 0 \\ -3 & 0 & 1 \\ -1 & 0 & 0 \end{bmatrix}, \quad \text{and} \quad \begin{bmatrix} -4 & 1 & 0 & 0 \\ -6 & 0 & 1 & 0 \\ -4 & 0 & 0 & 1 \\ -1 & 0 & 0 & 0 \end{bmatrix},$$

respectively.

2.3.2 ▪ Riccati Equation Design

Let Q be the positive definite symmetric solution of the Riccati equation

$$AQ + QA^T - QC^T CQ + BB^T = 0, \tag{2.34}$$

where $B = \text{col}(0, 0, \ldots, 0, 1)$ is a $\rho \times 1$ matrix, and take

$$H_o = QC^T. \tag{2.35}$$

The existence of Q and the fact that $A - H_o C = A - QC^T C$ is Hurwitz follow from the properties of the solution of Riccati equations because (A, B) is controllable and (A, C) is observable [91]. The eigenvalues of $A - QC^T C$ are assigned on the unit circle in a Butterworth pattern, that is,

$$\lambda_k = e^{j\theta_k}, \quad \text{where} \quad \theta_k = \frac{(2k+\rho-1)\pi}{2\rho} \quad \text{for} \quad k = 1, 2, \ldots, \rho. \tag{2.36}$$

Table 2.1 shows the eigenvalues of $A - H_o C$ for four different values of ρ.

2.3.3 ▪ Simulation Comparison

The choice of the observer eigenvalues determines the shape of the transient response of the estimation error. In the example to follow we compare a Lyapunov equation design, a Riccati equation design, and a pole-placement design.

[8] See [78, Theorem 4.6 and Exercise 4.22].

Table 2.1. *The eigenvalues of $A - H_o C$ in the Riccati equation design.*

ρ	Eigenvalues	θ
2	$-0.7071+0.7071j$	$3\pi/4$
	$-0.7071-0.7071j$	$5\pi/4$
3	$-0.5+0.8660j$	$2\pi/3$
	-1	π
	$-0.5-0.8660j$	$4\pi/3$
4	$-0.3827+0.9239j$	$5\pi/8$
	$-0.9239+0.3827j$	$7\pi/8$
	$-0.9239-0.3827j$	$9\pi/8$
	$-0.3827-0.9239j$	$11\pi/8$
5	$-0.3090+0.9511j$	$3\pi/5$
	$-0.8090+0.5878j$	$4\pi/5$
	-1	π
	$-0.8090-0.5878j$	$6\pi/5$
	$-0.3090-0.9511j$	$7\pi/5$

Example 2.1. The states of the three-dimensional system

$$\dot{x}_1 = x_2, \qquad \dot{x}_2 = x_3, \qquad \dot{x}_3 = \cos(t), \qquad y = x_1,$$

are estimated using the high-gain observer

$$\begin{aligned}
\dot{\hat{x}}_1 &= \hat{x}_2 + (\alpha_1/\varepsilon)(y - \hat{x}_1),\\
\dot{\hat{x}}_2 &= \hat{x}_3 + (\alpha_2/\varepsilon^2)(y - \hat{x}_1),\\
\dot{\hat{x}}_2 &= \cos(t) + (\alpha_3/\varepsilon^3)(y - \hat{x}_1).
\end{aligned}$$

Three different designs of $H_o = \text{col}(\alpha_1, \alpha_2, \alpha_3)$ are considered. The first one, $H_o = \text{col}(3, 3, 1)$, is a Lyapunov equation design that assigns all three eigenvalues at -1. The second one, $H_o = \text{col}(2, 2, 1)$, is a Riccati equation design that assigns complex eigenvalues on the unit circle in a Butterworth pattern. The third design, $H_o = \text{col}(3, 2.75, 0.75)$, assigns the eigenvalues at -0.5, -1,-1.5. Figure 2.1 compares the estimation errors obtained by the three designs when $\varepsilon = 0.01$. It can be seen that the two designs with real eigenvalues outperform the one with complex eigenvalues. Between these two designs, the one with distinct eigenvalues is slightly better than the one with multiple eigenvalues. These observations are typical for observers with dimension other than three. △

2.4 ▪ Reduced-Order Observer

Reconsider the system (2.12)–(2.15):

$$\dot{w} = f_0(w, x, u), \tag{2.37}$$

$$\dot{x}_i = x_{i+1} + \psi_i(x_1, \ldots, x_i, u) \qquad \text{for } 1 \le i \le \rho - 1, \tag{2.38}$$

$$\dot{x}_\rho = \phi(w, x, u), \tag{2.39}$$

$$y = x_1. \tag{2.40}$$

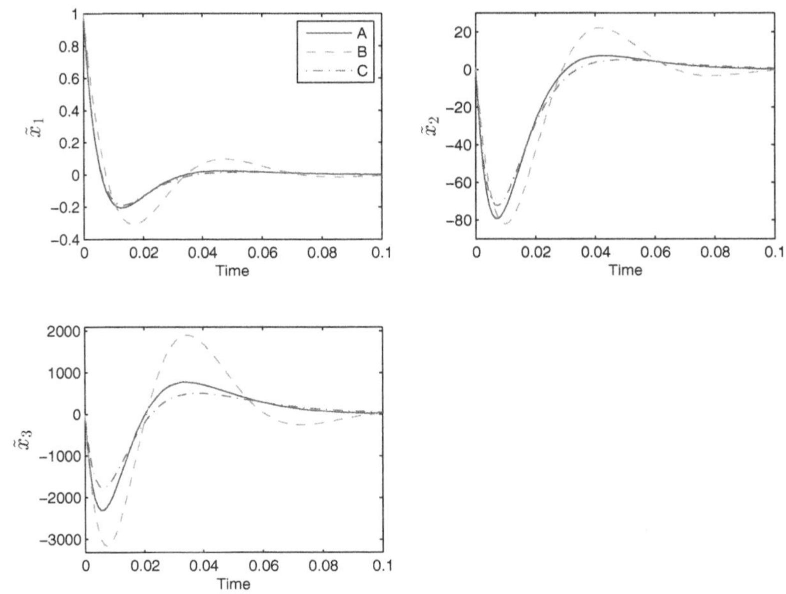

Figure 2.1. *Simulation of Example* 2.1. *Design A has three multiple eigenvalues at* −1; *design B has complex eigenvalues on the unit circle in a Butterworth pattern; design C has real eigenvalues at* −0.5, −1, −1.5.

The dimension ρ of the standard high-gain observer (2.17)–(2.18) is the same as the dimension of the vector x. Since the measured output is x_1, it is possible to design an observer of dimension $\rho-1$ to estimate x_2 to x_ρ. In this section we derive such reduced-order observer and calculate an upper bound on the estimation error. The derivation procedure is similar to the one used in linear systems.[9] We first assume that $\dot{y}-\psi_1(y,u)=x_2$ is the measured output and design an observer driven by $\dot{y}$, then we apply a change of variables to eliminate the dependence on $\dot{y}$.

Let

$$\xi_i = x_{i+1} \qquad \text{for } 1 \le i \le \rho-1.$$

Then

$$\dot{\xi}_i = \xi_{i+1} + \psi_{i+1}(y,\xi_1,\dots,\xi_i,u) \qquad \text{for } 1 \le i \le \rho-2, \tag{2.41}$$

$$\dot{\xi}_{\rho-1} = \phi(w,y,\xi_1,\dots,\xi_{\rho-1},u), \tag{2.42}$$

$$\dot{y}-\psi_1(y,u) = \xi_1, \tag{2.43}$$

which is similar to (2.38)–(2.40) except that the dimension of ξ is $\rho-1$ instead of ρ. A high-gain observer of dimension $\rho-1$ is given by

$$\begin{aligned}\dot{\hat{\xi}}_i = \hat{\xi}_{i+1} &+ \psi^s_{i+1}(y,\hat{\xi}_1,\dots,\hat{\xi}_i,u) \\ &+ \frac{\beta_i}{\varepsilon^i}[\dot{y}-\psi_1(y,u)-\hat{\xi}_1] \quad \text{for } 1 \le i \le \rho-2,\end{aligned} \tag{2.44}$$

$$\dot{\hat{\xi}}_{\rho-1} = \phi_0(y,\hat{\xi}_1,\dots,\hat{\xi}_{\rho-1},u) + \frac{\beta_{\rho-1}}{\varepsilon^{\rho-1}}[\dot{y}-\psi_1(y,u)-\hat{\xi}_1], \tag{2.45}$$

[9]See, for example, [9, Section 9.3.2] or [126, Chapter 15].

where ε is a sufficiently small positive constant, β_1 to $\beta_{\rho-1}$ are chosen such that the polynomial

$$s^{\rho-1}+\beta_1 s^{\rho-2}+\cdots+\beta_{\rho-2}s+\beta_{\rho-1} \tag{2.46}$$

is Hurwitz, and the functions ψ_2^s to $\psi_{\rho-1}^s$ and ϕ_0 are defined in Section 2.2. In particular, they satisfy (2.20), (2.21), and (2.23). The only difference is that the y argument of ψ_i^s is not saturated. With the change of variables

$$z_i=\hat{\xi}_i-\frac{\beta_i}{\varepsilon^i}y, \qquad 1\le i\le\rho-1, \tag{2.47}$$

we arrive at the reduced-order observer

$$\begin{aligned}\dot{z}_i &= z_{i+1}+\frac{\beta_{i+1}}{\varepsilon^{i+1}}y\\ &\quad +\psi_{i+1}^s\left(y,z_1+\frac{\beta_1}{\varepsilon}y,\ldots,z_i+\frac{\beta_i}{\varepsilon^i}y,u\right)\\ &\quad -\frac{\beta_i}{\varepsilon^i}\left[z_1+\frac{\beta_1}{\varepsilon}y+\psi_1(y,u)\right] \quad \text{for } 1\le i\le\rho-2, \end{aligned} \tag{2.48}$$

$$\begin{aligned}\dot{z}_{\rho-1} &= \phi_0\left(y,z_1+\frac{\beta_1}{\varepsilon}y,\ldots,z_{\rho-1}+\frac{\beta_{\rho-1}}{\varepsilon^{\rho-1}}y,u\right)\\ &\quad -\frac{\beta_{\rho-1}}{\varepsilon^{\rho-1}}\left[z_1+\frac{\beta_1}{\varepsilon}y+\psi_1(y,u)\right],\end{aligned} \tag{2.49}$$

$$\hat{x}_i=z_{i-1}+\frac{\beta_{i-1}}{\varepsilon^{i-1}}y \qquad \text{for } 2\le i\le\rho. \tag{2.50}$$

The bound on the estimation error is given in the following theorem.

Theorem 2.2. *Under the stated assumptions, there is $\varepsilon^*\in(0,1]$ such that for $0<\varepsilon\le\varepsilon^*$, the estimation error $\tilde{x}_i=x_i-\hat{x}_i$, for $2\le i\le\rho$, of the reduced-order high-gain observer* (2.48)–(2.50) *satisfies the bound*

$$|\tilde{x}_i|\le\max\left\{\frac{k}{\varepsilon^{i-1}}\,e^{-at/\varepsilon},\ \varepsilon^{\rho+1-i}cM\right\} \tag{2.51}$$

for some positive constants a, c, and k, where k depends on the initial conditions $x(0)$ and $z(0)$. ◇

Proof: Let

$$\eta_i=\frac{\xi_i-z_i-(\beta_i/\varepsilon^i)x_1}{\varepsilon^{\rho-1-i}} \qquad \text{for } 1\le i\le\rho-1. \tag{2.52}$$

It can be shown that $\eta=\text{col}(\eta_1,\eta_2,\ldots,\eta_{\rho-1})$ satisfies the equation

$$\varepsilon\dot{\eta}=E\eta+\varepsilon\Delta, \tag{2.53}$$

where

$$E=\begin{bmatrix}-\beta_1 & 1 & 0 & \cdots & 0\\ -\beta_2 & 0 & 1 & \cdots & 0\\ \vdots & & \ddots & & \vdots\\ -\beta_{\rho-2} & & & 0 & 1\\ -\beta_{\rho-1} & 0 & \cdots & \cdots & 0\end{bmatrix}, \qquad \Delta=\text{col}(\Delta_1,\ldots,\Delta_{\rho-1}),$$

$$\begin{aligned}
\Delta_i = {} & \frac{1}{\varepsilon^{\rho-1-i}}[\psi_{i+1}(x_1,\xi_1,\ldots,\xi_i,u) \\
& -\psi^s_{i+1}(x_1,\xi_1-\varepsilon^{\rho-2}\eta_1,\ldots,\xi_i-\varepsilon^{\rho-1-i}\eta_i,u)] \quad \text{for } 1\le i\le \rho-2,\\
\Delta_{\rho-1} = {} & \phi(w,x_1,\xi_1,\ldots,\xi_{\rho-1},u) \\
& -\phi_0(x_1,\xi_1-\varepsilon^{\rho-2}\eta_1,\ldots,\xi_{\rho-1}-\eta_{\rho-1},u).
\end{aligned}$$

The matrix E is Hurwitz by design because its characteristic polynomial is (2.46). Using (2.20), (2.21), and (2.23), it can be shown that

$$|\Delta_i| \le L_{i+1}\sum_{k=1}^{i}\varepsilon^{i-k}|\eta_k| \quad \text{for } 1\le i\le \rho-2,$$

$$|\Delta_{\rho-1}| \le L\sum_{k=1}^{\rho-1}\varepsilon^{\rho-1-k}|\eta_k| + M.$$

Thus, there exists a positive constant k_1 such that

$$\|\Delta\| \le k_1\|\eta\| + M.$$

Let $V=\eta^T P\eta$, where $P=P^T>0$ is the solution of the Lyapunov equation $PE+E^T P=-I$. The derivative $\dot V$ satisfies the inequality

$$\varepsilon\dot V \le -\|\eta\|^2 + 2k_1\varepsilon\|P\|\,\|\eta\|^2 + 2\varepsilon\|P\|M\|\eta\|.$$

Proceeding as in the proof Theorem 2.1, it can be shown that

$$\|\eta(t)\| \le \max\left\{be^{-at/\varepsilon}\|\eta(0)\|, \varepsilon cM\right\}$$

for some positive constants a, b, c. From (2.52) we see that $\|\eta(0)\|\le k_2/\varepsilon^{\rho-1}$, where $k_2>0$ depends on the initial conditions $x(0)$ and $z(0)$. From (2.47) and (2.52) we have

$$|\xi_i-\hat\xi_i| = \left|\varepsilon^{\rho-1-i}\eta_i\right| \le \max\left\{\frac{bk_2}{\varepsilon^i}e^{-at/\varepsilon}, \varepsilon^{\rho-i}cM\right\},$$

which yields (2.51) since $x_i-\hat x_i=\xi_{i-1}-\hat\xi_{i-1}$. □

While the bound on the estimation error for the reduced-order observer in Theorem 2.2 takes the same form as the one for the standard observer in Theorem 2.1, the standard observer has better performance, especially in the presence of measurement noise. When both observers are linear, the frequency response of the standard observer rolls off at high frequency, unlike the frequency response of the reduced-order observer, which stays flat at high frequency. This point is illustrated in Example 1.4 of Section 1.2.

2.5 ▪ Multi-output Systems

The high-gain observer of Section 2.2 can be extended to classes of multi-output systems. In the simplest such class, the system is represented by

$$\dot w = f_0(w,x,u), \tag{2.54}$$

$$\dot x^i_j = x^i_{j+1} \quad \text{for } 1\le j\le \rho_i-1,\ 1\le i\le q, \tag{2.55}$$

$$\dot x^i_{\rho_i} = \phi_i(w,x,u) \quad \text{for } 1\le i\le q, \tag{2.56}$$

$$y_i = x^i_1 \quad \text{for } 1\le i\le q, \tag{2.57}$$

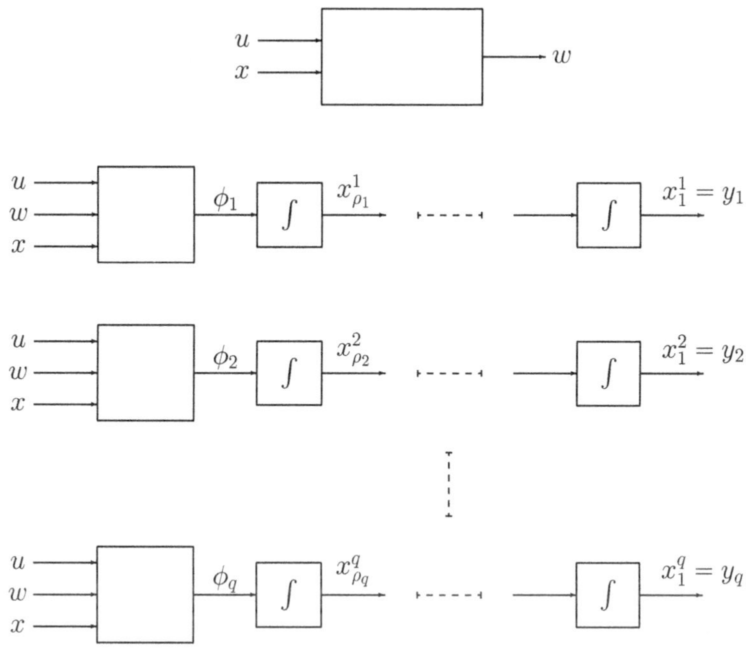

Figure 2.2. *Block diagram representation of the multi-output system* (2.54)–(2.57).

where

$$x^i = \mathrm{col}(x^i_1, x^i_2, \ldots, x^i_{\rho_i}),$$
$$x = \mathrm{col}(x^1, x^2, \ldots, x^q) \in R^\rho,$$
$$\rho = \rho_1 + \cdots + \rho_q.$$

The functions f_0, $\phi_1,\ldots, \phi_q$ are locally Lipschitz in their arguments, and $w(t) \in W \subset R^\ell$, $x(t) \in X \subset R^\rho$, and $u(t) \in U \subset R^m$ for all $t \geq 0$, for some compact sets W, X, and U. A block diagram of the system is shown in Figure 2.2. It has q chains of integrators with ρ_i integrators in each chain whose states are $x^i_1, x^i_2, \ldots x^i_{\rho_i}$. The measured outputs are the states of the first integrators of these chains. The q chains are coupled by the input to the last integrator in each chain, the function ϕ_i, which could depend on all integrator states as well as the state w and the input u. Without loss of generality, we assume that each chain has at least two integrators, that is, $\rho_i \geq 2$; a chain of a single integrator can be included in the equation $\dot{w} = f_0$.

The model (2.54)–(2.57) arises naturally in mechanical and electromechanical systems, where displacement variables are measured while their derivatives (velocities, accelerations, etc.) are not measured. For example, the m-link robot manipulator is modeled by [146].

$$M(q)\ddot{q} + C(q,\dot{q})\dot{q} + D\dot{q} + g(q) = u,$$

where q is an m-dimensional vector of generalized coordinates representing joint positions, u an m-dimensional control (torque) input, and $M(q)$ a symmetric positive definite inertia matrix. The terms $C(q,\dot{q})\dot{q}$, $D\dot{q}$, and $g(q)$ account for centrifugal/Coriolis forces, viscous damping, and gravity, respectively. If the measured variables are the joint positions, this model takes the foregoing form with $x_1^i = q_i$, $x_2^i = \dot{q}_i$, $\rho_i = 2$, for $i = 1,\ldots,m$, and the functions ϕ_i are given by

$$\text{col}(\phi_1,\ldots,\phi_m) = M^{-1}[u - C\dot{q} - D\dot{q} - g].$$

The multivariable normal form of a nonlinear system having a well-defined (vector) relative degree is a special case of (2.54)–(2.57) [64, Section 5.1]. An m-input–m-output nonlinear system of the form

$$\dot{\chi} = f(\chi) + \sum_{i=1}^{m} g_i(\chi)u_i, \qquad y_i = h_i(\chi) \quad \text{for } 1 \le i \le m$$

has vector relative degree $\{\rho_1, \rho_2, \ldots, \rho_m\}$ in an open set $\mathcal{R}$ if, for all $\chi \in \mathcal{R}$,

$$L_{g_j} L_f^{k-1} h_i(\chi) = 0 \quad \text{for } 1 \le k \le \rho_i - 1$$

for all $1 \le i \le m$ and $1 \le j \le m$, and the $m \times m$ matrix

$$\begin{bmatrix} L_{g_1} L_f^{\rho_1 -1} h_1 & \cdots & \cdots & L_{g_m} L_f^{\rho_1 -1} h_1 \\ L_{g_1} L_f^{\rho_2 -1} h_2 & \cdots & \cdots & L_{g_m} L_f^{\rho_2 -1} h_2 \\ \vdots & \vdots & \vdots & \vdots \\ L_{g_1} L_f^{\rho_m -1} h_m & \cdots & \cdots & L_{g_m} L_f^{\rho_m -1} h_m \end{bmatrix}$$

is nonsingular.

A high-gain observer for the system (2.54)–(2.57) is taken as

$$\dot{\hat{x}}_j^i = \hat{x}_{j+1}^i + \frac{\alpha_j^i}{\varepsilon^j}(y_i - \hat{x}_1^i) \qquad \text{for } 1 \le j \le \rho_i - 1, \tag{2.58}$$

$$\dot{\hat{x}}_{\rho_i}^i = \phi_{i0}(\hat{x}, u) + \frac{\alpha_{\rho_i}^i}{\varepsilon^{\rho_i}}(y_i - \hat{x}_1^i) \tag{2.59}$$

for $1 \le i \le q$, where ε is a sufficiently small positive constant and the positive constants α_j^i are chosen such that the polynomial

$$s^{\rho_i} + \alpha_1^i s^{\rho_i - 1} + \cdots + \alpha_{\rho_i - 1}^i s + \alpha_{\rho_i}^i$$

is Hurwitz for each $i = 1,\ldots,q$. The function $\phi_{i0}(x,u)$ is a nominal model of $\phi_i(w, x, u)$, which is required to satisfy the inequality

$$\|\phi_i(w,x,u) - \phi_{i0}(z,u)\| \le L_i \|x - z\| + M_i \tag{2.60}$$

for all $w \in W$, $x \in X$, $z \in R^{\rho}$, and $u \in U$.

Theorem 2.3. *Under the stated assumptions, there is $\varepsilon^* \in (0,1]$ such that for $0 < \varepsilon \le \varepsilon^*$, the estimation error $\tilde{x}_j^i = x_j^i - \hat{x}_j^i$, for $1 \le i \le q$ and $1 \le j \le \rho_i$, of the high-gain observer (2.58)–(2.59) satisfies the bound*

$$|\tilde{x}_j^i| \le \max\left\{ \frac{b}{\varepsilon^{\rho_{max} - \rho_i + j - 1}} \|\tilde{x}(0)\| \, e^{-at/\varepsilon}, \varepsilon^{\rho_i - j + 1} c M_{max} \right\} \tag{2.61}$$

for some positive constants a, b, c, where $\rho_{max} = \max\{\rho_1, \ldots, \rho_q\}$ *and* $M_{max} = \max\{M_1, \ldots, M_q\}$. ◇

Proof: With the scaled estimation errors

$$\eta_j^i = \frac{x_j^i - \hat{x}_j^i}{\varepsilon^{\rho_i - j}} \tag{2.62}$$

for $1 \le i \le q$ and $1 \le j \le \rho_i$, the observer dynamics are given by

$$\varepsilon \dot{\eta}^i = F_i \eta^i + \varepsilon B_i \delta_i(w, x, \hat{x}, u)$$

for $1 \le i \le q$, where

$$F_i = \begin{bmatrix} -\alpha_1^i & 1 & 0 & \cdots & 0 \\ -\alpha_2^i & 0 & 1 & \cdots & 0 \\ \vdots & & \ddots & & \vdots \\ -\alpha_{\rho_i - 1}^i & & & 0 & 1 \\ -\alpha_{\rho_i}^i & 0 & \cdots & \cdots & 0 \end{bmatrix}, \quad B_i = \begin{bmatrix} 0 \\ 0 \\ \vdots \\ 0 \\ 1 \end{bmatrix}, \quad \eta^i = \begin{bmatrix} \eta_1^i \\ \eta_2^i \\ \vdots \\ \vdots \\ \eta_{\rho_i}^i \end{bmatrix}$$

and

$$\delta_i(w, x, \hat{x}, u) = \phi_i(w, x, u) - \phi_{i0}(\hat{x}, u).$$

The matrix F_i is Hurwitz and, in view of (2.60),

$$|\delta_i| \le L_i \|x - \hat{x}\| + M_i \le L_i \sum_{j=1}^{q} \|x^j - \hat{x}^j\| + M_i \le L_i \sum_{j=1}^{q} \|\eta^j\| + M_i$$

for $\varepsilon \le 1$. Let $V = \sum_i^q (\eta^i)^T P_i \eta^i$, where P_i is the positive definite solution of the Lyapunov equation $P_i F_i + F_i^T P_i = -I$. Then

$$\begin{aligned} \varepsilon \dot{V} &= \sum_i^q \left[-(\eta^i)^T \eta^i + 2\varepsilon (\eta^i)^T P_i B_i \delta_i \right] \\ &\le \sum_i^q \left[-\|\eta^i\|^2 + 2\varepsilon \|\eta^i\| \|P_i B_i\| \left(L_i \sum_j^q \|\eta^j\| + M_i \right) \right] \\ &= -\varphi^T \varphi + \varepsilon \varphi^T N \varphi + 2\varepsilon M_{max} \pi^T \varphi, \end{aligned}$$

where $\varphi = \text{col}(\|\eta^1\|, \ldots, \|\eta^q\|)$, $\pi = \text{col}(\|P_1 B_1\|, \ldots, \|P_q B_q\|)$, and the elements of the symmetric matrix N are defined by $N_{ij} = \|P_i B_i\| L_i + \|P_j B_j\| L_j$. For $\varepsilon \|N\| \le \frac{1}{2}$,

$$\varepsilon \dot{V} \le -\tfrac{1}{2} \|\varphi\|^2 + 2\varepsilon M_{max} \|\pi\| \|\varphi\|.$$

Using the fact that $\|\varphi\| = \|\eta\|$, where $\eta = \text{col}(\eta^1, \ldots, \eta^q)$, we arrive at the inequality

$$\varepsilon \dot{V} \le -\tfrac{1}{2} \|\eta\|^2 + 2\varepsilon M_{max} \|\pi\| \|\eta\|.$$

Hence,

$$\varepsilon \dot{V} \le -\tfrac{1}{4} \|\eta\|^2 \qquad \forall \ \|\eta\| \ge 8\varepsilon M_{max} \|\pi\|,$$

and we conclude that [80, Theorem 4.5]

$$\|\eta(t)\| \leq \max\left\{be^{-at/\varepsilon}\|\eta(0)\|, \varepsilon c M_{max}\right\}$$

for some positive constants a, b, c. From (2.62) we see that $\|\eta(0)\| \leq \|\tilde{x}(0)\|/\varepsilon^{\rho_{max}-1}$ and $\tilde{x}^i_j = \varepsilon^{\rho_i - j}\eta^i_j$, which yield (2.61). □

2.6 ▪ Notes and References

High-gain observers can be designed for classes of nonlinear systems that are more general than the ones presented in Section 2.1 for single-output systems and Section 2.6 for multi-output systems. For such classes, the reader is referred to the book [49] and the papers [25, 41, 42, 51, 55, 56, 57]. The presentation of observer design in Section 2.2 follows closely the presentation in [80, Section 11.4]. The Lyapunov equation design of Section 2.3 is taken from [50], while the Riccati equation design can be found in [127]. Another design that is not included here is the H_∞ design that minimizes the H_∞ norm of the map from the uncertain input to the estimation error; see [116]. The reduced-order observer is discussed in [18], but more details are given in Section 2.4.

Chapter 3
Stabilization and Tracking

The separation principle is one of the fundamental results in linear control theory, whereby the design of a stabilizing output feedback controller can be achieved by separate designs of a stabilizing state feedback controller and an observer. Local versions of this result can be shown for nonlinear systems by using linearization. In this chapter we show a nonlocal, even semiglobal, separation principle for a class of nonlinear systems via the use of high-gain observers together with the global boundedness of the feedback controller in the estimated variables. The global boundedness property plays a crucial role in protecting against peaking in the observer. The separation principle we prove here goes beyond the separation principle in linear control theory because it does not only show stabilization under output feedback, but it also shows that the trajectories under output feedback approach the trajectories under state feedback, as the observer gain increases. This performance recovery property is valid only when the feedback controller is globally bounded in the estimated variables. In Section 3.1, we assume that the state feedback controller has this global boundedness property, which can be always achieved by saturating the control or the state estimates outside a compact set of interest. This saturation procedure is illustrated in examples and elaborated on in more detail in Section 3.2 in the context of robust stabilization.

In Section 3.2 we deal with the robust stabilization of minimum-phase nonlinear systems using output feedback. For relative-degree-one systems, the design of robust control is as simple as in state feedback. With high-gain observers, a system of relative degree higher than one can be treated as a system of relative degree one, which leads to straightforward design procedures. The robust control can be designed using sliding mode control, Lyapunov redesign, or high-gain feedback. In this chapter we use sliding mode control.[10]

The separation result of Section 3.1 is proved for the standard high-gain observer. When specialized to single-output systems, the result holds for a system of the form (2.1)–(2.4) with $\psi_i = 0$. In view of Theorem 2.1, it is not hard to see that a similar separation result holds if the standard observer is implemented as in (2.17)–(2.18) with functions ψ_i^s that satisfy (2.20) and (2.21). Moreover, in view of Theorem 2.2, a similar separation result holds for the reduced-order observer of Section 2.4. The proofs of such results are almost the same as the proof of Theorem 3.1 of Section 3.1.

[10]Throughout the book, sliding mode control is implemented with the signum function $\text{sgn}(s)$ replaced by $\text{sat}(s/\mu)$. By abuse of notation, we refer to such control as sliding mode control.

For minimum-phase nonlinear systems in the normal form, the tracking problem reduces to a stabilization problem when the state variables are chosen as the tracking error and its derivatives. Section 3.3 presents the design of output feedback tracking control using the same separation approach that is delineated in Sections 3.1 and 3.2.

3.1 ▪ Separation Principle

3.1.1 ▪ Problem Statement

Consider the multi-input–multi-output nonlinear system

$$\dot{w} = f_0(w, x, u), \tag{3.1}$$
$$\dot{x} = Ax + B\phi(w, x, u), \tag{3.2}$$
$$y = Cx, \tag{3.3}$$
$$z = \psi(w, x), \tag{3.4}$$

where $u \in R^m$ is the control input, $y \in R^q$ and $z \in R^s$ are measured outputs, and $w \in R^\ell$ and $x \in R^\rho$ constitute the state vector. The $\rho \times \rho$ matrix A, the $\rho \times q$ matrix B, and the $q \times \rho$ matrix C, given by

$$A = \text{block diag}[A_1, \ldots, A_q], \quad A_i = \begin{bmatrix} 0 & 1 & \cdots & \cdots & 0 \\ 0 & 0 & 1 & \cdots & 0 \\ \vdots & & & & \vdots \\ 0 & \cdots & \cdots & 0 & 1 \\ 0 & \cdots & \cdots & \cdots & 0 \end{bmatrix}_{\rho_i \times \rho_i},$$

$$B = \text{block diag}[B_1, \ldots, B_q], \quad B_i = \begin{bmatrix} 0 \\ 0 \\ \vdots \\ 0 \\ 1 \end{bmatrix}_{\rho_i \times 1},$$

$$C = \text{block diag}[C_1, \ldots, C_q], \quad C_i = \begin{bmatrix} 1 & 0 & \cdots & \cdots & 0 \end{bmatrix}_{1 \times \rho_i},$$

where $\rho = \rho_1 + \cdots + \rho_q$, represent q chains of integrators. Equations (3.2) and (3.3) are the same as (2.55) to (2.57) of Section 2.5 but are written in a more compact form. The functions f_0, ϕ, and ψ are locally Lipschitz in their arguments for $(w, x, u) \in D_w \times D_x \times R^m$, where $D_w \subset R^\ell$ and $D_x \subset R^\rho$ are domains that contain their respective origins. Moreover, $f_0(0,0,0) = 0$, $\phi(0,0,0) = 0$, and $\psi(0,0) = 0$. Our goal is to design an output feedback controller to stabilize the origin.

The model (3.1)–(3.4) arises naturally in mechanical and electromechanical systems, where displacement variables are measured, while their derivatives (velocities, accelerations, etc.) are not measured. The multivariable normal form of a nonlinear system having a well-defined (vector) relative degree is a special case of (3.1)–(3.3) [64, Section 5.1]. If y is the only measured variable, (3.4) is dropped. However, in many problems, we can measure some state variables in addition to those at the end of the chains of integrators. For example, a magnetic-levitation system is modeled by [80, Appendix A.8].

$$\begin{aligned}
\dot{x}_1 &= x_2,\\
\dot{x}_2 &= -bx_2 + 1 - \frac{4cx_3^2}{(1+x_1)^2},\\
\dot{x}_3 &= \frac{1}{T(x_1)}\left[-x_3 + u + \frac{\beta x_2 x_3}{(1+x_1)^2}\right],
\end{aligned}$$

where the dimensionless variables x_1, x_2, and x_3, are the position of a magnetic ball, its velocity, and the electric current in an electromagnet, respectively. Typically, we measure the ball position x_1 and the current x_3. The model fits the form (3.1)–(3.4) with (x_1, x_2) as the x component and x_3 as the w component. The measured outputs are $y = x_1$ and $z = x_3$.

Another source for the model (3.1)–(3.4) where (3.4) is significant arises in systems in which the dynamics are extended by adding integrators. Consider a single-input–single-output nonlinear system represented by the nth-order differential equation

$$y^{(n)} = f\left(y, y^{(1)}, \ldots, y^{(n-1)}, v, v^{(1)}, \ldots, v^{(m-1)}, v^{(m)}\right),$$

where v is the input, y is the output, and $f(\cdot)$ is a sufficiently smooth function in a domain of interest. This nonlinear input–output model reduces to the transfer function

$$H(s) = \frac{b_m s^m + b_{m-1}s^{m-1} + \cdots + b_0}{s^n + a_{n-1}s^{n-1} + \cdots + a_0}, \qquad b_m \neq 0,$$

for linear systems. We extend the dynamics of the system by adding a series of m integrators at the input side and define $u = v^{(m)}$ as the control input of the extended system. The dimension of the extended system is $n+m$. Taking the state variables as

$$w = \begin{bmatrix} v \\ v^{(1)} \\ \vdots \\ v^{(m-1)} \end{bmatrix}, \qquad x = \begin{bmatrix} y \\ y^{(1)} \\ \vdots \\ \vdots \\ y^{(n-1)} \end{bmatrix},$$

the state model of the extended system takes the form (3.1)–(3.4) with

$$\begin{aligned}
\dot{w} &= A_u w + B_u u,\\
\dot{x} &= Ax + B\phi(w, x, u),\\
y &= Cx,\\
z &= w,
\end{aligned}$$

where

$$A_u = \begin{bmatrix} 0 & 1 & \cdots & \cdots & 0 \\ 0 & 0 & 1 & \cdots & 0 \\ \vdots & & & & \vdots \\ 0 & \cdots & \cdots & 0 & 1 \\ 0 & \cdots & \cdots & \cdots & 0 \end{bmatrix}_{m\times m}, \qquad B_u = \begin{bmatrix} 0 \\ 0 \\ \vdots \\ 0 \\ 1 \end{bmatrix}_{m\times 1},$$

$$A=\begin{bmatrix} 0 & 1 & \cdots & \cdots & 0 \\ 0 & 0 & 1 & \cdots & 0 \\ \vdots & & & & \vdots \\ 0 & \cdots & \cdots & 0 & 1 \\ 0 & \cdots & \cdots & \cdots & 0 \end{bmatrix}_{n\times n}, B=\begin{bmatrix} 0 \\ 0 \\ \vdots \\ 0 \\ 1 \end{bmatrix}_{n\times 1}, C=\begin{bmatrix} 1 & 0 & \cdots & \cdots & 0 \end{bmatrix}_{1\times n},$$

$$\phi(w,x,u)=f(x_1,x_2,\ldots,x_n,w_1,w_2,\ldots,w_m,u).$$

The measured variables are y and the whole vector w.

It is seen in Section 2.2 that a high-gain observer can be designed for the system

$$\begin{aligned} \dot{w} &= f_0(w,x,u), \\ \dot{x}_i &= x_{i+1}+\psi_i(x_1,\ldots,x_i,u) \qquad \text{for } 1\le i\le \rho-1, \\ \dot{x}_\rho &= \phi(w,x,u), \\ y &= x_1 \end{aligned}$$

if the functions ψ_1 to $\psi_{\rho-1}$ are known. In feedback control we can allow uncertainty in these functions by applying a change of variables that transforms the system into the single-output case of (3.1)–(3.4). Suppose ψ_1 to $\psi_{\rho-1}$ are sufficiently smooth. The change of variables

$$z_1=x_1, \quad z_2=\dot{z}_1, \quad \ldots, \quad z_\rho=\dot{z}_{\rho-1}$$

transforms the system into the form

$$\begin{aligned} \dot{w} &= \bar{f}_0(w,z,u), \\ \dot{z}_i &= z_{i+1} \qquad \text{for } 1\le i\le \rho-1, \\ \dot{z}_\rho &= \bar{\phi}(w,z,u), \\ y &= z_1. \end{aligned}$$

It can be verified that

$$z_i = x_i + \bar{\psi}_i(x_1,\ldots,x_{i-1}) \qquad \text{for } 2\le i\le \rho$$

for some functions $\bar{\psi}_i$. The change of variables is a global diffeomorphism. By designing the output feedback controller in the z-coordinates, there is no need to know the change of variables. The functions $\bar{f}_0$ and $\bar{\phi}$ depend on the uncertain functions ψ_1 to $\psi_{\rho-1}$, but, as seen in Section 2.2, the high-gain observer does not require knowledge of $\bar{f}_0$ and is robust with respect to $\bar{\phi}$. Therefore, as long as a robust state feedback controller can be designed to tolerate the uncertainty in $\bar{f}_0$ and $\bar{\phi}$, the output feedback controller will be robust with respect to these two functions.

3.1.2 ▪ Main Result

We use a two-step approach to design the output feedback controller. First, a partial state feedback controller that uses measurements of x and z is designed to asymptotically stabilize the origin. Then a high-gain observer is used to estimate x from y. The state feedback controller is allowed to be a dynamical system of the form

$$\dot{\vartheta}=\Gamma(\vartheta,x,z), \qquad u=\gamma(\vartheta,x,z), \tag{3.5}$$

where γ and Γ are locally Lipschitz functions in their arguments over the domain of interest and globally bounded functions of x. Moreover, $\gamma(0,0,0)=0$ and $\Gamma(0,0,0)=0$. A static state feedback controller $u=\gamma(x,z)$ will be viewed as a special case of the foregoing equation by dropping the $\dot{\vartheta}$-equation. For convenience, we write the closed-loop system under state feedback as

$$\dot{\chi}=f(\chi), \tag{3.6}$$

where $\chi=\text{col}(w,x,\vartheta)$. The output feedback controller is taken as

$$\dot{\vartheta}=\Gamma(\vartheta,\hat{x},z), \qquad u=\gamma(\vartheta,\hat{x},z), \tag{3.7}$$

where $\hat{x}$ is generated by the high-gain observer

$$\dot{\hat{x}}=A\hat{x}+B\phi_0(z,\hat{x},u)+H(y-C\hat{x}). \tag{3.8}$$

The observer gain H is chosen as

$$H=\text{block diag}[H_1,\ldots,H_q], \;\; H_i=\begin{bmatrix} \alpha_1^i/\varepsilon \\ \alpha_2^i/\varepsilon^2 \\ \vdots \\ \alpha_{\rho_i-1}^i/\varepsilon^{\rho_i-1} \\ \alpha_{\rho_i}^i/\varepsilon^{\rho_i} \end{bmatrix}_{\rho_i\times 1}, \tag{3.9}$$

where ε is a small positive constant and the positive constants α_j^i are chosen such that the polynomial

$$s^{\rho_i}+\alpha_1^i s^{\rho_i-1}+\cdots+\alpha_{\rho_i-1}^i s+\alpha_{\rho_i}^i$$

is Hurwitz for $i=1,\ldots,q$. The function $\phi_0(z,x,u)$ is a nominal model of $\phi(w,x,u)$, which is required to be locally Lipschitz in its arguments over the domain of interest and globally bounded in x. Moreover, $\phi_0(0,0,0)=0$.

Theorem 3.1. *Consider the closed-loop system of the plant* (3.1)–(3.4) *and the output feedback controller* (3.7)–(3.8)*. Suppose the origin of* (3.6) *is asymptotically stable and* $\mathcal{R}$ *is its region of attraction. Let* $\mathcal{S}$ *be any compact set in the interior of* $\mathcal{R}$ *and* $\mathcal{Q}$ *be any compact subset of* R^ρ*. Suppose* $(\chi(0),\hat{x}(0))\in\mathcal{S}\times\mathcal{Q}$*. Then*

- *there exists* $\varepsilon_1^*>0$ *such that for every* $0<\varepsilon\le\varepsilon_1^*$*, the solutions* $(\chi(t),\hat{x}(t))$ *of the closed-loop system are bounded for all* $t\ge 0$*;*
- *given any* $\mu>0$*, there exist* $\varepsilon_2^*>0$ *and* $T_2>0$*, both dependent on* μ*, such that for every* $0<\varepsilon\le\varepsilon_2^*$*, the solutions of the closed-loop system satisfy*

$$\|\chi(t)\|\le\mu \quad\text{and}\quad \|\hat{x}(t)\|\le\mu \quad \forall\, t\ge T_2; \tag{3.10}$$

- *given any* $\mu>0$*, there exists* $\varepsilon_3^*>0$*, dependent on* μ*, such that for every* $0<\varepsilon\le\varepsilon_3^*$*, the solutions of the closed-loop system satisfy*

$$\|\chi(t)-\chi_r(t)\|\le\mu \quad \forall\, t\ge 0, \tag{3.11}$$

where χ_r *is the solution of* (3.6) *with* $\chi_r(0)=\chi(0)$*;*

- *if $\phi_0(z,x,u) = \phi(w,x,u)$, there exists $\varepsilon_4^* > 0$ such that for every $0 < \varepsilon \le \varepsilon_4^*$, the origin of the closed-loop system is asymptotically stable and $\mathscr{S} \times \mathscr{Q}$ is a subset of its region of attraction;*
- *if the origin of (3.6) is exponentially stable and $f_r(\chi)$ is continuously differentiable in some neighborhood of the origin, there exists $\varepsilon_5^* > 0$ such that for every $0 < \varepsilon \le \varepsilon_5^*$, the origin of the closed-loop system is exponentially stable and $\mathscr{S} \times \mathscr{Q}$ is a subset of its region of attraction.* ◇

The theorem shows a number of properties of the output feedback controller when ε is sufficiently small. First, it shows boundedness of the trajectories. Second, it shows practical stabilization because, by (3.10), the trajectories can be brought to an arbitrarily small neighborhood of the origin by choosing ε small enough. Third, (3.11) shows that the solution $\chi(t)$ under output feedback approaches the solution $\chi_r(t)$ under state feedback as ε tends to zero. Fourth, the output feedback controller recovers the region of attraction of the state feedback controller in the sense that the foregoing properties hold for any compact set in the interior of the region of attraction. Fifth, if the observer is implemented with no uncertainty in the function ϕ, that is, $\phi_0 = \phi$, then the controller recovers asymptotic stability of the origin. Finally, even when $\phi_0 \ne \phi$, the controller recovers exponential stability when the origin under state feedback is exponentially stable.[11] As a corollary of the theorem, it is clear that if the state feedback controller achieves global or semiglobal stabilization, then for sufficiently small ε, the output feedback controller achieves semiglobal stabilization or semiglobal practical stabilization because in this case the set $\mathscr{S} \times \mathscr{Q}$ can be chosen arbitrarily large.

Proof of Theorem 3.1: For the purpose of analysis, the observer dynamics are replaced by the equivalent dynamics of the scaled estimation error

$$\eta_j^i = \frac{x_j^i - \hat{x}_j^i}{\varepsilon^{\rho_i - j}}$$

for $1 \le i \le q$ and $1 \le j \le \rho_i$. Hence, $\hat{x} = x - D(\varepsilon)\eta$, where

$$\begin{aligned} \eta &= \text{col}(\eta_1^1, \dots, \eta_{\rho_1}^1, \dots, \eta_1^q, \dots, \eta_{\rho_q}^q), \\ D(\varepsilon) &= \text{block diag}[D_1, \dots, D_q], \\ D_i &= \text{diag}[\varepsilon^{\rho_i - 1}, \dots, 1]_{\rho_i \times \rho_i}. \end{aligned}$$

The closed-loop system can be represented by

$$\begin{aligned} \dot{w} &= f_0(w, x, \gamma(\vartheta, x - D(\varepsilon)\eta, z)), \\ \dot{x} &= Ax + B\phi(w, x, \gamma(\vartheta, x - D(\varepsilon)\eta, z)), \\ \dot{\vartheta} &= \Gamma(\vartheta, x - D(\varepsilon)\eta, z), \\ \varepsilon\dot{\eta} &= A_0\eta + \varepsilon B\delta(w, x, \vartheta, D(\varepsilon)\eta), \end{aligned}$$

where

$$\delta(w, x, \vartheta, D(\varepsilon)\eta) = \phi(w, x, \gamma(\vartheta, \hat{x}, z)) - \phi_0(z, \hat{x}, \gamma(\vartheta, \hat{x}, z))$$

[11] For convenience, recovery of asymptotic stability when $\phi \ne \phi_0$ is shown only for the exponentially stable case. See [18] for the more general case when the origin is asymptotically, but not exponentially, stable.

and $A_0 = \varepsilon D^{-1}(\varepsilon)(A - HC)D(\varepsilon)$ is a $\rho \times \rho$ Hurwitz matrix, independent of ε. For convenience, we rewrite the system in the singularly perturbed form

$$\dot{\chi} = F(\chi, D(\varepsilon)\eta) \tag{3.12}$$
$$\varepsilon\dot{\eta} = A_0\eta + \varepsilon B\Delta(\chi, D(\varepsilon)\eta) \tag{3.13}$$

where $F(\chi, 0) = f(\chi)$. The initial states are $\chi(0) = \text{col}(w(0), x(0), \vartheta(0)) \in \mathscr{S}$ and $\hat{x}(0) \in \mathscr{Q}$. Thus, $\eta(0) = D^{-1}(\varepsilon)[x(0) - \hat{x}(0)]$. As common in singular perturbation analysis [87], we define the reduced (slow) and boundary-layer (fast) subsystems of (3.12)–(3.13). Setting $\varepsilon = 0$ in (3.13) yields $\eta = 0$, and the reduced system

$$\dot{\chi} = f(\chi) \tag{3.14}$$

is the closed-loop system under state feedback. The boundary-layer system, obtained by applying the change of time variable $\tau = t/\varepsilon$ and then setting $\varepsilon = 0$, is given by

$$\frac{d\eta}{d\tau} = A_0\eta.$$

Since the origin of (3.14) is asymptotically stable and $\mathscr{R}$ is its region of attraction, by the converse Lyapunov theorem of asymptotic stability [78, Theorem 4.17], there is a smooth, positive definite function $V(\chi)$ and a continuous, positive definite function $U(\chi)$, both defined for all $\chi \in \mathscr{R}$, such that

$$V(\chi) \to \infty \text{ as } \chi \to \partial\mathscr{R},$$

$$\frac{\partial V}{\partial \chi} f(\chi) \leq -U(\chi) \quad \forall\, \chi \in \mathscr{R}$$

and for any $c > 0$, $\{V(\chi) \leq c\}$ is a compact subset of $\mathscr{R}$.

Boundedness: Let $\mathscr{S}$ be any compact set in the interior of $\mathscr{R}$. Choose positive constants b and c such that $c > b > \max_{\chi \in \mathscr{S}} V(\chi)$. Then

$$\mathscr{S} \subset \Omega_b = \{V(\chi) \leq b\} \subset \Omega_c = \{V(\chi) \leq c\} \subset \mathscr{R}.$$

For the boundary-layer system, the Lyapunov function $W(\eta) = \eta^T P_0 \eta$, where P_0 is the positive definite solution of the Lyapunov equation $P_0 A_0 + A_0^T P_0 = -I$, satisfies

$$\lambda_{\min}(P_0)\|\eta\|^2 \leq W(\eta) \leq \lambda_{\max}(P_0)\|\eta\|^2,$$

$$\frac{\partial W}{\partial \eta} A_0 \eta = -\|\eta\|^2.$$

Let $\Sigma = \{W(\eta) \leq \varrho\varepsilon^2\}$ and $\Lambda = \Omega_c \times \Sigma$. Due to the global boundedness of F and Δ in $\hat{x}$, there are positive constants k_1 and k_2 independent of ε such that

$$\|F(\chi, D(\varepsilon)\eta)\| \leq k_1, \quad \|\Delta(\chi, D(\varepsilon)\eta)\| \leq k_2$$

for all $\chi \in \Omega_c$ and $\eta \in R^{\rho}$. Moreover, for any $0 < \tilde{\varepsilon} < 1$, there is L_1, independent of ε, such that for all $(\chi, \eta) \in \Lambda$ and every $0 < \varepsilon \leq \tilde{\varepsilon}$,

$$\|F(\chi, D(\varepsilon)\eta) - F(\chi, 0)\| \leq L_1\|\eta\|.$$

We will always consider $\varepsilon \leq \tilde{\varepsilon}$. We start by showing that there exist positive constants ϱ and ε_1 (dependent on ϱ) such that the compact set Λ is positively invariant for every $0 < \varepsilon \leq \varepsilon_1$. This can be done by verifying that

$$\dot{V} \leq -U(\chi) + \varepsilon k_3$$

and

$$\dot{W} \leq -\frac{1}{\varepsilon}||\eta||^2 + 2||\eta||\,||P_0B||k_2$$

for all $(\chi, \eta) \in \Lambda$, where $k_3 = L_1L_2\sqrt{\varrho/\lambda_{\min}(P_0)}$ and L_2 is an upper bound for $||\partial V/\partial \chi||$ over Ω_c. Taking $\varrho = 16k_2^2||P_0B||^2\lambda_{\max}(P_0)$ and $\varepsilon_1 = \beta/k_3$, where $\beta = \min_{\chi \in \partial\Omega_c} U(\chi)$, it can be shown that, for every $0 < \varepsilon \leq \varepsilon_1$, $\dot{V} \leq 0$ for all $(\chi, \eta) \in \{V(\chi) = c\} \times \Sigma$ and $\dot{W} < 0$ for all $(\chi, \eta) \in \Omega_c \times \{W(\eta) = \varrho\varepsilon^2\}$. Hence, Λ is positively invariant.

Now consider the initial state $(\chi(0), \hat{x}(0)) \in \mathscr{S} \times \mathscr{Q}$. It can be verified that the corresponding initial error $\eta(0)$ satisfies $||\eta(0)|| \leq k/\varepsilon^{(\rho_{\max}-1)}$ for some nonnegative constant k dependent on $\mathscr{S}$ and $\mathscr{Q}$, where $\rho_{\max} = \max\{\rho_1, \ldots, \rho_m\}$. Since $\chi(0)$ is in the interior of Ω_c,

$$||\chi(t) - \chi(0)|| \leq k_1 t \tag{3.15}$$

as long as $\chi(t) \in \Omega_c$. Thus, there exists a finite time T_0, independent of ε, such that $\chi(t) \in \Omega_c$ for all $t \in [0, T_0]$. During this time interval,

$$\dot{W} \leq -\frac{1}{2\varepsilon}||\eta||^2 - \frac{1}{2\varepsilon}||\eta||^2 + 2k_2||P_0B||\,||\eta|| \leq -\frac{1}{2\varepsilon}||\eta||^2, \quad \text{for } W(\eta) \geq \varrho\varepsilon^2$$

because

$$W(\eta) \geq \varrho\varepsilon^2 \Rightarrow \lambda_{\max}(P_0)||\eta||^2 \geq 16k_2^2||P_0B||^2\lambda_{\max}(P_0)\varepsilon^2 \Leftrightarrow ||\eta|| \geq 4k_2||P_0B||\varepsilon.$$

Therefore,

$$\dot{W} \leq -\frac{1}{2\lambda_{\max}(P_0)\varepsilon}W \qquad \text{for } W(\eta) \geq \varrho\varepsilon^2.$$

By the comparison lemma [78, Lemma 3.4],

$$W(\eta(t)) \leq W(\eta(0))\exp(-\sigma_1 t/\varepsilon) \leq \frac{\sigma_2}{\varepsilon^{2(\rho_{\max}-1)}}\exp(-\sigma_1 t/\varepsilon), \tag{3.16}$$

where $\sigma_1 = 1/(2\lambda_{\max}(P_0))$ and $\sigma_2 = k^2\lambda_{\max}(P_0)$. The right-hand side of (3.16) equals $\varrho\varepsilon^2$ when

$$\frac{\sigma_2}{\varepsilon^{2(\rho_{\max}-1)}}\exp(-\sigma_1 t/\varepsilon) = \varrho\varepsilon^2 \quad \Leftrightarrow \quad t = \frac{\varepsilon}{\sigma_1}\ln\left(\frac{\sigma_2}{\varrho\,\varepsilon^{2\rho_{\max}}}\right).$$

Choose $\varepsilon_2 > 0$ small enough that

$$T(\varepsilon) \stackrel{\text{def}}{=} \frac{\varepsilon}{\sigma_1}\ln\left(\frac{\sigma_2}{\varrho\,\varepsilon^{2\rho_{\max}}}\right) \leq \tfrac{1}{2}T_0$$

for all $0 < \varepsilon \leq \varepsilon_2$. We note that ε_2 exists since, by l'Hôpital's rule, $T(\varepsilon)$ tends to zero as ε tends to zero. It follows that $W(\eta(T(\varepsilon))) \leq \varrho\varepsilon^2$ for every $0 < \varepsilon \leq \varepsilon_2$. Taking $\varepsilon_1^* = \min\{\tilde{\varepsilon}, \varepsilon_1, \varepsilon_2\}$ guarantees that, for every $0 < \varepsilon \leq \varepsilon_1^*$, the trajectory

$(\chi(t),\eta(t))$ enters Λ during the interval $[0,T(\varepsilon)]$ and remains there for all $t\geq T(\varepsilon)$. Consequently, the trajectory is bounded for all $t\geq T(\varepsilon)$. On the other hand, for $t\in[0,T(\varepsilon)]$, the trajectory is bounded by virtue of (3.15) and (3.16).

Ultimate boundedness: Next, we show (3.10). We know that, for every $0<\varepsilon\leq\varepsilon_1^*$, the solutions are inside Λ for all $t\geq T(\varepsilon)$, where Λ is $O(\varepsilon)$ in the direction of the variable η. Thus, we can find $\varepsilon_3=\varepsilon_3(\mu)\leq\varepsilon_1^*$ such that, for every $0<\varepsilon\leq\varepsilon_3$,

$$\|\eta(t)\|\leq\mu/2 \quad \forall\ t\geq T(\varepsilon_3)\overset{\text{def}}{=}\bar{T}(\mu). \tag{3.17}$$

Using the fact that $\dot{V}\leq-U(\chi)+\varepsilon k_3$ for all $(\chi,\eta)\in\Lambda$, we conclude that

$$\dot{V}\leq-\tfrac{1}{2}U(\chi) \quad \text{for } \chi\notin\{U(\chi)\leq 2k_3\varepsilon\overset{\text{def}}{=}\nu(\varepsilon).\} \tag{3.18}$$

Because $U(\chi)$ is positive definite and continuous, the set $\{U(\chi)\leq\nu(\varepsilon)\}$ is compact for sufficiently small ε. Let $c_0(\varepsilon)=\max_{U(\chi)\leq\nu(\varepsilon)}\{V(\chi)\}$; $c_0(\varepsilon)$ is nondecreasing and $\lim_{\varepsilon\to 0}c_0(\varepsilon)=0$. Consider the compact set $\{V(\chi)\leq c_0(\varepsilon)\}$. We have $\{U(\chi)\leq\nu(\varepsilon)\}\subset\{V(\chi)\leq c_0(\varepsilon)\}$. Choose $\varepsilon_4=\varepsilon_4(\mu)\leq\varepsilon_1^*$ small enough such that, for all $\varepsilon\leq\varepsilon_4$, the set $\{U(\chi)\leq\nu(\varepsilon)\}$ is compact, the set $\{V(\chi)\leq c_0(\varepsilon)\}$ is in the interior of Ω_c, and

$$\{V(\chi)\leq c_0(\varepsilon)\}\subset\{\|\chi\|\leq\mu/2\}. \tag{3.19}$$

Then, for all $\chi\in\Omega_c$ but $\chi\notin\{V(\chi)\leq c_0(\varepsilon)\}$, we have an inequality similar to (3.18). Therefore, we conclude that the set $\{V(\chi)\leq c_0(\varepsilon)\}\times\Sigma$ is positively invariant and every trajectory in $\Omega_c\times\Sigma$ reaches $\{V(\chi)\leq c_0(\varepsilon)\}\times\Sigma$ in finite time. In other words, given (3.19), there exists a finite time $\tilde{T}=\tilde{T}(\mu)$ such that for every $0<\varepsilon\leq\varepsilon_4$,

$$\|\chi(t)\|\leq\mu/2 \quad \forall\ t\geq\tilde{T}. \tag{3.20}$$

Take $\varepsilon_2^*=\varepsilon_2^*(\mu)=\min\{\varepsilon_3,\varepsilon_4\}$ and $T_2=T_2(\mu)=\max\{\bar{T},\tilde{T}\}$. Then, (3.10) follows from (3.17), (3.20), $\hat{x}=x-D(\varepsilon)\eta$, and $\|D(\varepsilon)\|=1$.

Performance recovery: To show (3.11), we divide the interval $[0,\infty)$ into three intervals, $[0,T(\varepsilon)]$, $[T(\varepsilon),T_3]$, and $[T_3,\infty)$, and show (3.11) for each interval. From the ultimate boundedness of $\chi(t)$, shown in (3.10), and the asymptotic stability of the origin of (3.14), we conclude that there exists a finite time $T_3\geq T(\varepsilon)$, independent of ε, such that, for every $0<\varepsilon\leq\varepsilon_2^*$, we have

$$\|\chi(t)-\chi_r(t)\|\leq\mu \quad \forall\ t\geq T_3. \tag{3.21}$$

From (3.15), we know that

$$\|\chi(t)-\chi(0)\|\leq k_1 t$$

during the interval $[0,T(\varepsilon)]$. Similarly, it can be shown that

$$\|\chi_r(t)-\chi(0)\|\leq k_1 t$$

during the same interval. Hence,

$$\|\chi(t)-\chi_r(t)\|\leq 2k_1T(\varepsilon) \quad \forall\ t\in[0,T(\varepsilon)].$$

Since $T(\varepsilon)\to 0$ as $\varepsilon\to 0$, there exists $0<\varepsilon_5\leq\varepsilon_2^*$ such that, for every $0<\varepsilon\leq\varepsilon_5$,

$$\|\chi(t)-\chi_r(t)\|\leq\mu \quad \forall\ t\in[0,T(\varepsilon)]. \tag{3.22}$$

Over the interval $[T(\varepsilon), T_3]$, the solution $\chi(t)$ satisfies

$$\dot{\chi} = F(\chi, D(\varepsilon)\eta(t)), \quad \text{with } \|\chi(T(\varepsilon)) - \chi_r(T(\varepsilon))\| \le \delta_1(\varepsilon),$$

where $D(\varepsilon)\eta$ is $O(\varepsilon)$ and $\delta_1(\varepsilon) \to 0$ as $\varepsilon \to 0$. Thus, by the continuous dependence of the solutions of differential equations on initial conditions and parameters [78, Theorem 3.5], we conclude that there exists $0 < \varepsilon_6 \le \varepsilon_2^*$ such that, for every $0 < \varepsilon \le \varepsilon_6$,

$$\|\chi(t) - \chi_r(t)\| \le \mu \quad \forall\, t \in [T(\varepsilon), T_3]. \tag{3.23}$$

Take $\varepsilon_3^* = \min\{\varepsilon_5, \varepsilon_6\}$. Then, (3.11) follows from (3.21)–(3.23).

Asymptotic stability: If $\phi_0(z, x, u) = \phi(w, x, u)$, it follows from the local Lipschitz property of ϕ that there is $r > 0$ such that for all $\|\chi\| \le r$ and $\|\eta\| \le r$, $\|\Delta(\chi, D(\varepsilon)\eta)\| \le L_3\|\eta\|$ for some $L_3 > 0$. On the same set, there are positive constants L_4 and L_5 such that $\|\partial V/\partial \chi\| \le L_4$ and $\|F(\chi, D(\varepsilon)\eta) - F(\chi, 0)\| \le L_5\|\eta\|$. Consider the composite Lyapunov function $V_1(\chi, \eta) = V(\chi) + \sqrt{W(\eta)}$:

$$\begin{aligned}
\dot{V}_1 &= \frac{\partial V_1}{\partial \chi} F(\chi, D(\varepsilon)\eta) + \frac{1}{2\sqrt{W}}\left[-\frac{1}{\varepsilon}\|\eta\|^2 + 2\eta^T P_0 B \Delta\right] \\
&\le \frac{\partial V_1}{\partial \chi} F(\chi, 0) + \frac{\partial V_1}{\partial \chi}[F(\chi, D(\varepsilon)\eta) - F(\chi, 0)] \\
&\quad - \frac{1}{2\varepsilon\sqrt{\lambda_{\max}(P_0)}}\|\eta\| + \frac{\|P_0 B\| L_3}{\sqrt{\lambda_{\min}(P_0)}}\|\eta\| \\
&\le -U(\chi) - \frac{1}{4\varepsilon\sqrt{\lambda_{\max}(P_0)}}\|\eta\| \\
&\quad - \left[\frac{1}{4\varepsilon\sqrt{\lambda_{\max}(P_0)}} - L_4 L_5 - \frac{\|P_0 B\| L_3}{\sqrt{\lambda_{\min}(P_0)}}\right]\|\eta\|.
\end{aligned}$$

Choose $\varepsilon_7 > 0$ such that

$$\frac{1}{4\varepsilon_7\sqrt{\lambda_{\max}(P_0)}} - L_4 L_5 - \frac{\|P_0 B\| L_3}{\sqrt{\lambda_{\min}(P_0)}} > 0.$$

Then, for $0 < \varepsilon \le \varepsilon_7$,

$$\dot{V}_1 \le -U(\chi) - \frac{1}{4\varepsilon\sqrt{\lambda_{\max}(P_0)}}\|\eta\|.$$

Hence, there is a neighborhood $\mathcal{N}_1$ of the origin, independent of ε, such that for every $0 < \varepsilon \le \varepsilon_7$, the origin is asymptotically stable and every trajectory in $\mathcal{N}_1$ converges to the origin as $t \to \infty$. By (3.10), there exists $\varepsilon_8 > 0$ such that for every $0 < \varepsilon \le \varepsilon_8$, the solutions starting in $\mathcal{S} \times \mathcal{Q}$ enter $\mathcal{N}_1$ in finite time. Hence, for every $0 < \varepsilon \le \varepsilon_4^* = \min\{\varepsilon_7, \varepsilon_8\}$, the origin is asymptotically stable and $\mathcal{S} \times \mathcal{Q}$ is a subset of the region of attraction.

Exponential stability: Finally, assuming that the origin of (3.6) is exponentially stable, it follows from the converse Lyapunov theorem of exponential stability [78, Theorem 4.14]. that there exists a continuously differentiable Lyapunov function $V_2(\chi)$

that satisfies the inequalities

$$b_1\|\chi\|^2 \le V_2(\chi) \le b_2\|\chi\|^2, \quad \frac{\partial V_2}{\partial \chi}F(\chi,0) \le -b_3\|\chi\|^2, \quad \left\|\frac{\partial V_2}{\partial \chi}\right\| \le b_4\|\chi\|,$$

over a ball $B_r \subset \mathscr{R}$ for some positive constants r, b_1, b_2, b_3, and b_4. Using the local Lipschitz property of F and Δ and the fact that $F(0,0)=0$ and $\Delta(0,0)=0$, it can be shown that the composite Lyapunov function $V_3(\chi,\eta)=V_2(\chi)+W(\eta)$ satisfies

$$\begin{aligned}\dot{V}_3 &= \frac{\partial V_2}{\partial \chi}F(\chi,0)+\frac{\partial V_2}{\partial \chi}[F(\chi,D(\varepsilon)\eta)-F(\chi,0)] \\ &\quad -\frac{1}{\varepsilon}\|\eta\|^2+2\eta^T P_0 B\Delta(\chi,D(\varepsilon)\eta) \\ &\le -b_3\|\chi\|^2+2\beta_1\|\chi\|\,\|\eta\|-\frac{1}{\varepsilon}\|\eta\|^2+\beta_2\|\eta\|^2 \\ &= -\mathscr{Y}^T Q\mathscr{Y},\end{aligned}$$

where

$$Q=\begin{bmatrix} b_3 & -\beta_1 \\ -\beta_1 & (1/\varepsilon)-\beta_2 \end{bmatrix}, \quad \mathscr{Y}=\begin{bmatrix} \|\chi\| \\ \|\eta\| \end{bmatrix}$$

for some nonnegative constants β_1 and β_2. The matrix Q is positive definite for sufficiently small ε. Hence, there is a neighborhood $\mathscr{N}_2$ of the origin, independent of ε, and $\varepsilon_9>0$ such that for every $0<\varepsilon\le\varepsilon_9$, the origin is exponentially stable and every trajectory in $\mathscr{N}_2$ converges to the origin as $t\to\infty$. By (3.10), there exists $\varepsilon_{10}>0$ such that for every $0<\varepsilon\le\varepsilon_{10}$, the solutions starting in $\mathscr{S}\times\mathscr{Q}$ enter $\mathscr{N}_2$ in finite time. Hence, for every $0<\varepsilon\le\varepsilon_5^*=\min\{\varepsilon_9,\varepsilon_{10}\}$, the origin is exponentially stable and $\mathscr{S}\times\mathscr{Q}$ is a subset of the region of attraction. □

Theorem 3.1 is proved for the full-order high-gain observer 3.8. A similar result can be proved for the reduced-order high-gain observer (2.48)–(2.50). The performance of the reduced-order observer is demonstrated in Example 3.3.

3.1.3 ▪ Examples

We use three examples to illustrate the separation principle and some reasonable intuition.

Example 3.1. We consider a second-order system having an exponentially unstable mode, together with a bounded linear controller that achieves a finite region of attraction. The system is

$$\dot{x}_1=x_2, \qquad \dot{x}_2=2x_1+10u, \qquad y=x_1.$$

The origin is stabilized by the globally bounded state feedback controller

$$u=\tanh(-x_1-x_2).$$

The output feedback controller is

$$u=\tanh(-\hat{x}_1-\hat{x}_2),$$

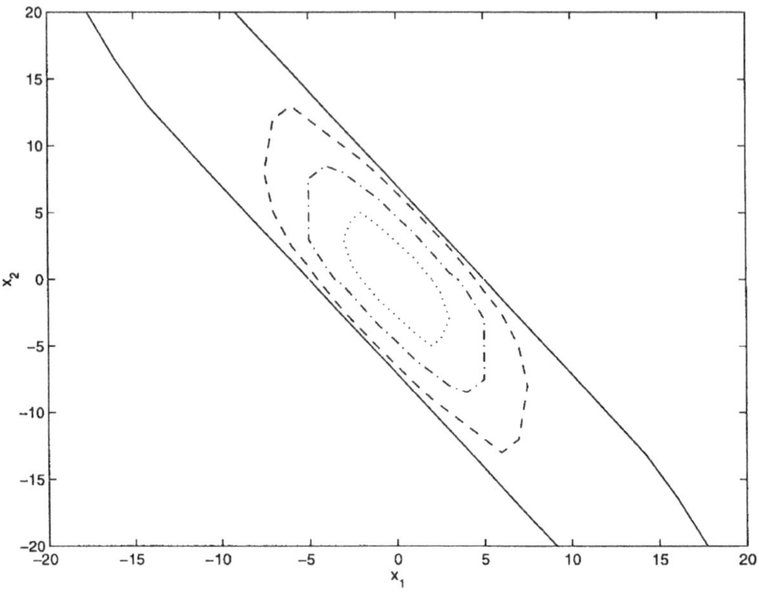

Figure 3.1. *Recovery of the region of attraction in Example* 3.1: $\varepsilon^\star = 0.007$ *(dashed),* $\varepsilon^\star = 0.057$ *(dash-dotted), and* $\varepsilon^\star = 0.082$ *(dotted).*

where the estimates $\hat{x}_1$ and $\hat{x}_2$ are provided by the high-gain observer

$$\dot{\hat{x}}_1 = \hat{x}_2 + (1/\varepsilon)(y - \hat{x}_1), \qquad \dot{\hat{x}}_2 = (1/\varepsilon^2)(y - \hat{x}_1).$$

We show how the output feedback controller recovers the region of attraction achieved under state feedback. Figure 3.1 shows the region of attraction under state feedback control (solid), in addition to three compact subsets that are recovered using the high-gain observer. In each case the compact subset is specified, then a design parameter $\varepsilon^\star$ is found through multiple simulations at different points of the subset such that for every $\varepsilon \le \varepsilon^\star$ the output feedback controller is able to recover the given subset; i.e., it is a subset of the region of attraction of the closed-loop system under output feedback. The bound $\varepsilon^\star$ is tight in the sense that for $\varepsilon > \varepsilon^\star$ there is a part of the given set that is not included in the region of attraction. The bounds $\varepsilon^\star$'s for these subsets are 0.082, 0.057, and 0.007, respectively, starting from the smallest subset. Notice that the bigger the subset, the smaller the bound $\varepsilon^\star$. In all cases we take $\hat{x}(0) = 0$.

Example 3.2. Consider the inverted pendulum-on-a-cart system, modeled by [80, Appendix A.11].

$$\begin{aligned}
\dot{x}_1 &= x_2,\\
\dot{x}_2 &= \frac{1}{\Delta(x_1)}\left[(m+M)mgL\sin x_1 - mL\cos x_1(u + mLx_2^2\sin x_1 - kx_4)\right],\\
\dot{x}_3 &= x_4,\\
\dot{x}_4 &= \frac{1}{\Delta(x_1)}\left[-m^2L^2 g\sin x_1\cos x_1 + (J + mL^2)(u + mLx_2^2\sin x_1 - kx_4)\right],
\end{aligned}$$

where m is the mass of the pendulum, M the mass of the cart, L the distance from the center of gravity to the pivot, J the moment of inertia of the pendulum with respect to the center of gravity, k a friction coefficient, x_1 the angular rotation of the pendulum from the upward vertical position (measured clockwise), x_3 the displacement of the pivot, g the acceleration due to gravity, and

$$\Delta(x_1) = (J + mL^2)(m + M) - m^2L^2\cos^2 x_1 \geq (J + mL^2)M + mJ > 0.$$

The model parameters are

$$m = 0.1, \quad M = 1, \quad k = 0.1, \quad J = 0.008, \quad L = 0.5, \quad g = 9.81.$$

The measured outputs are x_1 and x_3, and the goal is to stabilize the origin $x = 0$.

We start by the design of state feedback control. Assuming the domain of interest is limited to $|x_1| < \pi/2$, we take

$$u = -mLx_2^2 \sin x_1 + kx_4 + (m + M)g\tan x_1 - \frac{\Delta(x_1)}{mL\cos x_1}v$$

to simplify the foregoing equations to

$$\dot{x}_1 = x_2, \quad \dot{x}_2 = v, \quad \dot{x}_3 = x_4, \quad \dot{x}_4 = g\tan x_1 - \frac{J + mL^2}{mL\cos x_1}v.$$

The design of the auxiliary control v will follow a two-time-scale approach.[12] We take $v = -\beta_1(x_1 - \bar{x}_1) - \beta_2 x_2$ and design the positive feedback gains β_1 and β_2 such that the pendulum dynamics (x_1, x_2) are much faster than the cart dynamics (x_3, x_4), so that within a short period of time, x_1 reaches its quasi-steady-state $\bar{x}_1$ and x_2 and v reach zero. Then the slow cart dynamics are approximated by

$$\dot{x}_3 = x_4, \qquad \dot{x}_4 = g\tan(\bar{x}_1).$$

Taking $\bar{x}_1 = \tan^{-1}(-(\gamma_1 x_3 + \gamma_2 x_4)/g)$ results in

$$\dot{x}_3 = x_4, \qquad \dot{x}_4 = -\gamma_1 x_3 - \gamma_2 x_4,$$

where the positive feedback gains γ_1 and γ_2 are chosen much smaller than β_1 and β_2. This procedure can be validated by representing the closed-loop system in the singularly perturbed form. Taking

$$\gamma_1 = \mu^2 k_1, \ \gamma_2 = \mu k_2, \ \tau = \mu t, \ \eta_1 = \mu^2 x_3, \ \eta_2 = \mu x_4, \ \xi_1 = x_1 - \bar{x}_1, \ \xi_2 = x_2,$$

where k_1 and k_2 are positive constants and μ is a small positive constant, it can be shown that the closed-loop system takes the singularly perturbed form

$$\begin{aligned}
\frac{d\eta_1}{d\tau} &= \eta_2, \\
\frac{d\eta_2}{d\tau} &= -k_1\eta_1 - k_2\eta_2 + h_\eta(\cdot), \\
\mu\frac{d\xi_1}{d\tau} &= \xi_2 + \mu\, h_\xi(\cdot), \\
\mu\frac{d\xi_2}{d\tau} &= -\beta_1\xi_1 - \beta_2\xi_2,
\end{aligned}$$

[12]See [53], [92], and [147].

where h_η and h_ξ are smooth functions of their arguments and h_η vanishes at $\xi = 0$. The slow model

$$\frac{d\eta_1}{d\tau} = \eta_2, \quad \frac{d\eta_2}{d\tau} = -k_1\eta_1 - k_2\eta_2$$

and the fast model

$$\frac{d\xi_1}{dt} = \xi_2, \quad \frac{d\xi_2}{dt} = -\beta_1\xi_1 - \beta_2\xi_2$$

have exponentially stable equilibrium points at the origin. By singular perturbation analysis, it can be shown that the origin of the full system is exponentially stable for sufficiently small μ.[13] In summary, the state feedback control is given by

$$u = -mLx_2^2 \sin x_1 + kx_4 + (m+M)g \tan x_1 \\ - \frac{\Delta(x_1)}{mL\cos x_1}\left\{-\beta_1[x_1 - \tan^{-1}(-(\gamma_1 x_3 + \gamma_2 x_4)/g)] - \beta_2 x_2\right\}.$$

We take

$$\beta_1 = k_1 = 100, \beta_2 = k_2 = 10\sqrt{2}, \mu = 0.075 \Rightarrow \gamma_1 = 0.5625, \quad \gamma_2 = 1.0607.$$

In output feedback control, we use the high-gain observer

$$\begin{aligned}
\dot{\hat{x}}_1 &= \hat{x}_2 + (\alpha_1/\varepsilon)(x_1 - \hat{x}_1), \\
\dot{\hat{x}}_2 &= \phi_{01}(\hat{x}, u) + (\alpha_2/\varepsilon^2)(x_1 - \hat{x}_1), \\
\dot{\hat{x}}_3 &= \hat{x}_4 + (\alpha_1/\varepsilon)(x_3 - \hat{x}_3), \\
\dot{\hat{x}}_4 &= \phi_{02}(\hat{x}, u) + (\alpha_2/\varepsilon^2)(x_3 - \hat{x}_3).
\end{aligned}$$

The observer gains are taken as $\alpha_1 = 2$ and $\alpha_2 = 1$ to assign all the observer eigenvalues at $-1/\varepsilon$. We consider two observers: a linear observer where $\phi_{01} = \phi_{02} = 0$ and a nonlinear observer where ϕ_{01} and ϕ_{02} are given by

$$\begin{aligned}
\phi_{01}(\hat{x}, u) &= \frac{1}{\Delta(\hat{x}_1)}\left[(m+M)mgL\sin\hat{x}_1 - mL\cos\hat{x}_1(u + mL\hat{x}_2^2\sin\hat{x}_1 - k\hat{x}_4)\right], \\
\phi_{02}(\hat{x}, u) &= \frac{1}{\Delta(\hat{x}_1)}\left[-m^2L^2 g\sin\hat{x}_1\cos\hat{x}_1 + (J + mL^2)(u + mL\hat{x}_2^2\sin\hat{x}_1 - k\hat{x}_4)\right].
\end{aligned}$$

The state x in the feedback control is replaced by its estimate $\hat{x}$, and the control is saturated at ± 100. The functions ϕ_{01} and ϕ_{02} are made globally bounded by saturating $\hat{x}_2$ and $\hat{x}_4$ at ± 10. The saturation levels are determined by simulating the closed-loop system under state feedback for initial conditions in the domain of interest and determining the maximum values of $|u|$, $|x_2|$, and $|x_4|$. The saturation levels are higher than these maxima.

Figure 3.2 shows how the velocity estimate peaks due to the difference in the initial conditions between the position and its estimate. It also shows the control saturation during the peaking interval. This is done for a linear observer with $\varepsilon = 0.001$ and initial conditions $x(0) = \text{col}(1,0,1,0)$ and $\hat{x}(0) = 0$. Figure 3.3 shows how the output feedback controller recovers the trajectories achieved under state feedback. We use a linear high-gain observer with two values of ε: 0.03 and 0.005. The initial conditions

[13]See [92] for the details of the singular perturbation analysis.

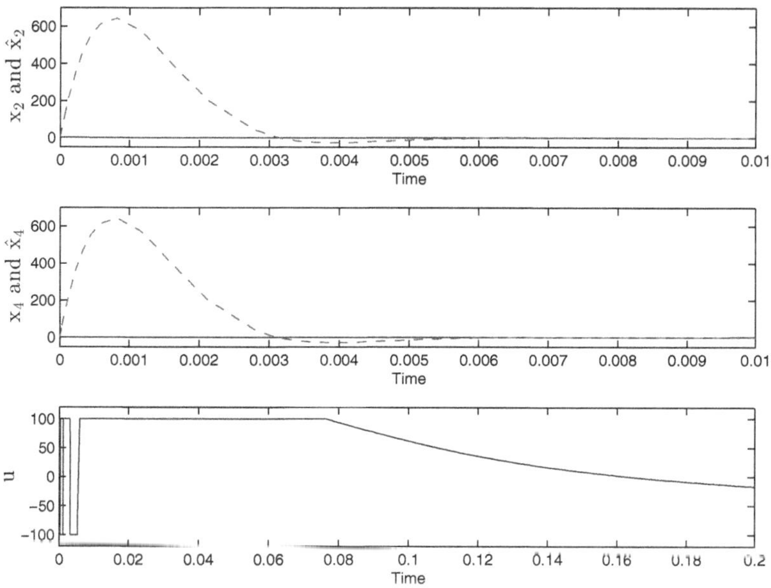

Figure 3.2. *Simulation of Example 3.2. Peaking of the velocity estimates and saturation of the control: the states x_2 and x_4 (solid) and their estimates $\hat{x}_2$ and $\hat{x}_4$ (dashed).*

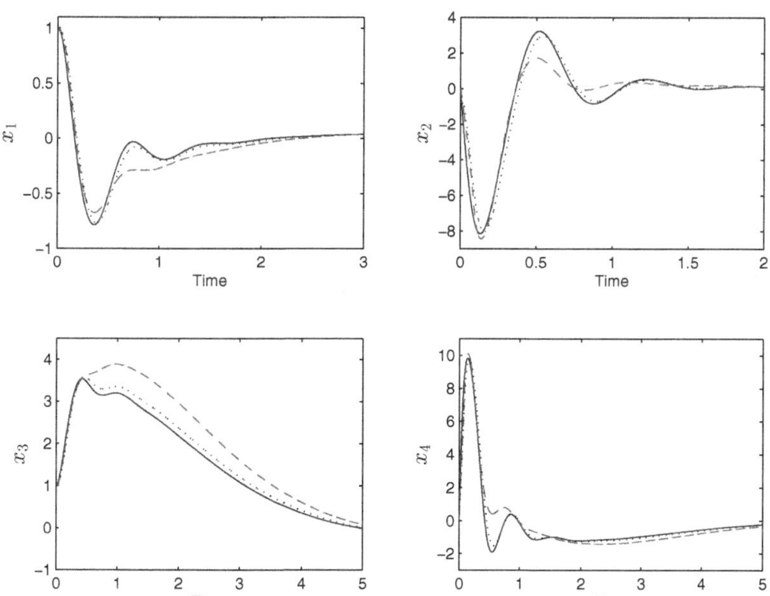

Figure 3.3. *Simulation of Example 3.2. Trajectory convergence: state feedback (solid); output feedback with $\varepsilon = 0.03$ (dashed) and $\varepsilon = 0.005$ (dotted).*

are $x(0) = \text{col}(1, 0, 1, 0)$ and $\hat{x}(0) = 0$. Intuitively we expect that a nonlinear observer that includes a model of the system's nonlinearities would outperform a linear one when the model is accurate and ε is not too small. Figure 3.4 shows that this intuition is justified. In Figures 3.4(a) and (b) we compare between a linear observer and a nonlinear one with $\varepsilon = 0.03$. The advantage of the nonlinear observer over the linear

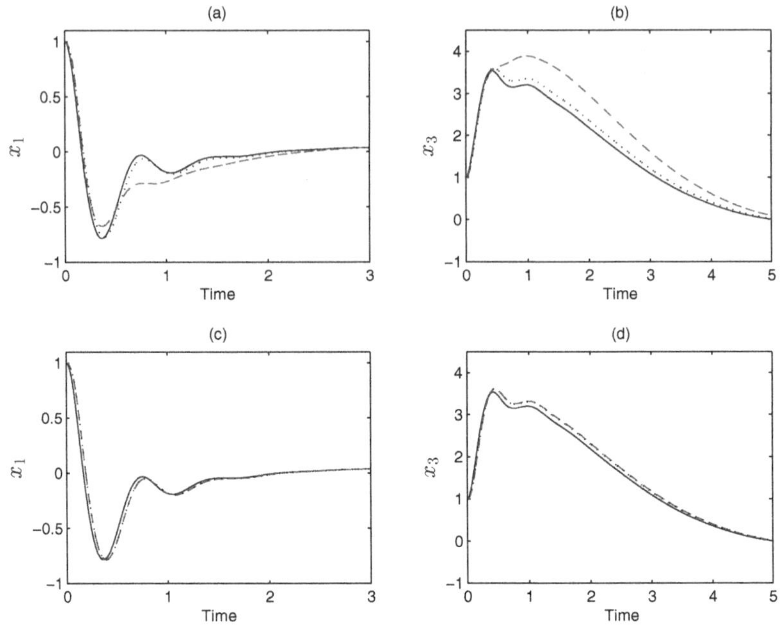

Figure 3.4. *Simulation of Example 3.2. Effect of nonlinearity in the observer: state feedback (solid); with linear observer (dashed); with nonlinear observer (dotted).* $\varepsilon = 0.03$ *in* (a) *and* (b) *and* $\varepsilon = 0.001$ *in* (c) *and* (d).

one is clear. In Figures 3.4(c) and (d) we repeat the comparison with $\varepsilon = 0.001$. This time we cannot differentiate between the performance of the two observers. As ε decreases, the effect of model uncertainty diminishes. In these two simulations the initial conditions are $x(0) = \text{col}(1,0,1,0)$ and $\hat{x}(0) = 0$.

Example 3.3. Consider the simplified PVTOL (Planar Vertical Take-off and Landing) aircraft, modeled by [14]

$$\begin{aligned}
\dot{x}_1 = x_2,\ \dot{x}_2 &= -u_1 \sin x_5 + \mu u_2 \cos x_5,\\
\dot{x}_3 = x_4,\ \dot{x}_4 &= u_1 \cos x_5 + \mu u_2 \sin x_5 - g,\\
\dot{x}_5 = x_6,\ \dot{x}_6 &= \lambda u_2,
\end{aligned}$$

where x_1, x_3, and x_5 are the horizontal coordinate, the vertical coordinate, and the inclination of the aircraft, respectively. A linearizing dynamic state feedback controller is given by

$$\begin{aligned}
&\dot{x}_7 = x_8,\ \dot{x}_8 = -v_1 \sin x_5 + v_2 \cos x_5 + x_6^2 x_7,\\
&u_1 = x_7 + \frac{\mu}{\lambda} x_6^2,\\
&u_2 = \frac{1}{\lambda x_7}(-v_1 \cos x_5 - v_2 \sin x_5 - 2x_6 x_8),
\end{aligned}$$

where v_1 and v_2 are the control inputs to the linearized system. The closed-loop system under the linearizing feedback has an equilibrium point at $x_{eq} = \text{col}(\bar{x}_1, 0, \bar{x}_3, 0, 0, 0,$

[14]The model and the state feedback design are taken from [107].

$g, 0)$, $\bar{u} = \text{col}(g, 0)$, and $\bar{v} = \text{col}(0, 0)$. The linearizing effect of this dynamic controller can be seen by applying the change of variables

$$\begin{aligned} \chi_1 &= x_1 - \bar{x}_1 - \tfrac{\mu}{\lambda}\sin x_5, & \chi_2 &= x_2 - \tfrac{\mu}{\lambda}x_6\cos x_5, \\ \chi_3 &= -x_7\sin x_5, & \chi_4 &= -x_8\sin x_5 - x_6x_7\cos x_5, \\ \chi_5 &= x_3 - \bar{x}_3 - \mu/\lambda + \tfrac{\mu}{\lambda}\cos x_5, & \chi_6 &= x_4 - \tfrac{\mu}{\lambda}x_6\sin x_5, \\ \chi_7 &= x_7\cos x_5 - g, & \chi_8 &= x_8\cos x_5 - x_6x_7\sin x_5 \end{aligned}$$

to obtain

$$\dot{\chi}_1 = \chi_2, \quad \dot{\chi}_2 = \chi_3, \quad \dot{\chi}_3 = \chi_4, \quad \dot{\chi}_4 = \nu_1,$$

$$\dot{\chi}_5 = \chi_6, \quad \dot{\chi}_6 = \chi_7, \quad \dot{\chi}_7 = \chi_8, \quad \dot{\chi}_8 = \nu_2.$$

In the new coordinates, the equilibrium point is at the origin $\chi = 0$. It can be stabilized by a linear state feedback control of the form

$$\nu_1 = -(k_1\chi_1 + k_2\chi_2 + k_3\chi_3 + k_4\chi_4), \tag{3.24}$$

$$\nu_2 = -(k_1\chi_5 + k_2\chi_6 + k_3\chi_7 + k_4\chi_8). \tag{3.25}$$

For the purpose of simulation, we take $\bar{x}_1 = \bar{x}_3 = 2$, $g = \lambda = 1$, and $\mu = 0.5$. The feedback gains $k_1 = 24$, $k_2 = 50$, $k_3 = 35$, and $k_4 = 10$ assign the closed-loop eigenvalues at $-1, -2, -3$, and -4.

Now suppose we only measure the position variables x_1, x_3, and x_5 and set $y_1 = x_1$, $y_2 = x_3$, and $y_3 = x_5$. We want to design an observer to estimate the velocity variables x_2, x_4, and x_6. Noting that the nonlinear functions $\sin x_5$ and $\cos x_5$ depend only on the measured output y_3, the system takes the form

$$\dot{x} = Ax + B\phi(u, y), \ \ y = Cx,$$

where

$$A = \text{block diag}[A_1, A_2, A_3], \quad A_i = \begin{bmatrix} 0 & 1 \\ 0 & 0 \end{bmatrix} \ \text{for } i = 1, 2, 3,$$

$$B = \text{block diag}[B_1, B_2, B_3], \quad B_i = \begin{bmatrix} 0 \\ 1 \end{bmatrix} \ \text{for } i = 1, 2, 3,$$

$$C = \text{block diag}[C_1, C_2, C_3], \quad C_i = \begin{bmatrix} 1 & 0 \end{bmatrix} \ \text{for } i = 1, 2, 3,$$

$$\phi(u, y) = \text{col}(-u_1\sin y_3 + \mu u_2\cos y_3,\ u_1\cos y_3 + \mu u_2\sin y_3 - g,\ \lambda u_2).$$

If the nonlinear function $\phi(u, y)$ is exactly known, we can design an observer that yields linear error dynamics [80, Section 11.3]. Assuming perfect knowledge of the parameters λ and μ, we design the reduced-order observer

$$\begin{aligned} \dot{z}_1 &= -(z_1 + y_1) - u_1\sin y_3 + \mu u_2\cos y_3, & \hat{x}_2 &= z_1 + y_1, \\ \dot{z}_2 &= -(z_2 + y_2) + u_1\sin y_3 + \mu u_2\sin y_3 - g, & \hat{x}_4 &= z_2 + y_2, \\ \dot{z}_3 &= -(z_3 + y_3) + \lambda u_2, & \hat{x}_6 &= z_3 + y_3, \end{aligned}$$

whose eigenvalues are assigned at -1. The estimation errors $\tilde{x}_2 = x_2 - \hat{x}_2$, $\tilde{x}_4 = x_4 - \hat{x}_4$, and $\tilde{x}_6 = x_6 - \hat{x}_6$ satisfy the equations $\dot{\tilde{x}}_i = -\tilde{x}_i$, for $i = 2, 4, 6$. The output feedback controller uses the estimates $\hat{x}_2$, $\hat{x}_4$, and $\hat{x}_6$ to replace x_2, x_4, and x_6, respectively. In view of the linear error dynamics, the output feedback control stabilizes the system at the desired equilibrium point. We refer to this control as the linear-error-observer controller.

On the other hand, if there is uncertainty in modeling the nonlinearity $\phi(u,y)$ or if we want the trajectories under output feedback to come arbitrarily close to the trajectories under state feedback, we use a high-gain observer. We design the reduced-order high-gain observer

$$\begin{aligned} \dot{z}_1 &= -(z_1+y_1)/\varepsilon - u_1\sin y_3 + \hat{\mu}u_2\cos y_3, & \hat{x}_2 &= (z_1+y_1)/\varepsilon, \\ \dot{z}_2 &= -(z_2+y_2)/\varepsilon + u_1\sin y_3 + \hat{\mu}u_2\sin y_3, & \hat{x}_4 &= (z_2+y_2)/\varepsilon, \\ \dot{z}_3 &= -(z_3+y_3)/\varepsilon + \hat{\lambda}u_2, & \hat{x}_6 &= (z_3+y_3)/\varepsilon, \end{aligned}$$

where $\hat{\lambda}$ and $\hat{\mu}$ are nominal values of λ and μ. The observer eigenvalues are assigned at $-1/\varepsilon$. In the simulation, we take $\hat{\lambda}=\lambda=1$ and $\hat{\mu}=\mu=0.5$. We saturate the dynamic feedback controller as follows: $\dot{x}_7 = 20\,\text{sat}(\cdot/20)$, $\dot{x}_8 = 200\,\text{sat}(\cdot/200)$, $u_1 = 40\,\text{sat}(\cdot/40)$, and $u_2 = 200\,\text{sat}(\cdot/200)$. These saturation levels have been determined from extensive simulations done to see the maximal values that the state trajectories would take when the initial state is in a region of interest around x_{eq}. The model nonlinearities are naturally globally bounded, so we do not need to saturate the nonlinearities in the observer. We refer to this control as the high-gain-observer controller. The difference between the two controllers is in the choice of the observer eigenvalues, $-1/\varepsilon$ versus -1, and the use of saturation in the high-gain-observer controller.

To compare the performance of the two controllers, we perform simulations with $x(0)=\text{col}(1,0,1,0,0,0,1,0)$ and $z(0)=0$. Figure 3.5 shows that as ε decreases, the trajectories under output feedback approach the ones under state feedback. The figure shows only x_1 to x_4, but the recovery holds for the entire state x_1 to x_8. Figure 3.6 illustrates that the trajectory recovery property does not hold for the linear-error observer controller. It might appear that if the observer eigenvalues of the linear-error observer are chosen as large as in the high-gain observer, the trajectory recovery prop-

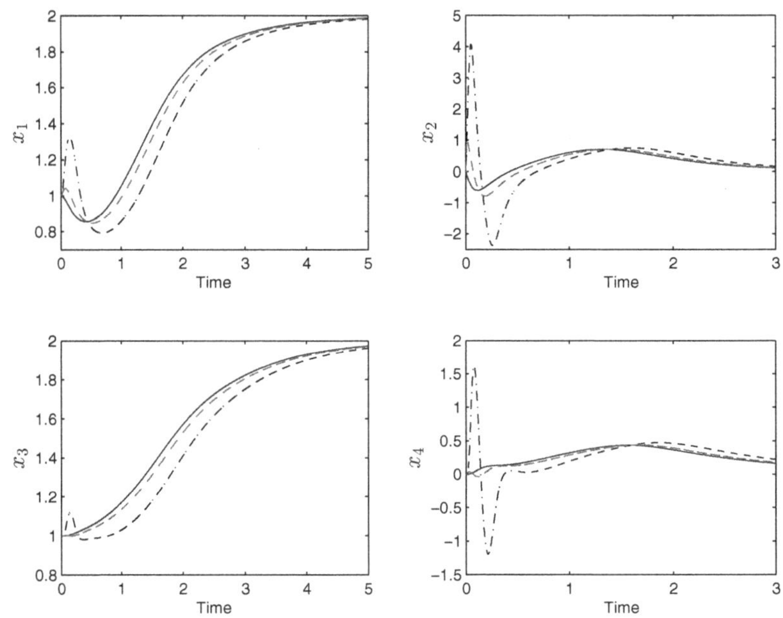

Figure 3.5. *Simulation of Example 3.3. Trajectory convergence: state feedback (solid); high-gain-observer output feedback with $\varepsilon = 0.01$ (dash-dotted) and $\varepsilon = 0.002$ (dashed).*

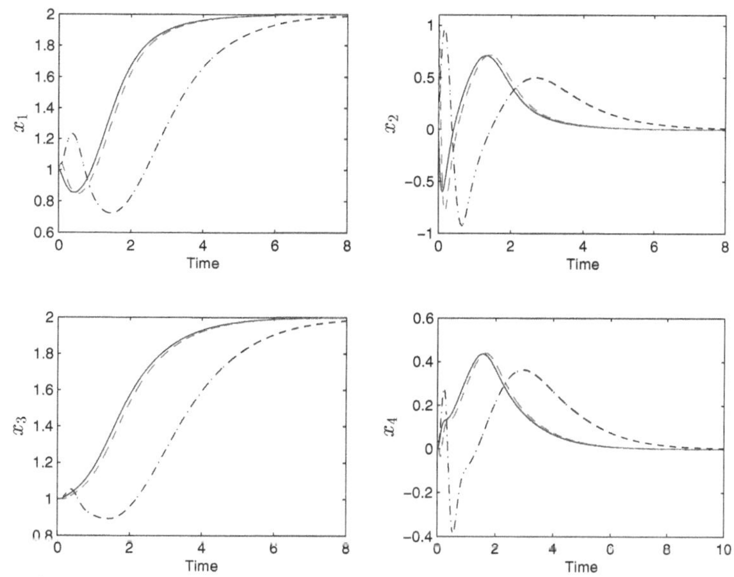

Figure 3.6. *Simulation of Example 3.3. High-gain versus linear-error observer: state feedback (solid); high-gain-observer controller with $\varepsilon = 0.002$ (dashed); linear-error-observer controller (dash-dotted).*

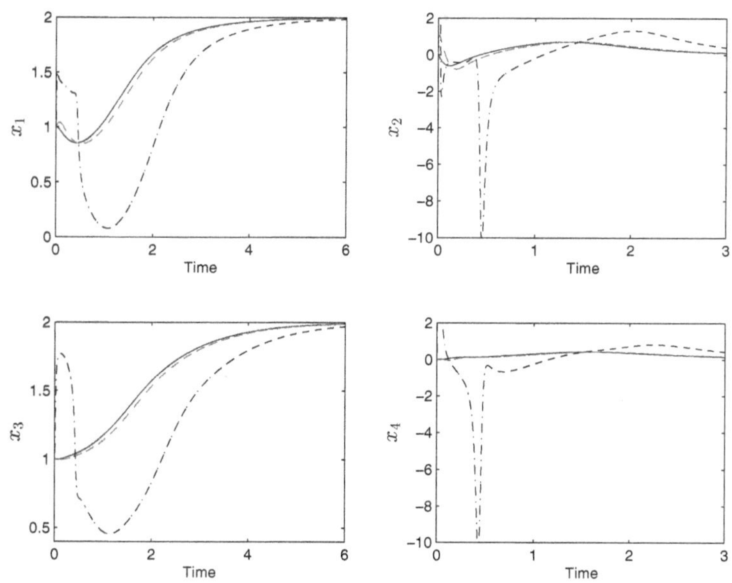

Figure 3.7. *Simulation of Example 3.3. High-gain versus linear-error observer: state feedback (solid), high-gain-observer controller with $\varepsilon = 0.002$ (dashed); linear-error-observer controller with eigenvalues at -500 (dash-dotted).*

erty will be realized. Figure 3.7 shows that this is not the case. In the simulation, the eigenvalues of the linear-error observer are assigned at -500, which coincide with the eigenvalues of the high-gain observer with $\varepsilon = 0.002$. This simulation emphasizes the fact that trajectory recovery is due to the combined effect of fast observer dynamics and global boundedness of the control input and the controller nonlinearities.

3.2 ▪ Robust Stabilization of Minimum-Phase Systems

In this section we consider the normal form of a single-input–single-output minimum-phase nonlinear system that has relative degree $\rho \geq 1$. Our goal is to design an output feedback controller to stabilize the origin in the presence of matched model uncertainty and time-varying disturbance. We start by considering the relative-degree-one case where a sliding mode controller can be designed without observers. Then, we show how to use a high-gain observer to reduce a relative-degree-higher-than-one system to a relative-degree-one system.

3.2.1 ▪ Relative-Degree-One System

When $\rho = 1$, the normal form is given by

$$\dot{\eta} = f_0(\eta, y), \qquad \dot{y} = a(\eta, y) + b(\eta, y)u + \delta(t, \eta, y, u), \tag{3.26}$$

where $\eta \in R^{n-1}$, $u \in R$, and $y \in R$. We assume that the functions f_0, a, and b are locally Lipschitz, $f_0(0,0) = 0$, $a(0,0) = 0$, and the origin of $\dot{\eta} = f_0(\eta, 0)$ is asymptotically stable.[15] Equation (3.26) includes a matched time-varying disturbance δ, which is assumed to be piecewise continuous and bounded in t and locally Lipschitz in (η, y, u) uniformly in t. Let $D \subset R^n$ be a domain that contains the origin and assume that for all $(\eta, y) \in D$, $b(\eta, y) \geq b_0 > 0$. Because the origin of $\dot{\eta} = f_0(\eta, 0)$ is asymptotically stable, the sliding surface is taken as $y = 0$. We assume that there is a (continuously differentiable) Lyapunov function $V(\eta)$ that satisfies the inequalities

$$\phi_1(\|\eta\|) \leq V(\eta) \leq \phi_2(\|\eta\|) \tag{3.27}$$

$$\frac{\partial V}{\partial \eta} f_0(\eta, y) \leq -\phi_3(\|\eta\|) \qquad \forall\ \|\eta\| \geq \phi_4(|y|) \tag{3.28}$$

for all $(\eta, y) \in D$, where ϕ_1 to ϕ_4 are class $\mathcal{K}$ functions. Inequality (3.28) implies regional input-to-state stability of the system $\dot{\eta} = f_0(\eta, y)$ with input y.[16] We take the control as

$$u = \psi(y) + v,$$

where ψ is locally Lipschitz with $\psi(0) = 0$. The function ψ could be zero or could be chosen to cancel known terms on the right-hand side of $\dot{y}$, which depend only on y. Thus,

$$\begin{aligned} y\dot{y} &= y\left[a(\eta, y) + b(\eta, y)\psi(y) + \delta(t, \eta, y, \psi(y) + v) + b(\eta, y)v\right] \\ &\leq b(\eta, y)\left[\left|\frac{a(\eta, y) + b(\eta, y)\psi(y) + \delta(t, \eta, y, \psi(y) + v)}{b(\eta, y)}\right| |y| + yv\right]. \end{aligned}$$

We assume that we know a locally Lipschitz function $\varrho(y) \geq 0$ and a constant $\kappa_0 \in [0, 1)$ such that the inequalities

$$\left|\frac{a(\eta, y) + b(\eta, y)\psi(y) + \delta(t, \eta, y, \psi(y))}{b(\eta, y)}\right| \leq \varrho(y) \tag{3.29}$$

---

[15] By definition, (3.26) is minimum phase when the origin of the zero dynamics $\dot{\eta} = f_0(\eta, 0)$ is asymptotically stable [64].

[16] See [80, Chapter 4] for the definition of regional input-to-state stability.

and

$$\left| \frac{\delta(t,\eta,y,\psi(y)+v) - \delta(t,\eta,y,\psi(y))}{b(\eta,y)} \right| \leq \kappa_0 \, |v| \tag{3.30}$$

hold for all $t \geq 0$, $(\eta, y) \in D$, and $v \in R$. Taking

$$v = -\beta(y) \operatorname{sat}\left(\frac{y}{\mu}\right), \tag{3.31}$$

where $\beta(y)$ is a locally Lipschitz function that satisfies

$$\beta(y) \geq \frac{\varrho(y)}{1-\kappa_0} + \beta_0, \tag{3.32}$$

for some $\beta_0 > 0$, ensures that

$$y\dot{y} \leq -\beta_0(1-\kappa_0)b(\eta,y)|y| \leq -\beta_0 b_0(1-\kappa_0)|y| \qquad \text{for } |y| \geq \mu. \tag{3.33}$$

Define the class $\mathcal{K}$ function ϕ by $\phi(r) = \phi_2(\phi_4(r))$ and suppose μ, $c > \mu$, and $c_0 \geq \phi(c)$ are chosen such that the set

$$\Omega = \{V(\eta) \leq c_0\} \times \{|y| \leq c\} \tag{3.34}$$

is compact and contained in D. Using (3.28) and (3.33) it can be shown that $\dot{V} \leq 0$ on the boundary $V = c_0$ and $y\dot{y} \leq 0$ on the boundary $|y| = c$. Therefore, the set Ω is positively invariant. By the same inequalities, the set

$$\Omega_\mu = \{V(\eta) \leq \phi(\mu)\} \times \{|y| \leq \mu\} \tag{3.35}$$

is positively invariant. Inside Ω, (3.33) shows that for $|y| \geq \mu$,

$$\frac{d}{dt}|y| = \frac{d}{dt}\sqrt{y^2} = \frac{y\dot{y}}{|y|} \leq -\beta_0 b_0(1-\kappa_0).$$

By integration,

$$|y(t)| \leq |y(t_0)| - \beta_0 b_0(1-\kappa_0)(t-t_0)$$

for $t \geq t_0$. Since the right-hand side of the preceding inequality reaches zero in finite time, we conclude that the trajectory enters the set $\{|y| \leq \mu\}$ in finite time. Once it is in, it cannot leave because $y\dot{y} < 0$ on the boundary $|y| = \mu$. Inside the set $\{V(\eta) \leq c\} \times \{|y| \leq \mu\}$,

$$V(\eta) \geq \phi(\mu) \;\Rightarrow\; \phi_2(\|\eta\|) \geq \phi_2(\phi_4(\mu)) \;\Leftrightarrow\; \|\eta\| \geq \phi_4(\mu).$$

It follows from (3.28) that

$$\dot{V} \leq -\phi_3(\|\eta\|) \leq -\phi_3(\phi_4(\mu)).$$

By integration,

$$V(\eta(t)) \leq V(\eta(t_0)) - \phi_3(\phi_4(\mu)),$$

which shows that $V(\eta(t))$ decreases monotonically and enters the set $\{V(\eta) \leq \phi(\mu)\}$ in finite time. In summary, for any initial state in Ω, the trajectory reaches the positively invariant set $\Omega_\mu = \{V(\eta) \leq \phi(\mu)\} \times \{|y| \leq \mu\}$ in finite time and stays in thereafter. If the assumptions hold globally and $V(\eta)$ is radially unbounded, the foregoing conclusion will hold for any initial state. Inside Ω_μ, $v = -\beta(y)y/\mu$. If the origin of $\dot{\eta} = f_0(\eta, 0)$ is exponentially stable, $f_0(\eta, 0)$ is continuously differentiable in some neighborhood of $\eta = 0$, and $\delta(t, 0, 0, 0) = 0$, we can use the converse Lyapunov theorem of exponential stability to show that, for sufficiently small μ, the origin $(\eta = 0, y = 0)$ of the closed-loop system is exponentially stable and Ω is a subset of its region of attraction [80, Section10.1]. If the assumptions hold globally and $V(\eta)$ is radially unbounded, the origin will be globally uniformly asymptotically stable.

3.2.2 ▪ Relative Degree Higher than One

The normal form of a system having relative degree $\rho > 1$ is given by

$$\dot{\eta} = f_0(\eta, \xi), \tag{3.36}$$

$$\dot{\xi}_i = \xi_{i+1} \quad \text{for } 1 \le i \le \rho - 1, \tag{3.37}$$

$$\dot{\xi}_\rho = a(\eta, \xi) + b(\eta, \xi)u + \delta(t, \eta, \xi, u), \tag{3.38}$$

$$y = \xi_1, \tag{3.39}$$

where $\eta \in R^{n-\rho}$, $\xi = \text{col}(\xi_1, \xi_2, \ldots, \xi_\rho)$, $u \in R$, and $y \in R$. Let $D \subset R^n$ be a domain that contains the origin and assume that for all $(\eta, \xi) \in D$, $u \in R$, and $t \ge 0$, the functions f_0, a, and b are locally Lipschitz, $f(0,0) = 0$, $a(0,0) = 0$, the matched disturbance δ is piecewise continuous and bounded in t and locally Lipschitz in (η, ξ, u) uniformly in t, and the origin of $\dot{\eta} = f_0(\eta, 0)$ is asymptotically stable.

From the properties of high-gain observers, presented in Section 2.2, and the separation principle of Section 3.1, we expect that if we can design a partial state feedback controller, in terms of ξ, which stabilizes or practically stabilizes the origin, we will be able to recover its performance by using a high-gain observer that estimates ξ. Based on this expectation we proceed to design a controller assuming that ξ is measured.

With measurement of ξ, we can convert the relative-degree-ρ system into a relative-degree-one system with respect to the output

$$s = k_1\xi_1 + k_2\xi_2 + \cdots + k_{\rho-1}\xi_{\rho-1} + \xi_\rho. \tag{3.40}$$

The normal form of this system is given by

$$\dot{z} = \bar{f}_0(z, s), \qquad \dot{s} = \bar{a}(z, s) + \bar{b}(z, s)u + \bar{\delta}(t, z, s, u), \tag{3.41}$$

where

$$z = \begin{bmatrix} \eta \\ \xi_1 \\ \vdots \\ \xi_{\rho-2} \\ \xi_{\rho-1} \end{bmatrix}, \quad \bar{f}_0(z, s) = \begin{bmatrix} f_0(\eta, \xi) \\ \xi_2 \\ \vdots \\ \xi_{\rho-1} \\ \xi_\rho \end{bmatrix}, \quad \bar{a}(z, s) = \sum_{i=1}^{\rho-1} k_i\xi_{i+1} + a(\eta, \xi),$$

$$\bar{b}(z, s) = b(\eta, \xi), \quad \bar{\delta}(t, z, s, u) = \delta(t, \eta, \xi, u),$$

with $\xi_\rho = s - \sum_{i=1}^{\rho-1} k_i\xi_i$. The functions $\bar{f}_0$, $\bar{a}$, $\bar{b}$, and $\bar{\delta}$ inherit the properties of f_0, a, b, and δ, respectively. In particular, $\bar{b}(z, s) \ge b_0 > 0$ for all $(\eta, \xi) \in D$. The zero dynamics of (3.41) are $\dot{z} = \bar{f}_0(z,0)$. This system can be written as the cascade connection

$$\dot{\eta} = f_0(\eta,\xi)\Big|_{\xi_\rho=-\sum_{i=1}^{\rho-1} k_i\xi_i}, \qquad \dot{\zeta} = F\zeta, \tag{3.42}$$

where

$$\zeta = \begin{bmatrix} \xi_1 \\ \xi_2 \\ \vdots \\ \xi_{\rho-1} \end{bmatrix}, \qquad F = \begin{bmatrix} 0 & 1 & 0 & \cdots & 0 \\ 0 & 0 & 1 & \cdots & 0 \\ \vdots & & & & \vdots \\ 0 & \cdots & & 0 & 1 \\ -k_1 & -k_2 & \cdots & -k_{\rho-2} & -k_{\rho-1} \end{bmatrix}.$$

Because the origin $\eta = 0$ of $\dot{\eta} = f_0(\eta,0)$ is asymptotically stable, the origin $z = 0$ of the cascade connection (3.42) will be asymptotically stable if the matrix F is Hurwitz [80, Appendix C.1]. Therefore, k_1 to $k_{\rho-1}$ in (3.40) are chosen such that the polynomial

$$\lambda^{\rho-1} + k_{\rho-1}\lambda^{\rho-2} + \cdots + k_2\lambda + k_1 \tag{3.43}$$

is Hurwitz.

In view of the converse Lyapunov theorem for asymptotic stability [78, Theorem 4.17] and the local Lipschitz property of $\bar{f}_0(z,s)$ with respect to s, it is reasonable to assume that there is a (continuously differentiable) Lyapunov function $V(z)$ that satisfies the inequalities

$$\phi_1(\|z\|) \le V(z) \le \phi_2(\|z\|), \tag{3.44}$$

$$\frac{\partial V}{\partial \eta}\bar{f}_0(z,s) \le -\phi_3(\|z\|) \qquad \forall\ \|z\| \ge \phi_4(|s|) \tag{3.45}$$

for all $(\eta,\xi) \in D$, where ϕ_1 to ϕ_4 are class $\mathcal{K}$ functions. We note that if $\rho = n$, the zero dynamics of (3.41) will be $\dot{\zeta} = F\zeta$.

Thus, we have converted a relative-degree-ρ system into a relative-degree-one system that satisfies the assumptions of Section 3.2.1. We proceed to design a sliding mode controller as in Section 3.2.1, with the exception that the functions ψ, ϱ, and β will be allowed to depend on the whole vector ξ and not only the output s since we are working under the assumption that ξ is measured. Let

$$u = \psi(\xi) + v$$

for some locally Lipschitz function ψ, with $\psi(0) = 0$, and suppose we know a locally Lipschitz function $\varrho(\xi) \ge 0$ and a constant $\kappa_0 \in [0,1)$ such that the inequalities

$$\left| \frac{\sum_{i=1}^{\rho-1} k_i\xi_{i+1} + a(\eta,\xi) + b(\eta,\xi)\psi(\xi) + \delta(t,\eta,\xi,\psi(\xi))}{b(\eta,\xi)} \right| \le \varrho(\xi) \tag{3.46}$$

and

$$\left| \frac{\delta(t,\eta,\xi,\psi(\xi)+v) - \delta(t,\eta,\xi,\psi(\xi))}{b(\eta,\xi)} \right| \le \kappa_0\,|v| \tag{3.47}$$

hold for all $t \ge 0$, $(\eta,\xi) \in D$, and $v \in R$. Similar to (3.31), take v as

$$v = -\beta(\xi)\,\mathrm{sat}\left(\frac{s}{\mu}\right),$$

where $\beta(\xi)$ is a locally Lipschitz function that satisfies

$$\beta(\xi) \geq \frac{\varrho(\xi)}{1-\kappa_0} + \beta_0 \tag{3.48}$$

for some $\beta_0 > 0$. The overall control is

$$u = \psi(\xi) - \beta(\xi)\,\mathrm{sat}\left(\frac{s}{\mu}\right) \stackrel{\text{def}}{=} \gamma(\xi). \tag{3.49}$$

Define the class $\mathcal{K}$ function ϕ by $\phi(r) = \phi_2(\phi_4(r))$ and suppose μ, $c > \mu$, and $c_0 \geq \phi(c)$ are chosen such that the set

$$\Omega = \{V(z) \leq c_0\} \times \{|s| \leq c\} \text{ with } c_0 \geq \phi(c) \tag{3.50}$$

is compact and contained in D. As in Section 3.2.1, we conclude that Ω is positively invariant, and for all initial states in Ω, the trajectories enter the positively invariant set

$$\Omega_\mu = \{V(z) \leq \phi(\mu)\} \times \{|s| \leq \mu\} \tag{3.51}$$

in finite time.

If $\delta(t,0,0,0) = 0$, the origin of $\dot{\eta} = f_0(\eta,0)$ is exponentially stable, and $f_0(\eta,0)$ is continuously differentiable in some neighborhood of $\eta = 0$, then for sufficiently small μ the origin $(z = 0, s = 0)$ of the closed-loop system is exponentially stable and Ω is a subset of its region of attraction. This can be shown as follows. By the converse Lyapunov theorem [78, Theorem 4.14], there is a Lyapunov function $V_0(\eta)$ that satisfies the inequalities

$$d_1\|\eta\|^2 \leq V_0(\eta) \leq d_2\|\eta\|^2, \quad \frac{\partial V_0}{\partial \eta} f_0(\eta,0) \leq -d_3\|\eta\|^2, \quad \left\|\frac{\partial V_0}{\partial \eta}\right\| \leq d_4\|\eta\|$$

in some neighborhood N_η of $\eta = 0$, where d_1 to d_4 are positive constants. By the local Lipschitz property of f_0, there is a neighborhood N of $(\eta,\xi) = (0,0)$ such that $\|f_0(\eta,\xi) - f_0(\eta,0)\| \leq L\|\xi\|$. Choose μ small enough that $\Omega_\mu \subset N_\eta \times N$. Consider the composite Lyapunov function

$$V = \alpha V_0(\eta) + \zeta^T P \zeta + \tfrac{1}{2}s^2, \tag{3.52}$$

where $P = P^T > 0$ is the solution of the Lyapunov equation $PF + F^T P = -I$ and α is a positive constant to be determined. Inside Ω_μ, it can be shown that $\dot{V} \leq -\mathcal{Y}^T Q \mathcal{Y}$, where

$$\mathcal{Y} = \begin{bmatrix} \|\eta\| \\ \|\zeta\| \\ |s| \end{bmatrix}, \quad Q = \begin{bmatrix} \alpha\ell_1 & -\alpha\ell_2 & -(\alpha\ell_3 + \ell_4) \\ -\alpha\ell_2 & 1 & -\ell_5 \\ -(\alpha\ell_3 + \ell_4) & -\ell_5 & (\ell_6/\mu - \ell_7) \end{bmatrix}$$

for some positive constants ℓ_1 to ℓ_7. Choose $\alpha < \ell_1/\ell_2^2$ to make the 2×2 principal minor positive, then choose μ small enough to make $\det(Q)$ positive. Hence, the origin $(\eta,\xi) = (0,0)$ is exponentially stable, and all trajectories in Ω_μ converge to the origin as time tends to infinity. Since all trajectories in Ω enter Ω_μ in finite time, Ω is a subset of the region of attraction.

To implement the partial state feedback controller (3.49) using a high-gain observer, we start by saturating the control outside the compact set Ω to obtain a globally bounded function $\gamma_s(\xi)$ such that $\gamma_s(\xi) = \gamma(\xi)$ for $\xi \in \Omega$. There are different ways to saturate the control. We can saturate each component of ξ in the functions ψ and β. Let

$$M_i > \max_{\Omega}\{|\xi_i|\}, \quad 1 \le i \le \rho,$$

and take ψ_s and β_s as the functions ψ and β with ξ_i replaced M_i sat(ξ_i/M_i). Alternatively, let

$$M_\psi > \max_{\Omega}\{|\psi(\xi)|\}, \qquad M_\beta > \max_{\Omega}\{|\beta(\xi)|\},$$

and take $\psi_s(\xi) = M_\psi \,\text{sat}(\psi(\xi)/M_\psi)$ and $\beta_s(\xi) = M_\beta \,\text{sat}(\beta(\xi)/M_\beta)$. In either case, the control is taken as

$$u = \psi_s(\xi) - \beta_s(\xi)\,\text{sat}\left(\frac{s}{\mu}\right).$$

We may also saturate the control signal itself. Let

$$M_u > \max_{\Omega}\{|\psi(\xi) - \beta(\xi)\,\text{sat}(s/\mu)|\}$$

and take the control as

$$u = M_u \,\text{sat}\left(\frac{\psi(\xi) - \beta(\xi)\,\text{sat}(s/\mu)}{M_u}\right).$$

The high-gain observer

$$\dot{\hat{\xi}}_i = \hat{\xi}_{i+1} + \frac{\alpha_i}{\varepsilon^i}(y - \hat{\xi}_1), \quad 1 \le i \le \rho - 1, \tag{3.53}$$

$$\dot{\hat{\xi}}_\rho = a_0(\hat{\xi}) + b_0(\hat{\xi})u + \frac{\alpha_\rho}{\varepsilon^\rho}(y - \hat{\xi}_1) \tag{3.54}$$

is used to estimate ξ by $\hat{\xi}$, where ε is a small positive constant; α_1 to α_ρ are chosen such that the polynomial

$$s^\rho + \alpha_1 s^{\rho-1} + \cdots + \alpha_{\rho-1}s + \alpha_\rho \tag{3.55}$$

is Hurwitz; $a_0(\xi)$ and $b_0(\xi)$ are locally Lipschitz, globally bounded functions of ξ, which serve as nominal models of $a(\eta,\xi)$ and $b(\eta,\xi)$, respectively; and $a_0(0) = 0$. The functions a_0 and b_0 are not allowed to depend on η because it is not estimated by the observer (3.53)–(3.54). The choice $a_0 = b_0 = 0$ results in a linear observer. This choice is typically used when no good models of a and b are available or when it is desired to use a linear observer. The output feedback controller is taken as

$$u = \gamma_s(\hat{\xi}), \tag{3.56}$$

where $\gamma_s(\hat{\xi})$ is given by

$$\psi_s(\hat{\xi}) - \beta_s(\hat{\xi})\,\text{sat}(\hat{s}/\mu) \quad \text{or} \quad M_u \,\text{sat}\left(\frac{\psi(\hat{\xi}) - \beta(\hat{\xi})\,\text{sat}(\hat{s}/\mu)}{M_u}\right)$$

and

$$\hat{s} = \sum_{i=1}^{\rho-1} k_i\hat{\xi}_i + \hat{\xi}_\rho.$$

The properties of the output feedback controller are stated in the following theorems.

Theorem 3.2. *Consider the system* (3.36)–(3.39) *and let* k_1 *to* $k_{\rho-1}$ *be chosen such that the polynomial* (3.43) *is Hurwitz. Suppose there exist* $V(z)$, $\varrho(\xi)$, *and* κ_0, *which satisfy* (3.44) *to* (3.47), *and* β *is chosen to satisfy* (3.48). *Let* Ω *and* Ω_μ *be defined by* (3.50) *and* (3.51), *respectively. Consider the high-gain observer* (3.53)–(3.54) *and the output feedback controller* (3.56). *Let* Ω_0 *be a compact set in the interior of* Ω *and* $\mathcal{Q}$ *be a compact subset of* R^ρ. *Suppose the initial states satisfy* $(\eta(0),\xi(0)) \in \Omega_0$ *and* $\hat{\xi}(0) \in \mathcal{Q}$. *Then there exists* ε^*, *dependent on* μ, *such that for all* $\varepsilon \in (0,\varepsilon^*)$, *the states* $(\eta(t),\xi(t),\hat{\xi}(t))$ *of the closed-loop system are bounded for all* $t \geq 0$ *and there is a finite time* T, *dependent on* μ, *such that* $(\eta(t),\xi(t)) \in \Omega_\mu$ *for all* $t \geq T$. *Moreover, if* $(\eta_r(t),\xi_r(t))$ *is the state of the closed-loop system under the state feedback controller* (3.49) *with* $\eta_r(0) = \eta(0)$ *and* $\xi_r(0) = \xi(0)$, *then given any* $\lambda > 0$, *there exists* $\varepsilon^{**} > 0$, *dependent on* μ *and* λ, *such that for all* $\varepsilon \in (0,\varepsilon^{**})$,

$$\|\eta(t)-\eta_r(t)\| \leq \lambda \quad \text{and} \quad \|\xi(t)-\xi_r(t)\| \leq \lambda \qquad \forall\, t \in [0,T]. \tag{3.57}$$

◇

Theorem 3.3. *Suppose all the assumptions of Theorems* 3.2 *are satisfied, the origin of* $\dot{\eta} = f_0(\eta,0)$ *is exponentially stable,* $f_0(\eta,0)$ *is continuously differentiable in some neighborhood* $\eta = 0$, *and* $\delta(t,0,0,0) = 0$. *Then, there exists* $\mu^* > 0$, *and for each* $\mu \in (0,\mu^*)$ *there exists* $\varepsilon^* > 0$, *dependent on* μ, *such that for all* $\varepsilon \in (0,\varepsilon^*)$, *the origin of the closed-loop system under the output feedback controller* (3.56) *is exponentially stable and* $\Omega_0 \times \mathcal{Q}$ *is a subset of its region of attraction.* ◇

Theorems 3.2 and 3.3 show that the output feedback controller (3.56) recovers the stabilization or practical stabilization properties of the state feedback controller (3.49) for sufficiently small ε. It also recovers its transient behavior, as shown by (3.57). We note that the inequalities of (3.57) will hold for all $t \geq 0$ when μ is chosen small enough. This is so because both $(\eta(t),\xi(t))$ and $(\eta_r(t),\xi_r(t))$ enter Ω_μ in finite time. By choosing μ small enough, we can ensure that $\|\eta(t)\| \leq \lambda/2$ and $\|\eta_r(t)\| \leq \lambda/2$ for all $t \geq T$ for some $T > 0$. Hence, $\|\eta(t)-\eta_r(t)\| \leq \lambda$ for all $t \geq T$. A similar argument holds for $\xi(t)-\xi_r(t)$.

Proof of Theorem 3.2: Write the trajectory under state feedback as $\chi_r = \text{col}(\eta_r,\xi_r)$ and the one under output feedback as $\chi = \text{col}(\eta,\xi)$. As in the proof of Theorem 3.1, represent the dynamics of the observer using the scaled estimation errors

$$\varphi_i = \frac{\xi_i - \hat{\xi}_i}{\varepsilon^{\rho-i}}, \qquad 1 \leq i \leq \rho, \tag{3.58}$$

to obtain

$$\varepsilon\dot{\varphi} = A_0\varphi + \varepsilon B_0 \Delta(t,\chi,\hat{\xi}),$$

where

$$A_0 = \begin{bmatrix} -\alpha_1 & 1 & 0 & \cdots & 0 \\ -\alpha_2 & 0 & 1 & \cdots & 0 \\ \vdots & & \ddots & & \vdots \\ -\alpha_{\rho-1} & & & 0 & 1 \\ -\alpha_\rho & 0 & \cdots & \cdots & 0 \end{bmatrix}, \qquad B_0 = \begin{bmatrix} 0 \\ 0 \\ \vdots \\ 0 \\ 1 \end{bmatrix},$$

and

$$\Delta(t,\chi,\hat{\xi}) = a(\eta,\xi) - a_0(\hat{\xi}) + [b(\eta,\xi) - b_0(\hat{\xi})]\gamma_s(\hat{\xi}) + \delta(t,\eta,\xi,\gamma_s(\hat{\xi})).$$

The matrix A_0 is Hurwitz by design. Because $a_0(\hat{\xi})$, $b_0(\hat{\xi})$, and $\gamma_s(\hat{\xi})$ are globally bounded in $\hat{\xi}$,

$$|\Delta(t,\chi,\hat{\xi})| \le L_1 \qquad \forall\, \chi \in \Omega,\ \hat{\xi} \in R^\rho,\ t \ge 0, \tag{3.59}$$

where throughout the proof L_i, for $i = 1,2,\dots$, denote positive constants independent of ε. Let P be the positive definite solution of the Lyapunov equation $PA_0 + A_0^T P = -I$, $W(\varphi) = \varphi^T P \varphi$, and $\Sigma = \{W(\varphi) \le k\varepsilon^2\}$. The first step of the proof is to show that the constant $k > 0$ in the definition of Σ can be chosen such that, for sufficiently small ε, the set $\Omega \times \Sigma$ is positively invariant; that is, $\chi(t_0) \in \Omega$ and $\varphi(t_0) \in \Sigma$ imply that $\chi(t) \in \Omega$ and $\varphi(t) \in \Sigma$ for all $t \ge t_0$. Using (3.59), it can be shown that, for all $\chi \in \Omega$,

$$\begin{aligned}\varepsilon \dot{W} &= -\varphi^T\varphi + 2\varepsilon\varphi^T P B_0 \Delta \le -\|\varphi\|^2 + 2\varepsilon L_1 \|PB_0\|\,\|\varphi\| \\ &\le -\tfrac{1}{2}\|\varphi\|^2 \quad \forall\, \|\varphi\| \ge 4\varepsilon L_1 \|PB_0\|.\end{aligned}$$

Using the inequalities

$$\lambda_{\min}(P)\|\varphi\|^2 \le \varphi^T P \varphi \le \lambda_{\max}(P)\|\varphi\|^2,$$

we arrive at

$$\varepsilon \dot{W} \le -\sigma W \qquad \forall\, W \ge \varepsilon^2 W_0, \tag{3.60}$$

where $\sigma = 1/(2\lambda_{\max}(P))$ and $W_0 = \lambda_{\max}(P)(4L_1\|PB_0\|)^2$. Taking $k = W_0$ shows that $\varphi(t)$ cannot leave Σ because $\dot{W}$ is negative on its boundary. On the other hand, for $\varphi \in \Sigma$, $\|\xi - \hat{\xi}\| \le L_2\varepsilon$. Using the Lipschitz property of β, ψ, and sat$(\cdot)$, it can be shown that $\|\gamma_s(\xi) - \gamma_s(\hat{\xi})\| \le L_3\varepsilon/\mu$, where μ appears because of the function sat(s/μ). Since $\gamma_s(\xi) = \gamma(\xi)$ in Ω, we have

$$\|\gamma(\xi) - \gamma_s(\hat{\xi})\| \le \frac{\varepsilon L_3}{\mu}. \tag{3.61}$$

Using (3.41), (3.46), (3.47), and (3.61), it can be shown that, for $|s| \ge \mu$,

$$\begin{aligned}s\dot{s} &= s\{\bar{a}(z,s) + \bar{b}(z,s)\gamma(\xi) + \bar{\delta}(t,z,s,\gamma(\xi)) + \bar{b}(z,s)[\gamma_s(\hat{\xi}) - \gamma(\xi)] \\ &\quad + \bar{\delta}(t,z,s,\gamma_s(\hat{\xi})) - \bar{\delta}(t,z,s,\gamma_s(\xi))\} \\ &\le b(\eta,\xi)[\varrho(\xi) + \kappa_0\,\beta(\xi) - \beta(\xi) + L_4\varepsilon/\mu]|s| \\ &\le b(\eta,\xi)[-\beta_0(1-\kappa_0) + L_4\varepsilon/\mu]|s|.\end{aligned}$$

Taking $\varepsilon \le \mu\beta_0(1-\kappa_0)/(2L_4)$ yields

$$s\dot{s} \le -\tfrac{1}{2} b_0\beta_0(1-\kappa_0)|s|.$$

With this inequality, the analysis of Section 3.2.1 carries over to show that the trajectory $(\eta(t), \xi(t))$ cannot leave Ω and enters Ω_μ in finite time.

The second step of the proof is to show that for all $\chi(0) \in \Omega_0$ and $\hat{\xi}(0) \in \mathcal{Q}$, the trajectory $(\chi(t), \varphi(t))$ enters $\Omega \times \Sigma$ within a time interval $[0, \tau(\varepsilon)]$, where $\tau(\varepsilon) \to 0$ as $\varepsilon \to 0$. Due to the scaling (3.58), the initial condition $\varphi(0)$ could be of the order of

$1/\varepsilon^{\rho-1}$. Because Ω_0 is in the interior of Ω and the control $\gamma_s(\hat{\xi})$ is globally bounded, there is time $T_1 > 0$, independent of ε, such that $\chi(t) \in \Omega$ for $t \in [0, T_1]$. During this time interval, (3.59) and consequently (3.60) hold. It follows that [80, Theorem 4.5]

$$W(t) \leq \max\left\{e^{-\sigma t/\varepsilon} W(0), \varepsilon^2 W_0\right\} \leq \max\left\{\frac{e^{-\sigma t/\varepsilon} L_5}{\varepsilon^{2(\rho-1)}}, \varepsilon^2 W_0\right\},$$

which shows that $\varphi(t)$ enters Σ within the time interval $[0, \tau(\varepsilon)]$, where

$$\tau(\varepsilon) = \frac{\varepsilon}{\sigma} \ln\left(\frac{L_5}{W_0 \varepsilon^{2\rho}}\right).$$

By l'Hôpital's rule it can be verified that $\lim_{\varepsilon \to 0} \tau(\varepsilon) = 0$. By choosing ε small enough, we can ensure that $\tau(\varepsilon) < T_1$. Thus, the trajectory $(\chi(t), \varphi(t))$ is bounded for all $t \geq 0$ and enters the set $\Omega_\mu \times \Sigma$ within $[0, \tau(\varepsilon)]$. Because $\gamma_s(\hat{\xi})$ is globally bounded and $\chi(0) = \chi_r(0)$,

$$\|\chi(t) - \chi_r(t)\| \leq L_6 \tau(\varepsilon) \qquad \forall\, t \in [0, \tau(\varepsilon)].$$

Over the interval $[\tau(\varepsilon), T]$, the $\dot{\chi}$-equation under output feedback is $O(\varepsilon)$ perturbation of the corresponding equation under state feedback. Therefore, (3.57) follows from the continuous dependence of solutions of differential equations on initial conditions and parameters [78, Theorem 3.4]. □

Proof of Theorem 3.3: Write the closed-loop system under state feedback as

$$\dot{\chi} = f(t, \chi, \gamma(\xi)). \tag{3.62}$$

Inside $\Omega_\mu \times \Sigma$, the closed-loop system under output feedback is given by

$$\dot{\chi} = f(t, \chi, \gamma(\xi - D\varphi)) \qquad \varepsilon\dot{\varphi} = A_0 \varphi + \varepsilon B_0 \Delta(t, \chi, \xi - D\varphi), \tag{3.63}$$

where D is a diagonal matrix whose ith diagonal element is $\varepsilon^{\rho-i}$ and $\Delta(t,0,0) = 0$. The origin of (3.62) is exponentially stable, and there is a Lyapunov function $V(\chi)$ that satisfies[17]

$$c_1\|\chi\|^2 \leq V(\chi) \leq c_2\|\chi\|^2, \quad \frac{\partial V}{\partial \chi} f(t, \chi, \gamma(\xi)) \leq -c_3\|\chi\|^2, \quad \left\|\frac{\partial V}{\partial \chi}\right\| \leq c_4\|\chi\|,$$

in Ω_μ, where c_1 to c_4 are positive constants. Taking $V_c(\chi, \varphi) = V(\chi) + W(\varphi)$ as a Lyapunov function candidate for (3.63). It can be shown that

$$\dot{V}_c \leq -c_3\|\chi\|^2 + c_5\|\chi\|\,\|\varphi\| - \frac{1}{\varepsilon}\|\varphi\|^2 + c_6\|\varphi\|^2$$

for some positive constant c_5 and c_6. The foregoing inequality can be written as

$$\dot{V}_c \leq -\tfrac{1}{2}c_3(\|\chi\|^2 + \|\varphi\|^2) - \tfrac{1}{2}\begin{bmatrix}\|\chi\| \\ \|\varphi\|\end{bmatrix}^T \begin{bmatrix} c_3 & -c_5 \\ -c_5 & (2/\varepsilon - c_3 - 2c_6)\end{bmatrix}\begin{bmatrix}\|\chi\| \\ \|\varphi\|\end{bmatrix}.$$

The matrix of the quadratic form can be made positive definite by choosing ε small enough. Then, $\dot{V}_c \leq -\frac{1}{2}c_3(\|\chi\|^2 + \|\varphi\|^2)$, which shows that the origin of (3.63) is

[17] See (3.52).

exponentially stable and all trajectories in $\Omega_\mu \times \Sigma$ converge to the origin. Since all trajectories with $\chi(0) \in \Omega_0$ and $\hat{\xi}(0) \in \mathcal{Q}$ enter $\Omega_\mu \times \Sigma$, we conclude that $\Omega_0 \times \mathcal{Q}$ is a subset of the region of attraction. □

Example 3.4. Consider the pendulum equation

$$\ddot{\theta} + \sin\theta + c_1\dot{\theta} = c_2 u,$$

where $0 \le c_1 \le 0.2$ and $0.5 \le c_2 \le 2$. To stabilize the pendulum at $(\theta = \pi, \dot{\theta} = 0)$, we take $x_1 = \theta - \pi$ and $x_2 = \dot{\theta}$, which yield the state equation

$$\dot{x}_1 = x_2, \qquad \dot{x}_2 = \sin x_1 - c_1 x_2 + c_2 u.$$

With $s = x_1 + x_2$, we obtain

$$\dot{s} = x_2 + \sin x_1 - c_1 x_2 + c_2 u.$$

It can be shown that

$$\left| \frac{(1-c_1)x_2 + \sin x_1}{c_2} \right| \le 2(|x_1| + |x_2|).$$

Therefore, the control is taken as

$$u = -2(|x_1| + |x_2| + 1)\, \text{sat}(s/\mu).$$

Inside the boundary layer $\{|s| \le \mu\}$, the system is given by

$$\dot{x}_1 = -x_1 + s, \qquad \dot{s} = (1-c_1)x_2 + \sin x_1 - 2c_2(|x_1| + |x_2| + 1)s/\mu$$

and has an equilibrium point at the origin. The derivative of $V = \frac{1}{2}x_1^2 + \frac{1}{2}s^2$ satisfies

$$\begin{aligned} \dot{V} &= -x_1^2 + x_1 s + (1-c_1)(s - x_1)s + s\sin x_1 - 2c_2(|x_1| + |x_2| + 1)s^2/\mu \\ &\le -x_1^2 + 1.2|x_1|\,|s| - (1/\mu\, -\, 1)s^2 = -\begin{bmatrix} |x_1| \\ |s| \end{bmatrix}^T \begin{bmatrix} 1 & -0.6 \\ -0.6 & (1/\mu\, -\, 1) \end{bmatrix} \begin{bmatrix} |x_1| \\ |s| \end{bmatrix}. \end{aligned}$$

For $\mu < 1/1.36 = 0.735$, the origin is exponentially stable. Take $\mu = 0.1$. The controller achieves global stabilization. Suppose now that we only measure θ. In preparation for using a high-gain observer, we saturate the state feedback control outside a compact set. With $V_1(x) = \frac{1}{2}x_1^2$, the set Ω is given by $\Omega = \{|x_1| \le c/k\} \times \{|s| \le c\}$, where $c > 0.1$ and $0 < k < 1$. Suppose that with $c = \pi$ and $k = 0.8$, Ω includes all the initial states of interest. Over Ω, we have $|x_1| \le 1.25\pi$ and $|x_2| \le 2.25\pi$. Therefore, the output feedback controller is taken as

$$u = -2\left[1.25\pi\, \text{sat}\left(\frac{|\hat{x}_1|}{1.25\pi} \right) + 2.25\pi\, \text{sat}\left(\frac{|\hat{x}_2|}{2.25\pi} \right) + 1 \right] \text{sat}\left(\frac{\hat{s}}{\mu} \right),$$

where $\hat{s} = \hat{x}_1 + \hat{x}_2$ and $\hat{x} = \text{col}(\hat{x}_1, \hat{x}_2)$ is determined by the high-gain observer

$$\dot{\hat{x}}_1 = \hat{x}_2 + \frac{2}{\varepsilon}(x_1 - \hat{x}_1), \qquad \dot{\hat{x}}_2 = \phi_0(\hat{x}, u) + \frac{1}{\varepsilon^2}(x_1 - \hat{x}_1),$$

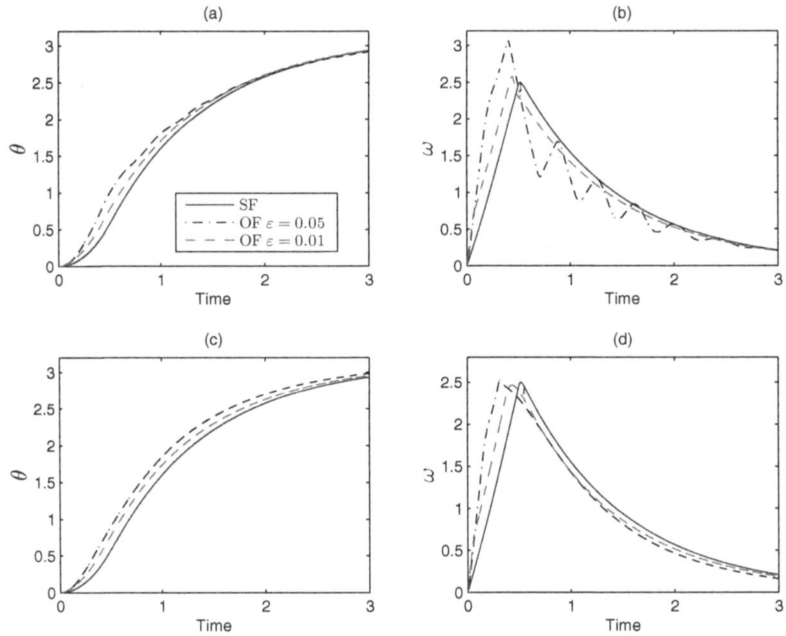

Figure 3.8. *Simulation of Example 3.4. Comparison of state feedback (SF) and output feedback (OF).* (a) *and* (b) *show* θ *and* $\omega = \dot{\theta}$ *for a linear high-gain observer with* $\phi_0 = 0$. (c) *and* (d) *show* θ *and* ω *for a nonlinear high-gain observer with* $\phi_0 = \sin\hat{x}_1 - 0.1\hat{x}_2 + 1.25u$.

with sufficiently small ε. We will consider two choices of ϕ_0. The first choice, $\phi_0 = 0$, yields a linear observer. The second choice, $\phi_0(\hat{x}) = \sin\hat{x}_1 - 0.1\hat{x}_2 + 1.25u$, duplicates the right-hand side of the state equation with 0.1 and 1.25 as nominal values of c_1 and c_2, respectively. In Figure 3.8, we compare the performance of the state and output feedback controllers for $\varepsilon = 0.05$ and $\varepsilon = 0.01$. The pendulum parameters are $c_1 = 0.01$ and $c_2 = 0.5$, and the initial conditions are $\theta(0) = \dot{\theta}(0) = \hat{x}_1(0) = \hat{x}_2(0) = 0$. Peaking is induced because $x_1(0) = -\pi \neq \hat{x}_1(0)$. Irrespective of the choice of ϕ_0, the simulation results show that the response under output feedback approaches the one under state feedback as ε decreases. For $\varepsilon = 0.01$, the inclusion of ϕ_0 in the observer has little effect on the response. However, for the larger value $\varepsilon = 0.05$, there is an advantage for including ϕ_0 in the observer. $\triangle$

3.3 ▪ Tracking

Consider a single-input–single-output system in the normal form

$$\dot{\eta} = f_0(\eta, \xi), \tag{3.64}$$

$$\dot{\xi}_i = \xi_{i+1} \qquad \text{for } 1 \le i \le \rho - 1, \tag{3.65}$$

$$\dot{\xi}_\rho = a(\eta, \xi) + b(\eta, \xi)u + \delta(t, \eta, \xi, u), \tag{3.66}$$

$$y = \xi_1, \tag{3.67}$$

where $\eta \in D_\eta \subset R^{n-\rho}$, $\xi = \text{col}(\xi_1, \ldots, \xi_\rho) \in D_\xi \subset R^\rho$, for some domains D_η and D_ξ. The functions f_0, a, and b are locally Lipschitz and $b(\eta, \xi) \ge b_0 > 0$ over $D_\eta \times D_\xi$.

The term δ is a matched time-varying disturbance that is assumed to be piecewise continuous and bounded in t and locally Lipschitz in (η, ξ, u) uniformly in t. When the relative degree $\rho = n$, η and its equation are dropped. The goal of the tracking problem is to design a feedback controller such that the output y asymptotically tracks a reference signal r, that is, $\lim_{t\to\infty}[y(t) - r(t)] = 0$, while ensuring boundedness of all state variables. We assume that the reference signal satisfies the following assumption.

Assumption 3.1. *$r(t)$ and its derivatives up to $r^{(\rho)}(t)$ are bounded for all $t \geq 0$, and the ρth derivative $r^{(\rho)}(t)$ is a piecewise continuous function of t. Moreover, $\mathcal{R} = \text{col}(r, \dot{r}, \ldots, r^{(\rho-1)}) \in D_\xi$ for all $t \geq 0$.*

The reference signal $r(t)$ could be specified, together with its derivatives, as given functions of time, or it could be the output of a *reference model* driven by a command input $u_c(t)$. In the latter case, the assumptions on r can be met by appropriately choosing the reference model. For example, for a relative-degree-two system, a reference model could be a second-order linear time-invariant system represented by the transfer function

$$\frac{\omega_n^2}{s^2 + 2\zeta\omega_n s + \omega_n^2},$$

where the positive constants ζ and ω_n are chosen to shape the reference signal $r(t)$ for a given input $u_c(t)$. The signal $r(t)$ is generated by the state model

$$\dot{y}_1 = y_2, \qquad \dot{y}_2 = -\omega_n^2 y_1 - 2\zeta\omega_n y_2 + \omega_n^2 u_c, \qquad r = y_1.$$

If $u_c(t)$ is piecewise continuous and bounded, then $r(t)$, $\dot{r}(t)$, and $\ddot{r}(t)$ will satisfy Assumption 3.1.

In the rest of this section we present two designs of the tracking controller. The first design is based on feedback linearization and applies to a special case of (3.64)–(3.67) where $\rho = n$, the functions a, and b are known, and $\delta = 0$. The second design uses sliding mode control and applies to the full model (3.64)–(3.67) even when f_0, a, and b are uncertain and $\delta \neq 0$. The latter design will require a bounded-input-bounded-state property of the system $\dot{\eta} = f_0(\eta, \xi)$.

3.3.1 ▪ Feedback Linearization

Consider a special case of (3.64)–(3.67) with no η and no disturbance, that is,

$$\dot{\xi}_i = \xi_{i+1} \qquad \text{for } 1 \leq i \leq n-1, \tag{3.68}$$

$$\dot{\xi}_n = a(\xi) + b(\xi)u, \tag{3.69}$$

$$y = \xi_1, \tag{3.70}$$

and suppose a and b are known. The change of variables

$$e_1 = \xi_1 - r, \quad e_2 = \xi_2 - r^{(1)}, \quad \ldots, \quad e_n = \xi_n - r^{(n-1)} \tag{3.71}$$

transforms the system (3.68)–(3.69) into the form

$$\dot{e}_i = e_{i+1} \qquad \text{for } 1 \leq i \leq n-1, \tag{3.72}$$

$$\dot{e}_n = a(\xi) + b(\xi)u - r^{(n)}. \tag{3.73}$$

In these coordinates, the goal of the design is to ensure that the vector $e = \text{col}(e_1, \ldots, e_n) = \xi - \mathscr{R}$ is bounded for all $t \geq 0$ and converges to zero as t tends to infinity. Boundedness of e implies boundedness of ξ since $\mathscr{R}$ is bounded. The limit $\lim_{t\to\infty} e(t) = 0$ ensures that the tracking error $e_1 = y - r$ converges to zero. A state feedback controller that achieves this goal will use e. A high-gain observer driven by e_1 will generate an estimate $\hat{e}$ of e, which replaces e in the control law after saturating the control to suppress the effect of the peaking phenomenon. Rewrite the $\dot{e}$-equation as

$$\dot{e} = Ae + B\left[a(\xi) + b(\xi)u - r^{(n)}\right],$$

where the pair (A, B) represents a chain of n integrators. By feedback linearization, the state feedback control

$$u = \frac{1}{b(\xi)}\left[-a(\xi) + r^{(n)} - Ke\right]$$

results in the linear system

$$\dot{e} = (A - BK)e, \tag{3.74}$$

in which $K = \begin{bmatrix} K_1 & K_2 & \cdots & K_n \end{bmatrix}$ is designed such that $A-BK$ is Hurwitz. Therefore, $e(t)$ is bounded and $\lim_{t\to\infty} e(t) = 0$. Consequently, $\xi = e + \mathscr{R}$ is bounded. Let P_1 be the symmetric positive definite solution of the Lyapunov equation

$$P_1(A - BK) + (A - BK)^T P_1 = -Q_1,$$

where $Q_1 = Q_1^T > 0$. Then, $\Omega = \{e^T P_1 e \leq c\}$ is a positively invariant compact set of the closed-loop system (3.74) for any $c > 0$. Choose $c > 0$ such that for every $e \in \Omega$, $\xi = e + \mathscr{R} \in D_\xi$. Suppose the initial error $e(0)$ belongs to $\mathscr{S}$, a compact set in the interior of Ω. In preparation for output feedback, saturate $a(\xi)$, $b(\xi)$, and Ke outside the compact set Ω. Take $M_i > \max_{e\in\Omega} |e_i|$ for $i = 1, \ldots, n$. Define $a_s(e, \mathscr{R})$ by

$$a_s(e, \mathscr{R}) = a(e + \mathscr{R})|_{e_i \to M_i\ \text{sat}(e_i/M_i)}$$

and $b_s(e, \mathscr{R})$ by

$$b_s(e, \mathscr{R}) = \max\left\{ b_0,\quad b(e + \mathscr{R})\Big|_{e_i \to M_i\ \text{sat}(e_i/M_i)} \right\}.$$

The functions a_s and b_s are globally bounded in e, $a_s = a$ and $b_s = b$ for $e \in \Omega$, and $b_s(e, \mathscr{R}) \geq b_0$ for all $e \in R^n$. The term $Ke = \sum_{i=1}^n K_i e_i$ is saturated as $\sum_{i=1}^n K_i M_i\ \text{sat}(e_i/M_i)$, which is equal to Ke for $e \in \Omega$.

Because the functions a and b are known, the high-gain observer and output feedback control can be taken as

$$\dot{\hat{e}} = A\hat{e} + B\left[a_s(\hat{e}, \mathscr{R}) + b_s(\hat{e}, \mathscr{R})u - r^{(n)}\right] + H(\varepsilon)(y - r - \hat{e}_1), \tag{3.75}$$

$$u = \frac{-a_s(\hat{e}, \mathscr{R}) + r^{(n)} - \sum_{i=1}^n K_i M_i\ \text{sat}(\hat{e}_i/M_i)}{b_s(\hat{e}, \mathscr{R})}, \tag{3.76}$$

where

$$H(\varepsilon) = \text{col}(\alpha_1/\varepsilon, \alpha_2/\varepsilon_2, \ldots, \alpha_n/\varepsilon^n)$$

with α_1 to α_n chosen such that the polynomial

$$s^n + \alpha_1 s^{n-1} + \cdots + \alpha_{n-1} s + \alpha_n$$

is Hurwitz and ε is a small positive constant. Substitution of u from (3.76) into (3.75) simplifies the observer equation to

$$\dot{\hat{e}} = A\hat{e} - B\sum_{i=1}^{n} K_i M_i \text{ sat}(\hat{e}_i/M_i) + H(\varepsilon)(e_1 - \hat{e}_1). \tag{3.77}$$

Theorem 3.4. *Suppose Assumption* 3.1 *is satisfied and consider the closed-loop system of the plant* (3.68)–(3.70) *and the output feedback controller* (3.76)–(3.77). *Let* $\mathcal{Q}$ *be any compact subset of* R^n. *Then, there exists* $\varepsilon^* > 0$ *such that for every* $0 < \varepsilon \leq \varepsilon^*$, *the solutions* $(e(t), \hat{e}(t))$ *of the closed-loop system, starting in* $\mathcal{S} \times \mathcal{Q}$, *are bounded for all* $t \geq 0$ *and* $\lim_{t\to\infty} e(t) = 0$. *Moreover, given any* $\lambda > 0$, *there exists* $\varepsilon^{**} > 0$, *dependent on* λ, *such that for every* $0 < \varepsilon \leq \varepsilon^{**}$,

$$\|e(t) - e_r(t)\| \leq \lambda \qquad \forall\, t \geq 0, \tag{3.78}$$

where e_r *is the solution of* (3.74) *with* $e_r(0) = e(0)$. ◇

Proof: With $\varphi_i = (e_i - \hat{e}_i)/\varepsilon^{n-i}$ for $i = 1, \ldots, n$, the closed-loop system is represented by

$$\dot{e} = (A - BK)e + B\delta_1(e, \hat{e}, \mathcal{R}, r^{(n)}),$$
$$\varepsilon\dot{\varphi} = A_0\varphi + \varepsilon B\delta_2(e, \hat{e}, \mathcal{R}, r^{(n)}),$$

where $\varphi = \text{col}(\varphi_1, \ldots, \varphi_n)$, $A_0 = A - H_o C$, $H_o = \text{col}(\alpha_1, \ldots \alpha_n)$,

$$\delta_1 = \frac{1}{b_s(\hat{e}, \mathcal{R})}\Big\{a(\xi)b_s(\hat{e}, \mathcal{R}) - a_s(\hat{e}, \mathcal{R})b(\xi) + [b(\xi) - b_s(\hat{e}, \mathcal{R})]r^{(n)}$$
$$-b(\xi)\sum_{i=1}^{n} K_i M_i \text{ sat}(\hat{e}_i/M_i) + b_s(\hat{e}, \mathcal{R})\sum_{i=1}^{n} K_i e_i\Big\},$$
$$\delta_2 = \delta_1 + \sum_{i}^{n} K_i M_i \text{ sat}(\hat{e}_i/M_i) - \sum_{i}^{n} K_i e_i,$$
$$\hat{e} = e - D(\varepsilon)\varphi,$$
$$D(\varepsilon) = \text{diag}[\varepsilon^{n-1}, \ldots, 1].$$

The matrix A_0 is Hurwitz by design. Similar to the boundedness part of the proof of Theorem 3.1, it can be shown that there exist $\varrho > 0$, $\varepsilon_1^* > 0$, and $T(\varepsilon) > 0$, with $\lim_{\varepsilon\to 0} T(\varepsilon) = 0$, such that for every $0 < \varepsilon \leq \varepsilon_1^*$, the solution $(e(t), \varphi(t))$ is bounded for all $t \geq 0$ and belongs to $\Lambda = \{e^T P_1 e \leq c\} \times \{\varphi^T P_o \varphi \leq \varepsilon^2 \varrho\}$ for all $t \geq T(\varepsilon)$, where P_0 is the solution of the Lyapunov equation $P_0 A_0 + A_0^T P_0 = -I$. Inside Λ,

$$|a(e+\mathcal{R})b_s(\hat{e}, \mathcal{R}) - a_s(\hat{e}, \mathcal{R})b(e+\mathcal{R})| = |a_s(e, \mathcal{R})b_s(\hat{e}, \mathcal{R}) - a_s(\hat{e}, \mathcal{R})b_s(e, \mathcal{R})| \leq L_1\|\varphi\|.$$

Similarly,

$$|b(e+\mathscr{R})-b_s(\hat{e},\mathscr{R})| \le L_2\|\varphi\|$$

and

$$\left|\sum_i^n K_i M_i \ \mathrm{sat}(\hat{e}_i/M_i)-\sum_i^n K_i e_i\right| \le L_3\|\varphi\|.$$

Thus, $|\delta_1| \le L_4\|\varphi\|$ and $|\delta_2| \le L_5\|\varphi\|$. Therefore, the derivative of $V_2 = e^T P_1 e + \varphi^T P_0 \varphi$ satisfies the inequality $\dot{V}_2 \le -\mathscr{Y}^T Q \mathscr{Y}$, where

$$Q = \begin{bmatrix} \lambda_{min}(Q_1) & -\beta_1 \\ -\beta_1 & (1/\varepsilon)-\beta_2 \end{bmatrix}, \quad \mathscr{Y} = \begin{bmatrix} \|e\| \\ \|\varphi\| \end{bmatrix}$$

for some nonnegative constants β_1 and β_2. The matrix Q is positive definite for sufficiently small ε. Hence, both $e(t)$ and $\varphi(t)$ tend to zero as t tends to infinity, and so is $\hat{e}(t)$. To complete the proof note that, since $\lim_{t\to\infty} e(t)=0$ and $\lim_{t\to\infty} e_r(t)=0$, there is $T_2 > 0$ such that (3.78) is satisfied for all $t \ge T_2$. Showing (3.78) on the compact interval $[0,T_2]$ is done as in the performance recovery part of the proof of Theorem 3.1. □

Example 3.5. Consider the pendulum equation

$$\dot{x}_1 = x_2, \qquad \dot{x}_2 = -\sin x_1 - c_1 x_2 + c_2 u, \qquad y = x_1,$$

where $c_1 = 0.015$ and $c_2 = 0.5$. We want the output y to track $r(t) = \cos t$. Taking

$$e_1 = x_1 - r = x_1 - \cos t, \qquad e_2 = x_2 - \dot{r} = x_2 + \sin t$$

yields

$$\dot{e}_1 = e_2, \qquad \dot{e}_2 = -\sin x_1 - c_1 x_2 + c_2 u + \cos t.$$

The state feedback control

$$u = 2(\sin x_1 + 0.015 x_2 - \cos t - 2e_1 - 3e_2)$$

assigns the eigenvalues of $A-BK$ at -1 and -2. The solution of the Lyapunov equation $P_1(A-BK)+(A-BK)^T P_1 = -I$ is $P_1 = \left[\begin{smallmatrix} 1.25 & 0.25 \\ 0.25 & 0.25 \end{smallmatrix}\right]$. Set $\Omega = \{e^T P_1 e \le 3\}$. It can be shown that $\max_{e\in\Omega} |e_1| = 1.7321$ and $\max_{e\in\Omega} |e_2| = 3.873$. We saturate e_1 at ± 2 and e_2 at ± 5. The observer gains α_1 and α_2 are chosen as 2 and 1, respectively, to assign the eigenvalues of $A-H_o C$ at -1 and -1. Thus, the observer and output feedback control are taken as

$$\begin{aligned}
\dot{\hat{e}}_1 &= \hat{e}_2 + (2/\varepsilon)(e_1 - \hat{e}_1), \\
\dot{\hat{e}}_2 &= -4 \ \mathrm{sat}(\hat{e}_1/2) - 15 \ \mathrm{sat}(\hat{e}_2/5) + (1/\varepsilon^2)(e_1 - \hat{e}_1), \\
u &= 2[\sin(\hat{e}_1 + \cos t) + 0.015(5 \ \mathrm{sat}(\hat{e}_2/5) - \sin t) - \cos t \\
&\quad -4 \ \mathrm{sat}(\hat{e}_1/2) - 15 \ \mathrm{sat}(\hat{e}_2/5)].
\end{aligned}$$

The term $\hat{e}_1$ in $\sin(\hat{e}_1 + \cos t)$ is not saturated because the sine function is bounded.

Figure 3.9 shows the response of the system under state and output feedback for two different values of ε with initial conditions $x(0) = 0$ and $\hat{e}(0) = 0$. Panel (a) compares the output under output feedback with $\varepsilon = 0.01$ to the reference signal. Panels (b) and (c) illustrate that the trajectories under output feedback approach the ones under state feedback as ε decreases. △

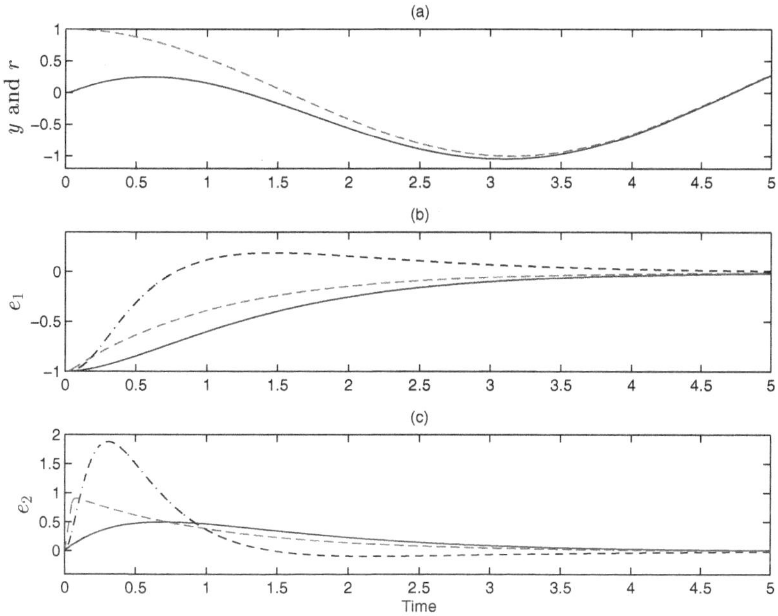

Figure 3.9. *Simulation of Example* 3.5. (a) *shows the reference r (dashed) and the output y under output feedback with $\varepsilon = 0.01$.* (b) *and* (c) *show the tracking errors e_1 and e_2 under state feedback (solid), output feedback with $\varepsilon = 0.1$ (dash-dotted), and output feedback with $\varepsilon = 0.01$ (dashed).*

3.3.2 ▪ Sliding Mode Control

Consider the system (3.64)–(3.67), where f_0, a and b could be uncertain. The change of variables

$$e_1 = \xi_1 - r, \quad e_2 = \xi_2 - r^{(1)}, \quad \ldots, \quad e_\rho = \xi_\rho - r^{(\rho-1)}$$

transforms the system into the form

$$\dot{\eta} = f_0(\eta, \xi), \tag{3.79}$$

$$\dot{e}_i = e_{i+1}, \qquad 1 \le i \le \rho - 1, \tag{3.80}$$

$$\dot{e}_\rho = a(\eta, \xi) + b(\eta, \xi)u + \delta(t, \eta, \xi, u) - r^{(\rho)}(t). \tag{3.81}$$

We start by designing a state feedback controller using sliding mode control. The sliding manifold is taken as $s = 0$, where

$$s = k_1 e_1 + \cdots + k_{\rho-1} e_{\rho-1} + e_\rho,$$

in which k_1 to $k_{\rho-1}$ are chosen such that the polynomial

$$\lambda^{\rho-1} + k_{\rho-1}\lambda^{\rho-2} + \cdots + k_1$$

is Hurwitz. The derivative of s is given by

$$\begin{aligned} \dot{s} &= \sum_{i=1}^{\rho-1} k_i e_{i+1} + a(\eta, \xi) + b(\eta, \xi)u + \delta(t, \eta, \xi, u) - r^{(\rho)}(t) \\ &\stackrel{\text{def}}{=} b(\eta, \xi)u + \Delta(t, \eta, e, \mathcal{R}, r^{(\rho)}, u). \end{aligned}$$

Suppose

$$\left|\frac{\Delta(t,\eta,e,\mathcal{R},r^{(\rho)},u)}{b(\eta,\xi)}\right| \leq \varrho\left(e,\mathcal{R},r^{(\rho)}\right)+\kappa_0\,|u|, \quad 0\leq\kappa_0<1,$$

for all $t \geq 0$, $u \in R$, and $(\eta,\xi) \in D_\eta \times D_\xi$, where ϱ and κ_0 are known, and let $\beta\left(e,\mathcal{R},r^{(\rho)}\right)$ be a locally Lipschitz function that satisfies

$$\beta\left(e,\mathcal{R},r^{(\rho)}\right)\geq \frac{\varrho\left(e,\mathcal{R},r^{(\rho)}\right)}{1-\kappa_0}+\beta_0 \qquad \text{with } \beta_0>0.$$

Then

$$u=-\beta\left(e,\mathcal{R},r^{(\rho)}\right)\ \text{sat}\left(\frac{s}{\mu}\right) \tag{3.82}$$

ensures that $s\dot{s} \leq -\beta_0 b_0(1-\kappa_0)|s|$ for $|s|\geq\mu$. Setting $\zeta=\text{col}(e_1,\ldots,e_{\rho-1})$, it can be seen that ζ satisfies the equation

$$\dot{\zeta}=(A_c-B_cK)\zeta+B_c s,$$

where $K=\begin{bmatrix}k_1 & k_2 & \ldots & k_{\rho-1}\end{bmatrix}$ and the pair (A_c,B_c) represents a chain of $(\rho-1)$ integrators. The matrix A_c-B_cK is Hurwitz by design. Let P be the solution of the Lyapunov equation $P(A_c-B_cK)+(A_c-B_cK)^TP=-I$. The derivative of $V=\zeta^TP\zeta$ satisfies the inequality

$$\dot{V}=-\zeta^T\zeta+2\zeta^TPB_cs\leq-(1-\theta)\|\zeta\|^2 \qquad \forall\ \|\zeta\|\geq 2\|PB_c\|\,|s|/\theta,$$

where $0<\theta<1$. Since $\zeta^TP\zeta\leq\lambda_{\max}(P)\|\zeta\|^2$, for any $\sigma\geq\mu$,

$$\{\|\zeta\|\leq 2\|PB_c\|\,\sigma/\theta\}\subset\{\zeta^TP\zeta\leq\lambda_{\max}(P)(2\|PB_c\|/\theta)^2\sigma^2\}.$$

Let $\rho_1=\lambda_{\max}(P)(2\|PB_c\|/\theta)^2$ and $\Omega_c=\{\zeta^TP\zeta\leq\rho_1c^2\}\times\{|s|\leq c\}$ with $c>\mu$. Because $e=\text{col}(\zeta,s-K\zeta)$, there is $\rho_2>0$ such that $\|e\|\leq\rho_2c$ for all $e\in\Omega_c$. Since $\xi=e+\mathcal{R}$ and $\mathcal{R}\in D_\xi$, by choosing c small enough we can ensure that for all $e\in\Omega_c$, $\xi\in D_\xi$ and $\|\xi\|\leq\rho_2c+\rho_3$ for some $\rho_3>0$. The set Ω_c is positively invariant because $s\dot{s}<0$ on the boundary $\{|s|=c\}$ and $\dot{V}<0$ on the boundary $\{V=\rho_1c^2\}$. Let $e(0)\in\Omega_c$. Within finite time, $e(t)$ enters the set $\Omega_\mu=\{\zeta^TP\zeta\leq\rho_1\mu^2\}\times\{|s|\leq\mu\}$, which is also positively invariant. Therefore, $\|e\|$ is ultimately bounded by $k\mu$ for some $k>0$. We turn now to the equation $\dot{\eta}=f_0(\eta,\xi)$ and impose the following assumption to ensure boundedness of η.

Assumption 3.2. *There is a continuously differentiable function $V_0(\eta)$, class $\mathcal{K}$ functions ϕ_1 and ϕ_2, and a nonnegative continuous nondecreasing function χ such that*

$$\phi_1(\|\eta\|)\leq V_0(\eta)\leq\phi_2(\|\eta\|),$$

$$\frac{\partial V_0}{\partial\eta}f_0(\eta,\xi)\leq 0 \qquad \forall\ \|\eta\|\geq\chi(\|\xi\|)$$

for all $(\eta,\xi)\in D_\eta\times D_\xi$. Moreover, there is $c_0\geq\phi_2(\chi(\rho_2c+\rho_3))$ such that $\{V_0(\eta)\leq c_0\}$ is compact and contained in D_η.

Assumption 3.2 ensures that the set $\{V_0(\eta) \le c_0\} \times \Omega_c$ is positively invariant because on the boundary $\{V_0 = c_0\}$,

$$\phi_2(\|\eta\|) \ge c_0 \ge \phi_2(\chi(\rho_2 c + \rho_3)) \Rightarrow \|\eta\| \ge \chi(\rho_2 c + \rho_3) \Rightarrow \|\eta\| \ge \chi(\|\xi\|) \Rightarrow \dot{V}_0 \le 0.$$

If all the assumptions hold globally, that is, $D_\eta \times D_\xi = R^n$, and ϕ_1 is class $\mathscr{K}_\infty$, the constants c and c_0 can be chosen arbitrarily large, and any compact set of R^n can be put in the interior of $\{V_0(\eta) \le c_0\} \times \Omega_c$.

For minimum-phase systems, where the origin of $\dot{\eta} = f_0(\eta, 0)$ is asymptotically stable, Assumption 3.2 holds locally [80, Lemma 4.7]. It will hold globally or regionally if $\dot{\eta} = f_0(\eta, \xi)$ is input-to-state stable or regionally input-to-state stable, respectively. However, the assumption might hold while the origin of $\dot{\eta} = f_0(\eta, 0)$ is not asymptotically stable, as in the system

$$\dot{\eta} = -\frac{|\xi|}{|\xi|+1}\eta + \xi,$$

where, with $\xi = 0$, the origin of $\dot{\eta} = 0$ is stable but not asymptotically stable. Assumption 3.2 is satisfied because the derivative of $V_0 = \frac{1}{2}\eta^2$ satisfies $\dot{V}_0 \le 0$ for $|\eta| \ge |\xi| + 1$.

In preparation for output feedback, we saturate the control outside the compact set Ω_c. Because the saturation function is bounded, we only need to saturate $\beta\left(e, \mathscr{R}, r^{(\rho)}\right)$. Take $M_i \ge \max_{e \in \Omega_c} |e_i|$ for $i = 1, \dots, \rho$ and define $\beta_s\left(e, \mathscr{R}, r^{(\rho)}\right)$ by

$$\beta_s\left(e, \mathscr{R}, r^{(\rho)}\right) = \beta\left(e, \mathscr{R}, r^{(\rho)}\right)\Big|_{e_i \to M_i \,\mathrm{sat}(e_i/M_i)}.$$

The function β_s is globally bounded in e and $\beta_s = \beta$ for all $e \in \Omega_c$. The linear high-gain observer

$$\dot{\hat{e}}_i = \hat{e}_{i+1} + \frac{\alpha_i}{\varepsilon^i}(y - r - \hat{e}_1), \quad 1 \le i \le \rho - 1, \tag{3.83}$$

$$\dot{\hat{e}}_\rho = \frac{\alpha_\rho}{\varepsilon^\rho}(y - r - \hat{e}_1) \tag{3.84}$$

estimates e by $\hat{e}$, where ε is a sufficiently small positive constant and α_1 to α_ρ are chosen such that the polynomial

$$\lambda^\rho + \alpha_1 \lambda^{\rho-1} + \cdots + \alpha_{\rho-1}\lambda + \alpha_\rho$$

is Hurwitz. The output feedback tracking controller is given by

$$u = -\beta_s\left(\hat{e}, \mathscr{R}, r^{(\rho)}\right)\,\mathrm{sat}\left(\frac{\hat{s}}{\mu}\right), \tag{3.85}$$

where $\hat{s} = k_1\hat{e}_1 + \cdots + k_{\rho-1}\hat{e}_{\rho-1} + \hat{e}_\rho$.

Theorem 3.5. *Suppose Assumptions* 3.1 *and* 3.2 *hold and consider the closed-loop system of the plant* (3.79)–(3.81) *and the output feedback controller* (3.83)–(3.85). *Let* $\mathscr{S} \in R^n$ *be a compact set in the interior of* $\{V_0(\eta) \le c_0\} \times \Omega_c$, *and let* $\mathscr{Q}$ *be any compact subset of* R^ρ. *Then, there exists* $\varepsilon^* > 0$ *(dependent on* μ*) and for every* $0 < \varepsilon \le \varepsilon^*$ *there is a finite time* T_1 *(dependent on* μ *and* ε*) such that the solutions* $(\eta(t), e(t), \hat{e}(t))$ *of the closed-loop system, starting in* $\mathscr{S} \times \mathscr{Q}$, *are bounded for all* $t \ge 0$ *and* $\|e(t)\| \le k\mu$ *for all* $t \ge T_1$.

Moreover, given any $\lambda > 0$ and $T_2 > 0$, there exists $\varepsilon_1^* > 0$, dependent on μ, λ, and T_2, such that for every $0 < \varepsilon \le \varepsilon_1^*$,

$$\left\| \begin{bmatrix} \eta(t) - \eta_r(t) \\ e(t) - e_r(t) \end{bmatrix} \right\| \le \lambda \qquad \forall\, t \in [0, T_2], \tag{3.86}$$

where (η_r, e_r) is the solution of closed-loop system under state feedback with $\eta_r(0) = \eta(0)$ and $e_r(0) = e(0)$. ◇

Theorem 3.5 shows that the output feedback controller recovers the tracking properties of the state feedback controller for sufficiently small ε. It also recovers its transient behavior, as shown by (3.86). We note that the inequality $\|e(t) - e_r(t)\| \le \lambda$ will hold for all $t \ge 0$ when μ is chosen small enough. This is so because both $e(t)$ and $e_r(t)$ enter the set Ω_μ in finite time. By choosing μ small enough, we can ensure that $\|e(t)\| \le \lambda/2$ and $\|e_r(t)\| \le \lambda/2$ for all $t \ge T_3$ for some $T_3 > 0$. Hence, $\|e(t) - e_r(t)\| \le \lambda$ for all $t \ge T_3$.

Proof of Theorem 3.5: With $\varphi_i = (e_i - \hat{e}_i)/\varepsilon^{\rho - i}$ for $i = 1, \dots, \rho$, the closed-loop system is represented by

$$\begin{aligned} \dot{\eta} &= f_0(\eta, \xi), \\ \dot{\zeta} &= (A_c - B_c K)\zeta + B_c s, \\ \dot{s} &= b(\eta, \xi)\bar{u} + \Delta(t, \eta, e, \mathscr{R}, r^{(\rho)}, \bar{u}) + \delta_1(t, \eta, e, \hat{e}, \mathscr{R}, r^{(\rho)}), \\ \varepsilon \dot{\varphi} &= A_0 \varphi + \varepsilon B \delta_2(t, \eta, e, \hat{e}, \mathscr{R}, r^{(\rho)}), \end{aligned}$$

where $\varphi = \text{col}(\varphi_1, \dots, \varphi_\rho)$, $A_0 = A - H_o C$, $H_o = \text{col}(\alpha_1, \dots \alpha_\rho)$, $B = \text{col}(0, \dots, 0, 1)$,

$$\begin{aligned} \bar{u} &= -\beta\left(e, \mathscr{R}, r^{(\rho)}\right) \text{sat}(s/\mu), \\ \delta_1 &= b(\eta, \xi)(u - \bar{u}) + \Delta(t, \eta, e, \mathscr{R}, r^{(\rho)}, u) - \Delta(t, \eta, e, \mathscr{R}, r^{(\rho)}, \bar{u}), \\ \delta_2 &= a(\eta, \xi) + b(\eta, \xi)u + \delta(t, \eta, \xi, u) - r^{(\rho)}. \end{aligned}$$

The matrix A_0 is Hurwitz by design. Similar to the boundedness part of the proof of Theorem 3.1, it can be shown that there exist $\varrho > 0$, $\varepsilon_1^* > 0$, and $T(\varepsilon) > 0$, with $\lim_{\varepsilon \to 0} T(\varepsilon) = 0$, such that for every $0 < \varepsilon \le \varepsilon_1^*$, the solution $(\eta(t), e(t), \varphi(t))$ is bounded for all $t \ge 0$ and belongs to $\Lambda = \{V_0(\eta) \le c_0\} \times \Omega_c \times \{\varphi^T P_o \varphi \le \varepsilon^2 \varrho\}$ for all $t \ge T(\varepsilon)$, where P_0 is the solution of the Lyapunov equation $P_0 A_0 + A_0^T P_0 = -I$. Inside Λ,

$$\begin{aligned} |u - \bar{u}| &= \left| \beta_s\left(\hat{e}, \mathscr{R}, r^{(\rho)}\right) \text{sat}\left(\frac{\hat{s}}{\mu}\right) - \beta\left(e, \mathscr{R}, r^{(\rho)}\right) \text{sat}\left(\frac{s}{\mu}\right) \right| \\ &= \left| \beta_s\left(\hat{e}, \mathscr{R}, r^{(\rho)}\right) \text{sat}\left(\frac{\hat{s}}{\mu}\right) - \beta_s\left(e, \mathscr{R}, r^{(\rho)}\right) \text{sat}\left(\frac{s}{\mu}\right) \right| \le \frac{L_1}{\mu}\|\varphi\|. \end{aligned}$$

Therefore, $|\delta_1| \le L_2\|\varphi\|/\mu$. Hence, for $|s| \ge \mu$, the derivative of $\frac{1}{2}s^2$ satisfies the inequalities

$$s\dot{s} \le -\beta_0 b_0 (1 - \kappa_0)|s| + L_3 \varepsilon |s|/\mu \le -\tfrac{1}{2}\beta_0 b_0 (1 - \kappa_0)|s|$$

for $\varepsilon \leq \frac{1}{2}\mu\beta_0 b_0(1-\kappa_0)/L_3$, which shows that $e(t)$ enters the positively invariant set Ω_μ in finite time. Thus, $\|e\|$ is ultimately bounded by $k\mu$. To prove (3.86), we note that the closed-loop system under state feedback can be written as

$$\dot{\chi} = f\left(t, \chi, \mathcal{R}, r^{(\rho)}\right),$$

where $\chi = \text{col}(\eta, \zeta, s)$, while the closed-loop system under output feedback can be written as

$$\dot{\chi} = f\left(t, \chi, \mathcal{R}, r^{(\rho)}\right) + \bar{B}\delta_1,$$
$$\varepsilon\dot{\varphi} = A_0\varphi + \varepsilon B\delta_2,$$

where $\bar{B} = \text{col}(0, \ldots, 0, 1)$ is $n \times 1$. As in the performance recovery part of the proof of Theorems 3.1, it can be shown that

$$\|\chi(t) - \chi_r(t)\| \leq 2k_1 T(\varepsilon) \qquad \forall\, t \in [0, T(\varepsilon)].$$

Since $T(\varepsilon) \to 0$ as $\varepsilon \to 0$, we can ensure that (3.86) is satisfied over $[0, T(\varepsilon)]$ by choosing ε small enough. On the compact time interval $[T(\varepsilon), T_2]$, (3.86) follows from continuous dependence of the solution of differential equations on initial conditions and parameters [78, Theorem 3.4]. □

Example 3.6. Reconsider the pendulum tracking problem

$$\dot{e}_1 = e_2, \qquad \dot{e}_2 = -\sin x_1 - c_1 x_2 + c_2 u + \cos t$$

from Example 3.5 and suppose that c_1 and c_2 are uncertain parameters that satisfy $0 \leq c_1 \leq 0.1$ and $0.5 \leq c_2 \leq 2$. Taking $s = e_1 + e_2$, we have

$$\dot{s} = e_2 - \sin x_1 - c_1 x_2 + c_2 u + \cos t = (1 - c_1)e_2 - \sin x_1 + c_1 \sin t + \cos t + c_2 u.$$

It can be verified that

$$\left|\frac{(1 - c_1)e_2 - \sin x_1 + c_1 \sin t + \cos t}{c_2}\right| \leq \frac{|e_2| + 2.1}{0.5} \leq 2|e_2| + 4.2.$$

We take $\beta = 2|e_2| + 5$, which yields the control

$$u = -(2|e_2| + 5)\,\text{sat}\left(\frac{e_1 + e_2}{\mu}\right).$$

From the equation $\dot{e}_1 = -e_1 + s$, we see that the derivative of $\frac{1}{2}e_1^2$ satisfies

$$e_1\dot{e}_1 = -e_1^2 + e_1 s \leq -(1 - \theta)e_1^2 \qquad \forall\, |e_1| \geq \frac{|s|}{\theta},$$

where $0 < \theta < 1$. With $1/\theta = 1.1$, $\Omega_c = \{|e_1| \leq 1.1c\} \times \{|s| \leq c\}$. We take $c = 2$. It can be seen that inside Ω_c, $|e_2| \leq |e_1| + |s| \leq 4.2$. We saturate $|e_2|$ at 5. Thus, the output feedback control is given by

$$u = -(2 \times 5\,\text{sat}(|\hat{e}_2|/5) + 5)\,\text{sat}\left(\frac{\hat{e}_1 + \hat{e}_2}{\mu}\right),$$

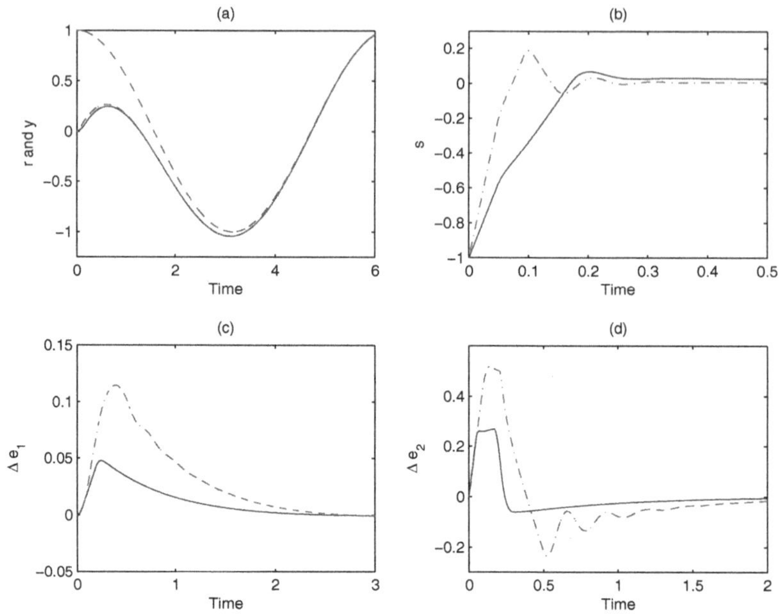

Figure 3.10. *Simulation of Example 3.6.* (a) *and* (b) *show y and s under output feedback control with $\varepsilon = 0.01$ for two different values of the parameters c_1 and c_2. The solid curves are for $c_1 = 0.015$ and $c_2 = 0.5$ and the dash-dotted curves for $c_1 = 0.03$ and $c_2 = 1$.* (a) *shows also the reference r (dashed).* (c) *and* (d) *show the components of Δe for $\varepsilon = 0.01$ (solid) and $\varepsilon = 0.03$ (dash-dotted).*

where $\hat{e}_1$ and $\hat{e}_2$ are provided by the high-gain observer

$$\dot{\hat{e}}_1 = \hat{e}_2 + (2/\varepsilon)(e_1 - \hat{e}_1), \quad \dot{\hat{e}}_2 = (1/\varepsilon^2)(e_1 - \hat{e}_1).$$

Simulation results with $\mu = 0.1$, $\varepsilon = 0.01$, and $x(0) = \hat{x}(0) = 0$ are shown in Figures 3.10(a) and (b) for $c_1 = 0.015$, $c_2 = 0.5$ (solid) and $c_1 = 0.03$, $c_2 = 1$ (dash-dotted). The dashed curve is the reference signal. The output responses in Figure 3.10(a) are almost indistinguishable. The plots in Figure 3.10(b) show that the trajectories enter the boundary layer $\{|s| \le \mu\}$ within the time interval $[0, 0.2]$. Figures 3.10(c) and (d) show $\Delta e = e(\text{under output feedback}) - e(\text{under state feedback})$ for two different values of ε; the rest of the parameters are $\mu = 0.1$, $c_1 = 0.015$, $c_2 = 0.5$. They illustrate the fact that $||\Delta e||$ decreases with ε. △

3.4 ▪ Notes and References

The separation principle of Section 3.1 is taken almost verbatim from Atassi and Khalil [18], but the proof of Theorem 3.1 is taken from Khalil [78], which gives a simplified version of the proof in [18]. One particular simplification is that asymptotic stability of the origin under output feedback when $\phi \neq \phi_0$ is shown when the origin under state feedback is exponentially stable. The result of [18] deals with the more general case when the origin is asymptotically but not exponentially stable. It shows asymptotic stability under an additional assumption that employs the center manifold theorem [78, Section 8.1]. The examples of Section 3.1.3 are taken from [18], but the design of

Example 3.2 is taken from [92]. It is worthwhile to note that Theorem 3.1 in the case $\phi_0 = \phi$ covers an earlier result on the separation principle by Teel and Praly [149], which was the first nonlinear separation principle based on high-gain observers.

An important generalization of Theorem 3.1 that is briefly reported in Atassi and Khalil [19] and described in more detail in Atassi's dissertation [17] is a separation principle for the stabilization of positively invariant sets for time-varying systems. The results of [19] include those of [18] as a special case because an equilibrium point is an invariant set, but they go way beyond that. Chapter 4 of [17] shows how the generalized separation principle applies to robust stabilization, tracking, regulation, and adaptive control.

Robust stabilization results using high-gain observers appeared early in the work of Esfandiari and Khalil [38] and Teel and Praly [150]. Section 3.2 is taken from a recent treatment in Khalil [80]. To simplify the presentation, we restricted attention to single-input–single-output systems in the normal form. More general systems are treated in [41, 150]. Section 3.3 is based on [80, Chapter 13].

Chapter 4

Adaptive Control

Adaptive control is one of the basic techniques to deal with uncertainty in the control of dynamical systems.[18] Starting in the 1990s, adaptive control techniques have been developed for classes of nonlinear systems.[19] Most of the results, however, have assumed state feedback. High-gain observers provide an effective tool to derive adaptive control assuming output feedback, and this is the subject of the current chapter.

The chapter starts in Section 4.1 by stating the tracking problem and defining the class of nonlinear systems. The design of feedback control is developed in two steps. State feedback control is derived in Section 4.2, followed by output feedback control in Section 4.3. The results are illustrated by examples in Section 4.4. The robustness of the controller to bounded disturbance is studied in Section 4.5. Finally, in Section 4.6 it is shown how the controller can be used when the nonlinearities of the systems are approximated by function approximators such as neural networks.

4.1 ▪ Problem Statement

Consider a single-input–single-output nonlinear system represented globally by the nth-order differential equation

$$y^{(n)} = f_0(\cdot) + \theta^T f(\cdot) + [g_0(\cdot) + \theta^T g(\cdot)]u^{(m)}, \tag{4.1}$$

where u is the control input, y is the measured output, the superscript (i) denotes the ith derivative, $m < n$, and f_0, f, g_0, and g are sufficiently smooth functions that could depend on $y, y^{(1)}, \ldots, y^{(n-1)}, u, u^{(1)}, \ldots, u^{(m-1)}$, for example,[20]

$$f_0(\cdot) = f_0(y, y^{(1)}, \ldots, y^{(n-1)}, u, u^{(1)}, \ldots, u^{(m-1)}).$$

The functions f_0, f, g_0, and g are known, while $\theta = \text{col}(\theta_1, \theta_2, \ldots, \theta_\ell)$ is a vector of unknown constant parameters that belongs to a known compact convex set $\Omega \subset R^\ell$. Let $\hat{\Omega} \subset R^\ell$ be a compact convex set that contains Ω in its interior. The control task is

[18] See, for example, [6], [15], [63], [110], [131], and [148].

[19] See, for example, [10], [90], [106], [137], [142], and [157].

[20] By sufficiently smooth we mean that the function is continuously differentiable a sufficient number of times, where the number is problem dependent.

to design an output feedback controller such that all variables of the closed-loop system are bounded and the output y asymptotically tracks a reference signal $r(t)$, that is,

$$\lim_{t\to\infty} |y(t)-r(t)| = 0.$$

We assume that $r(t)$ is bounded and has bounded derivatives up to the nth order and $r^{(n)}(t)$ is piecewise continuous. We augment a series of m integrators at the input side of the system and represent the extended system by a state-space model. The states of these integrators are $z_1 = u$, $z_2 = u^{(1)}$, up to $z_m = u^{(m-1)}$. The control input of the extended system is taken as $v = u^{(m)}$. With $x_1 = y$, $x_2 = y^{(1)}$, up to $x_n = y^{(n-1)}$, the extended state model is given by

$$\left.\begin{aligned} \dot{x}_i &= x_{i+1}, \quad 1 \le i \le n-1, \\ \dot{x}_n &= f_0(x,z) + \theta^T f(x,z) + [g_0(x,z) + \theta^T g(x,z)]v, \\ \dot{z}_i &= z_{i+1}, \quad 1 \le i \le m-1, \\ \dot{z}_m &= v, \\ y &= x_1, \end{aligned}\right\} \tag{4.2}$$

where

$$x = \text{col}(x_1, \dots, x_n), \qquad z = \text{col}(z_1, \dots, z_m).$$

We note that the state of this system comprises z_1 to z_m, which are available for feedback because they are the states of the augmented integrators, and x_1 to x_n, which can be estimated from y using a high-gain observer. Based on the properties of high-gain observers, we anticipate that the performance of a state feedback controller that uses the full state (x,z) can be recovered by an output feedback controller in which x is substituted by its estimate $\hat{x}$. Once v has been determined as a function of $(\hat{x}, z)$, it will be integrated m times to obtain the actual control u. Thus, the m integrators are actually part of the feedback controller.

The system is required to satisfy the following two assumptions.

Assumption 4.1.

$$g_0(x,z) + \theta^T g(x,z) \ge \bar{g} > 0 \quad \forall\, x \in R^n,\ z \in R^m,\ \theta \in \hat{\Omega}.$$

Assumption 4.2. *There exists a global diffeomorphism*

$$\begin{bmatrix} \eta \\ x \end{bmatrix} = \begin{bmatrix} T_1(x,z,\theta) \\ x \end{bmatrix} \stackrel{\text{def}}{=} T(x,z,\theta) \tag{4.3}$$

with $T_1(0,0,\theta) = 0$, *which transforms* (4.2) *into the global normal form*

$$\left.\begin{aligned} \dot{\eta} &= p(\eta, x, \theta), \\ \dot{x}_i &= x_{i+1}, \quad 1 \le i \le n-1, \\ \dot{x}_n &= f_0(x,z) + \theta^T f(x,z) + [g_0(x,z) + \theta^T g(x,z)]v, \\ y &= x_1. \end{aligned}\right\} \tag{4.4}$$

Assumption 4.1 implies that the system (4.2) has relative degree n. Under this assumption, the system is locally transformable into the normal form [64]. Assumption 4.2 requires the change of variables to hold globally. In the special case when g_0 and g are independent of z_m, the mapping T_1 can be taken as

$$\eta_i = z_i, \quad 1 \le i \le m-1,$$
$$\eta_m = z_m - \int_0^{x_n} \frac{1}{G(x_1, \ldots, x_{n-1}, \sigma, z_1, \ldots, z_{m-1}, \theta)}\, d\sigma,$$

where $G = g_0 + \theta^T g$.

Let

$$e_i = x_i - r^{(i-1)}, \quad 1 \le i \le n,$$
$$e = \mathrm{col}(e_1, \ldots, e_n),$$
$$\mathcal{R} = \mathrm{col}(r, r^{(1)}, \ldots, r^{(n-1)}).$$

The error vector e satisfies the equation

$$\dot{e} = Ae + B\left\{f_0(x,z) + \theta^T f(x,z) + [g_0(x,z) + \theta^T g(x,z)]v - r^{(n)}\right\}, \tag{4.5}$$

where

$$A = \begin{bmatrix} 0 & 1 & 0 & \cdots & \cdots & 0 \\ 0 & 0 & 1 & \cdots & \cdots & 0 \\ \vdots & & & & & \vdots \\ 0 & 0 & \cdots & \cdots & 1 & 0 \\ 0 & 0 & \cdots & \cdots & 0 & 1 \\ 0 & 0 & \cdots & \cdots & 0 & 0 \end{bmatrix}, \quad B = \begin{bmatrix} 0 \\ 0 \\ \vdots \\ 0 \\ 0 \\ 1 \end{bmatrix}.$$

If θ was known, a state feedback controller could have been taken as

$$v = \frac{-Ke + r^{(n)} - f_0(x,z) - \theta^T f(x,z)}{g_0(x,z) + \theta^T g(x,z)},$$

which results in the system

$$\dot{e} = (A - BK)e.$$

Choosing K such that $A_m = A - BK$ is Hurwitz ensures that $e(t)$ converges to zero as t tends to infinity. The boundedness of all variables will follow if the solution $\eta(t)$ of $\dot{\eta} = p(\eta, x, \theta)$ is bounded for every bounded $x(t)$, which is a stronger version of the minimum-phase assumption.[21] For later use with adaptive control, we make the following assumption.

Assumption 4.3. *The system $\dot{\bar{\eta}} = p(\bar{\eta}, \mathcal{R}, \theta)$ has a unique bounded steady-state solution $\eta_{ss}(t)$.*[22] *Moreover, with $\tilde{\eta} = \eta - \eta_{ss}$, there is a continuously differentiable function $V_1(t, \tilde{\eta})$, possibly dependent on θ, that satisfies the inequalities*

$$\rho_1 \|\tilde{\eta}\|^2 \le V_1(t, \tilde{\eta}) \le \rho_2 \|\tilde{\eta}\|^2, \tag{4.6}$$

$$\frac{\partial V_1}{\partial t} + \frac{\partial V_1}{\partial \tilde{\eta}}[p(\eta_{ss} + \tilde{\eta}, e + \mathcal{R}, \theta) - p(\eta_{ss}, \mathcal{R}, \theta)] \le -\rho_3 \|\tilde{\eta}\|^2 + \rho_4 \|\tilde{\eta}\|\, \|e\| \tag{4.7}$$

globally, where ρ_1 to ρ_4 are positive constants independent of $\mathcal{R}$ and θ.

[21] The minimum-phase assumption requires the equation $\dot{\eta} = p(\eta, 0, \theta)$ to have an asymptotically stable equilibrium point [64].

[22] See [64, Section 8.1] for the definition of the steady-state solution of a nonlinear system.

Inequalities (4.6) and (4.7) imply that the solution of $\dot{\eta} = p(\eta, x, \theta)$ is bounded for every bounded $x(t)$.

A special case of (4.1) arises when the functions f_0, f, g_0, and g depend only on $y, y^{(1)}, \ldots, y^{(n-1)}$. In this case no integrators are augmented with the system, and (4.2) takes the form

$$\left.\begin{aligned} \dot{x}_i &= x_{i+1}, \quad 1 \le i \le n-1, \\ \dot{x}_n &= f_0(x) + \theta^T f(x) + [g_0(x) + \theta^T g(x)]u, \\ y &= x_1, \end{aligned}\right\} \tag{4.8}$$

and Assumptions 4.2 and 4.3 are not needed.

The assumption that the model (4.1) depends linearly on the unknown parameters may require redefinition of physical parameters, as shown in the next example.

Example 4.1. A single-link manipulator with flexible joints and negligible damping can be modeled by [146].

$$I\ddot{q}_1 + MgL\sin q_1 + k(q_1 - q_2) = 0,$$
$$J\ddot{q}_2 - k(q_1 - q_2) = u,$$

where q_1 and q_2 are angular positions and u is a torque input. The physical parameters g, I, J, k, L, and M are all positive. Taking $y = q_1$ as the output, it can be verified that y satisfies the fourth-order differential equation

$$y^{(4)} = \frac{gLM}{I}(\dot{y}^2 \sin y - \ddot{y}\cos y) - \left(\frac{k}{I} + \frac{k}{J}\right)\ddot{y} - \frac{gkLM}{IJ}\sin y + \frac{k}{IJ}u.$$

Taking

$$\theta_1 = \frac{gLM}{I}, \quad \theta_2 = \frac{k}{I} + \frac{k}{J}, \quad \theta_3 = \frac{gkLM}{IJ}, \quad \theta_4 = \frac{k}{IJ}$$

we obtain

$$y^{(4)} = \theta_1(\dot{y}^2 \sin y - \ddot{y}\cos y) - \theta_2\ddot{y} - \theta_3 \sin y + \theta_4 u,$$

which takes the form (4.1) with $f_0 = 0$, $g_0 = 0$,

$$f = \begin{bmatrix} \dot{y}^2 \sin y - \ddot{y}\cos y \\ -\ddot{y} \\ -\sin y \\ 0 \end{bmatrix}, \quad g = \begin{bmatrix} 0 \\ 0 \\ 0 \\ 1 \end{bmatrix}, \quad \text{and} \quad \theta = \begin{bmatrix} \theta_1 \\ \theta_2 \\ \theta_3 \\ \theta_4 \end{bmatrix}.$$

Knowing bounds on the physical parameters, we can define a compact convex set Ω such that $\theta \in \Omega$. If Ω is defined such that θ_4 is bounded from below by a positive constant on a compact convex set $\hat{\Omega}$ that contains Ω in its interior, Assumption 4.1 will be satisfied. △

4.2 ▪ State Feedback Adaptive Control

We start by deriving a state feedback adaptive controller. Let $P = P^T > 0$ be the solution of the Lyapunov equation

$$PA_m + A_m^T P = -Q, \quad Q = Q^T > 0$$

and consider the Lyapunov function candidate

$$V = e^T Pe + \tfrac{1}{2}\tilde{\theta}^T \Gamma^{-1} \tilde{\theta},$$

where $\Gamma = \Gamma^T > 0$, $\tilde{\theta} = \hat{\theta} - \theta$, and $\hat{\theta}$ is an estimate of θ to be determined by the adaptive law. The derivative of V is given by

$$\dot{V} = -e^T Qe + 2e^T PB\Big[f_0 + \theta^T f + (g_0 + \theta^T g)v + Ke - r^{(n)}\Big] + \tilde{\theta}^T \Gamma^{-1}\dot{\hat{\theta}}.$$

By adding and subtracting $\hat{\theta}^T(f + gv)$, $\dot{V}$ is rewritten as

$$\dot{V} = -e^T Qe + 2e^T PB\Big[f_0 + \hat{\theta}^T f + (g_0 + \hat{\theta}^T g)v + Ke - r^{(n)}\Big] + \tilde{\theta}^T \Gamma^{-1}\Big[\dot{\hat{\theta}} - \Gamma(2e^T PB)(f + gv)\Big].$$

Taking

$$v = \frac{-Ke + r^{(n)} - f_0(e + \mathscr{R}, z) - \hat{\theta}^T f(e + \mathscr{R}, z)}{g_0(e + \mathscr{R}, z) + \hat{\theta}^T g(e + \mathscr{R}, z)} \overset{\text{def}}{=} \psi(e, z, \mathscr{R}, r^{(n)}, \hat{\theta}) \tag{4.9}$$

and setting

$$\varphi(e, z, \mathscr{R}, r^{(n)}, \hat{\theta}) = 2e^T PB\Big[f(e + \mathscr{R}, z) + g(e + \mathscr{R}, z)\psi(e, z, \mathscr{R}, r^{(n)}, \hat{\theta})\Big], \tag{4.10}$$

we end up with

$$\dot{V} = -e^T Qe + \tilde{\theta}^T \Gamma^{-1}[\dot{\hat{\theta}} - \Gamma\varphi].$$

Taking $\dot{\hat{\theta}} = \Gamma\varphi$ yields $\dot{V} = -e^T Qe$, which can be used to show that $e(t)$ converges to zero as t tends to infinity. However, we need to keep $\hat{\theta}$ in $\hat{\Omega}$. This can be achieved by adaptive laws with parameter projection,[23] which can be chosen such that

$$\tilde{\theta}^T \Gamma^{-1}(\dot{\hat{\theta}} - \Gamma\varphi) \le 0 \tag{4.11}$$

and $\hat{\theta} \in \hat{\Omega}$ for all $t \ge 0$. We give two choices of the adaptive law for the cases when Ω is a closed ball and a convex hypercube, respectively. In the first case, $\Omega = \{\theta^T\theta \le k\}$. Let $\Omega_\delta = \{\theta^T\theta \le k + \delta\}$, where $\delta > 0$ is chosen such that $\Omega_\delta \subset \hat{\Omega}$. The adaptive law is taken as $\dot{\hat{\theta}} = \Pi\,\Gamma\varphi$, where the $\ell \times \ell$ matrix Π is defined by

$$\Pi(\hat{\theta}, \varphi) = \begin{cases} I - \frac{(\hat{\theta}^T\hat{\theta} - k)}{\delta\hat{\theta}^T\Gamma\hat{\theta}}\Gamma\hat{\theta}\hat{\theta}^T & \text{if } \hat{\theta}^T\hat{\theta} > k \text{ and } \hat{\theta}^T\Gamma\varphi > 0, \\ I & \text{otherwise.} \end{cases} \tag{4.12}$$

With $\dot{\hat{\theta}} = \Pi\,\Gamma\varphi$, $\hat{\theta}$ cannot leave Ω_δ because, at the boundary $\hat{\theta}^T\hat{\theta} = k + \delta$, if $\hat{\theta}^T\Gamma\varphi > 0$,

$$\hat{\theta}^T\dot{\hat{\theta}} = \hat{\theta}^T\left[I - \frac{\delta}{\delta\hat{\theta}^T\Gamma\hat{\theta}}\Gamma\hat{\theta}\hat{\theta}^T\right]\Gamma\varphi = \left[1 - \frac{\delta\hat{\theta}^T\Gamma\hat{\theta}}{\delta\hat{\theta}^T\Gamma\hat{\theta}}\right]\hat{\theta}^T\Gamma\varphi = 0,$$

[23] See, for example, [63].

and if $\hat{\theta}^T\Gamma\varphi \le 0$,

$$\hat{\theta}^T\dot{\hat{\theta}} = \hat{\theta}^T\Gamma\varphi \le 0.$$

It can also be seen that with $\dot{\hat{\theta}} = \Pi\,\Gamma\varphi$, (4.11) is satisfied. This is obvious when $\Pi = I$ since $\dot{\hat{\theta}} = \Gamma\varphi$. Otherwise,

$$\begin{aligned}\tilde{\theta}^T\Gamma^{-1}(\dot{\hat{\theta}} - \Gamma\varphi) &= \tilde{\theta}^T\Gamma^{-1}(\Pi - I)\Gamma\varphi \\ &= \tilde{\theta}^T\Gamma^{-1}\left[-\frac{(\hat{\theta}^T\hat{\theta} - k)}{\delta\hat{\theta}^T\Gamma\hat{\theta}}\Gamma\hat{\theta}\hat{\theta}^T\right]\Gamma\varphi \\ &= -\left(\frac{\hat{\theta}^T\hat{\theta} - k}{\delta\hat{\theta}^T\Gamma\hat{\theta}}\right)\tilde{\theta}^T\hat{\theta}\hat{\theta}^T\Gamma\varphi \; < 0\end{aligned}$$

because, in this case, $\hat{\theta}^T\hat{\theta} > k$, $\hat{\theta}^T\Gamma\varphi > 0$, and

$$\tilde{\theta}^T\hat{\theta} = \hat{\theta}^T\hat{\theta} - \theta^T\hat{\theta} \ge ||\hat{\theta}||^2 - ||\theta||\,||\hat{\theta}|| = ||\hat{\theta}||(||\hat{\theta}|| - ||\theta||) > 0$$

since $||\theta||^2 \le k$ and $||\hat{\theta}||^2 > k$.

In the case of a convex hypercube,

$$\Omega = \{\theta \mid a_i \le \theta_i \le b_i,\ 1 \le i \le p\},$$

and Γ is taken as a positive diagonal matrix. Let

$$\Omega_\delta = \{\theta \mid a_i - \delta \le \theta_i \le b_i + \delta,\ 1 \le i \le p\},$$

where $\delta > 0$ is chosen such that $\Omega_\delta \subset \hat{\Omega}$. The adaptive law is taken as $\dot{\hat{\theta}}_i = \pi_i\gamma_i\varphi_i$, where $\gamma_i > 0$ and π_i is defined by

$$\pi_i(\hat{\theta}, \varphi) = \begin{cases} 1 + (b_i - \hat{\theta}_i)/\delta & \text{if } \hat{\theta}_i > b_i \text{ and } \varphi_i > 0, \\ 1 + (\hat{\theta}_i - a_i)/\delta & \text{if } \hat{\theta}_i < a_i \text{ and } \varphi_i < 0, \\ 1 & \text{otherwise.} \end{cases} \tag{4.13}$$

With $\Gamma = \text{diag}(\gamma_1, \ldots, \gamma_p)$ and $\Pi = \text{diag}(\pi_1, \ldots, \pi_p)$, the adaptive law can be rewritten as $\dot{\hat{\theta}} = \Pi\,\Gamma\varphi$. Under this law, $\hat{\theta}$ cannot leave Ω_δ because, at the boundary $\hat{\theta}_i = b_i + \delta$, if $\varphi_i > 0$,

$$\dot{\hat{\theta}}_i = \left(1 + \frac{b_i - \hat{\theta}_i}{\delta}\right)\gamma_i\varphi_i = \left(1 + \frac{b_i - b_i - \delta}{\delta}\right)\gamma_i\varphi_i = 0,$$

and if $\varphi_i \le 0$, $\dot{\hat{\theta}} = \gamma_i\varphi_i \le 0$. Similarly, at the boundary $\hat{\theta}_i = a_i - \delta_i$, if $\varphi_i < 0$,

$$\dot{\hat{\theta}}_i = \left(1 + \frac{\hat{\theta}_i - a_i}{\delta}\right)\gamma_i\varphi_i = \left(1 + \frac{a_i - \delta - a_i}{\delta}\right)\gamma_i\varphi_i = 0,$$

and if $\varphi_i \ge 0$, $\dot{\hat{\theta}} = \gamma_i\varphi_i \ge 0$. Next we verify (4.11):

$$\tilde{\theta}^T\Gamma^{-1}(\dot{\hat{\theta}} - \Gamma\varphi) = \sum_i^p \frac{1}{\gamma_i}\tilde{\theta}_i(\pi_i\gamma_i\varphi_i - \gamma_i\varphi_i) = \sum_i^p(\pi_i - 1)\tilde{\theta}_i\varphi_i.$$

Clearly (4.11) is satisfied if $\pi_i = 1$ for all i. If $\pi_i = 1 + (b_i - \hat{\theta}_i)/\delta$,

$$(\pi_i - 1)\tilde{\theta}_i \varphi_i = (b_i - \hat{\theta}_i)\tilde{\theta}_i \varphi_i/\delta < 0$$

because, in this case, $\hat{\theta}_i > b_i$, $\varphi_i > 0$, and $\tilde{\theta}_i = \hat{\theta}_i - \theta_i > 0$. If $\pi_i = 1 + (\hat{\theta}_i - a_i)/\delta$,

$$(\pi_i - 1)\tilde{\theta}_i \varphi_i = (\hat{\theta}_i - a_i)\tilde{\theta}_i \varphi_i/\delta < 0$$

because, in this case, $\hat{\theta}_i < a_i$, $\varphi_i < 0$, and $\tilde{\theta}_i = \hat{\theta}_i - \theta_i < 0$

In summary, taking the adaptive law as

$$\dot{\hat{\theta}} = \Pi(\hat{\theta}, \varphi)\Gamma\varphi, \tag{4.14}$$

where Π is defined by (4.12) or (4.13), ensures that (4.11) is satisfied and

$$\hat{\theta}(0) \in \Omega \Rightarrow \hat{\theta}(t) \in \Omega_\delta \quad \forall\, t \geq 0.$$

It can be verified that $\Pi(\hat{\theta}, \varphi)\,\Gamma\varphi$ is locally Lipschitz in its arguments.[24]

In preparation for output feedback, we saturate the control $v = \psi$ outside a compact set that contains (e, z) under state feedback. Suppose $\hat{\theta}(0) \in \Omega$, $e(0) \in E_0$, and $z(0) \in Z_0$, where E_0 and Z_0 are compact sets, which can be chosen to cover any given bounded initial conditions. Suppose further that Y_r is a compact subset of R^n and M_r is a positive constant such that $\mathscr{R} \in Y_r$ and $|r^{(n)}| \leq M_r$ for all $t \geq 0$. Let

$$c_1 \geq \max_{e \in E_0} e^T P e, \quad c_2 \geq \max_{\theta \in \Omega, \hat{\theta} \in \Omega_\delta} \tfrac{1}{2}(\hat{\theta} - \theta)^T \Gamma^{-1} (\hat{\theta} - \theta),$$

and $c_3 > c_1 + c_2$. It follows from (4.11) that $\dot{V} \leq -e^T Q e$. Hence, the set $\{V \leq c_3\}$ is positively invariant. Consequently, $e(t) \in E \stackrel{\text{def}}{=} \{e^T P e \leq c_3\}$ for all $t \geq 0$. Let $\rho = \max_{e \in E} \|e\|$. From Assumption 4.3, we see that

$$\dot{V}_1 \leq -\rho_3 \|\tilde{\eta}\|^2 + \rho_4 \rho \|\tilde{\eta}\|.$$

Thus, we can choose $c_4 > 0$ large enough so that the set $\{V_1(t, \tilde{\eta}) \leq c_4\}$ is positively invariant and that for all $e(0) \in E_0$ and $z(0) \in Z_0$, $\tilde{\eta}(0)$ is in the interior of $\{V_1(t, \tilde{\eta}) \leq c_4\}$. Since $\{V_1(t, \tilde{\eta}) \leq c_4\} \subset \{\rho_1 \|\tilde{\eta}\|^2 \leq c_4\}$, mapping $\{\rho_1 \|\tilde{\eta}\|^2 \leq c_4\}$ into the z-coordinates, we can find a compact set $Z \in R^m$ such that $z(t) \in Z$ for all $t \geq 0$. Take $S > \max |\psi(e, z, \mathscr{R}, z, r^{(n)}, \hat{\theta})|$, where the maximization is taken over all $e \in E = \{e^T P e \leq c_3\}$, $z \in Z$, $\mathscr{R} \in Y_r$, $|r(n)| \leq M_r$, and $\hat{\theta} \in \Omega_\delta$. Define the saturated control function ψ^s by

$$\psi^s(e, z, \mathscr{R}, r^{(n)}, \hat{\theta}) = S \,\text{sat}\left(\frac{\psi(e, z, \mathscr{R}, r^{(n)}, \hat{\theta})}{S} \right), \tag{4.15}$$

where $\text{sat}(y) = \text{sign}(y)\min\{|y|, 1\}$.

[24] A parameter projection that confines $\hat{\theta}$ to Ω will be discontinuous. The parameter projection used here is locally Lipschitz and confines $\hat{\theta}$ to $\Omega_\delta \supset \Omega$.

4.3 ▪ Output Feedback Adaptive Control

The output feedback controller is given by

$$\dot{\hat{\theta}} = \Pi(\hat{\theta}, \varphi(\hat{e}, z, \mathcal{R}, r^{(n)}, \hat{\theta}))\Gamma\varphi(\hat{e}, z, \mathcal{R}, r^{(n)}, \hat{\theta}), \tag{4.16}$$

$$v = \psi^s(\hat{e}, z, \mathcal{R}, r^{(n)}, \hat{\theta}), \tag{4.17}$$

where $\hat{e}$ is provided by the high-gain observer

$$\left.\begin{aligned} \dot{\hat{e}}_i &= \hat{e}_{i+1} + (\alpha_i/\varepsilon^i)(e_1 - \hat{e}_1) \quad \text{for } 1 \le i \le n-1, \\ \dot{\hat{e}}_n &= (\alpha_n/\varepsilon^n)(e_1 - \hat{e}_1), \end{aligned}\right\} \tag{4.18}$$

in which ε is a small positive constant and α_1 to α_n are chosen such that the polynomial

$$s^n + \alpha_1 s^{n-1} + \cdots + \alpha_{n-1}s + \alpha_n \tag{4.19}$$

is Hurwitz. The closed-loop system under output feedback control is given by (4.2), (4.16), (4.17), and (4.18).

To show convergence of e and $\hat{\theta}$ under output feedback control, we need the following assumption. Let

$$w_r(t) = f(\mathcal{R}, z_{ss}) + g(\mathcal{R}, z_{ss})\psi(0, z_{ss}, \mathcal{R}, r^{(n)}, \theta), \tag{4.20}$$

where z_{ss} is the unique solution of $\eta_{ss} = T_1(\mathcal{R}, z_{ss}, \theta)$ in which η_{ss} is defined in Assumption 4.3.

Assumption 4.4. *w_r satisfies one of the following three conditions:*

(a) *w_r is persistently exciting and $\dot{w}_r$ is bounded.*

(b) $w_r = 0$.

(c) *There is a constant nonsingular matrix M, possibly dependent on θ, such that*

$$M w_r(t) = \begin{bmatrix} w_a(t) \\ 0 \end{bmatrix},$$

where w_a is a persistently exciting and $\dot{w}_a$ is bounded.

Recall that w is persistently exciting if there are positive constants β, β_1, and β_2 such that [63].

$$\beta_1 I \le \int_t^{t+\beta} w(\tau)w^T(\tau)\, d\tau \le \beta_2 I \quad \forall\, t \ge 0.$$

In case (c) of Assumption 4.4, define vectors θ_a and $\hat{\theta}_a$, of the same dimension as w_a, by

$$(M^{-1})^T\theta = \begin{bmatrix} \theta_a \\ \theta_b \end{bmatrix}, \quad (M^{-1})^T\hat{\theta} = \begin{bmatrix} \hat{\theta}_a \\ \hat{\theta}_b \end{bmatrix}.$$

Theorem 4.1. *Consider the closed-loop system formed of* (4.2), (4.16), (4.17), *and* (4.18). *Suppose Assumptions* 4.1 *to* 4.4 *are satisfied,* $\hat{\theta}(0) \in \Omega$, $e(0) \in E_0$, $z(0) \in Z_0$, *and* $\hat{e}(0)$ *is* *bounded. Then there is $\varepsilon^* > 0$ such that for all $0 < \varepsilon < \varepsilon^*$, all variables of the closed-loop system are bounded and*

$$\lim_{t\to\infty} e(t) = 0, \quad \lim_{t\to\infty} [z(t) - z_{ss}(t)] = 0, \quad \lim_{t\to\infty} \hat{e}(t) = 0. \tag{4.21}$$

Moreover, if Assumption 4.4(a) is satisfied,

$$\lim_{t\to\infty} \|\hat{\theta}(t) - \theta(t)\| = 0. \tag{4.22}$$

If Assumption 4.4(c) is satisfied,

$$\lim_{t\to\infty} \|\hat{\theta}_a(t) - \theta_a(t)\| = 0. \tag{4.23}$$

Proof: We start by applying a change of variables to represent the closed-loop system in a more convenient form. Let $\tilde{\eta} = \eta - \eta_{ss} = T_1(x, z, \theta) - \eta_{ss}$ and

$$\xi_i = \frac{e_i - \hat{e}_i}{\varepsilon^{n-i}}, \quad 1 \le i \le n.$$

The closed-loop system is represented by

$$\begin{aligned}
\dot{e} &= A_m e + B[Ke + f_0 + \theta^T f + (g_0 + \theta^T g)\psi^s - r^{(n)}],\\
\dot{\hat{\theta}} &= \Pi(\hat{\theta}, \varphi)\Gamma\varphi,\\
\dot{\tilde{\eta}} &= p(\eta_{ss} + \tilde{\eta}, e + \mathscr{R}, \theta) - p(\eta_{ss}, \mathscr{R}, \theta),\\
\varepsilon\dot{\xi} &= (A - HC)\xi + \varepsilon B[f_0 + \theta^T f + (g_0 + \theta^T g)\psi^s - r^{(n)}],
\end{aligned}$$

where $H = \text{col}(\alpha_1, \ldots, \alpha_n)$, $\xi = \text{col}(\xi_1, \ldots, \xi_n)$, $\psi^s = \psi^s(\hat{e}, z, \mathscr{R}, r^{(n)}, \hat{\theta})$, and $\varphi = \varphi(\hat{e}, z, \mathscr{R}, r^{(n)}, \hat{\theta})$. The arguments of f_0, f, g_0, g, ψ^s, and φ are omitted for convenience. The characteristic polynomial of $(A - HC)$ is (4.19); hence $(A - HC)$ is Hurwitz. Let $V_2 = \xi^T P_2 \xi$, where P_2 is the solution of the Lyapunov equation $P_2(A - HC) + (A - HC)^T P_2 = -I$. We start by showing that there is a constant $c_5 > 0$ such that the set

$$\Sigma = \{V \le c_3\} \cap \{\hat{\theta} \in \Omega_\delta\} \times \{V_1 \le c_4\} \times \{V_2 \le c_5 \varepsilon^2\}$$

is positively invariant for sufficiently small ε. Inside Σ, $\|\xi\| = O(\varepsilon)$, and the saturation of ψ is not effective, that is, $\psi^s = \psi$. Therefore, e satisfies the equation

$$\dot{e} = A_m e - B\tilde{\theta}^T(f + g\psi) + \varepsilon\Delta_1(\cdot),$$

where $\|\Delta_1\| \le k_1$ for some k_1 independent of ε. Thus,

$$\begin{aligned}
\dot{V} &\le -e^T Q e + k\varepsilon \le -c_0 e^T P e + k\varepsilon\\
&= -c_0 V + \tfrac{1}{2} c_0 \tilde{\theta}^T \Gamma^{-1} \tilde{\theta} + k\varepsilon \le -c_0 V + c_0 c_7 + k\varepsilon
\end{aligned}$$

for some constant $k > 0$, where $c_0 = \lambda_{min}(Q)/\lambda_{max}(P)$. When $\varepsilon < c_0(c_3 - c_2)/k$, $\dot{V} < 0$ on the boundary $V = c_3$, which implies that $\{V \le c_3\} \cap \{\hat{\theta} \in \Omega_\delta\}$ is positively invariant. Inside this set, $e \in E$. As shown earlier, the choice of c_4 ensures that $\{V_1 \le c_4\}$ is positively invariant. The derivative of V_2 satisfies

$$\varepsilon\dot{V}_2 \le -k_2 V_2 + k_3 \varepsilon\sqrt{V_2} \le -\tfrac{1}{2} k_2 V_2 \quad \forall\; V_2 \ge (2k_3\varepsilon/k_2)^2. \tag{4.24}$$

Taking $c_5 > (2k_3/k_2)^2$ ensures that $\{V_2 \leq c_5\varepsilon^2\}$ is positively invariant, which completes the proof that Σ is positively invariant. Next we show that there is time $T(\varepsilon)$, with $lim_{\varepsilon\to 0}T(\varepsilon) = 0$, such that the trajectories of the closed-loop system are confined to Σ for all $t \geq T(\varepsilon)$. This follows from two observations. First, because the initial conditions $(e(0), \hat{\theta}(0), \tilde{\eta}(0))$ are in the interior of the set $\{V \leq c_3\} \cap \{\hat{\theta} \in \Omega_\delta\} \times \{V_1 \leq c_4\}$, v is bounded uniformly in ε due to saturation, and $\hat{\theta}$ is confined to Ω_δ by parameter projection, there is time $T_0 > 0$, independent of ε, such that $(e(t), \hat{\theta}(t), \tilde{\eta}(t)) \in \{V \leq c_3\} \cap \{\hat{\theta} \in \Omega_\delta\} \times \{V_1 \leq c_4\}$ for all $t \in [0, T_0]$. On the other hand, (4.24) shows that there time $T(\varepsilon)$, with $\lim_{\varepsilon\to 0}T(\varepsilon) = 0$, such that $V_2 \leq c_5\varepsilon^2$ for all $t \geq T(\varepsilon)$.[25] For sufficiently small ε, $T(\varepsilon) < T_0$. Thus, all the variables of the closed-loop system are in Σ for all $t \geq T(\varepsilon)$.

From this time on, the control saturation is not effective; that is, $\psi^s = \psi$, and the closed-loop system can be represented by

$$\left.\begin{aligned} \dot{e} &= A_m e - B\tilde{\theta}^T w(t) + B\Lambda_1, \\ \dot{\hat{\theta}} &= \Pi(\hat{\theta}, \varphi)\Gamma\varphi, \\ \dot{\tilde{\eta}} &= p(\eta_{ss} + \tilde{\eta}, e + \mathcal{R}, \theta) - p(\eta_{ss}, \mathcal{R}, \theta), \\ \varepsilon\dot{\xi} &= (A - HC)\xi - \varepsilon B[\tilde{\theta}^T w(t) - \Lambda_1 + Ke], \end{aligned}\right\} \tag{4.25}$$

where

$$w = f(\hat{e} + \mathcal{R}, z) + g(\hat{e} + \mathcal{R}, z)\psi(\hat{e}, z, \mathcal{R}, r^{(n)}, \hat{\theta}) \tag{4.26}$$

and

$$\begin{aligned} \Lambda_1 = K(e - \hat{e}) &+ f_0(e + \mathcal{R}, z) - f_0(\hat{e} + \mathcal{R}, z) + \theta^T[f(e + \mathcal{R}, z) - f(\hat{e} + \mathcal{R}, z)] \\ &+ [g_0(e + \mathcal{R}, z) - g_0(\hat{e} + \mathcal{R}, z)]\psi(\hat{e}, z, \mathcal{R}, r^{(n)}, \hat{\theta}) \\ &+ \theta^T[g(e + \mathcal{R}, z) - g(\hat{e} + \mathcal{R}, z)]\psi(\hat{e}, z, \mathcal{R}, r^{(n)}, \hat{\theta}). \end{aligned}$$

The following lemma relates w_r and w, defined by (4.20) and (4.26), respectively.

Lemma 4.1.

$$\tilde{\theta}^T w = \mathcal{G}\tilde{\theta}^T w_r - \tilde{\theta}^T \Lambda_2, \tag{4.27}$$

where

$$\mathcal{G} = \frac{g_0(\mathcal{R}, z_{ss}) + \theta^T g(\mathcal{R}, z_{ss})}{g_0(\mathcal{R}, z_{ss}) + \hat{\theta}^T g(\mathcal{R}, z_{ss})}$$

and

$$\begin{aligned} \Lambda_2 = f(\mathcal{R}, z_{ss}) &- f(\hat{e} + \mathcal{R}, z) \\ &+ g(\mathcal{R}, z_{ss})\psi(0, z_{ss}, \mathcal{R}, r^{(n)}, \hat{\theta}) - g(\hat{e} + \mathcal{R}, z)\psi(\hat{e}, z, \mathcal{R}, r^{(n)}, \hat{\theta}). \end{aligned}$$

◇

Proof of Lemma 4.1: Let $\bar{f}_0 = f_0(\mathcal{R}, z_{ss})$, $\bar{f} = f(\mathcal{R}, z_{ss})$, $\hat{f} = f(\hat{e} + \mathcal{R}, z)$, $\bar{g}_0 = g_0(\mathcal{R}, z_{ss})$, $\bar{g} = g(\mathcal{R}, z_{ss})$, $\hat{g} = g(\hat{e} + \mathcal{R}, z)$, $\bar{\psi} = \psi(0, z_{ss}, \mathcal{R}, r^{(n)}, \theta)$, $\tilde{\psi} = \psi(0, z_{ss}, \mathcal{R}, r^{(n)}, \hat{\theta})$, and $\hat{\psi} = \psi(\hat{e}, z_{ss}, \mathcal{R}, r^{(n)}, \hat{\theta})$:

$$w_r - w = \bar{f} + \bar{g}\bar{\psi} - \hat{f} - \hat{g}\hat{\psi} = \bar{f} - \hat{f} + \bar{g}(\bar{\psi} - \tilde{\psi}) + \bar{g}\tilde{\psi} - \hat{g}\hat{\psi}.$$

[25] See the proof of Theorem 3.1 for the details of this argument.

The term $\bar{\psi} - \tilde{\psi}$ can be written as

$$
\begin{aligned}
\bar{\psi} - \tilde{\psi} &= \frac{r^{(n)}\bar{g}_0 + r^{(n)}\hat{\theta}^T\bar{g} - \bar{f}_0\bar{g}_0 - \bar{f}_0\hat{\theta}^T\bar{g} - \theta^T\bar{f}\bar{g}_0 - \theta^T\bar{f}\hat{\theta}^T\bar{g}}{(\bar{g}_0 + \theta^T\bar{g})(\bar{g}_0 + \hat{\theta}^T\bar{g})} \\
&\quad + \frac{-r^{(n)}\bar{g}_0 - r^{(n)}\theta^T\bar{g} + \bar{f}_0\bar{g}_0 + \bar{f}_0\theta^T\bar{g} + \hat{\theta}^T\bar{f}\bar{g}_0 + \hat{\theta}^T\bar{f}\theta^T\bar{g}}{(\bar{g}_0 + \theta^T\bar{g})(\bar{g}_0 + \hat{\theta}^T\bar{g})} \\
&= \frac{r^{(n)}\tilde{\theta}^T\bar{g} - \bar{f}_0\tilde{\theta}^T\bar{g} + \tilde{\theta}^T\bar{f}\bar{g}_0 - \theta^T\bar{f}\hat{\theta}^T\bar{g} + \hat{\theta}^T\bar{f}\theta^T\bar{g}}{(\bar{g}_0 + \theta^T\bar{g})(\bar{g}_0 + \hat{\theta}^T\bar{g})} \\
&= \frac{\tilde{\theta}^T\bar{g}(r^{(n)} - \bar{f}_0 - \theta^T\bar{f}) + \tilde{\theta}^T\bar{f}(\bar{g}_0 + \theta^T\bar{g})}{(\bar{g}_0 + \theta^T\bar{g})(\bar{g}_0 + \hat{\theta}^T\bar{g})} \\
&= \frac{\tilde{\theta}^T(\bar{f} + \bar{g}\bar{\psi})}{(\bar{g}_0 + \hat{\theta}^T\bar{g})} = \frac{\tilde{\theta}^T w_r}{(\bar{g}_0 + \hat{\theta}^T\bar{g})}.
\end{aligned}
$$

Hence,

$$
w_r - w = \bar{f} - \hat{f} + \bar{g}\tilde{\psi} - \hat{g}\hat{\psi} + \bar{g}\frac{\tilde{\theta}^T w_r}{(\bar{g}_0 + \hat{\theta}^T\bar{g})} = \Lambda_2 + \bar{g}\frac{\tilde{\theta}^T w_r}{(\bar{g}_0 + \hat{\theta}^T\bar{g})}
$$

and

$$
\tilde{\theta}^T w = \frac{(\bar{g}_0 + \theta^T\bar{g})}{(\bar{g}_0 + \hat{\theta}^T\bar{g})}\tilde{\theta}^T w_r - \tilde{\theta}^T\Lambda_2 = \mathcal{G}\tilde{\theta}^T w_r - \tilde{\theta}^T\Lambda_2,
$$

which completes the proof of Lemma 4.1. □

In case (c) of Assumption 4.4, the equations of $\dot{e}$ and $\dot{\tilde{\theta}}$ can be written as

$$
\dot{e} = A_m e - B\mathcal{G}\tilde{\theta}^T M^{-1} M w_r + B(\Lambda_1 + \tilde{\theta}^T\Lambda_2) = A_m e - B\mathcal{G}\tilde{\theta}_a^T w_a + B(\Lambda_1 + \tilde{\theta}^T\Lambda_2),
$$

$$
\begin{bmatrix} \dot{\tilde{\theta}}_a \\ \dot{\tilde{\theta}}_b \end{bmatrix} = M^{-T}\Pi\,\Gamma\varphi \stackrel{\text{def}}{=} \begin{bmatrix} \varpi_a(\cdot) \\ \varpi_b(\cdot) \end{bmatrix},
$$

where ϖ_a has the same dimension as $\tilde{\theta}_a$. Cases (a) and (b) of Assumption 4.4 are extreme cases of case (c), where in case (a), $\tilde{\theta}_a = \tilde{\theta}$ and $\tilde{\theta}_b$ does not exist, while in case (b), $\tilde{\theta}_b = \tilde{\theta}$ and $\tilde{\theta}_a$ does not exist. We continue the proof with case (c). It can be shown that

$$
\begin{bmatrix} \dot{e} \\ \dot{\tilde{\theta}}_a \end{bmatrix} = \begin{bmatrix} A_m & -B\mathcal{G}w_a^T \\ 2\Gamma_1\mathcal{G}w_a B^T P & 0 \end{bmatrix} \begin{bmatrix} e \\ \tilde{\theta}_a \end{bmatrix} + \begin{bmatrix} \Lambda_e \\ \Lambda_\theta \end{bmatrix}, \tag{4.28}
$$

where

$$
M^{-T}\Gamma M^{-1} = \begin{bmatrix} \Gamma_1 & \Gamma_2 \\ \Gamma_2^T & \Gamma_4 \end{bmatrix},
$$

$$
\Lambda_e = B(\Lambda_1 + \tilde{\theta}^T\Lambda_2),
$$

$$
\Lambda_\theta = \varpi_a(\cdot) - 2\Gamma_1\mathcal{G}w_a B^T P e,
$$

and

$$K_{\mathscr{G}1} < \mathscr{G} \le K_{\mathscr{G}2}$$

for some positive constants $K_{\mathscr{G}1}$ and $K_{\mathscr{G}2}$ independent of ε. Since $\dot{\mathscr{R}}$, $\dot{z}_{ss}$, and $\dot{\hat{\theta}}$ are bounded,[26] it can be shown that $\dot{\mathscr{G}}$ is bounded. Since f, g_0, g, and ψ are locally Lipschitz functions in their arguments, Λ_e and Λ_θ satisfy the inequalities

$$\|\Lambda_e\| \le \delta_1\|e\| + \delta_2\|\xi\| + \delta_3\|\tilde{\eta}\|, \quad \|\Lambda_\theta\| \le \delta_4\|e\| + \delta_5\|\xi\| \tag{4.29}$$

for some nonnegative constants δ_1 to δ_5. Consider the system

$$\begin{bmatrix} \dot{e} \\ \dot{\tilde{\theta}}_a \end{bmatrix} = \begin{bmatrix} A_m & -B\mathscr{G}w_a^T \\ 2\Gamma_1\mathscr{G}w_a^T B^T P & 0 \end{bmatrix} \begin{bmatrix} e \\ \tilde{\theta}_a \end{bmatrix}. \tag{4.30}$$

Because w_a is persistently exciting and $\mathscr{G}$ is bounded from below and from above, $\mathscr{G}w_a$ is persistently exciting. Because $\dot{w}_a$ and $\dot{\mathscr{G}}$ are bounded, the derivative of $\mathscr{G}w_a$ is bounded. Using well-known results from adaptive control theory,[27] we conclude that the origin of (4.30) is exponentially stable. By the converse Lyapunov theorem, [78, Theorem 4.14], there is a Lyapunov function $V_3(t, e, \tilde{\theta}_a)$ whose derivative with respect to (4.28) satisfies

$$\begin{aligned} \dot{V}_3 \le {} & -\delta_6\|e\|^2 - \delta_7\|\tilde{\theta}_a\|^2 + \delta_8\|e\|\,\|\xi\| + \delta_9\|\tilde{\theta}_a\|\,\|\xi\| \\ & + \delta_{10}\|e\|^2 + \delta_{11}\|e\|\,\|\tilde{\theta}_a\| + \delta_{12}\|e\|\,\|\tilde{\eta}\| + \delta_{13}\|\tilde{\theta}_a\|\,\|\tilde{\eta}\| \end{aligned} \tag{4.31}$$

for some positive constants δ_6 and δ_7 and nonnegative constants δ_8 to δ_{13}. The derivative of V_2 with respect to

$$\varepsilon\dot{\xi} = (A - HC)\xi - \varepsilon B[\mathscr{G}\tilde{\theta}_a^T w_a + Ke] + \varepsilon\Lambda_e$$

satisfies

$$\dot{V}_2 \le -\frac{1}{\varepsilon}\|\xi\|^2 + \sigma_1\|\tilde{\theta}_a\|\,\|\xi\| + \sigma_2\|e\|\,\|\xi\| + \sigma_3\|\tilde{\eta}\|\,\|\xi\| + \sigma_4\|\xi\|^2 \tag{4.32}$$

for some nonnegative constants σ_1 to σ_4.

Construct the Lyapunov function candidate

$$W = \alpha V + \beta V_1 + V_2 + V_3,$$

where the positive constants α and β will be chosen later on. Using the inequality

$$\dot{V} \le -k_4\|e\|^2 + k_5\|e\|\,\|\xi\|$$

with $k_4 > 0$ and $k_5 > 0$, together with Assumption 4.3, (4.31), and (4.32), it can be shown that the derivative of W with respect to (4.25) satisfies

$$\dot{W} \le -\chi^T N \chi, \tag{4.33}$$

[26] The boundedness of $\dot{z}_{ss}$ follows from the boundedness of $\dot{\eta}_{ss}$, which is the case because η_{ss} satisfies the equation $\dot{\eta}_{ss} = p(\eta_{ss}, \mathscr{R}, \theta)$.

[27] See, for example, [75, Section 13.4].

where $\chi = \text{col}(\|e\|, \|\tilde{\theta}_a\|, \|\tilde{\eta}\|, \|\xi\|)$ and

$$N = \frac{1}{2}\begin{bmatrix} 2(\alpha k_4 + \delta_6 - \delta_{10}) & -\delta_{11} & -(\beta\rho_4 + \delta_{12}) & -(\alpha k_5 + \delta_8 + \sigma_2) \\ -\delta_{11} & 2\delta_7 & -\delta_{13} & -(\delta_9 + \sigma_1) \\ -(\beta\rho_4 + \delta_{12}) & -\delta_{13} & 2\beta\rho_3 & -\sigma_3 \\ -(\alpha k_5 + \delta_8 + \sigma_2 & -(\delta_9 + \sigma_1) & -\sigma_3 & 2(1 - \varepsilon\sigma_4)/\varepsilon \end{bmatrix}.$$

Choose β large enough to make

$$\begin{bmatrix} 2\delta_7 & -\delta_{13} \\ -\delta_{13} & 2\beta\rho_3 \end{bmatrix}$$

positive definite; then choose α large enough so that

$$\begin{bmatrix} 2(\alpha k_4 + \delta_6 - \delta_{10}) & -\delta_{11} & -(\beta\rho_4 + \delta_{12}) \\ -\delta_{11} & 2\delta_7 & -\delta_{13} \\ -(\beta\rho_4 + \delta_{12}) & -\delta_{13} & 2\beta\rho_3 \end{bmatrix}$$

is positive definite. Finally, choosing ε small enough makes N positive definite. Hence, $\lim_{t\to\infty} \chi(t) = 0$, [78, Theorem 8.4], which proves (4.21) and (4.23). In case (a) of Assumption 4.4, $\tilde{\theta} = \tilde{\theta}_a$, which proves (4.22). □

It should be noted that in cases (b) and (c) of Assumption 4.4, the proof does not imply exponential convergence because the right-hand side of (4.33) is only negative semidefinite. In particular, the right-hand side of (4.33) is a quadratic form of $(e, \tilde{\theta}_a, \tilde{\eta}, \xi)$, while the Lyapunov function W is a quadratic form of $(e, \tilde{\theta}, \tilde{\eta}, \xi)$ The construction of W is a key point in the proof. While the perturbation terms Λ_e and Λ_θ on the right-hand side of (4.28) satisfy the growth conditions (4.29), the constants δ_1 to δ_5 are not necessarily small. Consequently, we see in (4.31) that the right-hand side contains the positive term $\delta_{10}\|e\|^2$, which could dominate the negative term $-\delta_6\|e\|^2$. We overcome this difficulty by including αV in the composite Lyapunov function W and choose α large enough to ensure that the negative term $-\alpha k_4\|e\|^2$ dominates $\delta_{10}\|e\|^2$ and other cross-product terms. In case (a) of Assumption 4.4, $\tilde{\theta} = \tilde{\theta}_a$ and the proof shows that the closed-loop system has an exponentially stable equilibrium point at $(e, \tilde{\theta}, \tilde{\eta}, \xi) = 0$.

4.4 ▪ Examples

We use three examples to illustrate the adaptive controller of the previous section. The first example is a linear system that shows how the algorithm that is developed for nonlinear systems applies to linear systems as a special case. The example also elaborates on the persistence of excitation conditions of Assumption 4.4. The second example revisits the tracking problem of the pendulum that is treated in Example 3.5 by feedback linearization when the systems parameters are known. The same feedback strategy is used here but with unknown parameters. The example allows us to compare the use of feedback linearization with and without adaptation. The last example considers a nonlinear system with zero dynamics and illustrates the extension of the system by adding integrators at the input side.[28]

[28] The zero dynamics are the dynamics of the system when the output is identically zero.

Example 4.2. A linear system whose transfer function is

$$G(s) = \frac{c}{(s+a)(s+b)},$$

with $c > 0$, can be represented by the second-order differential equation

$$\ddot{y} = -\theta_1 y - \theta_2 \dot{y} + \theta_3 u,$$

where $\theta_1 = ab$, $\theta_2 = a + b$, and $\theta_3 = c$. We assume that the parameters a, b, and c are unknown but with known bounds. The differential equation can be represented by the state model (4.2) with no z variable and with $f_0 = 0$, $g_0 = 0$,

$$\theta = \begin{bmatrix} \theta_1 \\ \theta_2 \\ \theta_3 \end{bmatrix}, \quad f = \begin{bmatrix} -x_1 \\ -x_2 \\ 0 \end{bmatrix} = \begin{bmatrix} -y \\ -\dot{y} \\ 0 \end{bmatrix}, \quad \text{and} \quad g = \begin{bmatrix} 0 \\ 0 \\ 1 \end{bmatrix}.$$

With

$$A = \begin{bmatrix} 0 & 1 \\ 0 & 0 \end{bmatrix} \quad \text{and} \quad B = \begin{bmatrix} 0 \\ 1 \end{bmatrix},$$

the matrix $K = \begin{bmatrix} k_1 & k_2 \end{bmatrix}$ is chosen such that $A - BK$ is Hurwitz. The state feedback adaptive controller is given by

$$\psi(e, \mathcal{R}, \ddot{r}, \hat{\theta}) = \frac{-Ke + \ddot{r} - \hat{\theta}^T f}{\hat{\theta}^T g} = \frac{-k_1 e_1 - k_2 e_2 + \ddot{r} + \hat{\theta}_1(e_1 + r) + \hat{\theta}_2(e_2 + \dot{r})}{\hat{\theta}_3}.$$

To examine Assumption 4.4, consider

$$\psi(0, \mathcal{R}, \ddot{r}, \theta) = \frac{\ddot{r} + \theta_1 r + \theta_2 \dot{r}}{\theta_3},$$

which, by (4.20), yields

$$w_r = \begin{bmatrix} -r \\ -\dot{r} \\ (\ddot{r} + \theta_1 r + \theta_2 \dot{r})/\theta_3 \end{bmatrix}.$$

We consider three cases of the reference signal: $r = 1$, $r = \sin \omega t$, and $r = 1 + \sin \omega t$, where $\omega > 0$. In the first case,

$$w_r = \begin{bmatrix} -1 \\ 0 \\ \theta_1/\theta_3 \end{bmatrix} \quad \text{and} \quad \underbrace{\begin{bmatrix} -1 & 0 & 0 \\ 0 & 1 & 0 \\ \theta_1/\theta_3 & 0 & 1 \end{bmatrix}}_{M} w_r = \begin{bmatrix} 1 \\ 0 \\ 0 \end{bmatrix}.$$

Hence, case (c) of Assumption 4.4 is satisfied with $w_a = 1$, and

$$(M^{-1})^T \theta = \begin{bmatrix} 0 \\ \theta_2 \\ \theta_3 \end{bmatrix}, \quad (M^{-1})^T \hat{\theta} = \begin{bmatrix} -(\hat{\theta}_1 - \theta_1 \hat{\theta}_3/\theta_3) \\ \hat{\theta}_2 \\ \hat{\theta}_3 \end{bmatrix}.$$

Thus, $\theta_a = 0$ and $\hat{\theta}_a = -(\hat{\theta}_1 - \theta_1\hat{\theta}_3/\theta_3)$. Theorem 4.1 shows that $\lim_{t\to\infty}[\hat{\theta}_a(t) - \theta_a(t)] = 0$, that is, $\lim_{t\to\infty}[\hat{\theta}_1(t) - \theta_1\hat{\theta}_3(t)/\theta_3] = 0$. When $r = \sin\omega t$,

$$w_r = \begin{bmatrix} -\sin\omega t \\ -\omega\cos\omega t \\ (-\omega^2\sin\omega t + \theta_1\sin\omega t + \theta_2\omega\cos\omega t)/\theta_3 \end{bmatrix}$$

and

$$\underbrace{\begin{bmatrix} -1 & 0 & 0 \\ 0 & -1/\omega & 0 \\ (\theta_1 - \omega^2)/\theta_3 & \theta_2/\theta_3 & 1 \end{bmatrix}}_{M} w_r = \begin{bmatrix} \sin\omega t \\ \cos\omega t \\ 0 \end{bmatrix}.$$

Therefore, case (c) of Assumption 4.4 is satisfied with $w_a = \text{col}(\sin\omega t,\ \cos\omega t)$, which is persistently exciting because

$$\int_t^{t+\pi/\omega} \begin{bmatrix} \sin\omega\tau \\ \cos\omega\tau \end{bmatrix} \begin{bmatrix} \sin\omega\tau \\ \cos\omega\tau \end{bmatrix}^T d\tau = \tfrac{1}{2}(\pi/\omega)I$$

and

$$(M^{-1})^T\theta = \begin{bmatrix} -\omega^2 \\ 0 \\ \theta_3 \end{bmatrix}, \quad (M^{-1})^T\hat{\theta} = \begin{bmatrix} -\hat{\theta}_1 + (\theta_1 - \omega^2)\hat{\theta}_3/\theta_3 \\ -\omega\hat{\theta}_2 + \omega\theta_2\hat{\theta}_3/\theta_3 \\ \hat{\theta}_3 \end{bmatrix}.$$

Thus,

$$\theta_a = \begin{bmatrix} -\omega^2 \\ 0 \end{bmatrix} \quad \text{and} \quad \hat{\theta}_a = \begin{bmatrix} -\hat{\theta}_1 + (\theta_1 - \omega^2)\hat{\theta}_3/\theta_3 \\ -\omega\hat{\theta}_2 + \omega\theta_2\hat{\theta}_3/\theta_3 \end{bmatrix}.$$

Theorem 4.1 shows that $\lim_{t\to\infty}[\hat{\theta}_a(t) - \theta_a(t)] = 0$, that is,

$$\lim_{t\to\infty}[\hat{\theta}_1(t) - (\theta_1 - \omega^2)\hat{\theta}_3(t)/\theta_3 - \omega^2] = 0 \text{ and } \lim_{t\to\infty}[\hat{\theta}_2(t) - \theta_2\hat{\theta}_3(t)/\theta_3] = 0.$$

Finally, when $r = 1 + \sin\omega t$,

$$\begin{aligned} w_r &= \begin{bmatrix} -1 - \sin\omega t \\ -\omega\cos\omega t \\ (-\omega^2\sin\omega t + \theta_1 + \theta_1\sin\omega t + \theta_2\omega\cos\omega t)/\theta_3 \end{bmatrix} \\ &= \underbrace{\begin{bmatrix} -1 & -1 & 0 \\ 0 & 0 & -\omega \\ \theta_1/\theta_3 & (\theta_1 - \omega^2)/\theta_3 & \theta_2\omega/\theta_3 \end{bmatrix}}_{Q} \begin{bmatrix} 1 \\ \sin\omega t \\ \cos\omega t \end{bmatrix}. \end{aligned}$$

The vector $\bar{w}_r = \text{col}(1,\ \sin\omega t,\ \cos\omega t)$ is persistently exciting because

$$\int_t^{t+\pi/\omega} \bar{w}_r(\tau)\bar{w}_r^T(\tau)\, d\tau = (\pi/\omega)\,\text{diag}(1, \tfrac{1}{2}, \tfrac{1}{2}),$$

and Q is nonsingular since $\det(Q) = \omega^3/\theta_3 > 0$. Thus, w_r is persistently exciting as in case (a) of Assumption 4.4. It follows from Theorem 4.1 that $\lim_{t\to\infty}[\hat{\theta}_i(t) - \theta_i] = 0$ for $i = 1, 2, 3$.

To complete the design, suppose $a \in [-1,1]$, $b \in [-1,1]$, and $c \in [1,2]$. Then

$$\Omega = \{\theta \mid \theta_1 \in [-1,1],\ \theta_2 \in [-2,2],\ \theta_3 \in [1,2]\}.$$

With $\delta = 0.1$,

$$\Omega_\delta = \{\theta \mid \theta_1 \in [-1.1,1.1],\ \theta_2 \in [-2.1,2.1],\ \theta_3 \in [0.9,2.1]\}.$$

The matrix K is taken as $K = \begin{bmatrix} 2 & 3 \end{bmatrix}$ to assign the eigenvalues of $A_m = A - BK$ at -1, -2. The solution of the Lyapunov equation $PA_m + A_m^T P = -I$ is

$$P = \tfrac{1}{4}\begin{bmatrix} 5 & 1 \\ 1 & 1 \end{bmatrix} \quad \text{and} \quad PB = \tfrac{1}{4}\begin{bmatrix} 1 \\ 1 \end{bmatrix}.$$

Thus,

$$\psi = \frac{-2e_1 - 3e_2 + \ddot{r} + \hat{\theta}_1(e_1 + r) + \hat{\theta}_2(e_2 + \dot{r})}{\hat{\theta}_3}$$

and

$$\varphi = 2e^T PB(f + g\psi) = \tfrac{1}{2}(e_1 + e_2)\begin{bmatrix} -(e_1 + r) \\ -(e_2 + \dot{r}) \\ \psi \end{bmatrix}.$$

The adaptive law is $\dot{\hat{\theta}}_i = \pi_i \gamma_i \varphi_i$ for $i = 1,2,3$, where

$$\pi_1(\hat{\theta}_1, \varphi_1) = \begin{cases} 1 + (1 - \hat{\theta}_1)/0.1 & \text{if } \hat{\theta}_1 > 1 \text{ and } \varphi_1 > 0, \\ 1 + (\hat{\theta}_1 + 1)/0.1 & \text{if } \hat{\theta}_1 < -1 \text{ and } \varphi_1 < 0, \\ 1 & \text{otherwise,} \end{cases}$$

$$\pi_2(\hat{\theta}_2, \varphi_2) = \begin{cases} 1 + (2 - \hat{\theta}_2)/0.1 & \text{if } \hat{\theta}_2 > 2 \text{ and } \varphi_2 > 0, \\ 1 + (\hat{\theta}_2 + 2)/0.1 & \text{if } \hat{\theta}_2 < -2 \text{ and } \varphi_2 < 0, \\ 1 & \text{otherwise,} \end{cases}$$

and

$$\pi_3(\hat{\theta}_3, \varphi_3) = \begin{cases} 1 + (2 - \hat{\theta}_3)/0.1 & \text{if } \hat{\theta}_3 > 2 \text{ and } \varphi_3 > 0, \\ 1 + (\hat{\theta}_3 - 1)/0.1 & \text{if } \hat{\theta}_3 < 1 \text{ and } \varphi_3 < 0, \\ 1 & \text{otherwise.} \end{cases}$$

To saturate the control, we need to find the maximum of $|\psi|$ under state feedback. We calculate this maximum for all three cases of the reference r. Suppose $\hat{\theta}(0) \in \Omega$, $e(0) \in E_0 = \{|e_1| \le 1,\ |e_2| \le 1\}$, $\omega \le 2$, and $\gamma_i \ge 10$ for $i = 1,2,3$. The maximum of $e^T Pe$ over E_0 happens at one of the vertices and $c_1 = \max_{e \in E_0} e^T Pe = 2$. The maximum of $\frac{1}{2}(\hat{\theta} - \theta)^T \Gamma^{-1}(\hat{\theta} - \theta)$ satisfies

$$\begin{aligned} \max_{\theta \in \Omega, \hat{\theta} \in \Omega_\delta} \tfrac{1}{2}(\hat{\theta} - \theta)^T \Gamma^{-1}(\hat{\theta} - \theta) &= \tfrac{1}{2} \max_{\theta \in \Omega, \hat{\theta} \in \Omega_\delta} \sum_{i=1}^{3} \frac{(\hat{\theta}_i - \theta_i)^2}{\gamma_i} \\ &\le \tfrac{1}{2}\left[\frac{(2.1)^2}{\gamma_1} + \frac{(4.1)^2}{\gamma_2} + \frac{(1.1)^2}{\gamma_3}\right] \le 1.1215. \end{aligned}$$

We take $c_2 = 1.125$ and $c_3 = 3.2 > c_1 + c_2$. Now we compute the maximum of $|\psi|$ for $e^T Pe \le c_3$ and $\hat{\theta} \in \Omega_\delta$:

$$\begin{aligned}|\psi| &= \left| \frac{(-2+\hat{\theta}_1)e_1 + (-3+\hat{\theta}_2)e_2 + \ddot{r} + \hat{\theta}_1 r + \hat{\theta}_2 \dot{r}}{\hat{\theta}_3} \right| \\ &\le \frac{1}{0.9}[|(-2+\hat{\theta}_1)e_1 + (-3+\hat{\theta}_2)e_2| + |\ddot{r} + \hat{\theta}_1 r + \hat{\theta}_2 \dot{r}|].\end{aligned}$$

It can be verified that for all three cases of r,

$$|\ddot{r} + \hat{\theta}_1 r + \hat{\theta}_2 \dot{r}| \le 7.71.$$

On the other hand, the maximum of $|(-2+\hat{\theta}_1)e_1+(-3+\hat{\theta}_2)e_2|$ over $e^T Pe \le c_3$ is given by $\sqrt{c_3}\left\|\left[(-2+\hat{\theta}_1) \quad (-3+\hat{\theta}_2)\right]P^{-1/2}\right\|$, where $P^{-1/2}$ is the inverse of the square root matrix of P. Maximizing this norm over $\hat{\theta} \in \Omega_\delta$, we arrive at the bound $|\psi| \le 30.57$. We take $S = 32$. The output feedback controller is given by

$$\begin{aligned}u &= 32 \text{ sat}\left(\frac{-2\hat{e}_1 - 3\hat{e}_2 + \ddot{r} + \hat{\theta}_1(\hat{e}_1 + r) + \hat{\theta}_2(\hat{e}_2 + \dot{r})}{32\,\hat{\theta}_3} \right), \\ \dot{\hat{e}}_1 &= \hat{e}_2 + (2/\varepsilon)(y - r - \hat{e}_1), \quad \dot{\hat{e}}_2 = (1/\varepsilon^2)(y - r - \hat{e}_1), \\ \dot{\hat{\theta}}_1 &= \gamma_1 \pi_1(\hat{\theta}_1, \varphi_1)\varphi_1, \quad \varphi_1 = -\tfrac{1}{2}(\hat{e}_1 + \hat{e}_2)(\hat{e}_1 + r), \\ \dot{\hat{\theta}}_2 &= \gamma_2 \pi_2(\hat{\theta}_2, \varphi_2)\varphi_2, \quad \varphi_2 = -\tfrac{1}{2}(\hat{e}_1 + \hat{e}_2)(\hat{e}_2 + \dot{r}), \\ \dot{\hat{\theta}}_3 &= \gamma_3 \pi_3(\hat{\theta}_3, \varphi_3)\varphi_3, \quad \varphi_3 = \tfrac{1}{2}(\hat{e}_1 + \hat{e}_2)u,\end{aligned}$$

where the observer gains assign the observer eigenvalues at $-1/\varepsilon$ and $-1/\epsilon$.

Figures 4.1 to 4.4 show simulation results. The simulation is carried out with $a = 0.5$, $b = -1$, $c = 1.5$, $\omega = 2$, and $\varepsilon = 0.01$. Consequently, $\theta_1 = -0.5$, $\theta_2 = -0.5$, and $\theta_3 = 1.5$. The initial conditions of x, $\hat{e}$, and $\hat{\theta}$ are taken to be zero, except $\hat{\theta}_3(0) = 1$. Figures 4.1 and 4.2 are for the case $r = 1$. Figure 4.1 shows the tracking errors e_1 and e_2 and the control u in two different time scales. Zooming on the initial period shows the control saturating at 32 during the peaking period of the observer. Figure 4.2 shows the parameter estimates $\hat{\theta}_1$ to $\hat{\theta}_3$. As expected, they do not converge to the true values of θ_1 to θ_3, which are shown by the dashed lines. As we saw earlier, in this case $\hat{\theta}_1 - \theta_1\hat{\theta}_3/\theta_3 = \hat{\theta}_1 + \frac{1}{3}\hat{\theta}_3$ converges to zero as t tends to infinity, which is demonstrated in Figure 4.2(d). Figure 4.3 shows simulation results when $r = \sin 2t$. Figures 4.2(a) and (b) demonstrate that the tracking errors e_1 and e_2 converge to zero. Figures 4.3(c) and (d) show the parameter estimates $\hat{\theta}_1$ to $\hat{\theta}_3$, which do not converge to the true values of θ_1 to θ_3. We saw earlier that $\hat{\theta}_1 - (\theta_1 - \omega^2)\hat{\theta}_3/\theta_3 - \omega^2 = \hat{\theta}_1 + 3\hat{\theta}_3 - 4$ and $\hat{\theta}_2 - \theta_2\hat{\theta}_3/\theta_3 = \hat{\theta}_2 + \frac{1}{3}\hat{\theta}_3$ converge to zero, which is shown in Figures 4.3(e) and (f). Finally, Figure 4.4 shows simulation results when $r = 1 + \sin 2t$. As expected, the tracking errors e_1 and e_2 converge to zero, and the parameter estimates converge to the true parameters. △

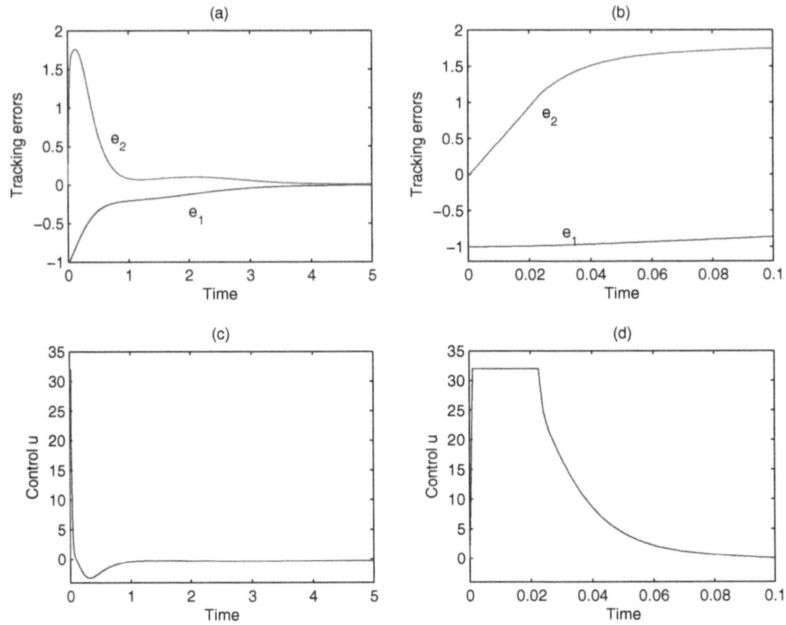

Figure 4.1. *Simulation of Example* 4.2 *with* $r = 1$. (a) *and* (b) *show the tracking error* e_1 *and* e_2 *on two different time scales.* (c) *and* (d) *do the same for the control signal* u.

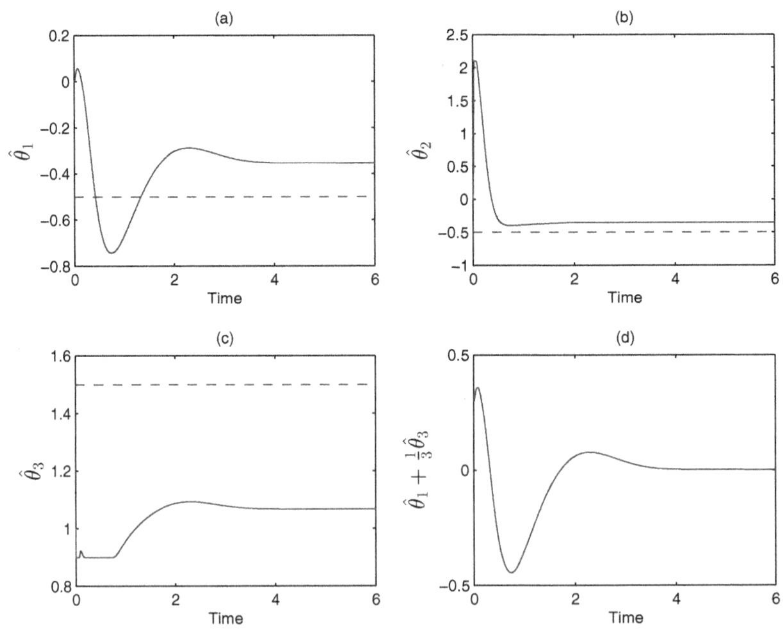

Figure 4.2. *Simulation of Example* 4.2 *with* $r = 1$. *The parameter estimates do not converge to the true values of the parameters, shown by the dashed lines, but* $\hat{\theta}_1 + \frac{1}{3}\hat{\theta}_3$ *converges to zero.*

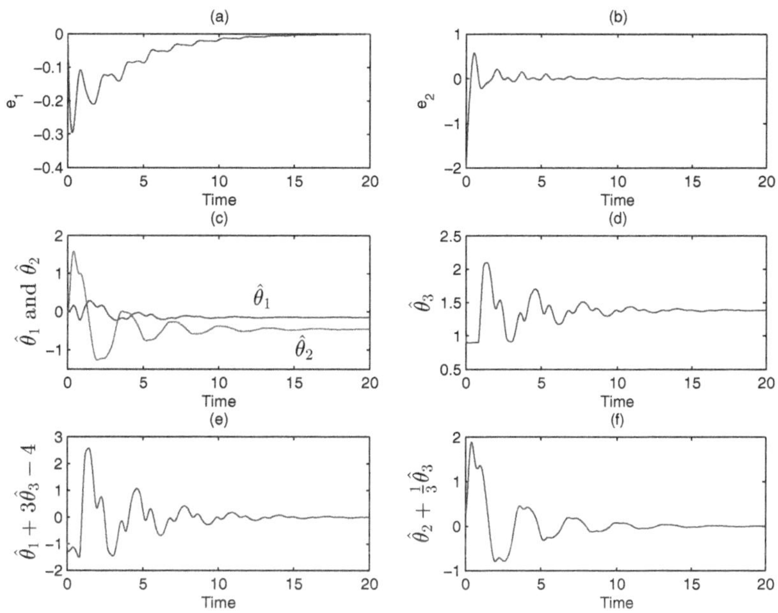

Figure 4.3. *Simulation of Example 4.2 with $r = \sin 2t$. The tracking errors e_1 and e_2 converge to zero. The parameter estimates do not converge to the true values of the parameters, but $\hat{\theta}_1 + 3\hat{\theta}_3 - 4$ and $\hat{\theta}_2 + \frac{1}{3}\hat{\theta}_3$ converge to zero.*

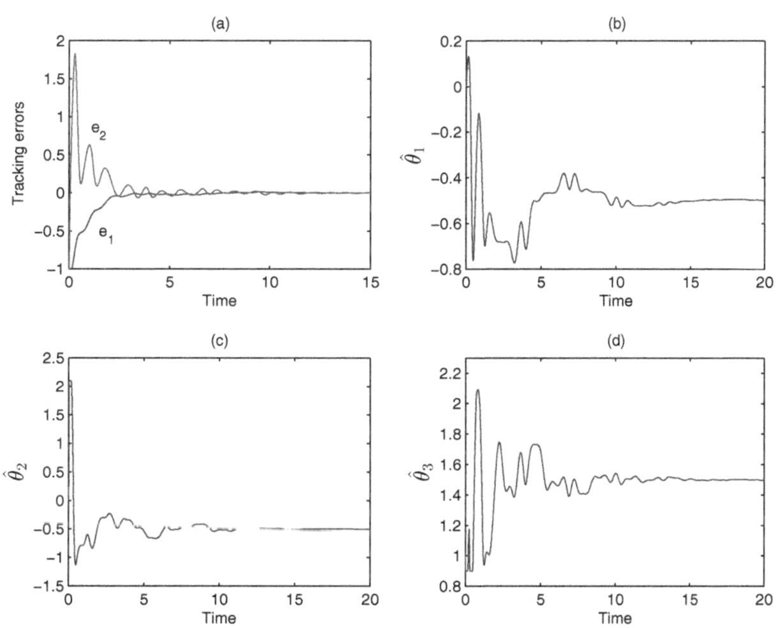

Figure 4.4. *Simulation of Example 4.2 with $r = 1 + \sin 2t$. The tracking errors e_1 and e_2 converge to zero, and the parameter estimates converge to the true values of the parameters.*

Example 4.3. Reconsider the tracking problem for the pendulum equation

$$\dot{x}_1 = x_2, \quad \dot{x}_2 = -\sin x_1 - \theta_1 x_2 + \theta_2 u, \quad y = x_1$$

from Example 3.5, where the goal is to have the output y track the reference $r(t) = \cos t$. In Example 3.5 feedback linearization is used when the parameters $\theta_1 = 0.015$ and $\theta_2 = 0.5$ are known. Suppose these parameters are unknown but we know that $\theta_1 \in [0, 0.1]$ and $\theta_2 \in [0.5, 2]$. With the change of variables

$$e_1 = x_1 - r = x_1 - \cos t, \qquad e_2 = x_2 - \dot{r} = x_2 + \sin t,$$

the system is represented by

$$\dot{e} = \underbrace{\begin{bmatrix} 0 & 1 \\ 0 & 0 \end{bmatrix}}_{A} e + \underbrace{\begin{bmatrix} 0 \\ 1 \end{bmatrix}}_{B} [-\sin(e_1 + \cos t) - \theta_1(e_2 - \sin t) + \theta_2 u + \cos t].$$

The matrix $K = \begin{bmatrix} 2 & 3 \end{bmatrix}$ assigns the eigenvalues of $A_m = A - BK$ at $-1, -2$. The solution of the Lyapunov equation $PA_m + A_m^T P = -I$ is

$$P = \tfrac{1}{4}\begin{bmatrix} 5 & 1 \\ 1 & 1 \end{bmatrix} \quad \text{and} \quad PB = \tfrac{1}{4}\begin{bmatrix} 1 \\ 1 \end{bmatrix}.$$

From (4.9) and (4.10), ψ and φ are given by

$$\psi = \frac{-2e_1 - 3e_2 + \sin(e_1 + \cos t) - \cos t + \hat{\theta}_1(e_2 - \sin t)}{\hat{\theta}_2},$$

$$\varphi = \tfrac{1}{2}(e_1 + e_2)\begin{bmatrix} -e_2 + \sin t \\ \psi \end{bmatrix}.$$

The sets Ω and Ω_δ, with $\delta = 0.05$, are given by

$$\Omega = \{\theta \mid \theta_1 \in [0, 0.1],\ \theta_2 \in [0.5, 2]\},$$

$$\Omega_\delta = \{\theta \mid \theta_1 \in [-0.05, 0.15],\ \theta_2 \in [0.45, 2.05]\}.$$

The adaptive law is $\dot{\hat{\theta}}_i = \pi_i \gamma_i \varphi_i$ for $i = 1, 2$, where

$$\pi_1(\hat{\theta}_1, \varphi_1) = \begin{cases} 1 + (0.1 - \hat{\theta}_1)/0.05 & \text{if } \hat{\theta}_1 > 0.1 \text{ and } \varphi_1 > 0, \\ 1 + \hat{\theta}_1/0.05 & \text{if } \hat{\theta}_1 < 0 \text{ and } \varphi_1 < 0, \\ 1 & \text{otherwise,} \end{cases}$$

$$\pi_2(\hat{\theta}_1, \varphi_1) = \begin{cases} 1 + (2 - \hat{\theta}_2)/0.05 & \text{if } \hat{\theta}_2 > 2 \text{ and } \varphi_2 > 0, \\ 1 + (\hat{\theta}_2 - 0.5)/0.05 & \text{if } \hat{\theta}_2 < 0.5 \text{ and } \varphi_2 < 0, \\ 1 & \text{otherwise.} \end{cases}$$

We take $\gamma_1 = \gamma_2 = 10$. Let $E_0 = \{|e_1| \le 1,\ |e_2| \le 1\}$. Then

$$c_1 = \max_{e \in E_0} e^T P e = 2, \quad c_2 = \max_{\theta \in \Omega, \hat{\theta} \in \Omega_\delta} \tfrac{1}{2}(\hat{\theta} - \theta)^T \Gamma^{-1}(\hat{\theta} - \theta) = 0.11461.$$

With $c_3 = 2.2$, $E = \{e^T P e \le 2.2\}$, and

$$\max_{e\in E,\hat{\theta}\in\Omega_\delta} \left| \frac{-2e_1 - 3e_2 + \sin(e_1 + \cos t) - \cos t + \hat{\theta}_1(e_2 - \sin t)}{\hat{\theta}_2} \right| = 22.7835.$$

Take $S = 25$. To examine the persistence of excitation condition, consider

$$w_r = \begin{bmatrix} \sin t \\ (\sin(\cos t) - \cos t - \theta_1 \sin t)/\theta_2 \end{bmatrix} = \begin{bmatrix} 1 & 0 \\ -\theta_1/\theta_2 & 1/\theta_2 \end{bmatrix} \begin{bmatrix} \sin t \\ \sin(\cos t) - \cos t \end{bmatrix}.$$

Using the fact that $(\sin t)^2$ and $(\sin(\cos t) - \cos t)^2$ are even functions while $\sin t(\sin(\cos t) - \cos t)$ is an odd function, it can be verified that w_r is persistently exciting. The output feedback controller is given by

$$u = 25 \operatorname{sat}\left(\frac{-2\hat{e}_1 - 3\hat{e}_2 + \sin(\hat{e}_1 + \cos t) - \cos t + \hat{\theta}_1(\hat{e}_2 - \sin t)}{25\,\hat{\theta}_2} \right),$$

$$\dot{\hat{e}}_1 = \hat{e}_2 + (2/\varepsilon)(e_1 - \hat{e}_1),$$

$$\dot{\hat{e}}_2 = (1/\varepsilon^2)(e_1 - \hat{e}_1),$$

$$\dot{\hat{\theta}}_1 = 10\,\pi_1(\hat{\theta}_1, \varphi_1)\varphi_1, \quad \varphi_1 = \tfrac{1}{2}(\hat{e}_1 + \hat{e}_2)(-\hat{e}_2 + \sin t),$$

$$\dot{\hat{\theta}}_2 = 10\,\pi_2(\hat{\theta}_2, \varphi_2)\varphi_2, \quad \varphi_2 = \tfrac{1}{2}(\hat{e}_1 + \hat{e}_2)u,$$

where the observer has multiple eigenvalues at $-1/\varepsilon$. The simulation results in Figures 4.5 and 4.6 are carried out with $\theta_1 = 0.015$, $\theta_2 = 0.5$, $x(0) = \hat{e}(0) = 0$, $\hat{\theta}_1(0) = 0$, $\hat{\theta}_2(0) = 1$, and $\varepsilon = 0.01$. Figures 4.5(a) and (b) show y and u under the adaptive controller. Figures 4.5(c) and (d) compare the tracking errors e_1 and e_2 between the feedback linearizing controller and the adaptive controller. Despite the parameter uncertainty, the performance of the of the adaptive controller is very the close to the performance of the feedback linearizing controller. Figure 4.6 confirms that the parameter estimates converge to the true values of the parameters. △

Example 4.4. Consider the tracking problem for the system shown in Figure 4.7 with

$$G(s) = \frac{\theta(s+1)}{s(s-1)} \quad \text{and} \quad \sigma(y) = \tanh(y),$$

where $\theta \in [1, 3]$ is an unknown parameter. Two different reference signals are considered: $r(t) = \cos t$ and $r(t) = 1$. A state model of the system is given by

$$\dot{\xi}_1 = \xi_2, \quad \dot{\xi}_2 = \xi_2 + \theta[u - \sigma(\xi_1 + \xi_2)], \quad y = \xi_1 + \xi_2.$$

The system has relative degree one since $\dot{y} = 2\xi_2 + \theta[u - \sigma(\xi_1 + \xi_2)]$ and is minimum phase since $y(t) \equiv 0 \Rightarrow \xi_2(t) \equiv -\xi_1(t) \Rightarrow \dot{\xi}_1 = -\xi_1$. It can be represented by the second-order differential equation

$$\ddot{y} = \dot{y} + \theta[u - \sigma(y)] + \theta[\dot{u} - \sigma'(y)\dot{y}],$$

where $\sigma'(y) = d\sigma/dy = 1/(\cosh(y))^2$. This equation is in the form (4.1) with $f_0 = \dot{y}$, $f = u - \sigma(y) - \sigma'(y)\dot{y}$, $g_0 = 0$, and $g = 1$. Assumption 4.1 is satisfied with $\bar{g} = 1$.

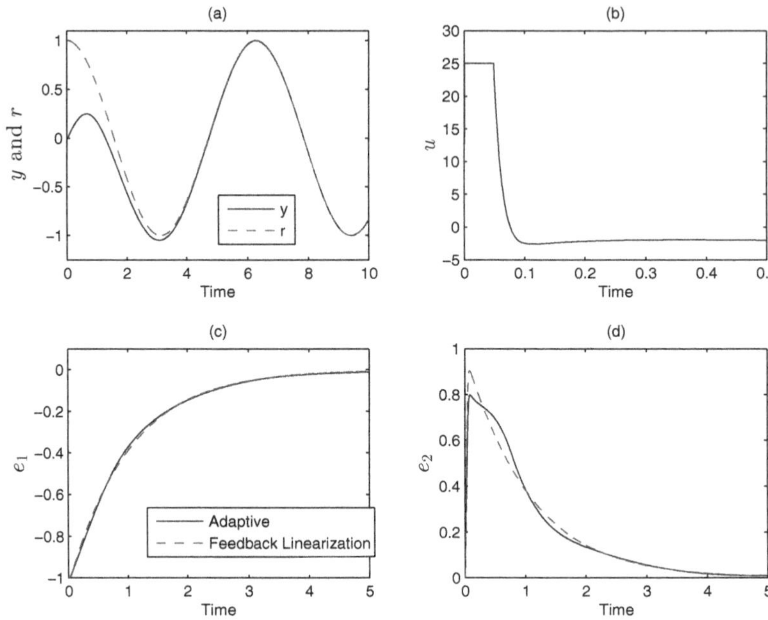

Figure 4.5. *Simulation of Example* 4.3. (a) *and* (b) *show* y *and* u *under the adaptive controller.* (c) *and* (d) *compare the tracking errors under the adaptive controller (solid) with the errors under the feedback linearization of Example* 3.5 *(dashed).*

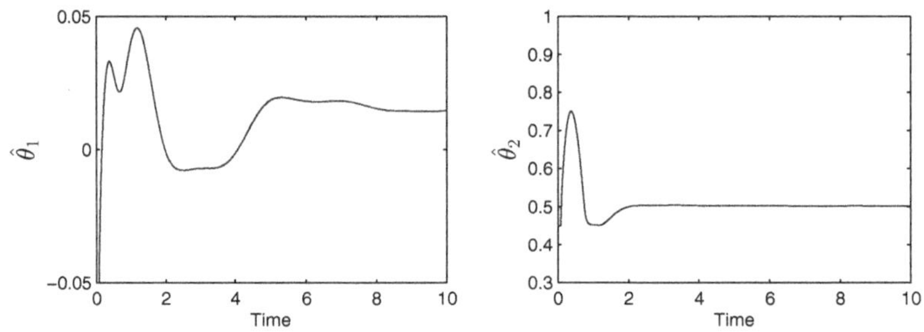

Figure 4.6. *Simulation of Example* 4.3. *Convergence of the parameter estimates.*

Extend the dynamics of the system by adding an integrator at the input and let $z = u$, $x_1 = y$, $x_2 = \dot{y}$, and $v = \dot{u}$ to arrive at the extended state model

$$\begin{aligned}
\dot{x}_1 &= x_2,\\
\dot{x}_2 &= x_2 + \theta[z - \sigma(x_1) - \sigma'(x_1)x_2] + \theta v,\\
\dot{z} &= v,\\
y &= x_1.
\end{aligned}$$

Assumption 4.2 is satisfied with $\eta = z - x_2/\theta$, which results in the $\dot{\eta}$-equation

$$\dot{\eta} = -\eta - (2/\theta)x_2 + \sigma(x_1) + \sigma'(x_1)x_2.$$

Let $e_1 = x_1 - r$ and $e_2 = x_2 - \dot{r}$. Then, $e = \text{col}(e_1, e_2)$ satisfies the equation

$$\dot{e} = Ae + B[e_2 + \dot{r} - \ddot{r} + \theta(z - \sigma(e_1 + r) - \sigma'(e_1 + r)(e_2 + \dot{r}) + v)],$$

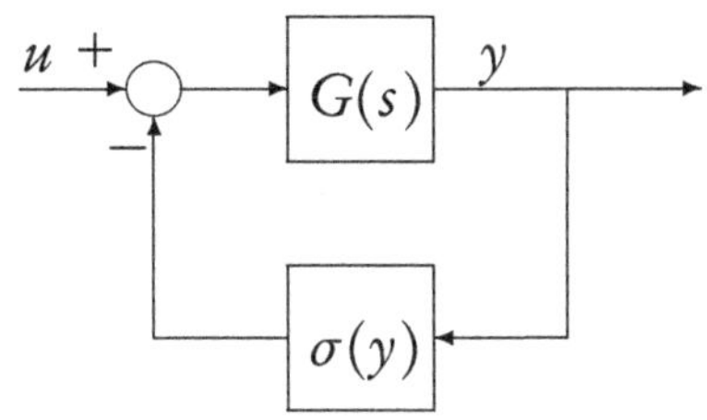

Figure 4.7. *Example* 4.4.

where

$$A=\begin{bmatrix}0 & 1\\ 0 & 0\end{bmatrix} \quad \text{and} \quad B=\begin{bmatrix}0\\ 1\end{bmatrix}.$$

The system $\dot{\bar{\eta}}=p(\bar{\eta},\mathscr{R},\theta)$ of Assumption 4.3 is given by

$$\dot{\bar{\eta}}=-\bar{\eta}-(2/\theta)\dot{r}+\sigma(r)+\sigma'(r)\dot{r}.$$

Since this is a linear exponentially stable system driven by a periodic input, it has a periodic steady-state solution η_{ss}, which satisfies the same equation, that is,

$$\dot{\eta}_{ss}=-\eta_{ss}-(2/\theta)\dot{r}+\sigma(r)+\sigma'(r)\dot{r}.$$

Then $\tilde{\eta}=\eta-\eta_{ss}$ satisfies the equation

$$\dot{\tilde{\eta}}=-\tilde{\eta}-(2/\theta)e_2+\sigma(e_1+r)-\sigma(r)+[\sigma'(e_1+r)-\sigma'(r)]\dot{r}+\sigma'(e_1+r)e_2.$$

Since $\sigma(y)=\tanh(y)$, $\sigma'(y)=1/(\cosh(y))^2$, and $\sigma''(y)=-2\tanh(y)/(\cosh(y))^2$, it can be seen that $|\sigma|\le 1$, $|\sigma'|\le 1$, $|\sigma''|\le 0.8$, $|\sigma(e_1+r)-\sigma(r)|\le |e_1|$, and $|\sigma'(e_1+r)-\sigma'(r)|\le 0.8|e_1|$. For the reference signals $r=\cos t$ and $r=1$, we have $|r|\le 1$, $|\dot{r}|\le 1$, and $|\ddot{r}|\le 1$. Writing $\dot{\tilde{\eta}}=-\tilde{\eta}+\delta_\eta$, it can be verified that $|\delta_\eta|\le 3.4986\|e\|$. With $V_1=\frac{1}{2}\tilde{\eta}^2$, we have

$$\dot{V}_1\le -|\tilde{\eta}|^2+3.4986\|e\|\,|\tilde{\eta}|.$$

Thus, Assumption 4.3 is satisfied with $V_1=\frac{1}{2}\tilde{\eta}^2$, $\rho_1=\rho_2=\frac{1}{2}$, $\rho_3=1$, and $\rho_4=3.4986$. We take $K=\begin{bmatrix}2 & 3\end{bmatrix}$ to assign the eigenvalues of $A_m=A-BK$ at -1 and -2. The solution of the Lyapunov equation $PA_m+A_m^TP=-I$ is

$$P=\frac{1}{4}\begin{bmatrix}5 & 1\\ 1 & 1\end{bmatrix} \quad \text{and} \quad 2e^TPB=\tfrac{1}{2}(e_1+e_2).$$

From (4.9) and (4.10), ψ and φ are given by

$$\begin{aligned}\psi&=\frac{-Ke-e_2-\dot{r}+\ddot{r}-\hat{\theta}[z-\sigma(e_1+r)-\sigma'(e_1+r)(e_2+\dot{r})]}{\hat{\theta}}\\ &=\frac{1}{\hat{\theta}}(-2e_1-4e_2-\dot{r}+\ddot{r})-[z-\sigma(e_1+r)-\sigma'(e_1+r)(e_2+\dot{r})],\end{aligned}$$

$$
\begin{aligned}
\varphi &= 2e^T PB[z - \sigma(e_1 + r) - \sigma'(e_1 + r)(e_2 + \dot{r}) + \frac{1}{\hat{\theta}}(-2e_1 - 4e_2 - \dot{r} + \ddot{r}) \\
&\quad - z + \sigma(e_1 + r) + \sigma'(e_1 + r)(e_2 + \dot{r})] \\
&= \frac{1}{2\hat{\theta}}(e_1 + e_2)(-2e_1 - 4e_2 - \dot{r} + \ddot{r}).
\end{aligned}
$$

To determine the saturation level, S, of ψ, suppose the initial error $e(0)$ belongs to $E_0 = \{|e_1| \le 1, |e_2| \le 1\}$. Set $c_1 = \max_{e \in E_0} e^T Pe = 2$. Taking $\gamma = 10$ and $\delta = 0.1$, the sets Ω and Ω_δ are defined by $\Omega = \{\theta \mid 1 \le \theta \le 3\}$ and $\Omega_\delta = \{\theta \mid 0.9 \le \hat{\theta} \le 3.1\}$, and

$$
\max_{\theta \in \Omega, \hat{\theta} \in \Omega_\delta} \frac{1}{2\gamma}(\theta - \hat{\theta})^2 = 0.2205.
$$

Take $c_2 = 0.221$ and $c_3 = 2.25 > c_1 + c_2$. Now, $e(t) \in E = \{e^T Pe \le c_3\}$ for all $t \ge 0$ and

$$
\max_{e \in E} \|e\| \le \sqrt{\frac{c_3}{\lambda_{\min}(P)}} = \sqrt{\frac{2.25}{0.191}} = 3.4322,
$$

$$
\max_{e \in E} |e_2| = \sqrt{c_3}\, \left\| \begin{bmatrix} 0 & 1 \end{bmatrix} P^{-1/2} \right\| = 3.3541,
$$

$$
\max_{e \in E} \left| \begin{bmatrix} 2 & 4 \end{bmatrix} e \right| = \sqrt{c_3}\, \left\| \begin{bmatrix} 2 & 4 \end{bmatrix} P^{-1/2} \right\| = 13.0977.
$$

It follows that $\dot{V}_1 \le -\tilde{\eta}^2 + 12.1|\tilde{\eta}|$, which shows that the set $\Xi = \{|\tilde{\eta}| \le 12.1\}$ is positively invariant. Since $\theta \in [1, 3]$ and $\sigma'(r) \in [0, 1]$, we have $-(2/\theta) + \sigma'(r) \in [-2, \frac{1}{3}]$. Therefore, $| - (2/\theta)\dot{r} + \sigma(r) + \sigma'(r)\dot{r}| \le 3$. It follows from the $\dot{\eta}_{ss}$-equation that $|\eta_{ss}(t)| \le 3$. Since $\tilde{\eta} = \eta - \eta_{ss} = z - \eta_{ss} - (e_2 + \dot{r})/\theta$, restricting $z(0)$ to $|z(0)| \le 7$ ensures that $\tilde{\eta}(0) \in \Xi$. Therefore, $|\tilde{\eta}(t)| \le 12.1$ for all $t \ge 0$. Hence, $|z| \le |\tilde{\eta}| + |\eta_{ss}| + (|e_2| + |\dot{r}|)/\theta \le 19.4541$. We can now calculate an upper bound on $|\psi|$ as

$$
|\psi| \le \left| \begin{bmatrix} 2 & 4 \end{bmatrix} e \right| + |\dot{r}| + |\ddot{r}| + |z| + |\sigma| + |\sigma'|(|e_2| + |\dot{r}|) \le 39.5495.
$$

We take $S = 41$. To examine the persistence of excitation condition, consider

$$
\begin{aligned}
w_r &= z_{ss} - \sigma(r) - \sigma'(r)\dot{r} + \frac{1}{\theta}(-\dot{r} + \ddot{r}) - z_{ss} + \sigma(r) + \sigma'(r)\dot{r} \\
&= \frac{1}{\theta}(-\dot{r} + \ddot{r}),
\end{aligned}
$$

where w_r is persistently exciting when $r = \cos t$ since $\int_t^{t+\pi} (\sin \tau - \cos \tau)^2 \, d\tau = \pi$, while $w_r = 0$ when $r = 1$. So, the parameter error $\hat{\theta} - \theta$ is guaranteed to converge to zero only when $r = \cos t$. The output feedback controller is given by

$$
\begin{aligned}
u &= \int v, \\
v &= 41 \,\mathrm{sat}\left(\frac{(-2\hat{e}_1 - 4\hat{e}_2 - \dot{r} + \ddot{r})/\hat{\theta} - u + \sigma(\hat{e}_1 + r) + \sigma'(\hat{e}_1 + r)(\hat{e}_2 + \dot{r})}{41} \right), \\
\dot{\hat{e}}_1 &= \hat{e}_2 + (2/\varepsilon)(e_1 - \hat{e}_1), \\
\dot{\hat{e}}_2 &= (1/\varepsilon^2)(e_1 - \hat{e}_1), \\
\dot{\hat{\theta}} &= 10\, \pi(\hat{\theta}, \varphi)\varphi, \quad \varphi = \tfrac{1}{2}(\hat{e}_1 + \hat{e}_2)(-2\hat{e}_1 - 4\hat{e}_2 - \dot{r} + \ddot{r})/\hat{\theta},
\end{aligned}
$$

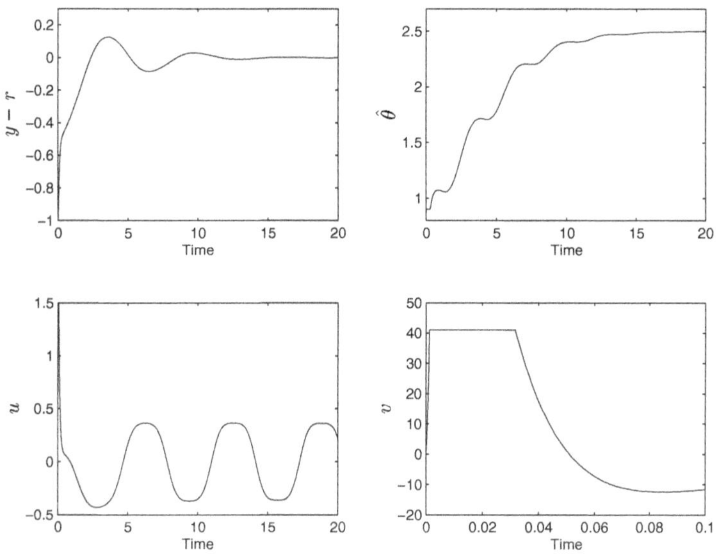

Figure 4.8. *Simulation of Example* 4.4 *when* $r = \cos t$.

where the observer eigenvalues are assigned at $-1/\varepsilon$ and

$$\pi(\hat{\theta},\varphi) = \begin{cases} 1+(3-\hat{\theta})/0.1 & \text{if } \hat{\theta} > 3 \text{ and } \varphi > 0, \\ 1+(\hat{\theta}-1)/0.1 & \text{if } \hat{\theta} < 1 \text{ and } \varphi < 0, \\ 1 & \text{otherwise.} \end{cases}$$

The simulation results of Figures 4.8 and 4.9 are carried out with $\theta = 2.5$, $\varepsilon = 0.01$, and the initial conditions $x(0) = \hat{e}(0) = 0$, $u(0) = 0$, and $\hat{\theta}(0) = 1.5$. Figure 4.8 shows the results when $r = \cos t$ and Figure 4.9 when $r = 1$. The plot of v is shown in a short time interval to demonstrate the saturation during the peaking period of the observer. As expected, in both cases the tracking error $y - r$ converges to zero. The parameter error $\hat{\theta} - \theta$ converges to zero only when $r = \cos t$. △

4.5 ▪ Robustness

In this section we study the robustness of the adaptive controller of Sections 4.2 and 4.3 with respect to bounded disturbance. Consider a perturbation of the system (4.2), represented by

$$\left.\begin{aligned} \dot{x}_i &= x_{i+1}, \quad 1 \le i \le n-1, \\ \dot{x}_n &= f_0(x,z) + \theta^T f(x,z) + [g_0(x,z) + \theta^T g(x,z)]v + \Delta(t,x,z), \\ \dot{z}_i &= z_{i+1}, \quad 1 \le i \le m-1, \\ \dot{z}_m &= v, \\ y &= x_1, \end{aligned}\right\} \tag{4.34}$$

where Δ is piecewise continuous and bounded in t and locally Lipschitz in (x,z) uniformly in t and f_0, f, g_0, g, and θ are the same as in Sections 4.1–4.3. Suppose Assumptions 4.1 and 4.2 are satisfied. The change of variables (4.3) transforms (4.34) into the form

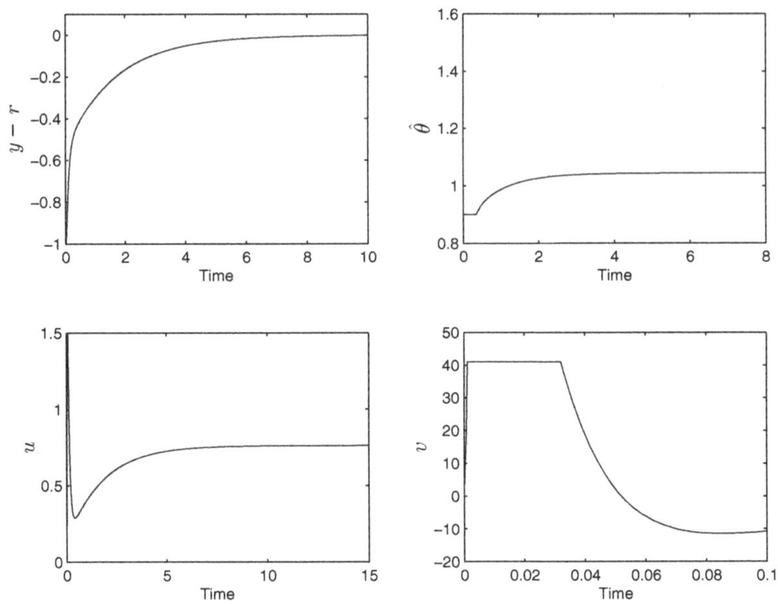

Figure 4.9. *Simulation of Example* 4.4 *when* $r = 1$.

$$\left.\begin{aligned} \dot{\eta} &= p(\eta,x,\theta)+\Delta_0(t,\eta,x,\theta),\\ \dot{x}_i &= x_{i+1}, \quad 1\le i\le n-1,\\ \dot{x}_n &= f_0(x,z)+\theta^T f(x,z)+[g_0(x,z)+\theta^T g(x,z)]v+\Delta(t,x,z),\\ y &= x_1, \end{aligned}\right\} \tag{4.35}$$

where $\Delta_0 = [\partial T_1/\partial x]B\Delta$. The error vector $e = x - \mathscr{R}$ satisfies the equation

$$\dot{e} = Ae + B\left\{f_0(x,z)+\theta^T f(x,z)+[g_0(x,z)+\theta^T g(x,z)]v - r^n + \Delta(t,x,z)\right\}. \tag{4.36}$$

As discussed in Section 4.1, to achieve boundedness of all variables we need to impose a minimum-phase assumption, which was done there by Assumption 4.3. In addition to ensuring that η is bounded for every bounded x, Assumption 4.3 also establishes the existence and exponential stability of a steady-state solution of the internal dynamics, which played a crucial role in showing convergence of e to zero. In the presence of Δ, we cannot show convergence of e to zero; the best we can hope for is to show that, after a transient period, e will be small in some sense. Therefore, we replace Assumption 4.3 with the less restrictive assumption, given next.[29]

Assumption 4.5. *There is a continuously differentiable function* $V_0(\eta)$, *possibly dependent on* θ, *class* $\mathscr{K}_\infty$ *functions* ϑ_1 *and* ϑ_2, *and class* $\mathscr{K}$ *functions* ϑ_3 *and* ϑ_4, *all independent of* θ, *such that*

$$\vartheta_1(\|\eta\|) \le V_0(\eta) \le \vartheta_2(\|\eta\|), \tag{4.37}$$

$$\frac{\partial V_0}{\partial \eta} p(\eta,x,\theta) \le -\vartheta_3(\|\eta\|) \qquad \forall\ \|\eta\| \ge \vartheta_4(\|x\|) \tag{4.38}$$

for all $\eta \in R^m$, $x \in R^n$, *and* $t \ge 0$.

[29] Assumption 4.5 is equivalent to input-to-state stability of $\dot{\eta} = p(\eta,x,\theta)$ with input x [145].

We start by determining the compact set of operation under state feedback when $\Delta = 0$. As in Section 4.2, suppose $\hat{\theta}(0) \in \Omega$, $e(0) \in E_0$, and $z(0) \in Z_0$ and define c_1, c_2, and $c_3 > c_1 + c_2$ as before. Let $E = \{e^T P e \le c_3\}$ and $c_x = \max_{e \in E, \mathscr{R} \in Y_r} \|e + \mathscr{R}\|$. Then, $\|x\| \le c_x$. For any $c_4 \ge \vartheta_2(\vartheta_4(c_x))$, $\{V_0(\eta) \le c_4\}$ is positively invariant because

$$V_0 = c_4 \Rightarrow \vartheta_2(\|\eta\|) \ge c_4 \Rightarrow \vartheta_2(\|\eta\|) \ge \vartheta_2(\vartheta_4(c_x)) \Rightarrow \|\eta\| \ge \vartheta_4(c_x) \Rightarrow \dot{V}_0 < 0.$$

By choosing c_4 large enough, we can ensure that for all $e(0) \in E_0$ and $z(0) \in Z_0$, $\eta(t) \in \{V_0(\eta) \le c_4\}$ for all $t \ge 0$. Mapping this set into the z-coordinates, we define a compact set Z such that $z(t) \in Z$ for all $t \ge 0$. Thus, the compact set of operation is given by $e \in E$, $z \in Z$, $\mathscr{R} \in Y_r$, $|r^{(n)}| \le M_r$, and $\hat{\theta} \in \Omega_\delta$. The only difference from the set of operation in Section 4.2 is the definition of Z, which is now based on Assumption 4.5 instead of Assumption 4.3. The saturated control is taken as in (4.15), where S is obtained by maximization of $|\psi|$ over the current set of operation.

When $\Delta \ne 0$, $\dot{V}$ satisfies the inequality

$$\dot{V} \le -e^T Q e + 2e^T P B \Delta.$$

Suppose

$$|\Delta(t, e + \mathscr{R}, z)| \le d \quad \forall\, t \ge 0,\ e \in E,\ \mathscr{R} \in Y_r,\ z \in Z.$$

Over the set $\{V \le c_3\} \cap \Omega_\delta \times Z \times Y_r$,

$$\begin{aligned} \dot{V} &\le -e^T Q e + k_a d \\ &\le -c_0 V + \tfrac{1}{2} c_0 \tilde{\theta}^T \Gamma^{-1} \tilde{\theta} + k_a d \le -c_0 V + c_0 c_2 + k_a d \end{aligned}$$

for some positive constant k_a, where $c_0 = \lambda_{\min}(Q)/\lambda_{\max}(P)$. When $d < \frac{1}{2} c_0 (c_3 - c_2)/k_a$, $\dot{V} \le -\frac{1}{2} c_0 c_3$ on the boundary $V = c_3$. On the other hand, $\dot{V}_0$ satisfies the inequality

$$\dot{V}_0 \le -\vartheta_3(\|\eta\|) + \frac{\partial V_0}{\partial \eta} \Delta_0(t, \eta, x, \theta).$$

Over the set $E \times \Omega_\delta \times \{V_0 \le c_4\} \times Y_r$,

$$\dot{V}_0 \le -\vartheta_3(\vartheta_2^{-1}(V_0)) + k_b d$$

for some positive constant k_b. For $d < \vartheta_3(\vartheta_2^{-1}(c_4))/k_b$, $\dot{V}_0 \le 0$ on the boundary $V_0 = c_4$. Thus, there is $d^* > 0$ such that for all $d \le d^*$ the sets $\{V \le c_3\}$ and $\{V_0 \le c_4\}$ remain positively invariant under the perturbation Δ. It follows that the compact set of operation under state feedback remains as $e \in E$, $z \in Z$, $\mathscr{R} \in Y_r$, $|r^{(n)}| \le M_r$, and $\hat{\theta} \in \Omega_\delta$.

The analysis of the output feedback controller proceeds as in the proof of Theorem 4.1 to show that the set

$$\Sigma = \{V \le c_3\} \cap \{\hat{\theta} \in \Omega_\delta\} \times \{V_0 \le c_4\} \times \{V_2 \le c_5 \varepsilon^2\}$$

is positively invariant and there is time $T(\varepsilon)$, with $\lim_{\varepsilon \to 0} T(\varepsilon) = 0$, such that the trajectories of the closed-loop system are confined to Σ for all $t \ge T(\varepsilon)$. The main difference arises in showing that $\dot{V} \leq 0$ on the boundary $V = c_3$. In the presence of the disturbance Δ, $\dot{V}$ satisfies the inequality

$$\begin{aligned}\dot{V} &\leq -e^T Q e + k_a d + k_c \varepsilon \\ &\leq -c_0 V + \tfrac{1}{2} c_0 \tilde{\theta}^T \Gamma^{-1} \tilde{\theta} + k_a d + k_c \varepsilon \\ &\leq -c_0 V + c_0 c_2 + k_a d + k_c \varepsilon\end{aligned}$$

for some positive constant k_a and k_c. When $d \leq \frac{1}{2} c_0(c_3 - c_2)/k_a$ and $\varepsilon < \frac{1}{2} c_0(c_3 - c_2)/k_c$, $\dot{V} \leq 0$ on the boundary $V = c_3$. Inside Σ, we have

$$\lim_{\tau\to\infty} \frac{1}{\tau} \int_0^\tau e^T(\sigma) Q e(\sigma)\, d\sigma \leq \lim_{\tau\to\infty} \frac{V(0) - V(\tau)}{\tau} + k_a d + k_c \varepsilon.$$

The first term on the right-hand side is zero due to the boundedness of V. Therefore,

$$\lim_{\tau\to\infty} \frac{1}{\tau} \int_0^\tau e^T(\sigma) Q e(\sigma)\, d\sigma \leq k_a d + k_c \varepsilon, \tag{4.39}$$

which shows that the mean-square tracking error is of the order $O(d) + O(\varepsilon)$. The upper bound on the mean-square tracking error in (4.39) will be small when both d and ε are small. While ε is a design parameter that can be chosen to be sufficiently small, d is an upper bound of the disturbance Δ, which we cannot control. When d is not small enough, we can use Lyapunov redesign to add a robustifying control component to make the mean-square error arbitrarily small, provided an upper bound on Δ is known.[30]

Suppose Δ satisfies the inequality

$$|\Delta(t,x,z)| \leq \beta(x,z) \qquad \forall\ t \geq 0,\ e \in E,\ \mathscr{R} \in Y_r,\ z \in Z,$$

for a known locally Lipschitz function β. Modify the feedback control function ψ to

$$\psi(e,z,\mathscr{R},r^{(n)},\hat{\theta}) = \frac{-Ke + r^{(n)} - f_0(e+\mathscr{R},z) - \hat{\theta}^T f(e+\mathscr{R},z) + \psi_r(e,z,\mathscr{R})}{g_0(e+\mathscr{R},z) + \hat{\theta}^T g(e+\mathscr{R},z)}, \tag{4.40}$$

where ψ_r is an additional control component to be determined. The derivative $\dot{V}$ satisfies the inequality

$$\dot{V} \leq -c_0 V + c_0 c_2 + \underbrace{2e^T PB}_{s}(\psi_r + \Delta).$$

Taking

$$\psi_r = -\beta\ \mathrm{sat}\left(\frac{\beta s}{\mu}\right), \tag{4.41}$$

where μ is a positive constant, we have

$$s(\psi_r + \Delta) \leq -\beta s\ \mathrm{sat}\left(\frac{\beta s}{\mu}\right) + \beta|s|.$$

When $\beta|s| \geq \mu$,

$$s(\psi_r + \Delta) \leq -\beta|s| + \beta|s| = 0,$$

[30]See [78, Section 14.2] or [80, Section 10.2] for an introduction to Lyapunov redesign.

and when $\beta|s| < \mu$,

$$s(\psi_r + \Delta) \le -\frac{(\beta s)^2}{\mu} + \beta|s| \le \frac{\mu}{4}.$$

Thus,

$$\dot{V} \le -c_0 V + c_0 c_2 + \frac{\mu}{4},$$

and the set $\{V \le c_3\}$ is positively invariant for sufficiently small μ. As for the set $\{V_0 \le c_4\}$, we still need $d < \vartheta_3(\vartheta_2^{-1}(c_4))/k_b$ because the inequality satisfied by $\dot{V}_0$ remains the same. Under output feedback, ψ is saturated outside the compact set of operation as before, e is replaced by $\hat{e}$, and

$$\dot{V} \le -c_0 V + c_0 c_2 + \frac{\mu}{4} + k_d \varepsilon$$

for some positive constant k_d. Thus, the set $\{V \le c_3\}$ is positively invariant for sufficiently small ε and μ. Repeating previous arguments, it can be shown that the trajectory is bounded and enters the positively invariant set Σ within a time interval $[0, T(\varepsilon)]$, where $\lim_{\varepsilon \to 0} T(\varepsilon) = 0$. Inside Σ,

$$\dot{V} \le -e^T Q e + \frac{\mu}{4} + k_d \varepsilon,$$

from which we see that the mean-square tracking error will be of the order $O(\mu) + O(\varepsilon)$. Thus, it can be made arbitrarily small by choice of ε and μ.

Example 4.5. Reconsider the pendulum tracking problem of Example 4.3 with constant disturbance d:

$$\dot{x}_1 = x_2, \quad \dot{x}_2 = -\sin x_1 - \theta_1 x_2 + \theta_2 u + d, \quad y = x_1,$$

where $|d| \le 0.1$. We consider two controllers. The first one is the same controller of Example 4.3. The second one includes the robustifying component ψ_r with $\beta = 0.2$ and the same saturation level $S = 25$; it was determined by simulation under state feedback that the saturation level is still adequate after adding ψ_r. Recalling the matrix P from Example 4.3, it can be seen that $s = 2e^T PB = \frac{1}{2}(e_1 + e_2)$. The simulation results of Figure 4.10 are carried out with the same parameters and initial conditions as in Example 4.3, with two values of the disturbance d and two values of the parameter μ. Figure 4.10(a) compares the tracking error e_1 under the controller of Example 4.3 for $d = 0.1$ and $d = 0.01$. It confirms that the error reduces with $|d|$. Figure 4.10(b) shows also e_1 in three cases: with no robustifying control (dashed), with $\psi_r = -0.2\,\text{sat}(s/0.1)$ (solid), and with $\psi_r = -0.2\,\text{sat}(s/0.01)$ (dotted). It demonstrates how reducing μ reduces the error. △

4.6 ▪ Approximation-Based Control

One approach to the design of feedback control for nonlinear systems with uncertain nonlinearities is the adaptive approximation-based control.[31] In this approach, a nonlinear function $f(x)$ is approximated by a function approximator of the form

$$f(x) \approx \sum_{i}^{N} \theta_i \phi_i(x) = \theta^T \phi(x), \tag{4.42}$$

[31] See [40] for a comprehensive introduction.

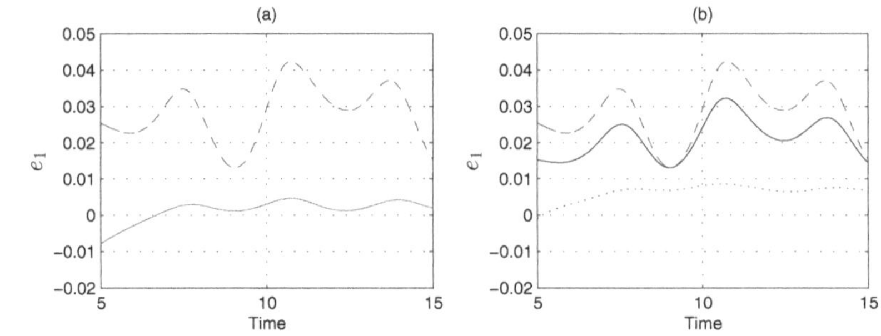

Figure 4.10. *Simulation of Example 4.5. (a) shows the tracking error e_1 at steady state for $d = 0.1$ (dashed) and $d = 0.01$ (solid) when no robustifying control is used. (b) shows e_1 for $d = 0.1$ in three cases: no robustifying control (dashed), robustifying control with $\mu = 0.02$ (solid), and robustifying control with $\mu = 0.002$ (dotted).*

where ϕ_1 to ϕ_N are known basis functions while θ_1 to θ_N are adjustable weights to be determined by an adaptive law.[32] Different approximation structures are available in the literature, including polynomials, splines, neural networks with radial basis functions, fuzzy approximation, and wavelets. All these methods share the universal approximation property; namely, in a given compact set of x, the approximation error

$$\Delta_f(x) = f(x) - \sum_{i}^{N} \theta_i \phi_i(x)$$

can be made arbitrarily small by choosing N sufficiently large.

In this section, we extend the adaptive controller of Sections 4.2 and 4.3 to approximation-based control of a single-input–single-output nonlinear system represented globally by the input-output model

$$y^{(n)} = F(\cdot) + G(\cdot)u^{(m)}, \tag{4.43}$$

where u is the control input, y is the measured output, $m < n$, and F and G are sufficiently smooth functions that could depend on $y, y^{(1)}, \dots, y^{(n-1)}, u, u^{(1)}, \dots, u^{(m-1)}$. The control task is to design an output feedback controller such that all variables of the closed-loop system are bounded and y asymptotically tracks a reference signal $r(t)$, which is bounded and has bounded derivatives up to the nth order and $r^{(n)}(t)$ is piecewise continuous. As in Section 4.1, we augment a series of m integrators at the input side of the system and represent the extended system by the state model

$$\left.\begin{aligned} \dot{x}_i &= x_{i+1}, \quad 1 \le i \le n-1, \\ \dot{x}_n &= F(x,z) + G(x,z)v, \\ \dot{z}_i &= z_{i+1}, \quad 1 \le i \le m-1, \\ \dot{z}_m &= v, \\ y &= x_1. \end{aligned}\right\} \tag{4.44}$$

The following three assumptions restate Assumptions 4.1, 4.2, and 4.5 for the current case.

[32]There are cases where the function approximator depends nonlinearly on the weights, but we limit our discussion to the case of linear dependence as shown in (4.42).

Assumption 4.6.

$$G(x,z) \geq \bar{g} > 0 \quad \forall\, x \in R^n,\ z \in R^m.$$

Assumption 4.7. *There exists a global diffeomorphism*

$$\begin{bmatrix} \eta \\ x \end{bmatrix} = \begin{bmatrix} T_1(x,z) \\ x \end{bmatrix} \stackrel{\text{def}}{=} T(x,z) \tag{4.45}$$

with $T_1(0,0)=0$*, which transforms* (4.44) *into the global normal form*

$$\left.\begin{aligned} \dot{\eta} &= p(\eta,x), \\ \dot{x}_i &= x_{i+1}, \quad 1 \leq i \leq n-1, \\ \dot{x}_n &= F(x,z)+G(x,z)v, \\ y &= x_1. \end{aligned}\right\} \tag{4.46}$$

Assumption 4.8. *There is a continuously differentiable function* $V_0(\eta)$*, class* $\mathcal{K}_\infty$ *functions* ϑ_1 *and* ϑ_2*, and class* $\mathcal{K}$ *functions* ϑ_3 *and* ϑ_4 *such that*

$$\vartheta_1(\|\eta\|) \leq V_0(\eta) \leq \vartheta_2(\|\eta\|), \tag{4.47}$$

$$\frac{\partial V_0}{\partial \eta} p(\eta,x) \leq -\vartheta_3(\|\eta\|) \qquad \forall\, \|\eta\| \geq \vartheta_4(\|x\|) \tag{4.48}$$

for all $\eta \in R^m$ *and* $x \in R^n$.

The functions $F(x,z)$ and $G(x,z)$ will be approximated for $x \in X$ and $z \in Z$, for some compact sets $X \subset R^n$ and $Z \subset R^m$. The construction of X and Z builds on our earlier discussion and makes use of Assumption 4.8. With $e, \mathcal{R}, A, B, K,$ and P as defined earlier, suppose $e(0)$ and $z(0)$ belong to the compact sets $E_0 \subset R^n$ and $Z_0 \subset R^m$, respectively. Take $c_1 \geq \max_{e \in E_0} e^T P e$ and $c_3 > c_1$. Define $E = \{e^T P e \leq c_3\}$ and $X = \{e + \mathcal{R} \mid e \in E,\ \mathcal{R} \in Y_r\}$. Let $c_x = \max_{x \in X} \|x\|$ and choose $c_4 > 0$ large enough such that $c_4 \geq \vartheta_2(\vartheta_4(c_x))$ and for all $e(0) \in E_0$ and $z(0) \in Z_0$, $\eta(0) \in \{V_0(\eta) \leq c_4\}$. Define a compact set Z such that for all $x \in X$ and $\eta \in \{V_0(\eta) \leq c_4\}$, $z \in Z$. The definition of X and Z is done in such a way that if we can ensure that $e(t) \in E$ for all $t \geq 0$, then $x(t) \in X$ and $z(t) \in Z$ for all $t \geq 0$.

Using a universal approximator, we represent the functions $F(x,z)$ and $G(x,z)$ over $X \times Z$ as

$$\begin{aligned} F(x,z) &= \theta_f^{*T} f(x,z) + \Delta_f(x,z), \\ G(x,z) &= \theta_g^{*T} g(x,z) + \Delta_g(x,z), \end{aligned}$$

where the weight vectors $\theta_f^* \in R^{\ell_1}$ and $\theta_g^* \in R^{\ell_2}$ are the optimal weights, defined by

$$\theta_f^* = \arg \min_{\theta_f \in R^{\ell_1}} \left\{ \sup_{x \in X, z \in Z} |F(x,z) - \theta_f^T f(x,z)| \right\},$$

$$\theta_g^* = \arg \min_{\theta_g \in R^{\ell_2}} \left\{ \sup_{x \in X, z \in Z} |G(x,z) - \theta_g^T g(x,z)| \right\},$$

and Δ_f and Δ_g are the approximation errors.

Assumption 4.9. *The vectors θ_f^* and θ_g^* belong to known compact convex sets $\Omega_f \subset R^{\ell_1}$ and $\Omega_g \subset R^{\ell_2}$, respectively.*

Assumption 4.10. $|\theta_g^T g(x,z)| \geq \bar{g} > 0$ *for all $x \in X$, $z \in Z$, and $\theta_g \in \hat{\Omega}_g$, where $\hat{\Omega}_g$ is a compact convex set that contains Ω_g in its interior.*

In view of the robustness results of the previous section, the adaptive controller can be designed as in Section 4.2 using the Lyapunov function

$$V = e^T P e + \tfrac{1}{2}\tilde{\theta}^T \Gamma^{-1} \tilde{\theta},$$

where $\tilde{\theta} = \text{col}(\hat{\theta}_f - \theta_f^*, \hat{\theta}_g - \theta_g^*)$, the feedback control

$$v = \frac{-Ke + r^{(n)} - \hat{\theta}_f^T f(x,z)}{\hat{\theta}_g^T g(x,z)} \stackrel{\text{def}}{=} \psi(e, z, \mathcal{R}, r^{(n)}, \hat{\theta}),$$

and the adaptive law

$$\dot{\hat{\theta}} = \Pi\, \Gamma\, \varphi,$$

where $\varphi = \text{col}(f, g\psi)$, $\hat{\theta} = \text{col}(\hat{\theta}_f, \hat{\theta}_g)$, and the parameter projection Π is defined as in Section 4.2 to ensure that $\hat{\theta} \in \Omega_\delta$. The only missing piece from the analysis of the previous sections is the requirement $c_3 > c_1 + c_2$, where

$$c_2 \geq \max_{\theta^* \in \Omega, \hat{\theta} \in \Omega_\delta} \tfrac{1}{2}\tilde{\theta}^T \Gamma^{-1} \tilde{\theta}.$$

In Section 4.2, c_1 and c_2 were determined first; then c_3 was chosen to satisfy $c_3 > c_1 + c_2$. In the current case c_1 and c_3 are chosen first because the choice of c_3 determines the sets X and Z, which are needed to define the function approximation. The condition $c_3 > c_1 + c_2$ is imposed by choosing the adaptation gain large enough to make $c_2 < c_3 - c_1$. Let

$$k_2 = \max_{\theta \in \Omega, \hat{\theta} \in \Omega_\delta} \tfrac{1}{2}\tilde{\theta}^T \tilde{\theta} \text{ and } \gamma = \lambda_{min}(\Gamma).$$

Then

$$\max_{\theta \in \Omega, \hat{\theta} \in \Omega_\delta} \tfrac{1}{2}\tilde{\theta}^T \Gamma^{-1} \tilde{\theta} \leq k_2/\gamma.$$

Set $c_2 = k_2/\gamma$. By choosing $\gamma > k_2/(c_3 - c_1)$, we can ensure that $c_2 < c_3 - c_1$. Let

$$d = \max |\Delta_f(e + \mathcal{R}, z) + \Delta_g(e + \mathcal{R}, z)\psi(e, z, \mathcal{R}, r^{(n)}, \hat{\theta})|,$$

where the maximization is taken over $e \in E$, $z \in Z$, $\mathcal{R} \in Y_r$, $|r^{(n)}| \leq M_r$, and $\hat{\theta} \in \Omega_\delta$. The derivative $\dot{V}$ satisfies the inequality

$$\begin{aligned}\dot{V} &\leq -e^T Q e + 2e^T PB(\Delta_f + \Delta_g \psi) \leq -e^T Q e + k_a d \\ &\leq -c_0 V + \tfrac{1}{2} c_0 \tilde{\theta}^T \Gamma^{-1} \tilde{\theta} + k_a d \leq -c_0 V + \frac{c_0 k_2}{\gamma} + k_a d\end{aligned}$$

for some positive constant k_a, where $c_0 = \lambda_{\min}(Q)/\lambda_{\max}(P)$. When $d \le \frac{1}{2}c_0(c_3 - k_2/\gamma)/k_a$, $\dot{V} \le -\frac{1}{2}c_0 c_3$ on the boundary $V = c_3$. As in Section 4.5, we can show that $\dot{V}_0 \le 0$ on the boundary $V_0 = c_4$ for $d < \vartheta_3(\vartheta_2(c_4))/k_b$. Thus, there exists $d^* > 0$ (dependent on γ) such that for all $d \le d^*$, the sets $\{V \le c_3\}$ and $\{V_0 \le c_4\}$ are positively invariant under the perturbation $\Delta_f + \Delta_g \psi$.

The output feedback control is given by

$$v = \psi^s(e, z, \mathcal{R}, r^{(n)}, \hat{\theta}),$$

where $\hat{e}$ is provided by the high-gain observer (4.18), $\psi^s = S\ \mathrm{sat}(\psi/S)$, and $S \ge \max|\psi(e, z, \mathcal{R}, r^{(n)}, \hat{\theta})|$ with the maximization taken over all $e \in E$, $z \in Z$, $\mathcal{R} \in Y_r$, $|r^{(n)}| \le M_r$ and $\hat{\theta} \in \Omega_\delta$. The analysis of the output feedback controller proceeds as in the proof of Theorem 4.1 to show that the set

$$\Sigma = \{V \le c_3\} \cap \{\hat{\theta} \in \Omega_\delta\} \times \{V_0 \le c_4\} \times \{V_2 \le c_5 \varepsilon^2\}$$

is positively invariant and there is time $T(\varepsilon)$, with $\lim_{\varepsilon \to 0} T(\varepsilon) = 0$, such that the trajectories of the closed-loop system are confined to Σ for all $t \ge T(\varepsilon)$. The main difference arises in showing that $\dot{V} \le 0$ on the boundary $V = c_3$. In the presence of the disturbance $\Delta_f + \Delta_g \psi$, $\dot{V}$ satisfies the inequality

$$\begin{aligned}
\dot{V} &\le -e^T Q e + k_a d + k_c \varepsilon \\
&\le -c_0 V + \tfrac{1}{2} c_0 \tilde{\theta}^T \Gamma^{-1} \tilde{\theta} + k_a d + k_c \varepsilon \\
&\le -c_0 V + \frac{c_0 k_2}{\gamma} + k_a d + k_c \varepsilon
\end{aligned}$$

for some positive constant k_a and k_c. When $d \le \frac{1}{2}c_0(c_3 - k_2/\gamma)/k_a$ and $\varepsilon < \frac{1}{2}c_0(c_3 - k_2/\gamma)/k_c$, $\dot{V} \le 0$ on the boundary $V = c_3$. Inside Σ, we have

$$\dot{V} \le -c_0 V + \frac{c_0 k_2}{\gamma} + k_a d + k_c \varepsilon,$$

which shows that $\|e\|$ is ultimately bounded with an ultimate bound of the order $O(\sqrt{d + \varepsilon + 1/\gamma})$. For sufficiently large γ and sufficiently small ε, the ultimate bound on $\|e\|$ will be of the order of $O(\sqrt{d})$. As a point of comparison with Section 4.5, note that by choosing γ large, we establish an ultimate bound on the error rather than a bound on the mean-square error.

When d is not small enough, we can use Lyapunov redesign to reduce the tracking error. The state feedback control is modified to

$$\psi(e, z, \mathcal{R}, r^{(n)}, \hat{\theta}) = \frac{-Ke + r^{(n)} - \hat{\theta}_f^T f(x, z) + \psi_r(e, z, \mathcal{R})}{\hat{\theta}_g^T g(x, z)}, \tag{4.49}$$

where ψ_r is the robustifying control component. Because $\Delta_f + \Delta_g v$ depends on v, we assume that it satisfies the inequality

$$|\Delta_f(x, z) + \Delta_g(x, z)\psi(e, z, \mathcal{R}, r^{(n)}, \hat{\theta})| \le \beta(e, z, \mathcal{R}, r^{(n)}, \hat{\theta}) + \kappa |\psi_r|$$

for a known locally Lipschitz function β and a known positive constant $\kappa < 1$ for all $e \in E$, $z \in Z$, $\mathscr{R} \in Y$, $|r^{(n)}| \le M_r$, $\hat{\theta} \in \Omega_\delta$, and $\psi_r \in R$. The derivative $\dot{V}$ satisfies the inequality

$$\dot{V} \le -c_0 V + c_0 \frac{k_2}{\gamma} + \underbrace{2e^T PB}_{s}(\psi_r + \Delta_f + \Delta_g \psi).$$

Taking

$$\psi_r = -\frac{\beta}{1-\kappa} \operatorname{sat}\left(\frac{\beta s}{\mu}\right), \tag{4.50}$$

where μ is a positive constant, we have

$$s(\psi_r + \Delta_f + \Delta_g \psi) \le -\frac{\beta}{1-\kappa} s \operatorname{sat}\left(\frac{\beta s}{\mu}\right) + \beta|s| + \kappa \frac{\beta}{1-\kappa}|s| \left|\operatorname{sat}\left(\frac{\beta s}{\mu}\right)\right|.$$

When $\beta|s| \ge \mu$,

$$s(\psi_r + \Delta_f + \Delta_g \psi) \le -\frac{\beta}{1-\kappa}|s| + \beta|s| + \kappa \frac{\beta}{1-\kappa}|s| = 0,$$

and when $\beta|s| < \mu$,

$$s(\psi_r + \Delta_f + \Delta_g \psi) \le -\frac{(\beta s)^2}{\mu(1-\kappa)} + \beta|s| + \frac{\kappa(\beta s)^2}{\mu(1-\kappa)} = -\frac{(\beta s)^2}{\mu} + \beta|s| \le \frac{\mu}{4}.$$

Thus,

$$\dot{V} \le -c_0 V + c_0 \frac{k_2}{\gamma} + \frac{\mu}{4},$$

and the set $\{V \le c_3\}$ is positively invariant for sufficiently small μ. As for the set $\{V_0 \le c_4\}$, we still need $d < \vartheta_3(\vartheta_2^{-1}(c_4))/k_b$ because the inequality satisfied by $\dot{V}_0$ remains the same. Under output feedback, the control is saturated over the compact set of operation as before, e is replaced by $\hat{e}$, and $\dot{V}$ satisfies the inequality

$$\dot{V} \le -c_0 V + c_0 \frac{k_2}{\gamma} + \frac{\mu}{4} + k_d \varepsilon$$

for some positive constant k_d. Thus, the set $\{V \le c_3\}$ is positively invariant for sufficiently small ε and μ. Repeating previous arguments, it can be shown that the trajectory is bounded and enters the positively invariant set Σ within a time interval $[0, T(\varepsilon)]$, where $\lim_{\varepsilon \to 0} T(\varepsilon) = 0$. Inside Σ,

$$\dot{V} \le -c_0 V + c_0 \frac{k_2}{\gamma} + \frac{\mu}{4} + k_d \varepsilon,$$

which shows that for sufficiently large γ and sufficiently small ε, the ultimate bound on $\|e\|$ is of the order $(\sqrt{\mu})$.

4.7 ▪ Notes and References

The problem statement of Section 4.1 and the controller of Sections 4.2 and 4.3 are taken from [74], but Theorem 4.1 and the robustness results of Section 4.5 follow [4]. Section 4.6 follows [134], which includes simulation results when the controller is implemented using radial basis function neural networks. Another approach to the use of high-gain observers in neural network control can be found in [52].

Chapter 5
Regulation

The regulation problem in feedback control is concerned with asymptotically regulating the tracking error to zero when the reference and disturbance signals are generated by a known model. Robust regulation is based on the internal model principle, which is briefly described in Section 5.1. Throughout the chapter, we deal with a class of nonlinear systems that can be transformed into the normal form, for which a high-gain observer can be designed as described in Chapter 2. In the special case when the reference and disturbance signals are constant, the internal model principle requires the use of integral action to ensure zero steady-state error in the presence disturbance and parameter uncertainty. Sections 5.2 and 5.3 deal with this special case. In Section 5.2 an integrator is augmented with the plant, and output feedback control is designed using a separation approach to stabilize the augmented system at an equilibrium point where the error is zero. Because the equilibrium point is dependent on disturbance and uncertain parameters, robust control is used. The inclusion of integral action comes usually at the expense of degrading the transient response, when compared with a controller that does not include integrator. The conditional integrator of Section 5.3 removes this drawback by ensuring that the transient response of the closed-loop system can be brought arbitrarily close to the transient response under state feedback sliding mode control without integrator.

In the more general case when the reference and disturbance signals have constant and sinusoidal components, the internal model principle requires the use of a servocompensator that duplicates the dynamics of an internal model, which generates the signals that are needed to achieve zero steady-state error. While the servocompensator can be designed by extending the approach of Section 5.2,[33] the next three sections present a conditional servocompensator approach that recovers the transient response under state feedback sliding mode control. The main design, presented in Section 5.4, requires precise knowledge of the internal model. The effect of uncertainty in the internal model is studied in the next two sections. Section 5.5 shows that when the internal model perturbation is small, the steady-state error will be of the order of the product of two parameters, one of which is a bound on the model perturbation, while the other is a design parameter. Section 5.6 presents an adaptive approach to learn the internal model, which ensures zero steady-state error.

[33] See [73] and [76].

5.1 ▪ Internal Model Principle

One of the fundamental tools in linear feedback control theory is the use of an internal model of the class of disturbance and reference signals in order to achieve asymptotic tracking and disturbance rejection for all signals in the specified class.[34] The class of signals is characterized by a linear model, called *exosystem*, which generates all the exogenous signals of interest. A device, called *servocompensator*, is augmented with the plant, and a stabilizing controller is designed to stabilize the augmented system. The block diagram of Figure 5.1 shows the feedback structure of the controller. The servocompensator is a linear model whose eigenvalues coincide with the eigenvalues of exosystem. It is driven by the regulation error. At steady state, the regulation error must be zero. This can be easily seen in the special case when the exogenous signals are constant. Constant signals are generated by a first-order model whose eigenvalue is at the origin. Therefore, the servocompensator in this case is an integrator. Since the closed-loop system is asymptotically stable, its state converges to an equilibrium point as time tends to infinity. At the equilibrium point, all internal signals are constant. For the integrator to maintain a constant output, its input must be zero. An important feature of the internal model approach is its robustness to parameter perturbations that do not destroy the stability of the system. For all such perturbations, the regulation error converges to zero. Once again, this can be easily seen in the special case of constant signals. The closed-loop system has an asymptotically stable equilibrium point that is determined by the exogenous signals and system parameters. But no matter what the location of the equilibrium point is, the input to integrator will be zero at equilibrium. This robustness property explains the abundance of integral control in automatic control systems.

The internal model principle has been extended to classes of nonlinear systems.[35] The results of this chapter present one particular extension that is built around the use of high-gain observers. In this approach the exosystem is a linear model that has distinct eigenvalues on the imaginary axis; thus, it generates constant or sinusoidal signals. There are three basic ingredients of the approach. First, by studying the dynamics of the system on an invariant manifold at which the regulation error is zero, called the *zero-error manifold*, a linear internal model is derived. The internal model generates not only the modes of the exosystem but also a number of higher-order harmonics induced by nonlinearities. Second, a separation approach is used to design a robust output feedback controller where a state feedback controller is designed first,

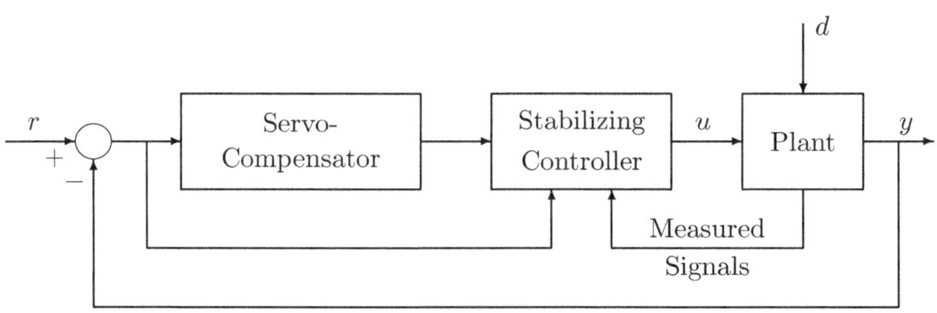

Figure 5.1. *Schematic diagram of feedback control with servocompensator.*

[34] See, for example, [33], [34], and [44].

[35] See [28], [30], [60], [69], [70], and [115], and the references therein.

and then a high-gain observer that estimates the derivatives of the regulation error is used to recover the performance achieved under state feedback. As with earlier chapters, the control is saturated outside a compact set of interest to overcome the effect of peaking. Third, to achieve regional or semiglobal stabilization of the augmented system (formed of the plant and the servocompensator), the state feedback design uses a strategy whereby a robust controller is designed as if the goal were to stabilize the origin. This controller brings the trajectories of the system to a small neighborhood of the origin in finite time. Near the origin, the robust controller, acting as a high-gain feedback controller, and the servocompensator stabilize the zero-error manifold. Three different robust control techniques have been used in the literature, namely, high-gain feedback, Lyapunov redesign, and sliding mode control.[36] The results of this chapter use a continuous implementation of sliding mode control.

5.2 ▪ Integral Control

Consider a single-input–single-output nonlinear system modeled by

$$\dot{x} = f(x,w) + g(x,w)u, \qquad e = h(x,w), \tag{5.1}$$

where $x \in R^n$ is the state, u is the control input, e is the output, and w is a constant vector of reference, disturbance, and system parameters that belongs to a compact set $W \subset R^l$. The functions f, g, and h are sufficiently smooth in a domain $D_x \subset R^n$. The goal is to design an output feedback controller such that all state variables are bounded and the output e is asymptotically regulated to zero.

Assumption 5.1. *For each $w \in W$, the system* (5.1) *has relative degree $\rho \le n$, for all $x \in D_x$, that is,*

$$L_g h = L_g L_f h = \cdots = L_g L_f^{\rho-2} h = 0, \quad L_g L_f^{\rho-1} h \neq 0,$$

and there is a diffeomorphism

$$\begin{bmatrix} \eta \\ \xi \end{bmatrix} = T(x)$$

in D_x, possibly dependent on w, that transforms (5.1) *into the normal form*[37]

$$\begin{aligned}
\dot{\eta} &= f_0(\eta,\xi,w),\\
\dot{\xi}_i &= \xi_{i+1}, \quad 1 \le i \le \rho-1,\\
\dot{\xi}_\rho &= a(\eta,\xi,w) + b(\eta,\xi,w)u,\\
e &= \xi_1.
\end{aligned}$$

Moreover, $b(\eta,\xi,w) \ge b_0 > 0$ for all $(\eta,\xi) \in T(D_x)$ and $w \in W$.

The relative degree assumption guarantees the existence of the change of variables (5.2) locally for each $w \in W$ [64]. Assumption 5.1 goes beyond that by requiring the change of variables to hold in a given region.

[36]See [80] for an introduction to these three techniques.

[37]For $\rho = n$, η and the $\dot{\eta}$-equation are dropped.

For the system to maintain the steady-state condition $e=0$, it must have an equilibrium point $(\eta,\xi)=(\bar{\eta},0)$ and a steady-state control $\bar{u}$ that satisfy the equations

$$0 = f_0(\bar{\eta},0,w),$$
$$0 = a(\bar{\eta},0,w)+b(\bar{\eta},0,w)\bar{u}.$$

Assumption 5.2. *For each $w\in W$, the equation $0=f_0(\bar{\eta},0,w)$ has a unique solution $\bar{\eta}$ such that $(\bar{\eta},0)\in T(D_x)$.*

Because $b\neq 0$, the steady-state control that maintains equilibrium is

$$\bar{u}=-\frac{a(\bar{\eta},0,w)}{b(\bar{\eta},0,w)}\stackrel{\text{def}}{=}\phi(w).$$

The change of variables $z=\eta-\bar{\eta}$ transforms the system into the form

$$\dot{z}=f_0(z+\bar{\eta},\xi,w)\stackrel{\text{def}}{=}\tilde{f}_0(z,\xi,w), \tag{5.2}$$
$$\dot{\xi}_i=\xi_{i+1} \qquad \text{for } 1\le i\le \rho-1, \tag{5.3}$$
$$\dot{\xi}_\rho=a_0(z,\xi,w)+b(\eta,\xi,w)[u-\phi(w)], \tag{5.4}$$

where $\tilde{f}_0(0,0,w)=0$ and $a_0(0,0,w)=0$. Let $\Gamma\subset R^n$ be a compact set, which contains the origin in its interior, such that $(z,\xi)\in\Gamma$ implies that $x\in D_x$ for each $w\in W$. The size of W may have to be restricted in order for Γ to exist.

The controller is designed using a separation approach. First, a partial state feedback controller is designed in terms of ξ to regulate the output. Then, a high-gain observer is used to recover the performance of the state feedback controller. Due to the uncertainty of the system, the state feedback controller is designed using sliding mode control.[38] The restriction to partial state feedback is possible when the system is minimum phase. The next two assumptions imply the minimum phase property.

Assumption 5.3. *For each $w\in W$, there is a Lyapunov function $V_1(z)$, possibly dependent on w, and class $\mathcal{K}$ functions γ_1 to γ_4, independent of w, such that*

$$\gamma_1(\|z\|)\le V_1(z)\le\gamma_2(\|z\|),$$
$$\frac{\partial V_1}{\partial z}\tilde{f}_0(z,\xi,w)\le-\gamma_3(\|z\|) \qquad \forall\ \|z\|\ge\gamma_4(\|\xi\|)$$

for all $(z,\xi)\in\Gamma$.

Assumption 5.4. *$z=0$ is an exponentially stable equilibrium point of $\dot{z}=\tilde{f}_0(z,0,w)$, uniformly in w.*

Assumption 5.3 implies that, with ξ as the driving input, the system $\dot{z}=\tilde{f}_0(z,\xi,w)$ is regionally input-to-state stable, uniformly in w.[39]

Augmenting the integrator

$$\dot{\sigma}=e \tag{5.5}$$

[38] Other robust control techniques such as Lyapunov redesign and high-gain feedback can be used. See [80, Chapter 10].

[39] See [144] or [80] for the definition of input-to-state stability.

with the system (5.2)–(5.4) yields

$$\begin{aligned}
\dot{z} &= \tilde{f}_0(z,\xi,w),\\
\dot{\sigma} &= \xi_1,\\
\dot{\xi}_i &= \xi_{i+1} \qquad \text{for } 1 \le i \le \rho-1,\\
\dot{\xi}_\rho &= a_0(z,\xi,w) + b(\eta,\xi,w)[u-\phi(w)],
\end{aligned}$$

which preserves the normal-form structure with a chain of $\rho+1$ integrators. Let

$$s = k_0\sigma + k_1\xi_1 + \cdots + k_{\rho-1}\xi_{\rho-1} + \xi_\rho,$$

where k_0 to $k_{\rho-1}$ are chosen such that the polynomial

$$\lambda^\rho + k_{\rho-1}\lambda^{\rho-1} + \cdots + k_1\lambda + k_0$$

is Hurwitz. Then

$$\dot{s} = \sum_{i=0}^{\rho-1} k_i\xi_{i+1} + a_0(z,\xi,w) + b(\eta,\xi,w)[u-\phi(w)] \stackrel{\text{def}}{=} \Delta(z,\xi,w) + b(\eta,\xi,w)u.$$

Let $\varrho(\xi)$ be a known locally Lipschitz function such that

$$\left|\frac{\Delta(z,\xi,w)}{b(\eta,\xi,w)}\right| \le \varrho(\xi)$$

for all $(z,\xi,w) \in \Gamma \times W$. The state feedback controller is taken as

$$u = -\beta(\xi)\,\text{sat}\left(\frac{s}{\mu}\right), \tag{5.6}$$

where $\beta(\xi)$ is a locally Lipschitz function that satisfies $\beta(\xi) \ge \varrho(\xi) + \beta_0$ with $\beta_0 > 0$. The functions ϱ and β are allowed to depend only on ξ rather than the full state (z,ξ). This is possible because the inequality $|\Delta/b| \le \varrho$ is required to hold over the compact set Γ where the z-dependent part of Δ/b can be bounded by a constant. Under the controller (5.6),

$$s\dot{s} = s\Delta - b\beta s\,\text{sat}\left(\frac{s}{\mu}\right) \le b\varrho|s| - b\beta s\,\text{sat}\left(\frac{s}{\mu}\right).$$

For $|s| \ge \mu$,

$$s\dot{s} \le b(\varrho|s| - \beta|s|) \le -b_0\beta_0|s|.$$

The closed-loop system is given by

$$\begin{aligned}
\dot{z} &= \tilde{f}_0(z,\xi,w),\\
\dot{\zeta} &= A_1\zeta + Bs,\\
\dot{s} &= -b(\eta,\xi,w)\beta(\xi)\,\text{sat}\left(\frac{s}{\mu}\right) + \Delta(z,\xi,w),
\end{aligned}$$

where $\zeta = \text{col}(\sigma, \xi_1, \dots, \xi_{\rho-1})$, $\xi = A_1\zeta + Bs$,

$$A_1 = \begin{bmatrix} 0 & 1 & \cdots & \cdots & 0 \\ 0 & 0 & 1 & \cdots & 0 \\ \vdots & & & & \vdots \\ 0 & \cdots & \cdots & 0 & 1 \\ -k_0 & \cdots & \cdots & \cdots & -k_{\rho-1} \end{bmatrix}, \quad \text{and} \quad B = \begin{bmatrix} 0 \\ 0 \\ \vdots \\ 0 \\ 1 \end{bmatrix}.$$

The matrix A_1 is Hurwitz by design. The inequality $s\dot{s} \leq -b_0\beta_0|s|$ shows that the set $\{|s| \leq c\}$ with $c > \mu$ is positively invariant because $s\dot{s} < 0$ on the boundary $|s| = c$. Let $V_2(\zeta) = \zeta^T P_1 \zeta$, where P_1 is the solution of the Lyapunov equation $P_1A_1 + A_1^T P_1 = -I$. The derivative of V_2 satisfies the inequalities

$$\dot{V}_2 \leq -\zeta^T\zeta + 2\|\zeta\|\,\|P_1B\|\,|s|,$$

which shows that the set $\{V_2 \leq c^2\rho_1\} \times \{|s| \leq c\}$ is positively invariant for $\rho_1 > 4\|P_1B\|^2\lambda_{\max}(P_1)$ because $\dot{V}_2 < 0$ on the boundary $V_2 = c^2\rho_1$. Inside this set, $\|\xi\| = \|A_1\zeta + Bs\| \leq c\|A_1\|\sqrt{\rho_1/\lambda_{\min}(P_1)} + c \stackrel{\text{def}}{=} c\rho_2$. The derivative of the Lyapunov function V_1 of Assumption 5.3 satisfies the inequality

$$\dot{V}_1 \leq -\gamma_3(\|z\|) \quad \forall\ \|z\| \geq \gamma_4(c\rho_2),$$

which shows that the set

$$\Omega = \{V_1(z) \leq c_0\} \times \{\zeta^T P_1 \zeta \leq \rho_1 c^2\} \times \{|s| \leq c\},$$

with $c_0 \geq \gamma_2(\gamma_4(c\rho_2))$, is positively invariant because $\dot{V}_1 < 0$ on the boundary $V_1 = c_0$. Similarly, it can be shown that the set

$$\Omega_\mu = \{V_1(z) \leq \gamma_2(\gamma_4(\mu\rho_2))\} \times \{\zeta^T P_1 \zeta \leq \rho_1\mu^2\} \times \{|s| \leq \mu\}$$

is positively invariant and every trajectory starting in Ω enters Ω_μ in finite time. The constants c and c_0 are chosen such that $(z, \xi) \in \Gamma$ for $(z, \zeta, s) \in \Omega$.

Under output feedback, the high-gain observer

$$\dot{\hat{\xi}}_i = \hat{\xi}_{i+1} + \frac{\alpha_i}{\varepsilon^i}(e - \hat{\xi}_1), \quad 1 \leq i \leq \rho - 1, \tag{5.7}$$

$$\dot{\hat{\xi}}_\rho = \frac{\alpha_\rho}{\varepsilon^\rho}(e - \hat{\xi}_1) \tag{5.8}$$

is used to estimate ξ by $\hat{\xi}$, where ε is a sufficiently small positive constant, and α_1 to α_ρ are chosen such that the polynomial

$$\lambda^\rho + \alpha_1\lambda^{\rho-1} + \cdots + \alpha_{\rho-1}\lambda + \alpha_\rho$$

is Hurwitz. To overcome the peaking phenomenon of high-gain observers, the control is saturated outside Ω. Because the saturation function is bounded, we only need to saturate $\beta(\hat{\xi})$. Denote the saturated function by $\beta_s(\hat{\xi})$. Then the output feedback controller is given by

$$u = -\beta_s(\hat{\xi})\ \text{sat}\left(\frac{k_0\sigma + k_1\hat{\xi}_1 + k_2\hat{\xi}_2 + \cdots + k_{\rho-1}\hat{\xi}_{\rho-1} + \hat{\xi}_\rho}{\mu}\right). \tag{5.9}$$

The integrator state σ is available for feedback because it is obtained by integrating the measured output ξ_1. In (5.9), $\hat{\xi}_1$ can be replaced by ξ_1.

Remark 5.1. *The precise information of the plant that is used in the design of the controller is its relative degree and the sign of its high-frequency gain b, which, without loss of generality, is assumed to be positive. Some additional rudimentary information will be needed to estimate bounds on design parameters. Such a controller is "universal" in the sense that it will stabilize any plant in a family of plants that share the relative degree and the sign of the high-frequency gain.* ◇

Remark 5.2. *For relative-degree-one systems, the controller* (5.9), *with $\hat{\xi}_1$ replaced by $\xi_1 = e$, takes the form*

$$u = -\beta(e)\,\text{sat}\left(\frac{k_0\sigma + k_1 e}{\mu}\right),$$

which is implemented without observer. ◇

Remark 5.3. *In the special case when $\beta = k$ (constant) and $\hat{\xi}_1$ is replaced by ξ_1, the controller is given by*

$$u = -k\,\text{sat}\left(\frac{k_0\sigma + k_1\xi_1 + k_2\hat{\xi}_2 + \cdots + k_{\rho-1}\hat{\xi}_{\rho-1} + \hat{\xi}_\rho}{\mu}\right). \tag{5.10}$$

When $\rho = 1$, the controller (5.10) *is a classical PI (proportional-integral) controller with high gain, followed by saturation. When $\rho = 2$, it is a classical PID (proportional-integral-derivative) controller with high gain, followed by saturation, where the derivative term is provided by the high-gain observer.* ◇

Theorem 5.1. *Suppose Assumptions* 5.1 *to* 5.4 *are satisfied and consider the closed-loop system formed of the system* (5.1), *the integrator* (5.5), *the observer* (5.7)–(5.8), *and the controller* (5.9). *Let Ψ be a compact set in the interior of Ω and suppose $(z(0), \zeta(0), s(0)) \in \Psi$ and $\hat{\xi}(0)$ is bounded. Then, $\mu^* > 0$ exists, and for each $\mu \in (0, \mu^*]$, $\varepsilon^* = \varepsilon^*(\mu)$ exists such that for each $\mu \in (0, \mu^*]$ and $\varepsilon \in (0, \varepsilon^*(\mu)]$, all state variables are bounded and $\lim_{t\to\infty} \xi(t) = 0$.* ◇

Proof: The closed-loop system under output feedback is given by

$$\dot{z} = \tilde{f}_0(z, \xi, w), \tag{5.11}$$

$$\dot{\zeta} = A_1\zeta + Bs, \tag{5.12}$$

$$\dot{s} = b(\eta, \xi, w)\psi(\sigma, \hat{\xi}, \mu) + \Delta(z, \xi, w), \tag{5.13}$$

$$\varepsilon\dot{\varphi} = A_0\varphi + \varepsilon B\left[\Delta(z, \xi, w) - \sum_{i=0}^{\rho-1} k_i\xi_{i+1} + b(\eta, \xi, w)\psi(\sigma, \hat{\xi}, \mu)\right], \tag{5.14}$$

where

$$\psi(\sigma, \xi, \mu) = -\beta_s(\xi)\,\text{sat}\left(\frac{s}{\mu}\right), \qquad \varphi_i = \frac{\xi_i - \hat{\xi}_i}{\varepsilon^{\rho-i}} \quad \text{for } 1 \le i \le \rho,$$

and

$$A_0 = \begin{bmatrix} -\alpha_1 & 1 & 0 & \cdots & 0 \\ -\alpha_2 & 0 & 1 & \cdots & 0 \\ \vdots & & \ddots & & \vdots \\ -\alpha_{\rho-1} & & & 0 & 1 \\ -\alpha_\rho & 0 & \cdots & \cdots & 0 \end{bmatrix}.$$

The matrix A_0 is Hurwitz by design. Equations (5.11)–(5.13) with $\psi(\sigma, \hat{\xi}, \mu)$ replaced by $\psi(\sigma, \xi, \mu)$ are the closed-loop system under state feedback. Let P_0 be the solution of the Lyapunov equation $P_0 A_0 + A_0^T P_0 = -I$, $V_3(\varphi) = \varphi^T P_0 \varphi$, and $\Sigma_\varepsilon = \{V_3(\varphi) \leq \rho_3 \varepsilon^2\}$, where the positive constant ρ_3 is to be determined. The proof proceeds in four steps:

Step 1: Show that there exist $\rho_3 > 0$, $\mu_1^* > 0$, and $\varepsilon_1^* = \varepsilon_1^*(\mu) > 0$ such that for each $\mu \in (0, \mu_1^*]$ and $\varepsilon \in (0, \varepsilon_1^*(\mu)]$ the set $\Omega \times \Sigma_\varepsilon$ is positively invariant.

Step 2: Show that for any bounded $\hat{\xi}(0)$ and any $(z(0), \zeta(0), s(0)) \in \Psi$, there exists $\varepsilon_2^* > 0$ such that for each $\varepsilon \in (0, \varepsilon_2^*]$ the trajectory enters the set $\Omega \times \Sigma_\varepsilon$ in finite time $T_1(\varepsilon)$, where $\lim_{\varepsilon \to 0} T_1(\varepsilon) = 0$.

Step 3: Show that there exists $\varepsilon_3^* = \varepsilon_3^*(\mu) > 0$ such that for each $\varepsilon \in (0, \varepsilon_3^*(\mu)]$ every trajectory in $\Omega \times \Sigma_\varepsilon$ enters $\Omega_\mu \times \Sigma_\varepsilon$ in finite time and stays therein for all future time.

Step 4: Show that there exists $\mu_2^* > 0$ and $\varepsilon_4^* = \varepsilon_4^*(\mu) > 0$ such that for each $\mu \in (0, \mu_2^*]$ and $\varepsilon \in (0, \varepsilon_4^*(\mu)]$ every trajectory in $\Omega_\mu \times \Sigma_\varepsilon$ converges to the equilibrium point $(z = 0, \zeta = \bar{\zeta}, s = \bar{s}, \varphi = 0)$ at which $\xi = 0$.

For the first step, calculate the derivative of V_3 on the boundary $V_3 = \rho_3 \varepsilon^2$:

$$\varepsilon \dot{V}_3 = -\varphi^T \varphi + 2\varepsilon \varphi^T P_0 B \left[\Delta(z, \xi, w) - \sum_{i=0}^{\rho-1} k_i \xi_{i+1} + b(\eta, \xi, w)\psi(\sigma, \hat{\xi}, \mu) \right].$$

Since $\psi(\sigma, \hat{\xi}, \mu)$ is globally bounded in $\hat{\xi}$, for all $(z, \zeta, s) \in \Omega$ there is $\ell_1 > 0$ such that $|\Delta(z, \xi, w) - \sum_{i=0}^{\rho-1} k_i \xi_{i+1} + b(\eta, \xi, w)\psi(\sigma, \hat{\xi}, \mu)| \leq \ell_1$. Therefore

$$\varepsilon \dot{V}_3 \leq -\|\varphi\|^2 + 2\varepsilon \ell_1 \|P_o B\| \, \|\varphi\| \leq -\tfrac{1}{2}\|\varphi\|^2 \quad \forall \, \|\varphi\| \geq 4\varepsilon \ell_1 \|P_0 B\|. \tag{5.15}$$

Taking $\rho_3 = \lambda_{\max}(P_0)(4\|P_0 B\| \ell_1)^2$ ensures that $\dot{V}_3 \leq -\frac{1}{2}\|\varphi\|^2$ for all $V_3 \geq \rho_3 \varepsilon^2$. Consequently, $\dot{V}_3 < 0$ on the boundary $V_3 = \rho_3 \varepsilon^2$. Consider, next, (5.13) as a perturbation of the corresponding equation under state feedback:

$$\dot{s} = b(\eta, \xi, w)\psi(\sigma, \xi, \mu) + \Delta(z, \xi, w) + b(\eta, \xi, w)[\psi(\sigma, \hat{\xi}, \mu) - \psi(\sigma, \xi, \mu)].$$

In $\Omega \times \Sigma_\varepsilon$, positive constants ℓ_2 and ℓ_3 exist such that

$$|\psi(\sigma, \hat{\xi}, \mu) - \psi(\sigma, \xi, \mu)| \leq \left(\ell_2 + \frac{\ell_3}{\mu}\right) \|\xi - \hat{\xi}\| \leq \left(\ell_2 + \frac{\ell_3}{\mu}\right) \|\varphi\|$$

for $\varepsilon \leq 1$. The factor $1/\mu$ is due to the Lipschitz constant of $\text{sat}(s/\mu)$. Since $\|\varphi\| \leq \ell_4 \varepsilon$ in Σ_ε,

$$s\dot{s} \leq b|s|[\varrho - \beta + \varepsilon \ell_4(\ell_2 + \ell_3/\mu)] \leq -\tfrac{1}{2} b_0 \beta_0 |s|$$

when $|s| \geq \mu$ and $\varepsilon \ell_4(\ell_2 + \ell_3/\mu) \leq \frac{1}{2}\beta_0$. Repeating the argument used with state feedback, it can be shown that Ω is positively invariant, which completes the first step. Because Ψ is in the interior of Ω and $\psi(\sigma, \hat{\xi}, \mu)$ is globally bounded in $\hat{\xi}$, there is time $T_0 > 0$, independent of ε, such that $(z(t), \zeta(t), s(t)) \in \Omega$ for $t \in [0, T_0]$. During this time, inequality (5.15) shows that

$$\dot{V}_3 \leq -\frac{1}{2\varepsilon \lambda_{\max}(P_0)} V_3.$$

Therefore, V_3 reduces to $\rho_3 \varepsilon^2$ within a time interval $[0, T_1(\varepsilon)]$ in which $\lim_{\varepsilon \to 0} T_1(\varepsilon) = 0$. For sufficiently small ε, $T_1(\varepsilon) < T_0$. The second step is complete. The third step follows from the fact that $s\dot{s} \leq -\frac{1}{2} b_0 \beta_0$ for $|s| \geq \mu$, which shows that s enters the set $\{|s| \leq \mu\}$ in finite time. The rest is similar to the analysis under state feedback because (5.11) and (5.12) are not altered under output feedback. For the final step, consider the system inside $\Omega_\mu \times \Sigma_\varepsilon$. There is an equilibrium point at $(z = 0, \zeta = \bar{\zeta}, s = \bar{s}, \varphi = 0)$, where $\bar{\zeta} = \text{col}(\bar{\sigma}, 0, \ldots, 0)$, and

$$\bar{s} = k_0 \bar{\sigma} = \frac{-\mu \phi(w)}{\beta(0)}.$$

Shifting the equilibrium point to the origin by the change of variables $\nu = \zeta - \bar{\zeta}$ and $p = s - \bar{s}$, the system takes the singularly perturbed form

$$\begin{aligned}
\dot{z} &= \tilde{f}_0(z, A_1 \nu + B p, w), \\
\dot{\nu} &= A_1 \nu + B p, \\
\mu \dot{p} &= -b(\eta, \xi, w)\beta(\xi) p + \mu \Delta_a(\cdot) + \mu \Delta_b(\cdot) + \mu b(\eta, \xi, w)[\psi(\sigma, \hat{\xi}, \mu) - \psi(\sigma, \xi, \mu)], \\
\varepsilon \dot{\varphi} &= A_0 \varphi + (\varepsilon/\mu) B\{-b(\eta, \xi, w)\beta(\xi) p + \mu \Delta_a(\cdot) \\
&\quad + \mu b(\eta, \xi, w)[\psi(\sigma, \hat{\xi}, \mu) - \psi(\sigma, \xi, \mu)]\},
\end{aligned}$$

where

$$\Delta_a = a_0(z, \xi, w) + b(\eta, \xi, w)\phi(w)\left[\frac{\beta(\xi) - \beta(0)}{\beta(0)}\right]$$

and $\Delta_b = \sum_{i=0}^{\rho-1} k_i \xi_{i+1}$. There are positive constants ℓ_5 to ℓ_9 such that

$$|\Delta_a| \leq \ell_5 \|z\| + \ell_6 \|\nu\| + \ell_7 |p|, \quad |\Delta_b| \leq \ell_8 \|\nu\| + \ell_9 |p|.$$

Moreover, $\mu |\psi(\sigma, \hat{\xi}, \mu) - \psi(\sigma, \xi, \mu)| \leq (\mu \ell_2 + \ell_3)\|\varphi\|$.

By Assumption 5.4, $z = 0$ is an exponentially stable equilibrium point of $\dot{z} = \tilde{f}_0(z, 0, w)$ uniformly in w. By the converse Lyapunov theorem [78, Lemma 9.8], there is a Lyapunov function $V_0(z)$, possibly dependent on w, and positive constant $\bar{c}_1$ to $\bar{c}_4$, independent of w, such that

$$\bar{c}_1 \|z\|^2 \leq V_0(z, w) \leq \bar{c}_2 \|z\|^2, \quad \frac{\partial V_0}{\partial z} \tilde{f}_0(z, 0, w) \leq -\bar{c}_3 \|z\|^2, \quad \left\|\frac{\partial V_0}{\partial z}\right\| \leq \bar{c}_4 \|z\|$$

in some neighborhood of $z = 0$. Consider the composite Lyapunov function

$$V = \alpha V_0 + \nu^T P_1 \nu + \frac{1}{2} p^2 + \varphi^T P_0 \varphi$$

with $\alpha > 0$. It can be shown that $\dot{V} \leq -\mathscr{Y}^T Q \mathscr{Y}$, where

$$\mathscr{Y} = \begin{bmatrix} \|z\| \\ \|v\| \\ |p| \\ \|\varphi\| \end{bmatrix}, \quad Q = \begin{bmatrix} \alpha c_1 & -\alpha c_2 & -(\alpha c_3 + c_4) & -c_5 \\ -\alpha c_2 & 1 & -c_6 & -c_7 \\ -(\alpha c_3 + c_4) & -c_6 & (c_8/\mu - c_9) & -(c_{10} + c_{11}/\mu) \\ -c_5 & -c_7 & -(c_{10} + c_{11}/\mu) & (1/\varepsilon - c_{12} - c_{13}/\mu) \end{bmatrix},$$

and c_1 to c_{13} are positive constants. Choose $\alpha < c_1/c_2^2$ to make the 2×2 principal minor of Q positive; then choose μ small enough to make the 3×3 principal minor positive. Finally, choose ε small enough to make the determinant of Q positive. Hence, the origin $(z = 0, v = 0, p = 0, \varphi = 0)$ is exponentially stable, and there is a neighborhood N of the origin, independent of μ and ε, such that all trajectories in N converge to the origin as $t \to \infty$. By choosing μ and ε small enough, it can be ensured that for all $(z, \zeta, s, \varphi) \in \Omega_\mu \times \Sigma_\varepsilon$, $(z, v, p, \varphi) \in N$. Thus, all trajectories in $\Omega_\mu \times \Sigma_\varepsilon$ converge to the equilibrium point $(z = 0, \zeta = \bar{\zeta}, s = \bar{s}, \varphi = 0)$. Consequently, all trajectories with $(z(0), \zeta(0), s(0)) \in \Psi$ and bounded $\hat{\xi}(0)$ converge to this equilibrium point because such trajectories enter $\Omega_\mu \times \Sigma_\varepsilon$ in finite time. □

Remark 5.4. *It is also possible to prove that the trajectories under output feedback approach the ones under state feedback as $\varepsilon \to 0$. The proof repeats steps from the proof of Theorem 3.1. It is not given here, but the trajectory recovery property is illustrated by simulation.* ◇

Remark 5.5. *If Assumptions 5.1 to 5.3 hold globally, that is, $D_x = T(D_x) = R^n$ and γ_1 is class $\mathscr{K}_\infty$, then the sets Γ and Ω can be chosen arbitrarily large. For any bounded $(z(0), \zeta(0), s(0))$, the conclusion of Theorem 5.1 will hold by choosing β large enough.* ◇

Example 5.1. The state model

$$\dot{x}_1 = x_2, \qquad \dot{x}_2 = -\sin x_1 - a x_2 + b u + d \cos x_1, \qquad y = x_1$$

represents a pendulum equation whose suspension point is subjected to a constant horizontal acceleration. It is desired to regulate y to a constant reference r using output feedback. Suppose a, b, d are uncertain parameters that satisfy $0 \leq a \leq 0.1$, $0.5 \leq b \leq 2$, and $0 \leq d \leq 0.5$. Define $\xi_1 = x_1 - r$ and $\xi_2 = x_2$ and augment the integrator $\dot{\sigma} = \xi_1$ with the system to obtain

$$\dot{\sigma} = \xi_1, \qquad \dot{\xi}_1 = \xi_2, \qquad \dot{\xi}_2 = -\sin x_1 - a x_2 + b u + d \cos x_1.$$

Taking $s = \sigma + 2\xi_1 + \xi_2$ assigns the eigenvalues of $\lambda^2 + 2\lambda + 1$ at -1, -1. Then

$$\dot{s} = \xi_1 + (2 - a)\xi_2 - \sin x_1 + b u + d \cos x_1.$$

Using the bounds on a, b, and d, it can be shown that

$$\left| \frac{\xi_1 + (2 - a)\xi_2 - \sin x_1 + d \cos x_1}{b} \right| \leq 2|\xi_1| + 4|\xi_2| + 3.$$

The state feedback controller is taken as

$$u = -k \,\mathrm{sat}\left(\frac{\sigma + 2\xi_1 + \xi_2}{\mu} \right),$$

which is a globally bounded function of ξ. The initial condition $\xi(0)$ has to be restricted to a compact set such that the inequality $2|\xi_1|+4|\xi_2|+3<k$ is satisfied for all t. This set can be determined by simulation or estimated by choosing $c>0$ such that the maximum of $2|\xi_1|+4|\xi_2|+3$ over Ω is less than k. The output feedback controller is given by

$$u=-k\ \text{sat}\left(\frac{\sigma+2\hat{\xi}_1+\hat{\xi}_2}{\mu}\right),$$

where $\hat{\xi}_1$ and $\hat{\xi}_2$ are provided by the high-gain observer

$$\dot{\hat{\xi}}_1=\hat{\xi}_2+\frac{2}{\varepsilon}(\xi_1-\hat{\xi}_1),\qquad \dot{\hat{\xi}}_2=\frac{1}{\varepsilon^2}(\xi_1-\hat{\xi}_1).$$

Simulation results with zero initial conditions, $r=\pi$, $\mu=0.1$, $k=5$, $a=0.03$, $b=1$, and $d=0.3$ are shown in Figure 5.2. Figures 5.2(a) and (b) show the output x_1 when $\varepsilon=0.01$. The convergence of x_1 to π is shown in Figure 5.2(b). Figures 5.2(c) and (d) illustrate the property that the trajectories under output feedback approach the ones under state feedback as ε decreases. They show the differences

$$\Delta x_i=x_i(\text{under output feedback})-x_i(\text{under state feedback})$$

for $i=1,2$ when $\varepsilon=0.01$, 0.005, and 0.001.

For comparison, consider a sliding mode controller without integral action. The design proceeds as in Section 3.2.2 with $s_1=\xi_1+\xi_2$,

$$\dot{s}_1=(1-a)\xi_2-\sin x_1+bu+d\cos x_1,$$

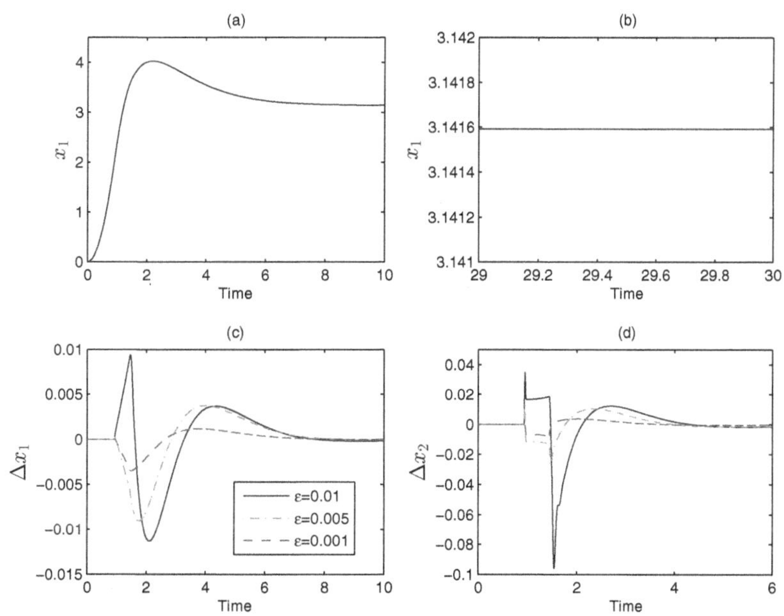

Figure 5.2. *Simulation of Example* 5.1. (a) *and* (b) *show the transient and steady-state responses of the output.* (c) *and* (d) *show the deviation of the state trajectories under output feedback from the ones under state feedback as* ε *decreases.*

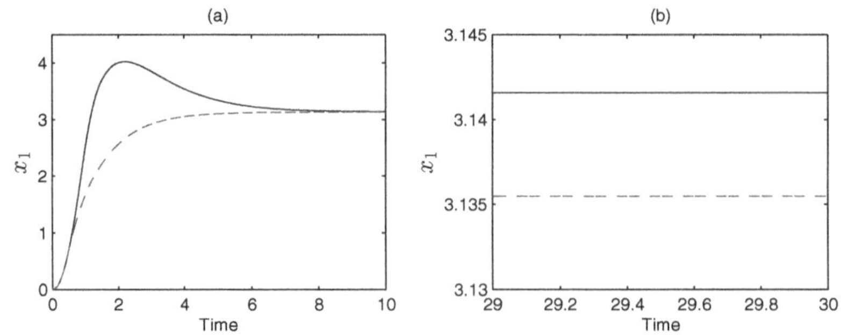

Figure 5.3. *Simulation of the output feedback controllers of Example* 5.1 *with (solid) and without (dashed) integral action.*

and

$$\left|\frac{(1-a)\xi_2 - \sin x_1 + d\cos x_1}{b}\right| \le \frac{|\xi_2| + 1 + 0.5}{0.5} = 2|\xi_2| + 3.$$

Considering a compact set of operation where $2|\xi_2| + 3 < k$, the state feedback controller is taken as

$$u = -k \text{ sat}\left(\frac{\xi_1 + \xi_2}{\mu}\right)$$

and the output feedback controller is

$$u = -k \text{ sat}\left(\frac{\hat{\xi}_1 + \hat{\xi}_2}{\mu}\right),$$

where $\hat{\xi}_1$ and $\hat{\xi}_2$ are provided by the same high-gain observer. Figure 5.3 compares the responses of the output feedback controllers with (solid) and without (dashed) integral action. The simulation is carried out using the same parameters as in Figure 5.2 with $\varepsilon = 0.01$. The controller without integral action results in a steady-state error due to the nonvanishing disturbance $d \cos x_1$, while the one with integral action regulates the error to zero. The inclusion of integral action comes at the expense of the transient response, which shows an overshoot that is not present in the case without integral action. $\triangle$

5.3 ▪ Conditional Integrator

The integral controller of the previous section achieves zero steady-state regulation error in the presence of constant reference and disturbance. This usually comes at the expense of degrading the transient response when compared with a controller that does not include integral action, as shown in Example 5.1. The conditional integrator of this section removes this drawback. It achieves zero steady-state error without degrading the transient response.

Reconsider the system (5.1) and suppose Assumptions 5.1 to 5.4 are satisfied. Once again, the goal is to design an output feedback controller such that all state variables are bounded and the output e is asymptotically regulated to zero. Similar to Section 5.2, the controller is designed using a separation approach. For the state feedback controller, start with (5.2) to (5.4) and let

$$s_1 = k_1\xi_1 + \cdots + k_{\rho-1}\xi_{\rho-1} + \xi_\rho,$$

where k_1 to $k_{\rho-1}$ are chosen such that the polynomial

$$\lambda^{\rho-1} + k_{\rho-1}\lambda^{\rho-2} + \cdots + k_2\lambda + k_1$$

is Hurwitz. Then

$$\dot{s}_1 = \sum_{i=1}^{\rho-1} k_i \xi_{i+1} + a_0(z,\xi,w) + b(\eta,\xi,w)[u - \phi(w)] \stackrel{\text{def}}{=} \Delta_1(z,\xi,w) + b(\eta,\xi,w)u.$$

Let $\varrho_1(\xi)$ be a known locally Lipschitz function such that

$$\left| \frac{\Delta_1(z,\xi,w)}{b(\eta,\xi,w)} \right| \le \varrho_1(\xi)$$

for all $(z,\xi,w) \in \Gamma \times W$. A state feedback sliding mode controller can be taken as

$$u = -\beta(\xi)\,\mathrm{sgn}(s_1), \tag{5.16}$$

where $\beta(\xi)$ is a locally Lipschitz function that satisfies $\beta(\xi) \ge \varrho_1(\xi) + \beta_0$ with $\beta_0 > 0$ and the signum function $\mathrm{sgn}(\cdot)$ is defined by

$$\mathrm{sgn}(s_1) = \begin{cases} 1 & \text{if } s_1 > 0, \\ -1 & \text{if } s_1 < 0. \end{cases}$$

Ideally, the control (5.16) can achieve zero steady-state error in the absence of unmolded dynamics and time delays. Its use will lead to control chattering, which can be eliminated by replacing $\mathrm{sgn}(s_1)$ by $\mathrm{sat}(s_1/\mu)$. The trajectories under the continuously implemented control approach the ones under the discontinuous control as μ decreases.[40] However, it cannot achieve zero steady-state error in the presence of nonvanishing disturbance. It can achieve only practical regulation, as the ultimate bound on the error will be of the order $O(\mu)$. Although, in theory, the error can be made arbitrarily small by reducing μ, a too small value of μ would induce chattering. Achieving zero steady-state error requires integral action, which is introduced through the conditional integrator

$$\dot{\sigma} = \gamma\left[-\sigma + \mu\,\mathrm{sat}\left(\frac{s}{\mu}\right)\right], \qquad |\sigma(0)| \le \mu, \tag{5.17}$$

$$u = -\beta(\xi)\,\mathrm{sat}\left(\frac{s}{\mu}\right) \tag{5.18}$$

where

$$s = \sigma + s_1 = \sigma + k_1\xi_1 + \cdots + k_{\rho-1}\xi_{\rho-1} + \xi_\rho,$$

and γ and μ are positive constants. From the inequality

$$\sigma\dot{\sigma} = \gamma\left[-\sigma^2 + \sigma\mu\,\mathrm{sat}\left(\frac{s}{\mu}\right)\right] \le \gamma(-\sigma^2 + \mu|\sigma|) \le 0 \;\; \forall\; |\sigma| \ge \mu,$$

it is seen that the set $\{|\sigma| \le \mu\}$ is positively invariant. Choosing $|\sigma(0)| \le \mu$ ensures that $|\sigma(t)| \le \mu$ for all $t \ge 0$. Inside the set $\{|s| \le \mu\}$, the $\dot{\sigma}$-equation becomes

$$\dot{\sigma} = \gamma s_1 = \gamma(k_1\xi_1 + \cdots + k_{\rho-1}\xi_{\rho-1} + \xi_\rho),$$

[40] This claim is shown in the proof of Theorem 5.2.

which provides integral action because at steady state $\dot{\sigma} = 0$ and $\xi_2 = \cdots = \xi_\rho = 0$. Hence, $\xi_1 = 0$.

The closed-loop system is given by

$$\begin{aligned}
\dot{\sigma} &= \gamma\left[-\sigma + \mu \operatorname{sat}\left(\frac{s}{\mu}\right)\right],\\
\dot{z} &= \tilde{f}_0(z,\xi,w),\\
\dot{q} &= A_2 q + B_2(s-\sigma),\\
\dot{s} &= -b(\eta,\xi,w)\beta(\xi)\operatorname{sat}\left(\frac{s}{\mu}\right) + \Delta_1(z,\xi,w) + \gamma\left[-\sigma + \mu \operatorname{sat}\left(\frac{s}{\mu}\right)\right],
\end{aligned}$$

where $q = \mathrm{col}(\xi_1,\ldots,\xi_{\rho-1})$, $\xi = \mathrm{col}(q, s-\sigma-Lq)$, $L = \begin{bmatrix} k_1 & \cdots & k_{\rho-1}\end{bmatrix}$,

$$A_2 = \begin{bmatrix} 0 & 1 & \cdots & \cdots & 0 \\ 0 & 0 & 1 & \cdots & 0 \\ \vdots & & & & \vdots \\ 0 & \cdots & \cdots & 0 & 1 \\ -k_1 & \cdots & \cdots & \cdots & -k_{\rho-1} \end{bmatrix}, \quad \text{and} \quad B_2 = \begin{bmatrix} 0 \\ 0 \\ \vdots \\ 0 \\ 1 \end{bmatrix}.$$

The matrix A_2 is Hurwitz by design. For $|s| \geq \mu$,

$$s\dot{s} \leq b\left(-\beta + \frac{|\Delta_1|}{b} + \frac{2\gamma\mu}{b}\right)|s| \leq b\left(-\beta_0 + \frac{2\gamma\mu}{b_0}\right)|s|.$$

For $\mu \leq \frac{1}{4} b_0\beta_0/\gamma$,

$$s\dot{s} \leq -\tfrac{1}{2} b\beta_0|s| \leq -\tfrac{1}{2} b_0\beta_0|s|,$$

which shows that the set $\{|s| \leq c\}$ is positively invariant for $c > \mu$ and s reaches the positively invariant set $\{|s| \leq \mu\}$ in finite time. Let $V_2(q) = q^T P_2 q$, where P_2 is the solution of the Lyapunov equation $P_2A_2 + A_2^T P_2 = -I$. For $|s| \leq c$, the derivative of V_2 satisfies the inequalities

$$\dot{V}_2 \leq -q^T q + 2\|q\|\,\|P_2B_2\|\,(|s| + |\sigma|) \leq -\|q\|^2 + 4\|P_2B_2\|\,\|q\|c,$$

which shows that the set $\{V_2 \leq \bar{\rho}_1 c^2\} \times \{|s| \leq c\}$ is positively invariant for $\bar{\rho}_1 > 16\|P_2B_2\|^2\lambda_{\max}(P_2)$ because $\dot{V}_2 < 0$ on the boundary $V_2 = \bar{\rho}_1 c^2$. Inside this set, $\|\xi\| = c(1+\|L\|)\sqrt{\bar{\rho}_1/\lambda_{\min}(P_2)} + 2c \stackrel{\text{def}}{=} \bar{\rho}_2 c$. The derivative of the Lyapunov function V_1 of Assumption 5.3 satisfies the inequality

$$\dot{V}_1 \leq -\gamma_3(\|z\|) \quad \forall\ \|z\| \geq \gamma_4(c\bar{\rho}_2),$$

which shows that the set

$$\Omega = \{V_1(z) \leq c_0\} \times \{q^T P_2 q \leq \bar{\rho}_1 c^2\} \times \{|s| \leq c\},$$

with $c_0 \geq \gamma_2(\gamma_4(c\bar{\rho}_2))$, is positively invariant because $\dot{V}_1 < 0$ on the boundary $V_1 = c_0$. Similarly, it can be shown that the set

$$\Omega_\mu = \{V_1(z) \leq \gamma_2(\gamma_4(\mu\bar{\rho}_2))\} \times \{q^T P_2 q \leq \bar{\rho}_1 \mu^2\} \times \{|s| \leq \mu\}$$

is positively invariant and every trajectory starting in $\{|\sigma| \le \mu\} \times \Omega$ enters $\{|\sigma| \le \mu\} \times \Omega_\mu$ in finite time. The constants c and c_0 are chosen such that $(z,\xi) \in \Gamma$ for $(z,q,s) \in \Omega$.

The output feedback controller is given by

$$\dot{\sigma} = \gamma \left[-\sigma + \mu \ \mathrm{sat}\left(\frac{\sigma + k_1\hat{\xi}_1 + \cdots + k_{\rho-1}\hat{\xi}_{\rho-1} + \hat{\xi}_\rho}{\mu} \right) \right], \tag{5.19}$$

$$u = -\beta_s(\hat{\xi}) \ \mathrm{sat}\left(\frac{\sigma + k_1\hat{\xi}_1 + \cdots + k_{\rho-1}\hat{\xi}_{\rho-1} + \hat{\xi}_\rho}{\mu} \right), \tag{5.20}$$

where β_s is obtained by saturating β outside Ω, and $\hat{\xi}$ is provided by the high-gain observer

$$\dot{\hat{\xi}}_i = \hat{\xi}_{i+1} + \frac{\alpha_i}{\varepsilon^i}(e - \hat{\xi}_1), \quad 1 \le i \le \rho - 1, \tag{5.21}$$

$$\dot{\hat{\xi}}_\rho = \frac{\alpha_\rho}{\varepsilon^\rho}(e - \hat{\xi}_1), \tag{5.22}$$

in which ε is a sufficiently small positive constant, and α_1 to α_ρ are chosen such that the polynomial

$$\lambda^\rho + \alpha_1\lambda^{\rho-1} + \cdots + \alpha_{\rho-1}\lambda + \alpha_\rho$$

is Hurwitz. In (5.19) and (5.20), $\hat{\xi}_1$ can be replaced with the measured error ξ_1.

Remark 5.6. *As in the previous section, the precise information of the plant that is used in the design of the controller is its relative degree and the sign of its high-frequency gain b.* ◇

Remark 5.7. *In the special case when $\beta = k$ (constant) and $\hat{\xi}_1$ is replaced by ξ_1, the controller has the structure of an integral controller with antiwindup scheme.*[41] *Figure 5.4 shows a block diagram representation of the controller for relative-degree-one systems. The difference between the input and output of the control saturation closes the loop around the*

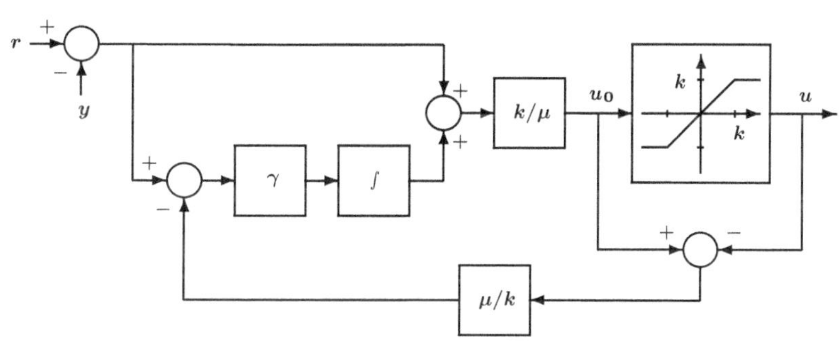

Figure 5.4. *The conditional integrator for relative-degree-one systems as a PI controller with antiwindup.*

[41] Antiwindup schemes are used to prevent integrator windup during control saturation. There are several such schemes. The scheme referred to here is due to [43]. See also [16, Section 8.5].

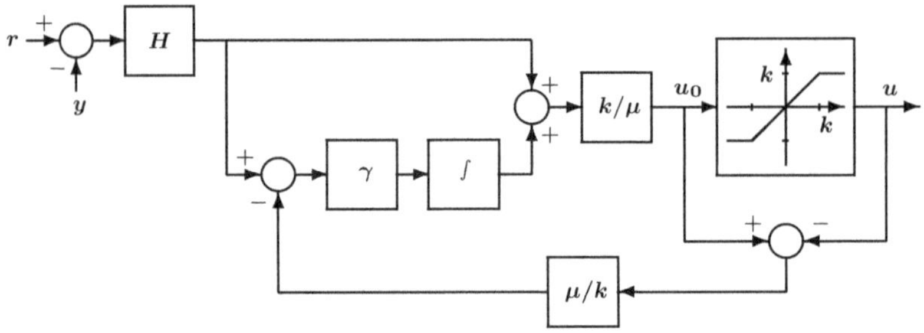

Figure 5.5. *The conditional integrator for relative-degree $\rho > 1$ as a PID$^{\rho-1}$ controller with antiwindup.*

integrator when the control saturates. The antiwindup scheme of Figure 5.4 *has two special features. First, the PI controller has high-gain k/μ. Second, the gain in the antiwindup loop μ/k is the reciprocal of the controller's high-gain. Figure* 5.5 *shows the block diagram when the relative degree is higher than one. The transfer function H represents the high-gain observer. When $\rho = 2$,*

$$H = k_1 + \frac{s}{(\varepsilon s)^2/\alpha_2 + (\varepsilon s)\alpha_1/\alpha_2 + 1},$$

and the controller takes the form of a PID controller with antiwindup. When $\rho = 3$,

$$H = k_1 + \frac{k_2 s + (1 + \varepsilon k_2 \alpha_2/\alpha_3)s^2}{(\varepsilon s)^3/\alpha_3 + (\varepsilon s)^2\alpha_2/\alpha_3 + (\varepsilon s)\alpha_1/\alpha_3 + 1},$$

and the controller is PID^2 with antiwindup. For $\rho > 3$, the controller is PID$^{\rho-1}$ with antiwindup. ◇

Theorem 5.2. *Suppose Assumptions* 5.1 *to* 5.4 *are satisfied and consider the closed-loop system formed of the system* (5.1), *the conditional integrator* (5.19), *the controller* (5.20), *and the observer* (5.21)–(5.22). *Let Ψ be a compact set in the interior of Ω and suppose $|\sigma(0)| \le \mu$, $(z(0), q(0), s(0)) \in \Psi$, and $\hat{\xi}(0)$ is bounded. Then $\mu^* > 0$ exists, and for each $\mu \in (0, \mu^*]$, $\varepsilon^* = \varepsilon^*(\mu)$ exists such that for each $\mu \in (0, \mu^*]$ and $\varepsilon \in (0, \varepsilon^*(\mu)]$, all state variables are bounded and $\lim_{t\to\infty} e(t) = 0$. Furthermore, let $\chi = (z, \xi)$ be part of the state of the closed-loop system under the output feedback controller, and let $\chi^* = (z^*, \xi^*)$ be the state of the closed-loop system under the state feedback sliding mode controller* (5.16), *with $\chi(0) = \chi^*(0)$. Then for every $\delta_0 > 0$ there is $\mu_1^* > 0$, and for each $\mu \in (0, \mu_1^*]$ there is $\varepsilon_1^* = \varepsilon_1^*(\mu) > 0$, such that for $\mu \in (0, \mu_1^*]$ and $\varepsilon \in (0, \varepsilon_1^*(\mu)]$,*

$$\|\chi(t) - \chi^*(t)\| \le \delta_0 \quad \forall\, t \ge 0. \tag{5.23}$$

◇

Proof: The closed-loop system under output feedback is given by

$$\dot{\sigma} = \gamma[-\sigma + \mu\ \mathrm{sat}(\hat{s}/\mu)], \tag{5.24}$$

$$\dot{z} = \tilde{f}_0(z, \xi, w), \tag{5.25}$$

$$\dot{q} = A_2 q + B_2(s - \sigma), \tag{5.26}$$

$$\dot{s} = b(\eta,\xi,w)\psi(\sigma,\hat{\xi},\mu) + \Delta_1(z,\xi,w) + \gamma[-\sigma + \mu\,\text{sat}(\hat{s}/\mu)], \tag{5.27}$$

$$\varepsilon\dot{\varphi} = A_0\varphi + \varepsilon B[a_0(z,\xi,w) - b(\eta,\xi,w)\phi(w) + b(\eta,\xi,w)\psi(\sigma,\hat{\xi},\mu)], \tag{5.28}$$

where

$$\hat{s} = \sigma + k_1\hat{\xi}_1 + \cdots + k_{\rho-1}\hat{\xi}_{\rho-1} + \hat{\xi}_\rho, \qquad \varphi_i = \frac{\xi_i - \hat{\xi}_i}{\varepsilon^{\rho-i}} \quad \text{for } 1 \le i \le \rho,$$

$\psi(\sigma,\xi,\mu) = -\beta_s(\xi)\,\text{sat}(s/\mu)$, and

$$A_0 = \begin{bmatrix} -\alpha_1 & 1 & 0 & \cdots & 0 \\ -\alpha_2 & 0 & 1 & \cdots & 0 \\ \vdots & & \ddots & & \vdots \\ -\alpha_{\rho-1} & & & 0 & 1 \\ -\alpha_\rho & 0 & \cdots & \cdots & 0 \end{bmatrix}.$$

The matrix A_0 is Hurwitz by design. The proof proceeds in four steps similar to the proof of Theorem 5.1. The first three steps are almost the same and will not be repeated here. For the fourth step, consider the system inside the set $\{|\sigma| \le \mu\} \times \Omega_\mu \times \Sigma_\varepsilon$, where $\Sigma_\varepsilon = \{\varphi^T P_0 \varphi \le \rho_3 \varepsilon^2\}$ and P_0 is the solution of the Lyapunov equation $P_0 A_0 + A_0^T P_0 = -I$. There is an equilibrium point at $(\sigma = \bar{\sigma}, z = 0, q = 0, s = \bar{s}, \varphi = 0)$, where

$$\bar{s} = \bar{\sigma} = \frac{-\mu\phi(w)}{\beta(0)}.$$

Shifting the equilibrium point to the origin by the change of variables $\vartheta = \sigma - \bar{\sigma}$ and $p = s - \bar{s}$, the system takes the singularly perturbed form

$$\begin{aligned}
\dot{\vartheta} &= \gamma\{-\vartheta + p + \mu[\text{sat}(\hat{s}/\mu) - \text{sat}(s/\mu)]\},\\
\dot{z} &= \tilde{f}_0(z, \text{col}(q, p - \vartheta - Lq), w),\\
\dot{q} &= A_2 q + B_2(p - \vartheta),\\
\mu\dot{p} &= -b(\eta,\xi,w)\beta(\xi)p + \mu\Delta_a(\cdot) + \mu\Delta_c(\cdot) + \mu b(\eta,\xi,w)[\psi(\sigma,\hat{\xi},\mu) - \psi(\sigma,\xi,\mu)]\\
&\quad + \gamma\mu^2[\text{sat}(\hat{s}/\mu) - \text{sat}(s/\mu)] + \gamma\mu(p - \vartheta),\\
\varepsilon\dot{\varphi} &= A_0\varphi + (\varepsilon/\mu)B\{-b(\eta,\xi,w)\beta(\xi)p + \mu\Delta_a(\cdot)\\
&\quad + \mu b(\eta,\xi,w)[\psi(\sigma,\hat{\xi},\mu) - \psi(\sigma,\xi,\mu)]\},
\end{aligned}$$

where

$$\Delta_a = a_0(z,\xi,w) + b(\eta,\xi,w)\phi(w)\left[\frac{\beta(\xi) - \beta(0)}{\beta(0)}\right] \quad \text{and} \quad \Delta_c = \sum_{i=1}^{\rho-1} k_i \xi_{i+1}.$$

There are positive constants ℓ_1 to ℓ_{10} such that

$$\begin{aligned}
\mu|\text{sat}(\hat{s}/\mu) - \text{sat}(s/\mu)| &\le \ell_1\|\varphi\|,\\
|\psi(\sigma,\hat{\xi},\mu) - \psi(\sigma,\xi,\mu)| &\le (\ell_2 + \ell_3/\mu)\|\varphi\|,\\
|\Delta_a| &\le \ell_4|\vartheta| + \ell_5\|z\| + \ell_6\|q\| + \ell_7|p|,\\
|\Delta_c| &\le \ell_8|\vartheta| + \ell_9\|q\| + \ell_{10}|p|.
\end{aligned}$$

By Assumption 5.4, $z=0$ is an exponentially stable equilibrium point of $\dot{z}=\tilde{f}_0(z,0,w)$. By the converse Lyapunov theorem [78, Lemma 9.8], there is a Lyapunov function $V_0(z)$, possibly dependent on w, that satisfies the inequalities

$$\bar{c}_1\|z\|^2 \leq V_0(z) \leq \bar{c}_2\|z\|^2, \quad \frac{\partial V_0}{\partial z}\tilde{f}_0(z,0,w) \leq -\bar{c}_3\|z\|^2, \quad \left\|\frac{\partial V_0}{\partial z}\right\| \leq \bar{c}_4\|z\|,$$

in some neighborhood of $z=0$, where $\bar{c}_1$ to $\bar{c}_4$ are positive constants independent of w. Consider the composite Lyapunov function

$$V = V_0 + \kappa_1\, q^T P_2 q + \frac{\kappa_2}{2\gamma}\vartheta^2 + \tfrac{1}{2}p^2 + \varphi^T P_0 \varphi,$$

with positive constants κ_1 and κ_2. It can be shown that $\dot{V} \leq -\mathcal{Y}^T Q \mathcal{Y}$, where $\mathcal{Y} = \text{col}(\|z\|, \|q\|, |\vartheta|, |p|, \|\varphi\|)$,

$$Q = \begin{bmatrix} c_1 & -c_2 & -c_3 & -c_4 & -c_5 \\ -c_2 & \kappa_1 & -\kappa_1 c_6 & -(\kappa_1 c_7 + c_8) & -c_9 \\ -c_3 & -\kappa_1 c_6 & \kappa_2 & -(\kappa_2 c_{10} + c_{11}) & -(\kappa_2 c_{12} + c_{13}) \\ -c_4 & -(\kappa_1 c_7 + c_8) & -(\kappa_2 c_{10} + c_{11}) & (c_{14}/\mu - c_{15}) & -(c_{16} + c_{17}/\mu) \\ -c_5 & -c_9 & -(\kappa_2 c_{12} + c_{13}) & -(c_{16} + c_{17}/\mu) & (1/\varepsilon - c_{18} - c_{19}/\mu) \end{bmatrix},$$

and c_1 to c_{19} are positive constants independent of κ_1, κ_2, μ, and ε. Choose κ_1 large enough to make the 2×2 principal minor of Q positive; then choose κ_2 large enough to make the 3×3 principal minor positive; then choose μ small enough to make the 4×4 principal minor positive; then choose ε small enough to make the determinant of Q positive. Hence, the origin ($\vartheta=0, z=0, q=0, p=0, \varphi=0$) is exponentially stable, and there is a neighborhood N of the origin, independent of μ and ε, such that all trajectories in N converge to the origin as $t\to\infty$. By choosing μ and ε small enough, it can be ensured that for all $(\sigma,z,q,s,\varphi) \in \{|\sigma|\leq\mu\}\times\Omega_\mu\times\Sigma_\varepsilon$, $(\vartheta,z,q,p,\varphi)\in N$. Thus, all trajectories in $\{|\sigma|\leq\mu\}\times\Omega_\mu\times\Sigma_\varepsilon$ converge to the equilibrium point ($\sigma=\bar{\sigma}, z=0, q=0, s=\bar{s}, \varphi=0$). Consequently, all trajectories with $(z(0),\zeta(0),s(0))\in\Psi$, $|\sigma(0)|\leq\mu$, and bounded $\hat{\xi}(0)$ converge to this equilibrium point because such trajectories enter $\{|\sigma|\leq\mu\}\times\Omega_\mu\times\Sigma_\varepsilon$ in finite time. This completes the proof that all the state variables are bounded and $\lim_{t\to\infty} e(t)=0$. The proof of (5.23) is done in two steps. In the first step, it is shown that the trajectories under the continuously implemented state feedback controller with conditional integrator (5.17)–(5.18) are $O(\mu)$ close to the trajectories under the state feedback sliding mode controller (5.16). In the second step, with fixed μ, it is shown that the trajectories under the output feedback controller (5.19)–(5.22) can be made arbitrarily close to the trajectories under the state feedback controller (5.17)–(5.18) by choosing ε small enough. The argument for the second step is similar to the argument used in the performance recovery part of the proof of Theorem 3.1 and will not be repeated here. For the first step, let $\chi^\dagger=(z^\dagger,\xi^\dagger)$ be part of the state of the closed-loop system under the controller (5.17)–(5.18), with $\chi^\dagger(0)=\chi^*(0)$. For the controller (5.16), $|s_1|$ is monotonically decreasing and reaches zero in finite time t_1. For the controller (5.17)–(5.18), $|s|=|\sigma+s_1|$ is monotonically decreasing and reaches the set $\{|s|\leq\mu\}$ in finite time t_2. Let $t_3=\min\{t_1,t_2\}$. If $t_3>0$, using $\text{sat}(s^\dagger(t)/\mu)=\text{sgn}(s^\dagger(t))=\text{sgn}(s_1^*(t))$ for $t\in[0,t_3]$, it can be shown that $\chi^\dagger(t)=\chi^*(t)$ for $t\in[0,t_3]$. Next, consider $\chi^\dagger(t)$ and $\chi^*(t)$ for $t\geq t_3$. Since $\chi^\dagger(t_3)=\chi^*(t_3)$, it must be true that $t_3=t_2\leq t_1$ because if $t_1<t_2$, it would be true that $s_1^*(t_1)=0$ and $|s^\dagger(t_1)|=|\sigma^\dagger(t_1)|\leq\mu$, which contradicts the claim that $t_1<t_2$. At

$t_3 = t_2$, $s^\dagger(t_3) = \mu$ and $|s_1^*(t_3)| = |s_1^\dagger(t_3)| \le |s^\dagger(t_3)| + |\sigma^\dagger(t_3)| \le 2\mu$. Since both $|s^\dagger|$ and $|s_1^*|$ are monotonically decreasing, $|s^\dagger(t) - s_1^*(t)| \le 3\mu$ for all $t \ge t_3$, which implies that $|s_1^\dagger(t) - s_1^*(t)| \le 4\mu$. If $t_3 = 0$, either $t_1 = 0$ or $t_2 = 0$. If $t_1 = 0$, $s_1^*(0) = 0$. This implies that $t_2 = 0$ because

$$s_1^*(0) = 0 \Rightarrow s_1^\dagger(0) = 0 \Rightarrow |s^\dagger(0)| = |\sigma(0)| \le \mu.$$

Then, for all $t \ge 0$, $s_1^*(t) \equiv 0$ and $|s^\dagger(t)| \le \mu$, which implies that $|s_1^*(t) - s_1^\dagger(t)| \le 2\mu$. If $t_2 = 0$,

$$|s^\dagger(0)| \le \mu \Rightarrow |s_1^\dagger(0)| \le 2\mu \Rightarrow |s_1^*(0)| \le 2\mu.$$

Hence, for all $t \ge 0$,

$$|s_1^*(t)| \le 2\mu \text{ and } |s_1^\dagger(t)| \le 2\mu \Rightarrow |s_1^*(t) - s_1^\dagger(t)| \le 4\mu.$$

Thus, in all cases, $s_1^\dagger(t) - s_1^*(t) = O(\mu)$ for all $t \ge 0$. The variable $q = \text{col}(\xi_1, \ldots, \xi_{\rho-1})$ satisfies the equations

$$\dot{q}^* = A_2 q^* + B_2 s_1^* \quad \text{and} \quad \dot{q}^\dagger = A_2 q^\dagger + B_2 s_1^\dagger$$

under the two controllers. Since A_2 is Hurwitz, continuity of the solution on the infinite time interval shows that $q^*(t) - q^\dagger(t) = O(\mu)$ [78, Theorem 9.1]. Hence, $\xi^*(t) - \xi^\dagger(t) = O(\mu)$ for all $t \ge 0$, which can be used to show that $z^*(t) - z^\dagger(t) = O(\mu)$ for all $t \ge 0$. □

Remark 5.8. *The proof delineates a procedure to tune the parameters of the controller* (5.19)–(5.22) *to shape the transient response. Starting with the sliding mode controller* (5.16), *the function β and the parameters k_1 to $k_{\rho-1}$ are chosen to shape the transient response. Then the state feedback controller with conditional integrator* (5.17)–(5.18) *is considered, and μ is reduced gradually until the transient response is close enough to the desired one. Finally, the output feedback controller* (5.19)–(5.22) *is considered, and ε is reduced gradually to bring the transient response towards the desired one.* ◇

Remark 5.9. *If Assumptions* 5.1 *to* 5.3 *hold globally, that is, $D_x = T(D_x) = R^n$ and γ_1 is class $\mathcal{K}_\infty$, then the sets Γ and Ω can be chosen arbitrarily large. For any bounded $(z(0), q(0), s(0))$, the conclusion of Theorem* 5.2 *will hold by choosing β large enough.* ◇

Example 5.2. Reconsider the pendulum regulation problem from Example 5.1. A state feedback sliding mode controller is taken as

$$u = -k \,\text{sgn}(\xi_1 + \xi_2).$$

A continuously implemented state feedback sliding mode controller with conditional integrator is taken as

$$\dot{\sigma} = -\sigma + \mu \,\text{sat}\left(\frac{\sigma + \xi_1 + \xi_2}{\mu}\right), \quad u = -k \,\text{sat}\left(\frac{\sigma + \xi_1 + \xi_2}{\mu}\right),$$

and its output feedback version is

$$\dot{\sigma} = -\sigma + \mu \,\text{sat}\left(\frac{\sigma + \hat{\xi}_1 + \hat{\xi}_2}{\mu}\right), \quad u = -k \,\text{sat}\left(\frac{\sigma + \hat{\xi}_1 + \hat{\xi}_2}{\mu}\right),$$

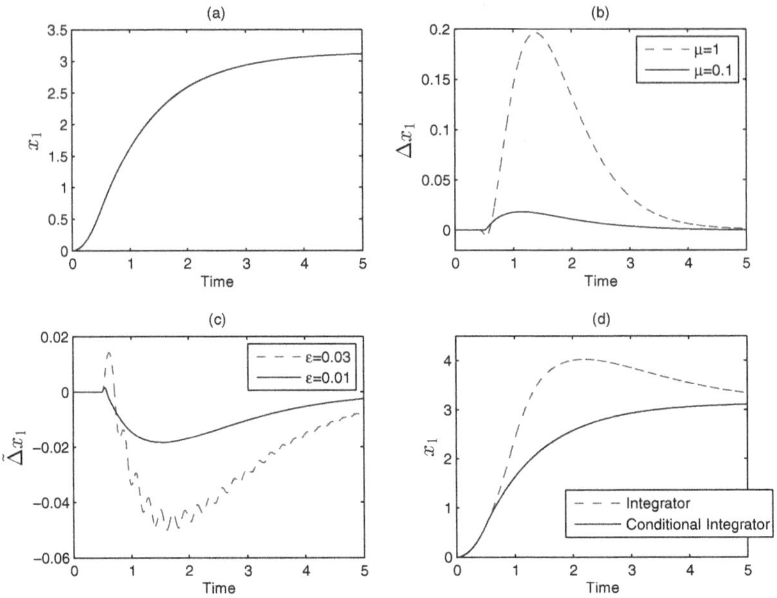

Figure 5.6. *Simulation of Example* 5.2. (a) *is the response of the sliding mode controller.* (b) *shows the difference between the responses of the sliding mode controller and the state feedback controller with conditional integrator.* (c) *shows the difference between the state and output feedback responses.* (d) *compares the responses of the traditional and conditional integrators.*

$$\dot{\hat{\xi}}_1 = \hat{\xi}_2 + \frac{2}{\varepsilon}(\xi_1 - \hat{\xi}_1), \qquad \dot{\hat{\xi}}_2 = \frac{1}{\varepsilon^2}(\xi_1 - \hat{\xi}_1).$$

Simulation results are shown in Figure 5.4 using zero initial conditions and the same parameters as in Example 5.1, namely, $r = \pi$, $k = 5$, $a = 0.03$, $b = 1$, and $d = 0.3$. Figure 5.6(a) shows the output response of the state feedback sliding mode controller. Figure 5.6(b) demonstrates how the response of the continuously implemented sliding mode controller with conditional integrator approaches the response of the sliding mode controller as μ decreases; Δx_1 is the difference between the two responses. Figure 5.6(c) demonstrates how the response of the output feedback controller with conditional integrator approaches its state feedback counterpart as ε decreases with $\mu = 0.1$; $\tilde{\Delta} x_1$ is the difference between the two responses. Finally, Figure 5.6(d) compares the response of the output feedback controller with conditional controller with the one designed in Example 5.1 using the traditional integrator; in both cases $\mu = 0.1$ and $\varepsilon = 0.01$. The advantage of the conditional integrator over the traditional one is clear. It avoids the degradation of the transient response associated with the traditional integrator because it recovers the response of the sliding mode controller. $\triangle$

5.4 ▪ Conditional Servocompensator

Consider a single-input–single-output nonlinear system modeled by

$$\dot{x} = f(x,\theta) + g(x,\theta)u + \delta(x,\theta,d), \tag{5.29}$$

$$y = h(x,\theta) + \gamma(\theta,d), \tag{5.30}$$

where $x \in R^n$ is the state, $u \in R$ is the control input, $y \in R$ is the measured output, $\theta \in \Theta$ is a vector of (possibly uncertain) system parameters, and $d \in R^l$ is a bounded time-varying disturbance. The functions f, g, h, and δ are sufficiently smooth in x in a domain $D_x \subset R^n$. The functions δ and γ vanish at $d = 0$, i.e., $\delta(x,\theta,0) = 0$ and $\gamma(\theta,0) = 0$ for all $x \in D_x$ and $\theta \in \Theta$. The output y is to be regulated to a bounded time-varying reference signal r:

$$\lim_{t\to\infty} |y(t) - r(t)| = 0.$$

Assumption 5.5. *For each $\theta \in \Theta$, the disturbance-free system,* (5.29)–(5.30) *with $d = 0$, has relative degree $\rho \le n$ in D_x, and there is a diffeomorphism*

$$\begin{bmatrix} \eta \\ \zeta \end{bmatrix} = T(x) \tag{5.31}$$

in D_x, possibly dependent on θ, that transforms (5.29)–(5.30), *with $d = 0$, into the normal form*[42]

$$\begin{aligned}
\dot{\eta} &= f_0(\eta,\zeta,\theta),\\
\dot{\zeta}_i &= \zeta_{i+1}, \quad 1 \le i \le \rho - 1,\\
\dot{\zeta}_\rho &= a(\eta,\zeta,\theta) + b(\eta,\zeta,\theta)u,\\
y &= \zeta_1.
\end{aligned}$$

Moreover, $b(\eta,\zeta,\theta) \ge b_0 > 0$ for all $(\eta,\zeta) \in T(D_x)$ and $\theta \in \Theta$.

Assumption 5.6. *The change of variables* (5.31) *transforms the disturbance-driven system* (5.29)–(5.30) *into the form*

$$\begin{aligned}
\dot{\eta} &= f_a(\eta,\zeta_1,\dots,\zeta_m,\theta,d),\\
\dot{\zeta}_i &= \zeta_{i+1} + \psi_i(\zeta_1,\dots,\zeta_i,\theta,d), \quad 1 \le i \le m-1,\\
\dot{\zeta}_i &= \zeta_{i+1} + \psi_i(\eta,\zeta_1,\dots,\zeta_i,\theta,d), \quad m \le i \le \rho-1,\\
\dot{\zeta}_\rho &= a(\eta,\zeta,\theta) + b(\eta,\zeta,\theta)u + \psi_\rho(\eta,\zeta,\theta,d),\\
y &= \zeta_1 + \gamma(\theta,d),
\end{aligned}$$

where $1 \le m \le \rho - 1$. The functions ψ_i vanish at $d = 0$.

For $m = 1$, the $\dot{\zeta}_i$-equations for $1 \le i \le m-1$ are dropped. In the absence of η, Assumption 5.6 is satisfied locally if the system (5.29)–(5.30) is observable uniformly in θ and d [49].

Assumption 5.7. *The disturbance and reference signals $d(t)$ and $r(t)$ are generated by the exosystem*

$$\dot{w} = S_0 w, \qquad \begin{bmatrix} d \\ r \end{bmatrix} = H_0 w, \tag{5.32}$$

where S_0 has distinct eigenvalues on the imaginary axis and $w(t)$ belongs to a compact set W.

This assumption says that $d(t)$ and $r(t)$ are linear combinations of constant and sinusoidal signals.

[42]For $\rho = n$, η and the $\dot{\eta}$-equation are dropped.

Define $\tau_1(\theta,w)$ to $\tau_m(\theta,w)$ by

$$\tau_1 = r-\gamma(\theta,d),$$
$$\tau_{i+1} = \frac{\partial \tau_i}{\partial w}S_0 w - \psi_i(\tau_1,\dots,\tau_i,\theta,d), \quad 1\le i\le m-1.$$

Assumption 5.8. *There exists a unique mapping $\tau_0(\theta,w)$ that solves the partial differential equation*

$$\frac{\partial \tau_0}{\partial w}S_0 w = f_a(\tau_0,\tau_1,\dots,\tau_m,\theta,d) \tag{5.33}$$

for all $\theta\in\Theta$ and $w\in W$.

Remark 5.10. *In the special case when the $\dot\eta$-equation takes the form*

$$\dot\eta = A\eta + f_b(\zeta_1,\dots,\zeta_m,\theta,d)$$

with a Hurwitz matrix A, (5.33) *is a linear partial differential of the form*

$$\frac{\partial \tau_0}{\partial w}S_0 w = A\tau_0 + f_c(\theta,w),$$

and its unique solution is given by

$$\tau_0(\theta,w) = \int_{-\infty}^{0} e^{-At} f_c(\theta, e^{S_0 t}w)\,dt. \qquad \diamond$$

Using $\tau_0(\theta,w)$, define $\tau_{m+1}(\theta,w)$ to $\tau_\rho(\theta,w)$ by

$$\tau_{i+1} = \frac{\partial \tau_i}{\partial w}S_0 w - \psi_i(\tau_0,\tau_1,\dots,\tau_i,\theta,d), \quad m\le i\le \rho-1.$$

The steady-state zero-error manifold is given by $\{\eta=\tau_0(\theta,w),\ \zeta=\tau(\theta,w)\}$ because η and ζ satisfy the equations of Assumption 5.6 and $\zeta_1=\tau_1=r-\gamma(\theta,d)$ implies that $y=r$. The steady-state value of the control input u on this manifold is given by

$$\phi(\theta,w) = \frac{1}{b(\tau_0,\tau,\theta)}\left[(\partial\tau_\rho/\partial w)S_0 w - a(\tau_0,\tau,\theta) - \psi_\rho(\tau_0,\tau,\theta,d)\right]. \tag{5.34}$$

Assumption 5.9. *There are known real numbers $c_0,\dots,c_{q-1}$ such that the polynomial*

$$p^q + c_{q-1}p^{q-1} + \cdots + c_1 p + c_0$$

has distinct roots on the imaginary axis and $\phi(\theta,w)$ satisfies the differential equation

$$\phi^{(q)} + c_{q-1}\phi^{(q-1)} + \cdots + c_1\phi^{(1)} + c_0\phi = 0. \tag{5.35}$$

Therefore, $\phi(\theta,w)$ is generated by the linear internal model

$$\dot\Phi = S\Phi, \qquad \phi = H\Phi, \tag{5.36}$$

where

$$S = \begin{bmatrix} 0 & 1 & \cdots & \cdots & 0 \\ 0 & 0 & 1 & \cdots & 0 \\ \vdots & & & & \vdots \\ 0 & \cdots & \cdots & 0 & 1 \\ -c_0 & \cdots & \cdots & \cdots & -c_{q-1} \end{bmatrix}, \quad \Phi = \begin{bmatrix} \phi \\ \phi^{(1)} \\ \vdots \\ \phi^{(q-2)} \\ \phi^{(q-1)} \end{bmatrix}, \quad \textit{and} \quad H^T = \begin{bmatrix} 1 \\ 0 \\ \vdots \\ \vdots \\ 0 \end{bmatrix}.$$

This assumption reflects the fact that for nonlinear systems, the internal model must reproduce not only the sinusoidal signals generated by the exosystem but also higher-order harmonics induced by the nonlinearities. Because the model has finite dimension, there can be only a finite number of harmonics. The assumption is satisfied when the system has polynomial nonlinearities.

Example 5.3. Consider the system

$$\dot{x}_1 = x_2, \quad \dot{x}_2 = \theta_1(x_1 - x_1^3) + \theta_2 u, \quad y = x_1,$$

where θ_1 and θ_2 are unknown. It is required to regulate y to $r(t) = \alpha \sin(\omega_0 t + \theta_0)$, where α and θ_0 are unknown but ω_0 is known. The signal r is generated by the exosystem

$$\dot{w} = \begin{bmatrix} 0 & \omega_0 \\ -\omega_0 & 0 \end{bmatrix} w, \quad w(0) = \begin{bmatrix} \alpha \sin \theta_0 \\ \alpha \cos \theta_0 \end{bmatrix}, \quad r = w_1.$$

The steady-state control $\phi(\theta, w) = [-(\theta_1 + \omega_0^2) w_1 + \theta_1 w_1^3]/\theta_2$ satisfies the differential equation

$$\phi^{(4)} + 10\,\omega_0^2 \phi^{(2)} + 9\,\omega_0^4 \phi = 0.$$

The eigenvalues of

$$S = \begin{bmatrix} 0 & 1 & 0 & 0 \\ 0 & 0 & 1 & 0 \\ 0 & 0 & 0 & 1 \\ -9\,\omega_0^4 & 0 & -10\,\omega_0^2 & 0 \end{bmatrix}$$

are $\pm j\omega_0$, $\pm 3j\omega_0$. The internal model generates the first and third harmonics. △

The change of variables

$$z = \eta - \tau_0(\theta, w), \qquad \xi_i = y^{(i-1)} - r^{(i-1)} \text{ for } 1 \le i \le \rho, \tag{5.37}$$

transforms the system (5.29)–(5.30) into the form

$$\dot{z} = \tilde{f}_0(z, \xi, \theta, w), \tag{5.38}$$

$$\dot{\xi}_i = \xi_{i+1}, \quad 1 \le i \le \rho - 1, \tag{5.39}$$

$$\dot{\xi}_\rho = a_0(z, \xi, \theta, w) + b(\eta, \zeta, \theta)[u - \phi(\theta, w)], \tag{5.40}$$

$$e = \xi_1, \tag{5.41}$$

where $e = y - r$ is the measured regulation error, $\tilde{f}_0(0, 0, \theta, w) = 0$, and $a_0(0, 0, \theta, w) = 0$. In these new variables the zero-error manifold is $\{z = 0,\ \xi = 0\}$. Let $\Gamma \subset R^n$ be a compact set, which contains the origin in its interior, such that $(z,\xi)\in\Gamma$ implies that $(\eta,\zeta)\in T(D_x)$ for $\theta\in\Theta$ and $w\in W$. The existence of Γ may require restricting the sizes of Θ and W.

Assumption 5.10. *There is a Lyapunov function $V_1(z,\theta,w)$ for the system $\dot{z}=\tilde{f}_0(z,\xi,\theta,w)$ that satisfies the inequalities*

$$\gamma_1(\|z\|)\le V_1(z,\theta,w)\le\gamma_2(\|z\|),$$

$$\frac{\partial V_1}{\partial z}\tilde{f}_0(z,\xi,\theta,w)+\frac{\partial V_1}{\partial w}S_0 w\le-\gamma_3(\|z\|)\qquad\forall\ \|z\|\ge\gamma_4(\|\xi\|)$$

for all $(z,\xi,\theta,w)\in\Gamma\times\Theta\times W$, where γ_1 to γ_4 are class $\mathcal{K}$ functions independent of θ and w.

Assumption 5.11. *In some neighborhood of $z=0$, there is a Lyapunov function $V_0(z,\theta,w)$ that satisfies the inequalities*

$$\bar{c}_1\|z\|^2\le V_0(z,\theta,w)\le\bar{c}_2\|z\|^2,$$

$$\frac{\partial V_0}{\partial z}\tilde{f}_0(z,0,\theta,w)+\frac{\partial V_0}{\partial w}S_0 w\le-\bar{c}_3\|z\|^2,$$

$$\left\|\frac{\partial V_0}{\partial z}\right\|\le\bar{c}_4\|z\|$$

for some positive constants $\bar{c}_1$ to $\bar{c}_4$, independent of θ and w.

As in the conditional integrator of the previous section, the state feedback control design starts by considering the sliding mode controller

$$u=-\beta(\xi)\,\mathrm{sgn}(s_1),\tag{5.42}$$

where

$$s_1=k_1\xi_1+\cdots+k_{\rho-1}\xi_{\rho-1}+\xi_\rho$$

and k_1 to $k_{\rho-1}$ are chosen such that the polynomial

$$\lambda^{\rho-1}+k_{\rho-1}\lambda^{\rho-2}+\cdots+k_2\lambda+k_1$$

is Hurwitz. From the equation

$$\begin{aligned}\dot{s}_1&=\sum_{i=1}^{\rho-1}k_i\xi_{i+1}+a_0(z,\xi,\theta,w)+b(\eta,\zeta,\theta)[u-\phi(\theta,w)]\\&\stackrel{\text{def}}{=}\Delta_1(z,\xi,\theta,w)+b(\eta,\zeta,\theta)u\end{aligned}$$

it can be seen that the condition $s\dot{s}<-\beta_0|s|$ is ensured when the locally Lipschitz function $\beta(\xi)$ satisfies $\beta(\xi)\ge\varrho_1(\xi)+\beta_0$ with $\beta_0>0$, where $\varrho_1(\xi)$ is an upper bound on $|\Delta_1(z,\xi,\theta,w)/b(\eta,\zeta,\theta)|$ for all $(z,\xi,\theta,w)\in\Gamma\times\Theta\times W$. A continuously implemented sliding mode controller with a conditional servocompensator is taken as

$$u=-\beta(\xi)\,\mathrm{sat}\left(\frac{s}{\mu}\right),\tag{5.43}$$

$$\dot{\sigma} = F\sigma + \mu G \,\text{sat}\left(\frac{s}{\mu}\right), \tag{5.44}$$

where

$$s = \Lambda^T \sigma + s_1 = \Lambda^T \sigma + k_1 \xi_1 + \cdots + k_{\rho-1}\xi_{\rho-1} + \xi_\rho, \tag{5.45}$$

F is Hurwitz, the pair (F, G) is controllable, and Λ is the unique vector that assigns the eigenvalues of $(F + G\Lambda^T)$ at the eigenvalues of S.[43] The closed-loop system is given by

$$\begin{aligned}
\dot{w} &= S_0 w, \\
\dot{\sigma} &= F\sigma + \mu G \,\text{sat}\left(\frac{s}{\mu}\right), \\
\dot{z} &= \tilde{f}_0(z, \xi, \theta, w), \\
\dot{q} &= A_2 q + B_2(s - \Lambda^T \sigma), \\
\dot{s} &= -b(\eta, \zeta, \theta)\beta(\xi)\,\text{sat}\left(\frac{s}{\mu}\right) + \Delta_1(\cdot) + \Lambda^T\left[F\sigma + \mu G\,\text{sat}\left(\frac{s}{\mu}\right)\right],
\end{aligned}$$

where $q = \text{col}(\xi_1, \ldots, \xi_{\rho-1})$, $\xi = \text{col}(q, s - \Lambda^T\sigma - Lq)$, $L = \begin{bmatrix} k_1 & \cdots & k_{\rho-1}\end{bmatrix}$,

$$A_2 = \begin{bmatrix} 0 & 1 & \cdots & \cdots & 0 \\ 0 & 0 & 1 & \cdots & 0 \\ \vdots & & & & \vdots \\ 0 & \cdots & \cdots & 0 & 1 \\ -k_1 & \cdots & \cdots & \cdots & -k_{\rho-1} \end{bmatrix}, \quad \text{and} \quad B_2 = \begin{bmatrix} 0 \\ 0 \\ \vdots \\ 0 \\ 1 \end{bmatrix}.$$

The matrices F and A_2 are Hurwitz by design. Let $V_\sigma = \sigma^T P \sigma$, where P is the solution of the Lyapunov equation $PF + F^T P = -I$. From the inequality

$$\dot{V}_\sigma = -\sigma^T\sigma + 2\mu\sigma^T PG\,\text{sat}\left(\frac{s}{\mu}\right) \le -\|\sigma\|^2 + 2\mu\|PG\|\,\|\sigma\|,$$

it follows that the set $\Xi = \{V_\sigma \le \rho_0 \mu^2\}$ with $\rho_0 = 4\|PG\|^2\lambda_{\max}(P)$ is positively invariant because $\dot{V}_\sigma \le 0$ on the boundary $V_\sigma = \rho_0\mu^2$. Therefore, $\sigma(0) \in \Xi$ implies that $\sigma(t) = O(\mu)$ for all $t \ge 0$. For $|s| \ge \mu$,

$$\begin{aligned}
s\dot{s} &\le -b\beta|s| + |\Delta_1||s| + |s|\,\|\Lambda\|(\mu\|F\|\sqrt{\rho_0/\lambda_{\min}(P)} + \mu\|G\|) \\
&= b\left(-\beta + \frac{|\Delta_1|}{b} + \frac{k\mu}{b_0}\right)|s| \le b\left(-\beta_0 + \frac{k\mu}{b}\right)|s|,
\end{aligned}$$

where $k = \|\Lambda\|(\|F\|\sqrt{\rho_0/\lambda_{\min}(P)} + \|G\|)$. For $\mu \le \frac{1}{2}b_0\beta_0/k$,

$$s\dot{s} \le -\tfrac{1}{2}b\beta_0|s| \le -\tfrac{1}{2}b_0\beta_0|s|,$$

which shows that the set $\{|s| \le c\}$ is positively invariant for $c > \mu$, and s reaches the positively invariant set $\{|s| \le \mu\}$ in finite time. Let $V_2(q) = q^T P_2 q$, where P_2 is the solution of the Lyapunov equation $P_2 A_2 + A_2^T P_2 = -I$. For $|s| \le c$,

$$\dot{V}_2 \le -q^T q + 2\|q\|\,\|P_2 B_2\|\,(|s| + |\Lambda^T\sigma|) \le -\|q\|^2 + 2\|P_2B_2\|\,\|q\|(c + \ell\mu),$$

[43] Λ is unique because G has one column. See [9, Lemma 9.10].

where $\ell\mu = \max_{\sigma\in\Xi} \|\Lambda^T\sigma\|$. For $\mu \le c/\ell$,

$$\dot{V}_2 \le -\|q\|^2 + 4\|P_2B_2\|\,\|q\|c,$$

which shows that the set $\{V_2 \le \bar{\rho}_1 c^2\} \times \{|s| \le c\}$ is positively invariant for $\bar{\rho}_1 > 16\|P_2B_2\|^2\lambda_{\max}(P_2)$ because $\dot{V}_2 < 0$ on the boundary $V_2 = \bar{\rho}_1 c^2$. Inside this set, $\|\xi\| = c(1+\|L\|)\sqrt{\bar{\rho}_1/\lambda_{\min}(P_2)} + 2c \stackrel{\text{def}}{=} \bar{\rho}_2 c$. The derivative of the Lyapunov function V_1 of Assumption 5.10 satisfies the inequality

$$\dot{V}_1 \le -\gamma_3(\|z\|) \quad \forall\ \|z\| \ge \gamma_4(c\bar{\rho}_2),$$

which shows that the set

$$\Omega = \{V_1(z) \le c_0\} \times \{q^T P_2 q \le \bar{\rho}_1 c^2\} \times \{|s| \le c\},$$

with $c_0 \ge \gamma_2(\gamma_4(c\bar{\rho}_2))$, is positively invariant because $\dot{V}_1 < 0$ on the boundary $V_1 = c_0$. Similarly, it can be shown that the set

$$\Omega_\mu = \{V_1(z) \le \gamma_2(\gamma_4(\mu\bar{\rho}_2))\} \times \{q^T P_2 q \le \bar{\rho}_1 \mu^2\} \times \{|s| \le \mu\}$$

is positively invariant and every trajectory starting in $\Xi \times \Omega$ enters $\Xi \times \Omega_\mu$ in finite time. The constants c and c_0 are chosen such that $(z,\xi) \in \Gamma$ for $(z,q,s) \in \Omega$.

Inside the set $\{|s| \le \mu\}$, $\mu\ \text{sat}(s/\mu) = s$, and the conditional servocompensator reduces to

$$\dot{\sigma} = (F + G\Lambda^T)\sigma + G\left(\sum_{i=1}^{\rho-1} k_i \xi_i + \xi_\rho\right).$$

The equation

$$MS - FM = GH$$

has a unique solution M, which is nonsingular.[44] Then

$$MSM^{-1} = F + GHM^{-1}.$$

Hence, the eigenvalues of $(F + GHM^{-1})$ are located at the eigenvalues of S. Since Λ is the unique vector that assigns the eigenvalues of $(F + G\Lambda^T)$ at the eigenvalues of S, it follows that $\Lambda^T = HM^{-1}$. Setting

$$\bar{\sigma} = -\frac{\mu M\Phi}{\beta(0)}$$

it can be seen that

$$\dot{\bar{\sigma}} = (F + G\Lambda^T)\bar{\sigma} \tag{5.46}$$

and

$$\bar{s} \stackrel{\text{def}}{=} \Lambda^T\bar{\sigma} = -\frac{\mu\Lambda^T M\Phi}{\beta(0)} = -\frac{\mu H\Phi}{\beta(0)} = -\frac{\mu\phi(\theta,w)}{\beta(0)}. \tag{5.47}$$

Therefore, $\{\sigma = \bar{\sigma}, z = 0, \xi = 0\}$ is an invariant manifold where $s = \bar{s}$ and $e = 0$. Assumption 5.11 shows that the manifold is exponentially stable.

[44] See [112]. The existence and uniqueness of the solution follows from the fact that F and S have no common eigenvalues. M is nonsingular because the pair (F,G) is controllable and the pair (S,H) is observable.

The output feedback controller is given by

$$\dot{\sigma} = F\sigma + \mu G \,\text{sat}\left(\frac{\Lambda^T\sigma + k_1\hat{\xi}_1 + \cdots + k_{\rho-1}\hat{\xi}_{\rho-1} + \hat{\xi}_\rho}{\mu}\right), \tag{5.48}$$

$$u = -\beta_s(\hat{\xi})\,\text{sat}\left(\frac{\Lambda^T\sigma + k_1\hat{\xi}_1 + \cdots + k_{\rho-1}\hat{\xi}_{\rho-1} + \hat{\xi}_\rho}{\mu}\right), \tag{5.49}$$

where β_s is obtained by saturating β outside Ω, and $\hat{\xi}$ is provided by the high-gain observer

$$\dot{\hat{\xi}}_i = \hat{\xi}_{i+1} + \frac{\alpha_i}{\varepsilon^i}(e - \hat{\xi}_1), \quad 1 \leq i \leq \rho - 1, \tag{5.50}$$

$$\dot{\hat{\xi}}_\rho = \frac{\alpha_\rho}{\varepsilon^\rho}(e - \hat{\xi}_1), \tag{5.51}$$

in which ε is a sufficiently small positive constant, and α_1 to α_ρ are chosen such that the polynomial

$$\lambda^\rho + \alpha_1\lambda^{\rho-1} + \cdots + \alpha_{\rho-1}\lambda + \alpha_\rho$$

is Hurwitz. In (5.48) and (5.49), $\hat{\xi}_1$ can be replaced with the measured error ξ_1

Remark 5.11. *The precise information that is used in the design of the controller is the relative degree ρ, the fact that b is positive, and the eigenvalues of the internal model* (5.36). ◇

Theorem 5.3. *Suppose Assumptions* 5.5 *to* 5.11 *are satisfied and consider the closed-loop system formed of the system* (5.29)–(5.30)*, the conditional servocompensator* (5.48)*, the controller* (5.49)*, and the observer* (5.50)–(5.51)*. Let Ψ be a compact set in the interior of Ω and suppose $\sigma(0) \in \Xi$, $(z(0), q(0), s(0)) \in \Psi$, and $\hat{\xi}(0)$ is bounded. Then $\mu^* > 0$ exists, and for each $\mu \in (0, \mu^*]$, $\varepsilon^* = \varepsilon^*(\mu)$ exists such that for each $\mu \in (0, \mu^*]$ and $\varepsilon \in (0, \varepsilon^*(\mu)]$ all state variables are bounded and $\lim_{t\to\infty} e(t) = 0$. Furthermore, let $\chi = (z, \xi)$ be part of the state of the closed-loop system under the output feedback controller, and let $\chi^* = (z^*, \xi^*)$ be the state of the closed-loop system under the state feedback sliding mode controller* (5.42)*, with $\chi(0) = \chi^*(0)$. Then, for every $\delta_0 > 0$, there is $\mu_1^* > 0$, and for each $\mu \in (0, \mu_1^*]$, there is $\varepsilon_1^* = \varepsilon_1^*(\mu) > 0$ such that for $\mu \in (0, \mu_1^*]$ and $\varepsilon \in (0, \varepsilon_1^*(\mu)]$,*

$$\|\chi(t) - \chi^*(t)\| \leq \delta_0 \quad \forall\, t \geq 0. \tag{5.52}$$

◇

Proof: The closed-loop system under output feedback is given by

$$\dot{w} = S_0 w, \tag{5.53}$$

$$\dot{\sigma} = F\sigma + \mu G\,\text{sat}\left(\frac{\hat{s}}{\mu}\right), \tag{5.54}$$

$$\dot{z} = \tilde{f}_0(z, \xi, \theta, w), \tag{5.55}$$

$$\dot{q} = A_2 q + B_2(s - \Lambda^T\sigma), \tag{5.56}$$

$$\dot{s} = b(\eta,\zeta,\theta)\psi(\sigma,\hat{\xi},\mu) + \Delta_1(z,\xi,\theta,w) + \Lambda^T\left[F\sigma + \mu G \,\mathrm{sat}\left(\frac{\hat{s}}{\mu}\right)\right], \tag{5.57}$$

$$\begin{aligned}\varepsilon\dot{\varphi} &= A_0\varphi + \varepsilon B[a_0(z,\xi,\theta,w) - b(\eta,\zeta,\theta)\phi(\theta,w)\\ &\quad + b(\eta,\zeta,\theta)\psi(\sigma,\hat{\xi},\mu)],\end{aligned} \tag{5.58}$$

where

$$\hat{s} = \Lambda^T\sigma + k_1\hat{\xi}_1 + \cdots + k_{\rho-1}\hat{\xi}_{\rho-1} + \hat{\xi}_\rho, \qquad \varphi_i = \frac{\xi_i - \hat{\xi}_i}{\varepsilon^{\rho-i}} \quad \text{for } 1 \le i \le \rho,$$

$\psi(\sigma,\xi,\mu) = -\beta_s(\xi)\,\mathrm{sat}(s/\mu)$, and

$$A_0 = \begin{bmatrix} -\alpha_1 & 1 & 0 & \cdots & 0\\ -\alpha_2 & 0 & 1 & \cdots & 0\\ \vdots & & \ddots & & \vdots\\ -\alpha_{\rho-1} & & & 0 & 1\\ -\alpha_\rho & 0 & \cdots & \cdots & 0\end{bmatrix}.$$

The matrix A_0 is Hurwitz by design. Equations (5.53) to (5.57) with $\hat{s}$ and $\psi(\sigma,\hat{\xi},\mu)$ replaced by s and $\psi(\sigma,\xi,\mu)$, respectively, are the closed-loop system under state feedback. Let P_0 be the solution of the Lyapunov equation $P_0A_0 + A_0^TP_0 = -I$, $V_3(\varphi) = \varphi^TP_0\varphi$, and $\Sigma_\varepsilon = \{V_3(\varphi) \le \rho_3\varepsilon^2\}$, where the positive constant ρ_3 is to be determined. The proof proceeds in four steps:

Step 1: Show that there exist $\rho_3 > 0$, $\mu_1^* > 0$ and $\varepsilon_1^* = \varepsilon_1^*(\mu) > 0$ such that for each $\mu \in (0,\mu_1^*]$ and $\varepsilon \in (0,\varepsilon_1^*(\mu)]$ the set $\Xi \times \Omega \times \Sigma_\varepsilon$ is positively invariant.

Step 2: Show that for $\sigma(0) \in \Xi$, $(z(0),q(0),s(0)) \in \Psi$, and any bounded $\hat{\xi}(0)$, there exists $\varepsilon_2^* > 0$ such that for each $\varepsilon \in (0,\varepsilon_2^*]$ the trajectory enters the set $\Xi\times\Omega\times\Sigma_\varepsilon$ in finite time $T_1(\varepsilon)$, where $\lim_{\varepsilon\to 0} T_1(\varepsilon) = 0$.

Step 3: Show that there exists $\varepsilon_3^* = \varepsilon_3^*(\mu) > 0$ such that for each $\varepsilon \in (0,\varepsilon_3^*(\mu)]$ every trajectory in $\Xi\times\Omega\times\Sigma_\varepsilon$ enters $\Xi\times\Omega_\mu\times\Sigma_\varepsilon$ in finite time and stays therein for all future time.

Step 4: Show that there exists $\mu_2^* > 0$ and $\varepsilon_4^* = \varepsilon_4^*(\mu) > 0$ such that for each $\mu \in (0,\mu_2^*]$ and $\varepsilon \in (0,\varepsilon_4^*(\mu)]$ every trajectory in $\Xi\times\Omega_\mu\times\Sigma_\varepsilon$ approaches the invariant manifold $\{\sigma = \bar{\sigma}, z = 0, \xi = 0, \varphi = 0\}$ as $t \to \infty$, where $e = 0$ on the manifold.

For the first step, calculate the derivative of V_3 on the boundary $V_3 = \rho_3\varepsilon^2$:

$$\varepsilon\dot{V}_3 = -\varphi^T\varphi + 2\varepsilon\varphi^TP_0B[a_0(z,\xi,\theta,w) - b(\eta,\zeta,\theta)\phi(\theta,w) + b(\eta,\zeta,\theta)\psi(\sigma,\hat{\xi},\mu)].$$

Since $\psi(\sigma,\hat{\xi},\mu)$ is globally bounded in $\hat{\xi}$, for all $(\sigma,z,q,s) \in \Xi\times\Omega$ there is $\ell_1 > 0$ such that

$$|a_0(z,\xi,\theta,w) - b(\eta,\zeta,\theta)\phi(\theta,w) + b(\eta,\zeta,\theta)\psi(\sigma,\hat{\xi},\mu)| \le \ell_1.$$

Therefore,

$$\varepsilon\dot{V}_3 \le -\|\varphi\|^2 + 2\varepsilon\ell_1\|P_0B\|\,\|\varphi\| \le -\tfrac{1}{2}\|\varphi\|^2 \quad \forall\ \|\varphi\| \ge 4\varepsilon\ell_1\|P_0B\|. \tag{5.59}$$

Taking $\rho_3 = \lambda_{\max}(P_0)(4\|P_0B\|\ell_1)^2$ ensures that $\dot{V}_3 \le -\frac{1}{2}\|\varphi\|^2$ for all $V_3 \ge \rho_3\varepsilon^2$. Consequently, $\dot{V}_3 < 0$ on the boundary $V_3 = \rho_3\varepsilon^2$. Next, consider (5.57):

$$\dot{s} = b(\eta,\zeta,\theta)\psi(\sigma,\xi,\mu) + \Delta_1(z,\xi,\theta,w) + \Lambda^T\left[F\sigma + \mu G \operatorname{sat}\left(\frac{\hat{s}}{\mu}\right)\right] \\ + b(\eta,\zeta,\theta)[\psi(\sigma,\hat{\xi},\mu) - \psi(\sigma,\xi,\mu)].$$

In $\Xi \times \Omega \times \Sigma_\varepsilon$, positive constants ℓ_2 and ℓ_3 exist such that

$$|\psi(\sigma,\hat{\xi},\mu) - \psi(\sigma,\xi,\mu)| \le \left(\ell_2 + \frac{\ell_3}{\mu}\right)\|\xi - \hat{\xi}\| \le \left(\ell_2 + \frac{\ell_3}{\mu}\right)\|\varphi\|$$

for $\varepsilon \le 1$. The factor $1/\mu$ is due to the Lipschitz constant of $\operatorname{sat}(s/\mu)$. Since $\|\varphi\| \le \ell_0\varepsilon$ in Σ_ε and $|\Lambda^T[F\sigma + \mu G \operatorname{sat}(\hat{s}/\mu)]| \le \gamma\mu$ for $\sigma \in \Xi$,

$$s\dot{s} \le b\left[-\beta + \frac{|\Delta_1|}{b} + \frac{\gamma\mu}{b_0} + \varepsilon\ell_0\left(\ell_2 + \frac{\ell_3}{\mu}\right)\right]|s| \le -\tfrac{1}{2}b_0\beta_0|s|$$

when $|s| \ge \mu$, $\varepsilon\ell_0(\ell_2 + \ell_3/\mu) \le \frac{1}{4}\beta_0$, and $\gamma\mu/b_0 \le \frac{1}{4}\beta_0$. Repeating the argument used with state feedback, it can be shown that Ω is positively invariant, which completes the first step. Because Ψ is in the interior of Ω and $\psi(\sigma,\hat{\xi},\mu)$ is globally bounded in $\hat{\xi}$, there is time $T_0 > 0$, independent of ε, such that $(z(t), q(t), s(t)) \in \Omega$ for $t \in [0, T_0]$. During this time, (5.59) shows that

$$\dot{V}_3 \le -\frac{1}{2\varepsilon\lambda_{\max}(P_0)}V_3.$$

Therefore, V_3 reduces to $\rho_3\varepsilon^2$ within a time interval $[0, T_1(\varepsilon)]$ in which $\lim_{\varepsilon\to 0} T_1(\varepsilon) = 0$. For sufficiently small ε, $T_1(\varepsilon) < T_0$. The second step is complete. The third step follows from the fact that $s\dot{s} \le -\frac{1}{2}b_0\beta_0$ for $|s| \ge \mu$, which shows that s enters the set $\{|s| \le \mu\}$ in finite time. The rest is similar to the analysis under state feedback because (5.55) and (5.56) are not altered under output feedback. For the final step, consider the system inside $\Xi \times \Omega_\mu \times \Sigma_\varepsilon$. When $\varphi = 0$, equations (5.53) to (5.57) coincide with the corresponding equations under state feedback. On the other hand, when $\sigma = \bar{\sigma}$, $z = 0$, and $\xi = 0$, equation (5.58) has an equilibrium point at $\varphi = 0$ because $a_0(0,0,\theta,w) = 0$ and $\psi(\bar{\sigma},0,\mu) = -\beta(0)\bar{s}/\mu = \phi(\theta,w)$. Therefore, $\{\sigma = \bar{\sigma}, z = 0, \xi = 0, \varphi = 0\}$ in an invariant manifold. With the change of variables $\vartheta = \sigma - \bar{\sigma}$ and $p = s - \bar{s}$, the system takes the singularly perturbed form

$$\begin{aligned}
\dot{w} &= S_0 w,\\
\dot{\vartheta} &= F\vartheta + Gp + \mu G[\operatorname{sat}(\hat{s}/\mu) - \operatorname{sat}(s/\mu)],\\
\dot{z} &= \tilde{f}_0(z, \operatorname{col}(q, p - \Lambda^T\vartheta - Lq), \theta, w),\\
\dot{q} &= A_2 q + B_2(p - \Lambda^T\vartheta),\\
\mu\dot{p} &= -b(\eta,\zeta,\theta)\beta(\xi)p + \mu\Delta_a(\cdot) + \mu\Delta_c(\cdot) + \mu b(\eta,\zeta,\theta)[\psi(\sigma,\hat{\xi},\mu) - \psi(\sigma,\xi,\mu)]\\
&\quad + \mu\Lambda^T\{F\vartheta + Gp + \mu G[\operatorname{sat}(\hat{s}/\mu) - \operatorname{sat}(s/\mu)]\},\\
\varepsilon\dot{\varphi} &= A_0\varphi + (\varepsilon/\mu)B[-b(\eta,\zeta,\theta)\beta(\xi)p + \mu\Delta_a(\cdot)\\
&\quad + \mu b(\eta,\zeta,\theta)[\psi(\sigma,\hat{\xi},\mu) - \psi(\sigma,\xi,\mu)],
\end{aligned}$$

where

$$\Delta_a = a_0(z,\xi,\theta,w) + b(\eta,\zeta,\theta)\phi(\theta,w)\left[\frac{\beta(\xi)-\beta(0)}{\beta(0)}\right] \text{ and } \Delta_c = \sum_{i=1}^{\rho-1} k_i \xi_{i+1}.$$

There are positive constants ℓ_1 to ℓ_{10} such that

$$\begin{aligned}\mu|\text{sat}(\hat{s}/\mu) - \text{sat}(s/\mu)| &\le \ell_1\|\varphi\|,\\ |\psi(\sigma,\hat{\xi},\mu) - \psi(\sigma,\xi,\mu)| &\le (\ell_2 + \ell_3/\mu)\|\varphi\|,\\ |\Delta_a| &\le \ell_4|\vartheta| + \ell_5\|z\| + \ell_6\|q\| + \ell_7|p|,\\ |\Delta_c| &\le \ell_8|\vartheta| + \ell_9\|q\| + \ell_{10}|p|.\end{aligned}$$

Consider the composite Lyapunov function

$$V = V_0 + \kappa_1 q^T P_2 q + \kappa_2 \vartheta^T P \vartheta + \tfrac{1}{2}p^2 + \varphi^T P_0 \varphi,$$

with positive constants κ_1 and κ_2. It can be shown that $\dot{V} \le -\mathcal{Y}^T Q \mathcal{Y}$, where $\mathcal{Y} = \text{col}(\|z\|, \|q\|, |\vartheta|, |p|, \|\varphi\|)$,

$$Q = \begin{bmatrix} c_1 & -c_2 & -c_3 & -c_4 & -c_5 \\ -c_2 & \kappa_1 & -\kappa_1 c_6 & -(\kappa_1 c_7 + c_8) & -c_9 \\ -c_3 & -\kappa_1 c_6 & \kappa_2 & -(\kappa_2 c_{10} + c_{11}) & -(\kappa_2 c_{12} + c_{13}) \\ -c_4 & -(\kappa_1 c_7 + c_8) & -(\kappa_2 c_{10} + c_{11}) & (c_{14}/\mu - c_{15}) & -(c_{16} + c_{17}/\mu) \\ -c_5 & -c_9 & -(\kappa_2 c_{12} + c_{13}) & -(c_{16} + c_{17}/\mu) & (1/\varepsilon - c_{18} - c_{19}/\mu) \end{bmatrix},$$

and c_1 to c_{19} are positive constants independent of κ_1, κ_2, μ, and ε. Choose κ_1 large enough to make the 2×2 principal minor positive; then choose κ_2 large enough to make the 3×3 principal minor positive; then choose μ small enough to make the 4×4 principal minor positive; then choose ε small enough to make the determinant of Q positive. Hence, the origin $(\vartheta = 0, z = 0, q = 0, p = 0, \varphi = 0)$ is exponentially stable, and there is a neighborhood N of the origin, independent of μ and ε, such that all trajectories in N converge to the origin as $t \to \infty$. By choosing μ and ε small enough, it can be ensured that for all $(\sigma, z, q, s, \varphi) \in \Xi \times \Omega_\mu \times \Sigma_\varepsilon$, $(\vartheta, z, q, p, \varphi) \in N$. Thus, all trajectories in $\Xi \times \Omega_\mu \times \Sigma_\varepsilon$ approach the invariant manifold $(\sigma = \bar{\sigma}, z = 0, \xi = 0, \varphi = 0)$. Consequently, all trajectories with $\sigma(0) \in \Xi$, $(z(0), q(0), s(0)) \in \Psi$, and bounded $\hat{\xi}(0)$ approach this manifold because such trajectories enter $\Xi \times \Omega_\mu \times \Sigma_\varepsilon$ in finite time. This completes the proof that all the state variables are bounded and $\lim_{t\to\infty} e(t) = 0$. The proof of (5.52) is done as in the proof of Theorem 5.2. □

Remark 5.12. *If Assumptions* 5.5, 5.6, *and* 5.10 *hold globally, that is,* $D_x = T(D_x) = R^n$, *and* γ_1 *is class* $\mathcal{K}_\infty$, *then the sets* Γ *and* Ω *can be chosen arbitrarily large. For any bounded* $(z(0), q(0), s(0))$, *the conclusion of Theorem* 5.3 *will hold by choosing* β *large enough.* ◇

Example 5.4. Consider the system

$$\dot{x}_1 = -\theta_1 x_1 + x_2^2 + d, \quad \dot{x}_2 = x_3, \quad \dot{x}_3 = -\theta_2 x_1 x_2 + u, \quad y = x_2,$$

where $\theta_1 > 0$ and θ_2 are unknown parameters, d is a constant disturbance, and the reference signal is $r = \alpha \sin(\omega_0 t + \theta_0)$ with known frequency ω_0 and unknown amplitude and phase. Start by verifying the assumptions. Assumptions 5.5 and 5.6 are satisfied globally with

$$\eta = x_1, \quad \zeta_1 = x_2, \quad \zeta_2 = x_3.$$

The exosystem of Assumption 5.7 is given by

$$\dot{w} = \begin{bmatrix} 0 & \omega_0 & 0 \\ -\omega_0 & 0 & 0 \\ 0 & 0 & 0 \end{bmatrix} w, \quad w(0) = \begin{bmatrix} \alpha \sin\theta_0 \\ \alpha\cos\theta_0 \\ d \end{bmatrix}, \quad r = w_1, \quad d = w_3.$$

With $\tau_1 = w_1$ and $\tau_2 = \omega_0 w_2$, Assumption 5.8 is satisfied with

$$\tau_0(\theta, w) = \frac{\theta_1^2 + 2\omega_0^2}{\theta_1(\theta_1^2 + 4\omega_0^2)} w_1^2 - \frac{2\omega_0}{\theta_1^2 + 4\omega_0^2} w_1 w_2 + \frac{2\omega_0^2}{\theta_1(\theta_1^2 + 4\omega_0^2)} w_2^2 + \frac{1}{\theta_1} w_3.$$

The steady-state control is given by

$$\begin{aligned}
\phi(\theta, w) &= -\omega_0^2 w_1 + \theta_2 \tau_0(\theta, w) w_1 \\
&= -\omega_0^2 w_1 + \frac{\theta_2}{\theta_1} w_1 w_3 \\
&\quad + \frac{\theta_2}{\theta_1(\theta_1^2 + 4\omega_0^2)} \left[(\theta_1^2 + 2\omega_0^2) w_1^3 - 2\theta_1 \omega_0 w_1^2 w_2 + 2\omega_0^2 w_1 w_2^2 \right].
\end{aligned}$$

Assumption 5.9 is satisfied, as ϕ satisfies the differential equation

$$\phi^{(4)} + 10\,\omega_0^2 \phi^{(2)} + 9\,\omega_0^4 \phi = 0.$$

Hence, the internal model (5.36) is given by

$$S = \begin{bmatrix} 0 & 1 & 0 & 0 \\ 0 & 0 & 1 & 0 \\ 0 & 0 & 0 & 1 \\ -9\,\omega_0^4 & 0 & -10\,\omega_0^2 & 0 \end{bmatrix}, \quad H = \begin{bmatrix} 1 & 0 & 0 & 0 \end{bmatrix}.$$

It is worthwhile to note that while finding the internal model goes through the elaborate procedure of finding τ_0 and ϕ and verifying the differential equation satisfied by ϕ, the model can be intuitively predicted. If x_2 is to be a sinusoidal signal, then it can be seen from the $\dot{x}_1$-equation that the steady state of x_1 will have constant and second harmonic terms. Then the product $x_1 x_2$ will have first and third harmonics. Finally, the $\dot{x}_3$-equation shows that the steady-state control will have first and third harmonics, which results in the internal model. With the change of variables

$$z = \eta - \tau_0 = x_1 - \tau_0, \quad \xi_1 = y - r = x_2 - w_1, \quad \xi_2 = \dot{y} - \dot{r} = x_3 - \omega_0 w_2,$$

the system is represented by

$$\begin{aligned}
\dot{z} &= -\theta_1 z + \xi_1^2 + 2\xi_1 w_1, \\
\dot{\xi}_1 &= \xi_2, \\
\dot{\xi}_2 &= -\theta_2[(z + \tau_0)\xi_1 + z w_1] + u - \phi(\theta, w).
\end{aligned}$$

Assumption 5.10 is satisfied globally with $V_1 = \frac{1}{2}z^2$. Assumption 5.11 is satisfied with $V_0 = \frac{1}{2}z^2$. The output feedback controller with conditional servocompensator is taken as[45]

$$\dot{\sigma} = F\sigma + \mu G \text{ sat}\left(\frac{\Lambda^T\sigma + \hat{\xi}_1 + \hat{\xi}_2}{\mu}\right),$$

$$u = -20 \text{ sat}\left(\frac{\Lambda^T\sigma + \hat{\xi}_1 + \hat{\xi}_2}{\mu}\right),$$

$$\dot{\hat{\xi}}_1 = \hat{\xi}_2 + (2/\varepsilon)(e - \hat{\xi}_1), \qquad \dot{\hat{\xi}}_2 = (1/\varepsilon^2)(e - \hat{\xi}_1),$$

where $\mu = 0.1$, $\varepsilon = 0.001$, and $e = y - r$. The pair (F, G) is taken in the controllable canonical form as

$$F = \varsigma \begin{bmatrix} 0 & 1 & 0 & 0 \\ 0 & 0 & 1 & 0 \\ 0 & 0 & 0 & 1 \\ -1.5 & -6.25 & -8.75 & -5 \end{bmatrix}, \quad G = \varsigma \begin{bmatrix} 0 \\ 0 \\ 0 \\ 1 \end{bmatrix},$$

where ς is a positive parameter. The eigenvalues of F/ς are -0.5, -1, -1.5, and -2. The vector Λ that assigns the eigenvalues of $F + G\Lambda^T$ at the eigenvalues of S is given by

$$\Lambda^T = \begin{bmatrix} -9(\omega_0/\varsigma)^4 + 1.5, & 6.25, & -10(\omega_0/\varsigma)^2 + 8.75, & 5 \end{bmatrix}.$$

The scaling parameter ς is chosen such that Λ does not have large values for large ω_0. Assuming $\omega_0 \leq 3$, ς is taken as $\varsigma = 3$. The simulation results of Figure 5.7 are for $\theta_1 = 3$, $\theta_2 = 4$, $\omega_0 = 2.5$ rad/sec, $d = 0.1$, and $\varepsilon = 0.001$. The initial conditions are $x_1(0) = x_2(0) = 1$, $x_3(0) = \hat{\xi}_1(0) = \hat{\xi}_2(0) = 0$. Figure 5.7(a) shows the regulation error e under the output feedback controller. It shows also the error under the state feedback sliding mode controller

$$u = -20 \text{ sgn}(\xi_1 + \xi_2).$$

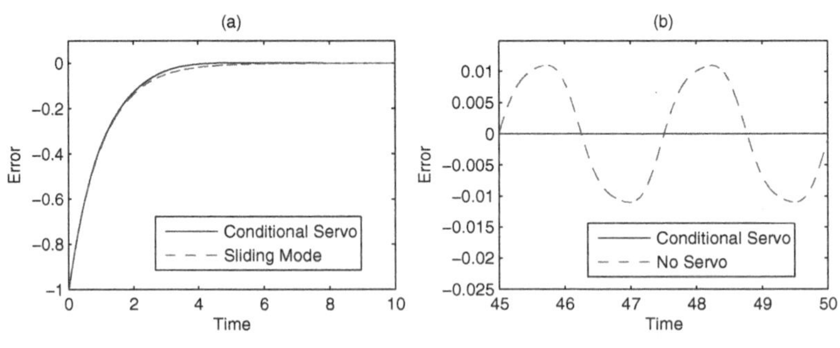

Figure 5.7. *Simulation of Example* 5.4. (a) *compares the regulation error e under the output feedback controller with conditional servocompensator to the error under the state feedback sliding mode controller.* (b) *shows the difference between the steady-state errors of the controllers with and without servocompensator.*

[45]The gain 20 is determined by simulation.

As expected, the two responses are very close. Figure 5.7(b) shows the advantage of including a servocompensator by comparing with the controller

$$u = -20 \,\mathrm{sat}\left(\frac{\hat{\xi}_1 + \hat{\xi}_2}{\mu}\right),$$

which does not include a servocompensator. This controller does not achieve zero steady-state error. It can only guarantee $O(\mu)$ steady-state error. △

5.5 ▪ Internal Model Perturbation

The servocompensator design of the previous section requires precise knowledge of the internal model (5.36), which is equivalent to knowing the frequency components of the steady-state control. In this section we study the effect of internal model perturbations on the steady-state error. Let $\bar{\phi}(\theta, w)$ be the steady-state control on the zero-error manifold, as defined by (5.34). The function $\bar{\phi}(\theta, w)$ may not satisfy (5.35), but an approximation of it, $\phi(\theta, w)$, could do so with known coefficients c_0 to c_{q-1}, which define the internal model that is used in the design. There are two sources for such perturbation. First, $\bar{\phi}(\theta, w)$ may not have a finite number of harmonics. Second, it may have a finite number of harmonics, but the frequencies are not precisely known.

Assumption 5.12.

$$b(\lambda, \pi, \theta)|\phi(\theta, w) - \bar{\phi}(\theta, w)| \leq \delta \quad \forall\, (\theta, w) \in \Theta \times W. \tag{5.60}$$

Theorem 5.4. *Under the assumptions of Theorem 5.3, if Assumption 5.12 is satisfied, then there exist positive constants μ^*, δ^*, ℓ, and T, and for each $\mu \in (0, \mu^*]$, there is a positive constant $\varepsilon^* = \varepsilon^*(\mu)$ such that for each $\mu \in (0, \mu^*]$, $\varepsilon \in (0, \varepsilon^*]$, and $\delta \in (0, \delta^*]$,*

$$|e(t)| \leq \ell \mu \delta \quad \forall\, t \geq T. \qquad \diamond$$

Proof: The closed-loop system is a perturbation of equations (5.53) to (5.58) in which (5.57) and (5.58) are perturbed by $b(\eta, \zeta, \theta)\tilde{\phi}(\theta, w)$ and $\varepsilon B b(\eta, \zeta, \theta)\tilde{\phi}(\theta, w)$, respectively, where $\tilde{\phi} = \phi - \bar{\phi}$. Provided δ is sufficiently small, the four steps of the proof of Theorem 5.3 can be repeated to show that every trajectory in $\Xi \times \Omega \times \Sigma_\varepsilon$ enters $\Xi \times \Omega_\mu \times \Sigma_\varepsilon$ in finite time and stays therein for all future time. Inside $\Xi \times \Omega_\mu \times \Sigma_\varepsilon$ the system can be represented in the form

$$\begin{aligned}
\dot{w} &= S_0 w,\\
\dot{\mathscr{X}} &= f_1(\mathscr{X}, p, \theta, w) + G_1 h_1(p, N(\varepsilon)\varphi, \mu),\\
\mu \dot{p} &= -b(\eta, \zeta, \theta)\beta(\xi)p + \mu f_2(\mathscr{X}, p, \theta, w) + h_2(\mathscr{X}, p, N(\varepsilon)\varphi, \theta, w)\\
&\quad + \mu b(\lambda, \pi, \theta)\tilde{\phi}(\theta, w),\\
\varepsilon \dot{\varphi} &= A_0 \varphi - (\varepsilon/\mu) B b(\eta, \zeta, \theta)\beta(\xi)p + (\varepsilon/\mu) h_3(\mathscr{X}, p, N(\varepsilon)\varphi, \theta, w)\\
&\quad + \varepsilon f_3(\mathscr{X}, p, \theta, w) + \varepsilon B b(\lambda, \pi, \theta)\tilde{\phi}(\theta, w),
\end{aligned}$$

where $\mathscr{X} = \mathrm{col}(z, q, \vartheta)$, $\vartheta = \sigma - \bar{\sigma}$, $p = s - \bar{s}$, and $N(\varepsilon)$ is a polynomial function of ε. All functions on the right-hand side are locally Lipschitz in their arguments, and the functions h_1 to h_3 satisfy the inequalities $|h_i| \leq \ell_i \|\varphi\|$ with some positive constants ℓ_1 to ℓ_3, independent of ε and μ. The function f_1 is given by

$$f_1 = \begin{bmatrix} \tilde{f}_0(z, \mathrm{col}(q, p - \Lambda^T \vartheta - Lq), \theta, w) \\ A_2 q + +B_2(p - \Lambda^T \vartheta) \\ F\vartheta + Gp \end{bmatrix}.$$

Because A_2 and F are Hurwitz and the origin of $\dot{z} = \tilde{f}(z, 0, \theta, w)$ is exponentially stable, the origin of $\dot{\mathscr{Z}} = f_1(\mathscr{Z}, 0, \theta, w)$ is exponentially stable, and a Lyapunov function for it can be constructed in the form

$$V_4(\mathscr{Z}, \theta, w) = \kappa_1 V_0(z, \theta, w) + \kappa_2 q^T P_2 q + \vartheta^T P \vartheta$$

with sufficiently small κ_2 and κ_1 / κ_2 [80, Appendix C]. It can be verified that V_4 satisfies the inequalities

$$\tilde{c}_1 \|\mathscr{Z}\|^2 \leq V_4(\mathscr{Z}, \theta, w) \leq \tilde{c}_2 \|\mathscr{Z}\|^2,$$

$$\frac{\partial V_4}{\partial \mathscr{Z}} f_1(\mathscr{Z}, 0, \theta, w) + \frac{\partial V_4}{\partial w} S_0 w \leq -\tilde{c}_3 \|\mathscr{Z}\|^2,$$

$$\left\| \frac{\partial V_4}{\partial \mathscr{Z}} \right\| \leq \tilde{c}_4 \|\mathscr{Z}\|$$

in a neighborhood of $\mathscr{Z} = 0$ with positive constants $\tilde{c}_1$ to $\tilde{c}_4$. Let $W_1 = \sqrt{V_4}$, $W_2 = \sqrt{p^2}$, and $W_3 = \sqrt{\varphi^T P_0 \varphi}$. By calculating upper bounds of $D^+ W_1$, $D^+ W_2$, and $D^+ W_3$,[46] it can be shown that

$$\begin{aligned} D^+ W_1 &\leq -b_1 W_1 + k_1 W_2 + k_2 W_3, \\ D^+ W_2 &\leq -(b_2/\mu) W_2 + k_3 W_1 + k_4 W_2 + (k_5/\mu) W_3 + \delta, \\ D^+ W_3 &\leq -(b_3/\varepsilon) W_3 + k_6 W_1 + k_7 W_2 + (k_8/\mu) W_2 + (k_9/\mu) W_3 + k_{10}\delta, \end{aligned}$$

where b_1 to b_3 and k_1 to k_{10} are positive constants independent of ε and μ. Rewrite the foregoing scalar differential inequalities as the vector differential inequality

$$D^+ W \leq \mathscr{A} W + \mathscr{B} \delta,$$

where

$$W = \begin{bmatrix} W_1 \\ W_2 \\ W_3 \end{bmatrix}, \quad \mathscr{A} = \begin{bmatrix} -b_1 & k_1 & k_2 \\ k_3 & -(b_2/\mu - k_4) & k_5/\mu \\ k_6 & (k_7 + k_8/\mu) & -(b_3/\varepsilon - k_9/\mu) \end{bmatrix}, \quad \text{and} \quad \mathscr{B} = \begin{bmatrix} 0 \\ 1 \\ k_{10} \end{bmatrix}.$$

For sufficiently small μ and ε/μ, the matrix $\mathscr{A}$ is Hurwitz and quasi monotone. Application of the comparison method shows that[47]

$$W(t) \leq U(t) \quad \forall\, t \geq 0,$$

[46] See [78, Section 3.4] for the definition of $D^+ W$ and [78, Section 9.3] for an example of calculating an upper bound on $D^+ W$.

[47] See [125, Chapter IX] for the definition quasi-monotone matrices and the comparison method for vector differential inequalities.

where U is the solution of the differential equation

$$\dot{U} = \mathscr{A}U + \mathscr{B}\delta, \quad U(0) = W(0). \tag{5.61}$$

Because $\mathscr{A}$ is Hurwitz, equation (5.61) has an exponentially stable equilibrium point at $\bar{U} = \text{col}(\bar{u}_1, \bar{u}_2, \bar{u}_3)$ that satisfies the algebraic equation

$$0 = \mathscr{A}\bar{U} + \mathscr{B}\delta.$$

It can be verified that there is a constant $k_{11} > 0$, independent of ε and μ, such that for sufficiently small μ and ε/μ, $|\bar{u}_1| \leq k_{11}\delta$. Therefore, $W_1(t)$ is ultimately bounded by $k_{12}\delta$ for $k_{12} > k_{11}$. The proof is completed by noting that e is a component of $\mathscr{Z}$ and $\|\mathscr{Z}\| \leq \sqrt{V_4/\tilde{c}_1} = W_1/\sqrt{\tilde{c}_1}$ □

Remark 5.13. *The proof uses linear-type Lyapunov functions and the vector Lyapunov function method rather than a quadratic-type Lyapunov function, which could have been constructed as a linear combination of V_4, p^2, and $\varphi^T \Gamma_0 \varphi$. The quadratic-type function would yield a bound on the error of the order $O(\sqrt{\mu\delta})$, which is more conservative than $O(\mu\delta)$ for small $\mu\delta$.* ◇

The next two examples illustrate the $O(\mu\delta)$ bound on the steady-state error. In the first example the internal model perturbation is due to uncertainty in the parameters of the model, while in the second example the steady-state control does not satisfy (5.35), but an approximation of it does so.

Example 5.5. Reconsider Example 5.4 where the conditional servocompensator is designed assuming the frequency of the reference signal is 2.5 rad/sec when the actual frequency $\omega_0 \neq 2.5$. The feedback controller is the same as in Example 5.4, and the simulation is carried out using the same parameters and initial conditions. The simulation results are shown in Figure 5.8. Figures 5.8(a) and (b) compare the regulation error e for $\omega_0 = 3$ and $\omega_0 = 2.7$ rad/sec when $\mu = 0.1$. It is observed first that the frequency error has a little effect on the transient response. This is expected because the transient response is basically the response under the sliding mode controller $u = -20\ \text{sgn}(\xi_1 + \xi_2)$. As for the steady-state response, the error decreases as the frequency approaches the nominal frequency of 2.5 rad/sec. Figures 5.8(c) and (d) compare the regulation error e for $\mu = 0.1$ and $\mu = 0.01$ with fixed frequency $\omega_0 = 3$ rad/sec. Once again, the change in μ has a little effect on the transient response, as both values are small enough to bring the transient response close to that of the sliding mode control. The steady-state error decreases with decreasing μ. Both cases demonstrate the fact that the steady-state error is of the order $O(\mu\delta)$. △

Example 5.6. Consider the system

$$\dot{x}_1 = -\theta_1 x_1 + x_2^2 + d, \quad \dot{x}_2 = x_3, \quad \dot{x}_3 = -\theta_2 x_1 x_2 + \theta_3 \sin x_2 + u, \quad y = x_2,$$

where $\theta_1 > 0$, θ_2, and θ_3 are unknown parameters, d is a constant disturbance, and the reference signal is $r = \alpha \sin(\omega_0 t + \theta_0)$ with known frequency ω_0 and unknown amplitude and phase. This is the same problem considered in Examples 5.4 and 5.5 with

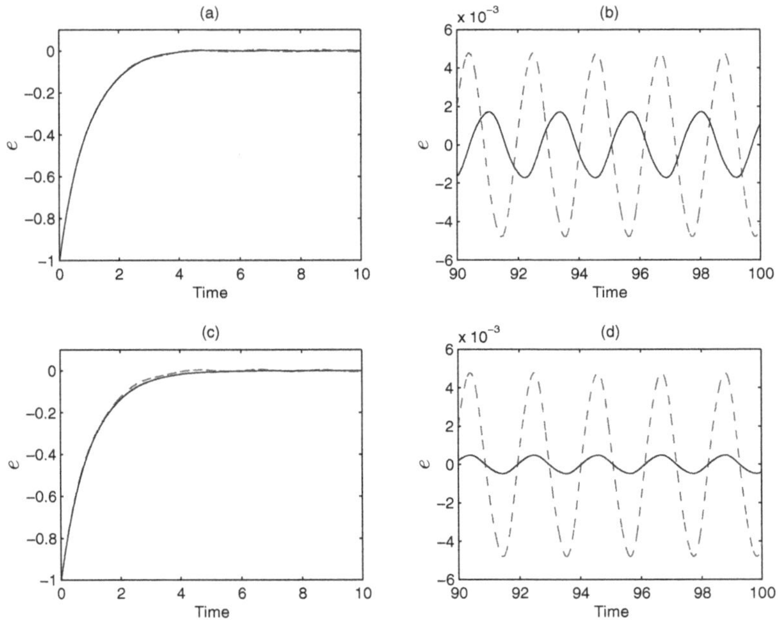

Figure 5.8. *Simulation of Example* 5.5. (a) *and* (b) *show the transient and steady-state regulation error e for* $\omega_0 = 3$ *rad/sec (dashed) and* $\omega_0 = 2.7$ *rad/sec (solid) when* $\mu = 0.1$. (c) *and* (d) *show the transient and steady state regulation error e for* $\omega_0 = 3$ *rad/sec when* $\mu = 0.1$ *(dashed) and* $\mu = 0.01$ *(solid).*

the additional term $\theta_3 \sin x_2$ in the $\dot{x}_3$-equation. The only change from Example 5.4 is in the steady-state control on the zero-error manifold, which is given by

$$\bar{\phi}(\theta, w) = -\theta_3 \sin w_1 - \omega_0^2 w_1 + \frac{\theta_2}{\theta_1} w_1 w_3 + \frac{\theta_2}{\theta_1(\theta_1^2 + 4\omega_0^2)} \left[(\theta_1^2 + 2\omega_0^2) w_1^3 - 2\theta_1 \omega_0 w_1^2 w_2 + 2\omega_0^2 w_1 w_2^2 \right].$$

The function $\bar{\phi}$ does not satisfy Assumption 5.33 due to transcendental function $\sin(\cdot)$, which generates an infinite number of harmonics of the sinusoidal reference signal. The sinusoidal function can be approximated by its truncated Taylor series

$$\sin w_1 \approx \sum_{i=1}^{n} \frac{(-1)^{i-1} w_1^{2i-1}}{(2i-1)!},$$

and the approximation error decreases as n increases. The approximate function

$$\phi(\theta, w) = -\theta_3 \sum_{i=1}^{n} \frac{(-1)^{i-1} w_1^{2i-1}}{(2i-1)!} - \omega_0^2 w_1 + \frac{\theta_2}{\theta_1} w_1 w_3 + \frac{\theta_2}{\theta_1(\theta_1^2 + 4\omega_0^2)} \left[(\theta_1^2 + 2\omega_0^2) w_1^3 + -2\theta_1 \omega_0 w_1^2 w_2 + 2\omega_0^2 w_1 w_2^2 \right]$$

satisfies Assumption 5.33 because it is a polynomial function of w. Two approximations are considered with $n = 3$ and $n = 5$. In the first case, ϕ satisfies the equation

$$\phi^{(4)} + 10\,\omega_0^2 \phi^{(2)} + 9\,\omega_0^4 \phi = 0$$

as in Example 5.4. In this case, the same controller of Example 5.4 is used, that is,

$$\dot{\sigma} = F\sigma + \mu G \,\mathrm{sat}\left(\frac{\Lambda^T\sigma + \hat{\xi}_1 + \hat{\xi}_2}{\mu}\right),$$

$$u = -20 \,\mathrm{sat}\left(\frac{\Lambda^T\sigma + \hat{\xi}_1 + \hat{\xi}_2}{\mu}\right),$$

$$\dot{\hat{\xi}}_1 = \hat{\xi}_2 + (2/\varepsilon)(e - \hat{\xi}_1), \qquad \dot{\hat{\xi}}_2 = (1/\varepsilon^2)(e - \hat{\xi}_1),$$

$$F = \varsigma \begin{bmatrix} 0 & 1 & 0 & 0 \\ 0 & 0 & 1 & 0 \\ 0 & 0 & 0 & 1 \\ -1.5 & -6.25 & -8.75 & -5 \end{bmatrix}, \quad G = \varsigma \begin{bmatrix} 0 \\ 0 \\ 0 \\ 1 \end{bmatrix},$$

$$\Lambda^T = \begin{bmatrix} -9(\omega_0/\varsigma)^4 + 1.5, & 6.25, & -10(\omega_0/\varsigma)^2 + 8.75, & 5 \end{bmatrix},$$

and $\varsigma = 3$. In the case $n = 5$, ϕ satisfies the equation

$$\phi^{(6)} + 35\omega_0^2\phi^{(4)} + 259\omega_0^4\phi^{(2)} + 225\omega_0^6 = 0,$$

and the internal model (5.36) is given by

$$S = \begin{bmatrix} 0 & 1 & 0 & 0 & 0 & 0 \\ 0 & 0 & 1 & 0 & 0 & 0 \\ 0 & 0 & 0 & 1 & 0 & 0 \\ 0 & 0 & 0 & 0 & 1 & 0 \\ 0 & 0 & 0 & 0 & 0 & 0 \\ -225\,\omega_0^6 & 0 & -259\,\omega_0^4 & 0 & -35\,\omega_0^2 & 0 \end{bmatrix}, \quad H = \begin{bmatrix} 1 & 0 & 0 & 0 & 0 & 0 \end{bmatrix}.$$

The controller is the same as in the previous case except for F, G, and Λ, which are taken as

$$F = \varsigma \begin{bmatrix} 0 & 1 & 0 & 0 & 0 & 0 \\ 0 & 0 & 1 & 0 & 0 & 0 \\ 0 & 0 & 0 & 1 & 0 & 0 \\ 0 & 0 & 0 & 0 & 1 & 0 \\ 0 & 0 & 0 & 1 & 0 & 1 \\ -11.25 & -55.125 & -101.5 & -91.875 & -43.75 & -10.5 \end{bmatrix}, \quad G = \varsigma \begin{bmatrix} 0 \\ 0 \\ 0 \\ 0 \\ 0 \\ 1 \end{bmatrix},$$

$$\Lambda^T = \begin{bmatrix} \Lambda_1, & 55.125, & \Lambda_3, & 91.875, & \Lambda_5, & 10.5 \end{bmatrix},$$

$$\Lambda_1 = 11.25 - 225 * (\omega_0/\varsigma)^6, \quad \Lambda_3 = 101.5 - 259 * (\omega_0/\varsigma)^4, \quad \Lambda_5 = 43.75 - 35 * (\omega_0/\varsigma)^2,$$

and $\varsigma = 3$. The eigenvalues of F/ς are -0.5, -1, -1.5, -2, -2.5, and -3. Simulation results are shown in Figures 5.9 and 5.10 for the parameters $\theta_1 = 3$, $\theta_2 = 4$, $\theta_3 = 1$, $\omega_0 = 2.5$, $d = 0.1$, and $\varepsilon = 10^{-4}$ and the initial conditions $x_1(0) = x_2(0) = 1$, $x_3(0) = \hat{\xi}_1(0) = \hat{\xi}_2(0) = 0$. In Figure 5.9, μ is fixed at 0.1, while δ is reduced by going from $n = 3$ to $n = 5$. In Figure 5.10, n is fixed at 3, while μ is reduced from 0.1 to 0.01. The first observation is that in all cases the transient response of the regulation error is almost the same. For the steady-state error, it is seen that reducing δ reduces the error. It is interesting to note that in the case $n = 3$, the approximation of the sinusoidal function maintains the first and third harmonics and neglects the

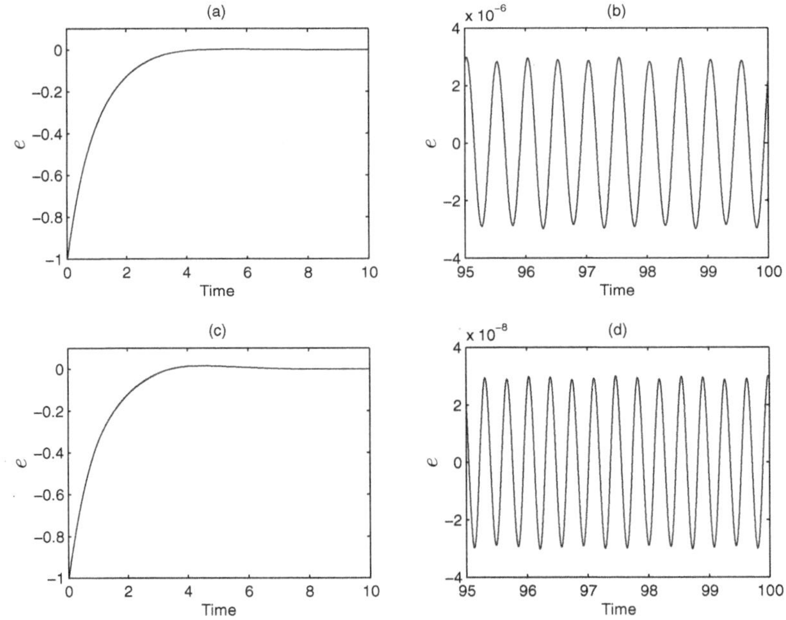

Figure 5.9. *Simulation of Example* 5.6. (a) *and* (b) *show the transient and steady-state regulation error e for the case $n = 3$, while* (c) *and* (d) *show the error for the case $n = 5$. In both cases, $\mu = 0.1$.*

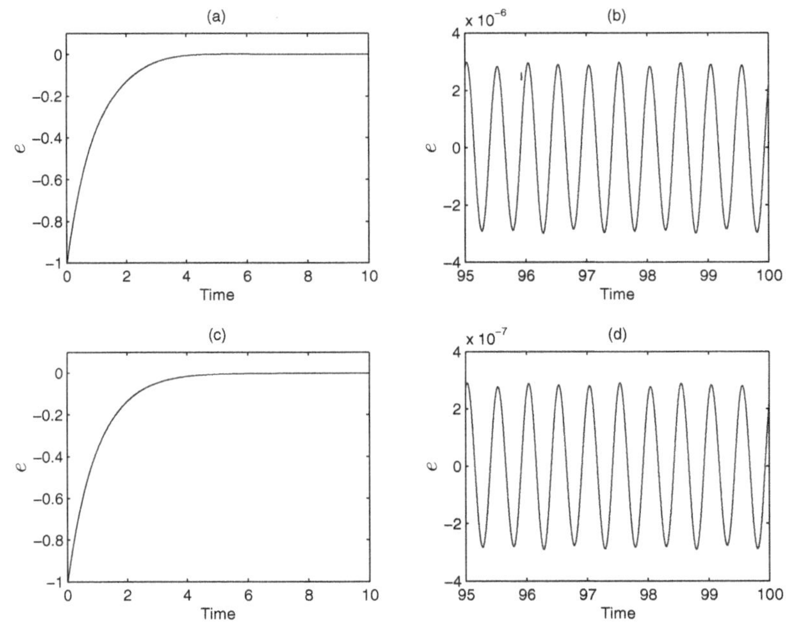

Figure 5.10. *Simulation of Example* 5.6. (a) *and* (b) *show the transient and steady-state regulation error e for the case $n = 3$ when $\mu = 0.1$, while* (c) *and* (d) *show the error when $\mu = 0.01$.*

higher-order harmonics, with the fifth harmonic being the most significant neglected one. Figure 5.9(b) shows that the error oscillates at the fifth harmonic frequency (12.5 rad/sec.). In the case $n = 5$, the approximation maintains the first, third, and fifth harmonics and neglects the higher-order harmonics, with the seventh harmonic being the most significant neglected one. Figure 5.9(d) shows that the error oscillates at the seventh harmonic frequency (17.5 rad/sec.). With n fixed, reducing μ decreases the steady-state error. △

5.6 ▪ Adaptive Internal Model

The controller of Section 5.4 requires precise knowledge of the eigenvalues of S because they are used to calculate Λ. When they are not known, Λ is replaced by $\hat{\Lambda}$, and an adaptive law is used to adjust $\hat{\Lambda}$ in real time. Two cases will be considered: *full-parameter adaptation* and *partial-parameter adaptation*. In the first case, all q elements of $\hat{\Lambda}$ are adapted. In the second case, the number of adapted parameters is equal to the number of complex modes of S, which is $q/2$ when q is even and $(q-1)/2$ when it is odd. This case corresponds to a special choice of the pair (F,G) in the controllable canonical form

$$F_c = \begin{bmatrix} 0 & 1 & \cdots & \cdots & 0 \\ 0 & 0 & 1 & \cdots & 0 \\ \vdots & & & & \vdots \\ 0 & \cdots & \cdots & 0 & 1 \\ * & \cdots & \cdots & \cdots & * \end{bmatrix}, \quad G_c = \begin{bmatrix} 0 \\ 0 \\ \vdots \\ 0 \\ 1 \end{bmatrix},$$

where F_c is Hurwitz. Because the matrix S is in the companion form and has simple eigenvalues on the imaginary axis, the vector Λ that assigns the eigenvalues of $(F_c + G_c\Lambda^T)$ at those of S will have only $q/2$ (or $(q-1)/2$) elements that depend on the eigenvalues of S. For example, if S has eigenvalues at $\pm j\omega_1$ and $\pm j\omega_2$, the last row of S is

$$\begin{bmatrix} -\omega_1^2\omega_2^2 & 0 & -(\omega_1^2+\omega_2^2) & 0 \end{bmatrix}.$$

One concern with this choice of F is that some components of Λ could be very large when the eigenvalues of S are large,[48] as it can be seen from the foregoing example of S when the frequencies ω_1 and ω_2 are large. To address this concern, F and G are chosen as $F = \varsigma F_c$ and $G = \varsigma G_c$ for some positive constant ς. It can be seen that the coefficient of s^{q-i} in the characteristic equation of $(F + G\Lambda^T)$ is $-\varsigma^i(f_i + \Lambda_i)$, where f_i is the ith element in the last row of F_c and Λ_i the ith element of Λ. If β_i is the coefficient of s^{q-i} in the characteristic equation of S, then $\Lambda_i = -f_i - \beta_i/\varsigma^i$. Knowing the range of values of the eigenvalues of S, the scaling factor ς can be chosen to control the range of the elements of Λ. For example, when the eigenvalues of S are $\pm j\omega_1$ and $\pm j\omega_2$, its characteristic equation is

$$s^4 + (\omega_1^2+\omega_2^2)s^2 + \omega_1^2\omega_2^2 = 0.$$

Choosing $\varsigma^2 = \max(\omega_1^2+\omega_2^2)$ ensures that

$$\frac{\omega_1^2\omega_2^2}{\varsigma^4} < \tfrac{1}{2}, \quad \frac{(\omega_1^2+\omega_2^2)}{\varsigma^2} \le 1.$$

[48]This concern is raised in [133].

To handle the full- and partial-parameter adaptation cases simultaneously, let m be the number of adapted parameters and define m-dimensional vectors λ and ν by

$$\lambda = E\Lambda \quad \text{and} \quad \nu = E\sigma, \tag{5.62}$$

where in the full-parameter case $m = q$, and E is the identity matrix, while in the partial-parameter case λ contains the unknown elements of Λ with $m = q/2$ when q is even and $m = (q-1)/2$ when it is odd. The known elements of Λ are stacked in a vector $\lambda_r = E_r\Lambda$. It can be verified that $\Lambda = E^T\lambda + E_r^T\lambda_r$, where in the full-parameter case the term $E_r^T\lambda_r$ does not exist and $\Lambda = \lambda$.

The forthcoming adaptive law will estimate λ by $\hat{\lambda}$. For notational convenience, define $\hat{\Lambda} = E^T\hat{\lambda} + E_r^T\lambda_r$. This notation applies for both the full- and partial-parameter adaptation cases.

To derive the adaptive law, consider the state feedback control

$$u = -\beta(\xi)\,\text{sat}\left(\frac{\hat{\Lambda}\sigma + s_1}{\mu}\right),$$

$$\dot{\sigma} = F\sigma + \mu G\,\text{sat}\left(\frac{\hat{\Lambda}\sigma + s_1}{\mu}\right),$$

where

$$s_1 = k_1\xi_1 + \cdots + k_{\rho-1}\xi_{\rho-1} + \xi_\rho,$$

which is taken from Section 5.4 with Λ replaced by its estimate $\hat{\Lambda}$. Repeating the analysis of Section 5.4 it is not hard to see that the trajectories enter the set $\Xi \times \Omega_\mu$ in finite time. Inside this set, the closed-loop system is given by

$$\begin{aligned}
\dot{w} &= S_0 w,\\
\dot{\sigma} &= F\sigma + G(\hat{\Lambda}^T\sigma + s_1),\\
\dot{z} &= \tilde{f}_0(z,\xi,\theta,w),\\
\dot{q} &= A_2 q + B_2 s_1,\\
\dot{s}_1 &= -b(\eta,\zeta,\theta)\beta(\xi)(\hat{\Lambda}^T\sigma + s_1)/\mu + \Delta_1(\cdot),
\end{aligned}$$

where $q = \text{col}(\xi_1,\ldots,\xi_{\rho-1})$. In the foregoing equations, $\hat{\Lambda}^T\sigma$ can be written as $\hat{\Lambda}^T\sigma = \Lambda^T\sigma + \tilde{\lambda}^T\nu$, where $\tilde{\lambda} = \hat{\lambda} - \lambda$. Let

$$\varrho = \int_0^{\xi_\rho} \frac{1}{\beta(\xi_1,\ldots,\xi_{\rho-1},y)\tilde{b}(z,\xi_1,\ldots,\xi_{\rho-1},y,\theta,w)}\,dy,$$

where $\tilde{b}(z,\xi,\theta,w) = b(\eta,\zeta,\theta)$. The function ϱ is well defined because $\beta b \geq \beta_0 b_0 > 0$. With the change of variables

$$\vartheta = (\sigma - \bar{\sigma})/\mu + G\varrho, \tag{5.63}$$

the closed-loop system takes the form[49]

$$\begin{aligned}
\dot{w} &= S_0 w,\\
\dot{\vartheta} &= F\vartheta + f_1(z,q,s_1,\theta,w),\\
\dot{z} &= \tilde{f}_0(z,0,\theta,w) + f_2(z,q,s_1,\theta,w),\\
\dot{q} &= A_2 q + B_2 s_1,\\
\dot{s}_1 &= -b(\eta,\zeta,\theta)\beta(\xi)s_1/\mu - b(\eta,\zeta,\theta)\beta(\xi)\tilde{\lambda}^T\nu/\mu + f_3(\vartheta,z,q,s_1,\theta,w),
\end{aligned}$$

in which f_1 to f_3 are locally Lipschitz functions that satisfy $f_1(0,0,0,\theta,w) = 0$, $f_2(z,0,0,\theta,w) = 0$, and $f_3(0,0,0,0,\theta,w) = 0$. Consider the Lyapunov function candidate

$$V_a = \vartheta^T P\vartheta + \kappa_1 V_0 + \kappa_2 q^T P_2 q + V_5 + \frac{\mu}{2\gamma}\tilde{\lambda}^T\tilde{\lambda}, \tag{5.64}$$

where V_0, P, and P_2 are defined in Section 5.4; γ, κ_1, and κ_2 are positive constants; and V_5 is defined by

$$V_5 = \int_0^{s_1} \frac{y}{\beta(\xi_1,\dots,\xi_{\rho-1},y - \sum_{i=1}^{\rho-1} k_i\xi_i)\tilde{b}(z,\xi_1,\dots,\xi_{\rho-1},y - \sum_{i=1}^{\rho-1} k_i\xi_i,\theta,w)}\,dy.$$

It can be seen that

$$\frac{s_1^2}{2b_m\beta_m} \le V_5 \le \frac{s_1^2}{2b_0\beta_0},$$

where b_m and β_m are upper bounds on b and β in the set Ω_μ. The derivative of V_a satisfies the inequality

$$\dot{V}_a \le -\mathcal{Y}_a^T Q_a \mathcal{Y}_a - \frac{1}{\mu}s_1\tilde{\lambda}^T\nu + \frac{\mu}{\gamma}\tilde{\lambda}^T\dot{\lambda},$$

where

$$\mathcal{Y}_a = \text{col}(\|\vartheta\|,\|z\|,\|q\|,|s_1|), \tag{5.65}$$

$$Q_a = \begin{bmatrix} 1 & -c_2 & -c_3 & -c_4\\ -c_2 & \kappa_1 c_5 & -\kappa_1 c_6 & -(\kappa_1 c_7 + c_8)\\ -c_3 & -\kappa_1 c_6 & \kappa_2 & -(\kappa_2 c_9 + c_{10})\\ -c_4 & -(\kappa_1 c_7 + c_8) & -(\kappa_2 c_9 + c_{10}) & 1/\mu - c_{11}\end{bmatrix}, \tag{5.66}$$

and c_1 to c_{11} are positive constants independent of κ_1, κ_2, and μ. Choose κ_1 large enough to make the 2×2 principal minor of Q_a positive; then choose κ_2 large enough to make the 3×3 principal minor positive; then choose μ small enough to make the determinant of Q_a positive. The adaptive law $\dot{\lambda} = (\gamma/\mu^2)s_1\nu$ results in $\dot{V}_a \le -\mathcal{Y}_a^T Q_a \mathcal{Y}_a$, which shows that $\lim_{t\to\infty}\mathcal{Y}_a(t) = 0$ [78, Theorem 8.4]; hence, $\lim_{t\to\infty} e(t) = 0$ because $e = \xi_1 = q_1$. Under output feedback, s_1 is replaced by its estimate $\hat{s}_1$ provided by a high-gain observer; therefore, parameter projection is used, as in Chapter 4, to ensure that $\hat{\lambda}$ remains bounded. Knowing upper bounds on the unknown frequencies of the internal model, it is possible to determine constants a_i and b_i such that λ belongs to the set

$$\Upsilon = \{\lambda \mid a_i \le \lambda_i \le b_i,\ 1 \le i \le m\}.$$

[49]Equations (5.46) and (5.47) are used in arriving at these equations.

Let

$$\Upsilon_\delta = \{\lambda \mid a_i - \delta \le \lambda_i \le b_i + \delta,\ 1 \le i \le m\},$$

where $\delta > 0$. The adaptive law is taken as $\dot{\hat{\lambda}}_i = (\gamma/\mu^2)\mathcal{P}_i(\hat{\lambda}_i, s_1 \nu_i) s_1 \nu_i$, where

$$\mathcal{P}_i(\hat{\lambda}_i, s_1\nu_i) = \begin{cases} 1 + (b_i - \hat{\lambda}_i)/\delta & \text{if } \hat{\lambda}_i > b_i \text{ and } s_1\nu_i > 0, \\ 1 + (\hat{\lambda}_i - a_i)/\delta & \text{if } \hat{\lambda}_i < a_i \text{ and } s_1\nu_i < 0, \\ 1 & \text{otherwise,} \end{cases} \tag{5.67}$$

with $\hat{\lambda}(0) \in \Upsilon$. Similar to Section 4.2, it can be verified that the adaptive law with parameter projection ensures that $\hat{\lambda}$ cannot leave Υ_δ and

$$-\frac{1}{\mu} s_1 \tilde{\lambda}^T \nu + \frac{\mu}{\gamma} \tilde{\lambda}^T \dot{\hat{\lambda}} \le 0.$$

Furthermore, because the adaptive law is derived based on Lyapunov analysis inside $\{|s| \le \mu\}$, it is modified to keep $\hat{\lambda}$ constant outside $\{|s| \le 2\mu\}$. Let

$$\Pi(s,\mu) = \begin{cases} 0 & \text{if } \ |s| \ge 2\mu, \\ 1 & \text{if } \ |s| \le 1.5\mu, \\ 4 - 2|s|/\mu & \text{if } \ 1.5\mu < |s| < 2\mu \end{cases} \tag{5.68}$$

and take the adaptive law as

$$\dot{\hat{\lambda}}_i = (\gamma/\mu^2)\Pi(s,\mu)\mathcal{P}_i(\hat{\lambda}_i, s_1\nu_i) s_1 \nu_i \quad \text{for } 1 \le i \le m. \tag{5.69}$$

Inside $\{|s| \le \mu\}$, $\Pi = 1$; hence the Lyapunov analysis is not affected by Π. The transition of Π from zero to one is done over an interval to ensure that the adaptive law is locally Lipschitz.

The foregoing Lyapunov analysis shows convergence of the regulation error to zero. Convergence of $\hat{\lambda}$ to λ depends on the persistence of excitation of the vector $\bar{\nu} = E\bar{\sigma}$. The q-dimensional vector $\bar{\sigma}$ is persistently exciting if all q modes of S are excited by the initial condition $w(0)$. Due to observability of the pair (S, H), a mode of S is excited if and only if it is present in ϕ. When $\bar{\sigma}$ is persistently exciting, so is $\bar{\nu}$ because E has full rank. Depending on the initial condition $w(0)$, some modes of S may be absent from ϕ.[50] The following lemma deals with the case when $\bar{\nu}$ is not persistently exciting.

Lemma 5.1. *When $\bar{\nu} \ne 0$, there is a nonsingular matrix L, possibly dependent on θ and w, such that*

$$L\bar{\nu} = \begin{bmatrix} \bar{\nu}_a \\ 0 \end{bmatrix}, \tag{5.70}$$

where $\bar{\nu}_a$ is an m_0-dimensional persistently exciting vector with $m_0 \le m$. When $m_0 = m$, L is the identity matrix and $\bar{\nu}_a = \bar{\nu} = \bar{\sigma}$. When $\bar{\nu} = 0$, (5.70) holds with $\bar{\nu}_a = 0$. ◇

Proof: Consider the case when S has r pairs of pure imaginary eigenvalues and a real eigenvalue at the origin, that is, $q = 2r + 1$. The case when S does not have a zero

[50] If all the modes of S can be excited by come choice of $w(0)$, the internal model (5.36) is called the minimal internal model [100].

eigenvalue follows in a straightforward way. For convenience, the cases of full- and partial-parameter adaptations are treated separately. Let P_d be a nonsingular matrix that transforms S into its real modal form, that is, $P_d^{-1}SP_d = S_d$, where S_d is a block diagonal matrix that has a zero diagonal element corresponding to the zero eigenvalues and r diagonal blocks of the form $\left[\begin{smallmatrix} 0 & \omega_i \\ -\omega_i & 0 \end{smallmatrix}\right]$ corresponding to the r complex modes. Applying the change of variables $\Phi_d = P_d^{-1}\Phi$ to (5.36) results in

$$\dot{\Phi}_d = S_d\Phi_d, \quad \phi = HP_d\Phi_d.$$

In the full-parameter adaptation case, assume, without loss of generality, that the elements of Φ_d are ordered such that the first m_0 elements contain the modes of S that are present in ϕ. Partition Φ_d as $\Phi_d = \mathrm{col}(\Phi_a, \Phi_b)$, where the dimension of Φ_a is m_0. It follows from the observability of the pair (S, H) that $\Phi_b = 0$. Hence,

$$\bar{\sigma} = -\frac{\mu}{\beta(0)}M\Phi = -\frac{\mu}{\beta(0)}MP_d\begin{pmatrix}\Phi_a \\ 0\end{pmatrix}.$$

The $q \times q$ matrix MP_d is nonsingular. Taking $L = (MP_d)^{-1}$ yields

$$L\bar{\sigma} = -\frac{\mu}{\beta(0)}\begin{pmatrix}\Phi_a \\ 0\end{pmatrix}.$$

The m_0-dimensional vector Φ_a is persistently exciting. In the partial-parameter adaptation case, $F = \varsigma F_c$, $G = \varsigma G_c$, and $m = r$. It can be verified that $F + G\Lambda^T = DSD^{-1}$, where $D = \mathrm{diag}(\varsigma^q, \varsigma^{q-1}, \ldots, \varsigma)$. The equation $MS - FM = GH$ implies that $MS - (F + G\Lambda^T)M = 0$. Substitution of $(F + G\Lambda^T) = DSD^{-1}$ and $S = P_dS_dP_d^{-1}$ in the foregoing equation yields $M_dS_d - S_dM_d = 0$, where $M_d = P_d^{-1}D^{-1}MP_d$ is a block diagonal matrix with diagonal blocks compatible with the diagonal blocks of S_d. Since S is in the companion form, P_d takes the form

$$\begin{bmatrix} 1 & 1 & 0 & \cdots \\ 0 & 0 & \omega_1 & \cdots \\ 0 & -\omega_1^2 & 0 & \cdots \\ 0 & 0 & -\omega_1^3 & \cdots \\ 0 & \omega_1^4 & 0 & \cdots \\ \vdots & \vdots & \vdots & \cdots \end{bmatrix},$$

where in writing P_d it is assumed, without loss of generality, that the diagonal blocks of S_d are ordered such that the first element is zero, followed by $m_0 \le r$ blocks that contain the complex modes of S that are present in ϕ. The $r \times q$ matrix E of (5.62) takes the form

$$E = \begin{bmatrix} 0 & 1 & 0 & 0 & \cdots & 0 & 0 & 0 \\ 0 & 0 & 0 & 1 & \cdots & 0 & \cdots & 0 \\ \cdots & & & & & & & \cdots \\ 0 & 0 & \cdots & & & \cdots & 1 & 0 \end{bmatrix},$$

and $\bar{\nu} = E\bar{\sigma} = -(\mu/\beta(0))EMP_d\Phi_d = -(\mu/\beta(0))EDP_dM_d\Phi_d$. Using the forms of E and P_d and the fact that D is diagonal and M_d is block diagonal, it can be seen that the first column of EMP_d is zero, columns i and $i+1$ are linearly dependent for $i =$

$2,4,\ldots,q-1$, and $\text{rank}(EMP_d)=r$. Let $(EMP_d)_{\text{red}}$ be an $r\times r$ matrix that is formed of the first r linearly independent columns of EMP_d and define an r-dimensional vector Φ_{pd} by stacking the elements of Φ_d corresponding to the columns maintained in $(EMP_d)_{\text{red}}$. If $m_0=r$, the vectors Φ_{pd} and $\bar{\nu}=-(\mu/\beta(0))(EP_dM)_{\text{red}}\,\Phi_{pd}$ are persistently exciting. If $m_0<r$, $\Phi_{pd}=\text{col}(\Phi_{pa},0)$, where Φ_{pa} is an m_0-dimensional persistently exciting vector. Taking $L=((EMP_d)_{\text{red}})^{-1}$ yields $L\bar{\nu}=-(\mu/\beta(0))$ $\text{col}(\Phi_{pa},0)$. □

Remark 5.14. *In the partial-parameter adaptation case, the vector $\bar{\nu}$ will be persistently exciting if all complex modes of S are excited even if the zero-eigenvalue mode is not excited.* ◇

The output feedback controller is given by

$$\dot{\sigma}=F\sigma+\mu G\,\text{sat}\left(\frac{\hat{s}}{\mu}\right),\tag{5.71}$$

$$u=-\beta_s(\hat{\xi})\,\text{sat}\left(\frac{\hat{s}}{\mu}\right),\tag{5.72}$$

$$\dot{\hat{\lambda}}_i=(\gamma/\mu^2)\Pi(\hat{s},\mu)\mathscr{P}_i(\hat{\lambda}_i,\hat{s}_1\nu_i)\hat{s}_1\nu_i\quad\text{for } 1\le i\le m,\tag{5.73}$$

where $\hat{s}=\hat{\Lambda}^T\sigma+\hat{s}_1$, $\hat{s}_1=k_1\hat{\xi}_1+\cdots+k_{\rho-1}\hat{\xi}_{\rho-1}+\hat{\xi}_\rho$, $\hat{\Lambda}=E^T\hat{\lambda}+E_r^T\lambda_r$, $\hat{\lambda}=\text{col}(\hat{\lambda}_1,\ldots,\hat{\lambda}_m)$, $\nu=E\sigma$, γ is a positive constant, $\hat{\xi}$ is provided by the high-gain observer (5.50)–(5.51), $\mathscr{P}_i$ is defined by (5.67), and Π is defined by (5.68). This is the same controller (5.48)–(5.51) of Section 5.4 with Λ replaced by $\hat{\Lambda}$.

Theorem 5.5. *Suppose Assumptions* 5.5 *to* 5.11 *are satisfied and consider the closed-loop system formed of the system* (5.29)–(5.30), *the conditional servocompensator* (5.71), *the controller* (5.72), *the adaptive law* (5.73), *and the observer* (5.50)–(5.51). *Let Ψ be a compact set in the interior of Ω and suppose $\sigma(0)\in\Xi$, $(z(0),q(0),s(0))\in\Psi$, $\hat{\lambda}(0)\in\Upsilon$, and $\hat{\xi}(0)$ is bounded. Then $\mu^*>0$ exists, and for each $\mu\in(0,\mu^*]$, $\varepsilon^*=\varepsilon^*(\mu)$ exists such that for each $\mu\in(0,\mu^*]$ and $\varepsilon\in(0,\varepsilon^*(\mu)]$, all state variables are bounded and $\lim_{t\to\infty}e(t)=0$. Furthermore, let $\chi=(z,\xi)$ be part of the state of the closed-loop system under the output feedback controller, and let $\chi^*=(z^*,\xi^*)$ be the state of the closed-loop system under the state feedback sliding mode controller* (5.42), *with $\chi(0)=\chi^*(0)$. Then, for every $\delta_0>0$, there is $\mu_1^*>0$, and for each $\mu\in(0,\mu_1^*]$, there is $\varepsilon_1^*=\varepsilon_1^*(\mu)>0$ such that for $\mu\in(0,\mu_1^*]$ and $\varepsilon\in(0,\varepsilon_1^*(\mu)]$,*

$$\|\chi(t)-\chi^*(t)\|\le\delta_0\quad\forall\,t\ge 0.\tag{5.74}$$

Finally, if $\bar{\nu}$ is persistently exciting, then $\lim_{t\to\infty}\hat{\lambda}(t)=\lambda$. ◇

Proof: Similar to the case of no adaption (Theorem 5.3), it can be shown that, for sufficiently small μ and ε, the set $\Xi\times\Omega\times\Sigma_\varepsilon$ is positively invariant and the trajectory $(\vartheta(t),z(t),q(t),s(t),\varphi(t))$ enters $\Xi\times\Omega_\mu\times\Sigma_\varepsilon$ in finite time. On the other hand, due to parameter projection $\hat{\lambda}(t) \in \Upsilon_\delta$ for all $t \geq 0$. Inside $\Xi \times \Omega_\mu \times \Sigma_\varepsilon$ the closed-loop system is represented by[51]

$$\begin{aligned}
\dot{w} &= S_0 w,\\
\dot{\sigma} &= F\sigma + G(\hat{\Lambda}^T\sigma + s_1) + \mu G[\text{sat}(\hat{s}/\mu) - \text{sat}(s/\mu)],\\
\dot{z} &= \tilde{f}_0(z, \xi, \theta, w),\\
\dot{q} &= A_2 q + B_2 s_1,\\
\dot{s}_1 &= -b(\eta, \zeta, \theta)\beta(\xi)(\hat{\Lambda}^T\sigma + s_1)/\mu + \Delta_1(\cdot)\\
&\quad + b(\eta, \zeta, \theta)[-\beta_s(\hat{\xi})\,\text{sat}(\hat{s}/\mu) + \beta_s(\xi)\,\text{sat}(s/\mu)],\\
\varepsilon\dot{\varphi} &= A_0\varphi + \varepsilon B[a_0(z, \xi, \theta, w) - b(\eta, \zeta, \theta)\phi(\theta, w) - b(\eta, \zeta, \theta)\beta_s(\hat{\xi})\,\text{sat}(\hat{s}/\mu)],\\
\dot{\hat{\lambda}} &= (\gamma/\mu^2)\Pi(\hat{s}, \mu)\mathscr{P}(\hat{\lambda}, \hat{s}_1\nu)\hat{s}_1\nu,
\end{aligned}$$

where $\mathscr{P}$ is a diagonal matrix whose diagonal elements are $\mathscr{P}_1$ to $\mathscr{P}_m$. Inside $\Xi \times \Omega_\mu \times \Sigma_\varepsilon$, $|s| \leq \mu$ and $|s - \hat{s}| = O(\varepsilon)$. For sufficiently small ε, $|\hat{s}| \leq 1.5\mu$; hence, $\Pi(\hat{s}, \mu) = 1$. Applying the change of variables (5.63), the system can be represented in the form

$$\begin{aligned}
\dot{w} &= S_0 w,\\
\dot{\vartheta} &= F\vartheta + f_1(z, q, s_1, \theta, w) + h_1(z, q, s_1, \varphi, \theta, w)/\mu,\\
\dot{z} &= \tilde{f}_0(z, 0, \theta, w) + f_2(z, q, s_1, \theta, w),\\
\dot{q} &= A_2 q + B_2 s_1,\\
\dot{s}_1 &= -b(\eta, \zeta, \theta)\beta(\xi)s_1/\mu - b(\eta, \zeta, \theta)\beta(\xi)\tilde{\lambda}^T\nu/\mu + f_3(\vartheta, z, q, s_1, \theta, w)\\
&\quad + h_2(z, q, s_1, \varphi, \tilde{\lambda}, \theta, w)/\mu,\\
\varepsilon\dot{\varphi} &= A_0\varphi + (\varepsilon/\mu)B[-b(\eta, \zeta, \theta)\beta(\xi)s_1/\mu - b(\eta, \zeta, \theta)\beta(\xi)\tilde{\lambda}^T\nu/\mu]\\
&\quad + f_4(\vartheta, z, q, s_1, \theta, w) + h_3(\vartheta, z, q, s_1, \varphi, \tilde{\lambda}, \theta, w)/\mu,\\
\dot{\hat{\lambda}} &= (\gamma/\mu^2)\mathscr{P}(\hat{\lambda}, \hat{s}_1\nu)\hat{s}_1\nu \;=\; (\gamma/\mu^2)\mathscr{P}(\hat{\lambda}, s_1\nu)s_1\nu + h_4(\vartheta, q, s_1, \varphi, \tilde{\lambda}, \theta, w)/\mu,
\end{aligned}$$

where h_1 to h_4 are locally Lipschitz functions that vanish at $\varphi = 0$, and f_1 to f_4 are locally Lipschitz functions that satisfy $f_1(0, 0, 0, \theta, w) = 0$, $f_2(z, 0, 0, \theta, w) = 0$, $f_3(0, 0, 0, 0, \theta, w) = 0$, and $f_4(0, 0, 0, 0, \theta, w) = 0$. By repeating the state feedback analysis, it can be seen that the derivative of V_a of (5.64) satisfies

$$\dot{V}_a \leq -\mathscr{Y}_a^T Q_a \mathscr{Y}_a + (\ell_1 + \mu\ell_2)\|\mathscr{Y}_a\|\,\|\varphi\|/\mu + \ell_3\|\tilde{\lambda}^T\nu\|\,\|\varphi\|/\mu, \tag{5.75}$$

where $\mathscr{Y}_a$ and Q_a are defined by (5.65) and (5.66), respectively, and ℓ_1 to ℓ_3 are positive constants. In arriving at (5.75) the term $-s_1\tilde{\lambda}^T\nu + +\tilde{\lambda}^T\mathscr{P}\hat{s}_1\nu$ is written as

$$\begin{aligned}
-s_1\tilde{\lambda}^T\nu + \tilde{\lambda}^T\mathscr{P}(\hat{\lambda}, \hat{s}_1\nu)\hat{s}_1\nu &= -\hat{s}_1\tilde{\lambda}^T\nu + +\tilde{\lambda}^T\mathscr{P}(\hat{\lambda}, \hat{s}_1\nu)\hat{s}_1\nu + (\hat{s}_1 - s_1)\tilde{\lambda}^T\nu\\
&\leq (\hat{s}_1 - s_1)\tilde{\lambda}^T\nu,
\end{aligned}$$

where $-\hat{s}_1\tilde{\lambda}^T\nu + \tilde{\lambda}^T\mathscr{P}(\hat{\lambda}, \hat{s}_1\nu)\hat{s}_1\nu \leq 0$ according to the adaptive law. As it was shown earlier, the matrix Q_a is positive definite for sufficiently large κ_1 and κ_2 and sufficiently small μ.

[51] See the proof of Theorem 5.3 for the definition of variables that are not defined here.

By Lemma 5.1 and (5.63),

$$\tilde{\lambda}^T \nu = \tilde{\lambda}^T L^{-1} L E \sigma = \tilde{\lambda}^T L^{-1} L E(\mu\vartheta + \bar{\sigma} - \mu G \varrho]$$
$$= \begin{bmatrix} \tilde{\lambda}_a^T & \tilde{\lambda}_b^T \end{bmatrix} \begin{bmatrix} \mu\vartheta_a - \bar{\nu}_a - \mu G_a \varrho \\ \mu\vartheta_b - \mu G_b \varrho \end{bmatrix}, \tag{5.76}$$

where

$$\begin{bmatrix} \vartheta_a \\ \vartheta_b \end{bmatrix} = LE\vartheta, \quad \begin{bmatrix} G_a \\ G_b \end{bmatrix} = LEG, \quad \text{and} \quad \begin{bmatrix} \tilde{\lambda}_a \\ \tilde{\lambda}_b \end{bmatrix} = L^{-T}\tilde{\lambda}.$$

The $\dot{s}_1$- and $\dot{\tilde{\lambda}}$-equations can be written as

$$\begin{bmatrix} \dot{s}_1 \\ \dot{\tilde{\lambda}}_a \end{bmatrix} = (b\beta/\mu) \begin{bmatrix} -1 & -\bar{\nu}_a^T \\ \gamma\bar{\nu}_a & 0 \end{bmatrix} \begin{bmatrix} s_1 \\ \tilde{\lambda}_a \end{bmatrix}$$
$$+ \begin{bmatrix} -b\beta\tilde{\lambda}_a^T(\vartheta_a - G_a\varrho) - b\beta\tilde{\lambda}_b^T(\vartheta_b - G_b\varrho) + f_3 + h_2/\mu \\ -\gamma b\beta\bar{\nu}_a s_1/\mu + \gamma N_a \mathscr{P} s_1 \nu/\mu^2 + N_a h_4/\mu \end{bmatrix}, \tag{5.77}$$

where $b = b(\eta, \zeta, \theta)$, $\beta = \beta(\xi)$, and $L^{-T} = \text{col}(N_a, N_b)$. The right-hand side of (5.77) vanishes at $(\vartheta = 0, z = 0, q = 0, s_1 = 0, \varphi = 0, \tilde{\lambda}_a = 0)$ regardless of $\tilde{\lambda}_b$, which is bounded due to parameter projection. Therefore, $\tilde{\lambda}_b$ is treated as bounded time-varying disturbance. Because $\bar{\nu}_a$ is persistently exciting, the origin of the systems

$$\begin{bmatrix} \dot{s}_1 \\ \dot{\tilde{\lambda}}_a \end{bmatrix} = (b\beta/\mu) \begin{bmatrix} -1 & -\bar{\nu}_a^T \\ \gamma\bar{\nu}_a & 0 \end{bmatrix} \begin{bmatrix} s_1 \\ \tilde{\lambda}_a \end{bmatrix}$$

is exponentially stable [75, Section 13.4]. By the converse Lyapunov theorem [78, Theorem 4.14], there is a Lyapunov function V_6 whose derivative along the system (5.77) satisfies the inequality

$$\dot{V}_6 \le -k_1|s_1|^2 - k_2||\tilde{\lambda}_a||^2 + k_3||\mathscr{Y}_a||^2 + k_4||\mathscr{Y}_a||\,||\tilde{\lambda}_a|| + k_5||\mathscr{Y}_a||\,||\varphi|| + k_6||\tilde{\lambda}_a||\,||\varphi||,$$

where, from now on, the positive constants k_i could depend on μ but are independent of ε. Using (5.76) in (5.75) shows that

$$\dot{V}_a \le -k_7||\mathscr{Y}_a||^2 + k_8||\mathscr{Y}_a||\,||\varphi|| + k_9||\tilde{\lambda}_a||\,||\varphi||.$$

For the $\dot{\varphi}$-equation, it can be shown that the derivative of $V_3 = \varphi^T P_0 \varphi$ satisfies the inequality

$$\dot{V}_3 \le -||\varphi||^2/\varepsilon + k_{10}||\varphi||^2 + k_{11}||\tilde{\lambda}_a||\,||\varphi|| + k_{12}||\mathscr{Y}_a||\,||\varphi||.$$

Using $W = \alpha V_a + V_3 + V_6$ with $\alpha > 0$ as a Lyapunov function candidate for the closed-loop system, it can be shown that

$$\dot{W} \le -\mathscr{Y}^T Q \mathscr{Y},$$

where

$$\mathscr{Y} = \begin{bmatrix} ||\mathscr{Y}_a|| \\ ||\tilde{\lambda}_a|| \\ ||\varphi|| \end{bmatrix} \quad \text{and} \quad Q = \frac{1}{2}\begin{bmatrix} 2(\alpha k_7 - k_3) & -k_4 & -(k_{13} + \alpha k_8) \\ -k_4 & 2k_2 & -(k_{14} + \alpha k_9) \\ -(k_{13} + \alpha k_8) & -(k_{14} + \alpha k_9) & 2(1/\varepsilon - k_{10}) \end{bmatrix},$$

where $k_{13} = k_5 + k_{12}$ and $k_{14} = k_6 + k_{11}$. Choose α large enough to make the 2×2 principal minor of Q positive; then choose ε small enough to make Q positive definite. Thus, $\lim_{t \to \infty} \|\mathscr{Y}(t)\| = 0$ [78, Theorem 8.4], which implies that $\lim_{t \to \infty} e(t) = 0$ and $\lim_{t \to \infty} \tilde{\lambda}_a(t) = 0$. If $\bar{v}$ is persistently exciting, it follows from Lemma 5.1 that $\bar{v}_a = \bar{v}$. Hence, $\tilde{\lambda} = \tilde{\lambda}_a$. Therefore, $\lim_{t \to \infty} \hat{\lambda}(t) = \lambda$. The proof of (5.74) is done as in the proof of Theorem 5.2, where μ is reduced first to bring the trajectories under the state feedback controller with conditional servocompensator close to the trajectories under the sliding mode controller, then ε is reduced to bring the trajectories under output feedback close to the ones under state feedback. □

Remark 5.15. *The proof shows that when $\bar{v}$ is not persistently exciting, partial parameter convergence is achieved as* $\lim_{t \to \infty} \tilde{\lambda}_a(t) = 0$. ◇

The performance of the conditional servocompensator with adaptive internal model is illustrated by two examples. Example 5.7 reconsiders Example 5.4 and compares the adaptive internal model with the known one. Example 5.8 illustrates the effect of the persistence of excitation condition. Both examples use partial parameter adaptation.

Example 5.7. Consider the system

$$\dot{x}_1 = -\theta_1 x_1 + x_2^2 + d, \quad \dot{x}_2 = x_3, \quad \dot{x}_3 = -\theta_2 x_1 x_2 + u, \quad y = x_2$$

from Example 5.4, where $\theta_1 > 0$ and θ_2 are unknown parameters, d is a constant disturbance, and the reference signal $r = \alpha \sin(\omega_0 t + \theta_0)$ has unknown frequency ω_0, in addition to unknown amplitude and phase. Building on Example 5.4, the output feedback controller is taken as

$$\dot{\sigma} = F\sigma + \mu G \,\text{sat}\left(\frac{\hat{\Lambda}^T \sigma + \hat{\xi}_1 + \hat{\xi}_2}{\mu}\right),$$

$$u = -20 \,\text{sat}\left(\frac{\hat{\Lambda}^T \sigma + \hat{\xi}_1 + \hat{\xi}_2}{\mu}\right),$$

$$\dot{\hat{\xi}}_1 = \hat{\xi}_2 + (2/\varepsilon)(e - \hat{\xi}_1), \qquad \dot{\hat{\xi}}_2 = (1/\varepsilon^2)(e - \hat{\xi}_1),$$

where the only change is that Λ^T is replaced by

$$\hat{\Lambda}^T = \begin{bmatrix} \hat{\Lambda}_1, & 6.25, & \hat{\Lambda}_3, & 5 \end{bmatrix}.$$

Assuming that $\omega_0 \in [1.2, 3]$, the scaling factor $\varsigma = 3$ remains the same as in Example 5.4. Using the expressions $\Lambda_1 = -9(\omega_0/\varsigma)^4 + 1.5$ and $\Lambda_3 = -10(\omega_0/\varsigma)^2 + 8.75$, it can be verified that

$$-7.5 \le \Lambda_1 \le 1.27 \quad \text{and} \quad -1.25 \le \Lambda_3 \le 7.15.$$

The sets Υ and δ are taken as

$$\Upsilon = \{(\Lambda_1, \Lambda_3) \mid -7.5 \le \Lambda_1 \le 1.3, \ -1.25 \le \Lambda_3 \le 7.7\}, \quad \delta = 0.1.$$

With $\hat{\lambda}_1 = \hat{\Lambda}_1$ and $\hat{\lambda}_2 = \hat{\Lambda}_3$, the matrix E of (5.62) is given by

$$E = \begin{bmatrix} 1 & 0 & 0 & 0 \\ 0 & 0 & 1 & 0 \end{bmatrix}.$$

Therefore $\nu_1 = \sigma_1$ and $\nu_2 = \sigma_3$. The adaptive law is given by

$$\dot{\hat{\lambda}}_i = (\gamma/\mu^2)\Pi\mathscr{P}_i\hat{s}_1\nu_i \quad \text{for } i = 1,2,$$

where $\hat{s}_1 = \hat{\xi}_1 + \hat{\xi}_2$,

$$\mathscr{P}_1 = \begin{cases} 1+(1.3-\hat{\lambda}_1)/0.1 & \text{if } \hat{\lambda}_1 > 1.3 \text{ and } \hat{s}_1\nu_1 > 0, \\ 1+(\hat{\lambda}_1+7.5)/0.1 & \text{if } \hat{\lambda}_1 < -7.5 \text{ and } \hat{s}_1\nu_1 < 0, \\ 1 & \text{otherwise,} \end{cases}$$

$$\mathscr{P}_2 = \begin{cases} 1+(7.7-\hat{\lambda}_2)/0.1 & \text{if } \hat{\lambda}_2 > 7.7 \text{ and } \hat{s}_1\nu_2 > 0, \\ 1+(\hat{\lambda}_2+1.25)/0.1 & \text{if } \hat{\lambda}_2 < -1.25 \text{ and } \hat{s}_1\nu_2 < 0, \\ 1 & \text{otherwise,} \end{cases}$$

and Π is defined by (5.68). The simulation is carried out with $\gamma = 10^5$, $\varepsilon = 10^{-4}$, and $\hat{\lambda}(0) = \hat{\lambda}_2(0) = 0$. All other parameters and initial conditions are the same as in Example 5.4. The simulation results are shown in Figure 5.11. Figures 5.11(a) and (b) show the regulation error e and the control signal u for the adaptive internal model (solid) and the known internal model (dashed). It can be seen that the error trajectories are very close and that the control trajectories are almost indistinguishable. Recall from Example 5.4 that the trajectories under the known internal model are very close to the ones under sliding mode control, so the same observation applies to the trajectories under the adaptive internal model. Figures 5.11(c) and (d) show the convergence of the parameter errors to zero, which is valid because all modes of the internal model are excited; hence the signal $\bar{\nu}$ is persistently exciting. △

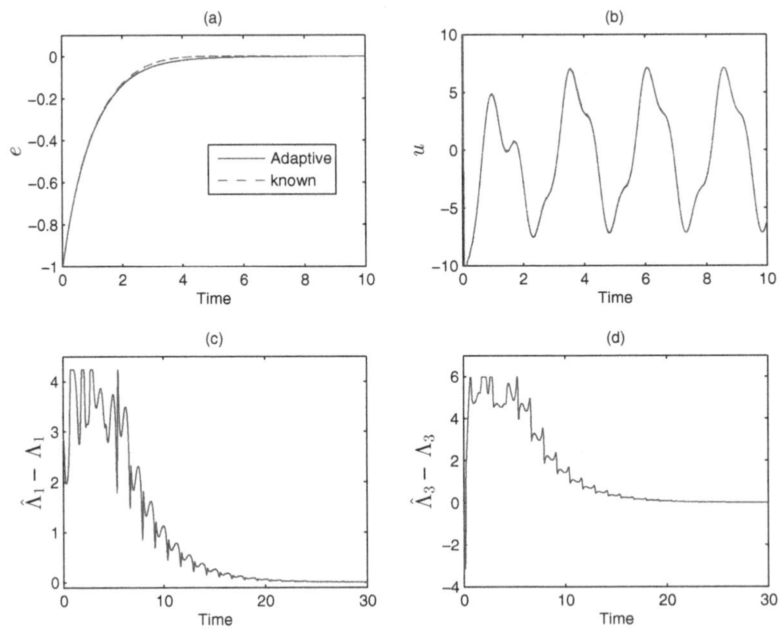

Figure 5.11. *Simulation of Example* 5.7. (a) *and* (b) *show the regulation error e and the control signal u for the adaptive internal model (solid) and the known internal model (dashed).* (c) *and* (d) *show the parameter errors for the adaptive internal model.*

Example 5.8. Consider the system

$$\dot{x}_1 = -\theta_1 x_1 + x_2^2 + d, \quad \dot{x}_2 = x_3, \quad \dot{x}_3 = -\theta_2 x_1 + u, \quad y = x_2,$$

where $\theta_1 > 0$ and θ_2 are unknown parameters, and $d = \alpha_1 \sin(\omega_1 t + \phi_1) + \alpha_2 \sin(\omega_2 t + \phi_2)$ is a disturbance input. The output y is to be regulated to a constant reference r. Assumptions 5.5 and 5.6 are satisfied globally with

$$\eta = x_1, \quad \zeta_1 = x_2, \quad \zeta_2 = x_3.$$

The exosystem of Assumption 5.7 is given by

$$\dot{w} = \begin{bmatrix} 0 & 0 & 0 & 0 & 0 \\ 0 & 0 & \omega_1 & 0 & 0 \\ 0 & -\omega_1 & 0 & 0 & 0 \\ 0 & 0 & 0 & 0 & \omega_2 \\ 0 & 0 & 0 & -\omega_2 & 0 \end{bmatrix} w, \quad r = w_1, \quad d = w_2 + w_4.$$

It can be verified that $\tau_1 = w_1$, $\tau_2 = 0$, the solution of (5.33) is

$$\tau_0 = \frac{1}{\theta_1} w_1^2 + \frac{1}{\theta_1^2 + \omega_1^2}(\theta_1 w_2 - \omega_1 w_3) + \frac{1}{\theta_1^2 + \omega_2^2}(\theta_1 w_4 - \omega_2 w_5),$$

and $\phi = \theta_2 \tau_0$ satisfies the differential equation

$$\phi^{(5)} + (\omega_1^2 + \omega_2^2)\phi^{(3)} + \omega_1^2 \omega_2^2 \phi = 0.$$

Thus, the internal model is given by

$$S = \begin{bmatrix} 0 & 1 & 0 & 0 & 0 \\ 0 & 0 & 1 & 0 & 0 \\ 0 & 0 & 0 & 1 & 0 \\ 0 & 0 & 0 & 0 & 1 \\ 0 & -\omega_1^2\omega_2^2 & 0 & -(\omega_1^2 + \omega_2^2) & 0 \end{bmatrix}, \quad H = \begin{bmatrix} 1 & 0 & 0 & 0 & 0 \end{bmatrix}.$$

With the change of variables

$$z = x_1 - \tau_0, \quad \xi_1 = x_2 - w_1, \quad \xi_2 = x_3,$$

the system is represented by

$$\begin{aligned} \dot{z} &= -\theta_1 z + \xi_1^2 + 2\xi_1 w_1, \\ \dot{\xi}_1 &= \xi_2, \\ \dot{\xi}_2 &= -\theta_2 z + u - \phi(\theta, w). \end{aligned}$$

With known frequencies ω_1 and ω_2, the output feedback controller is taken as[52]

$$\begin{aligned} \dot{\sigma} &= F\sigma + \mu G \,\text{sat}\left(\frac{\Lambda^T \sigma + \hat{\xi}_1 + \hat{\xi}_2}{\mu}\right), \\ u &= -20 \,\text{sat}\left(\frac{\Lambda^T \sigma + \hat{\xi}_1 + \hat{\xi}_2}{\mu}\right), \\ \dot{\hat{\xi}}_1 &= \hat{\xi}_2 + (2/\varepsilon)(e - \hat{\xi}_1), \qquad \dot{\hat{\xi}}_2 = (1/\varepsilon^2)(e - \hat{\xi}_1), \end{aligned}$$

[52]The gain 20 is determined by simulation.

where $e = y - r$,

$$F = \varsigma \begin{bmatrix} 0 & 1 & 0 & 0 & 0 \\ 0 & 0 & 1 & 0 & 0 \\ 0 & 0 & 0 & 1 & 0 \\ 0 & 0 & 0 & 0 & 1 \\ -3.75 & -17.125 & -28.125 & -21.25 & -7.5 \end{bmatrix} \quad \text{and} \quad G = \varsigma \begin{bmatrix} 0 \\ 0 \\ 0 \\ 0 \\ 1 \end{bmatrix}.$$

The eigenvalues of F/ς are -0.5, -1, -1.5, -2, and -2.5. The vector Λ that assigns the eigenvalues of $F + G\Lambda^T$ at the eigenvalues of S is

$$\Lambda^T = \begin{bmatrix} 3.75 & (17.125 - \omega_1^2\omega_2^2/\varsigma^4) & 28.125 & (21.25 - (\omega_1^2 + \omega_2^2)/\varsigma^2) & 7.5 \end{bmatrix}.$$

Assuming that ω_1 and ω_2 belong to the interval $[1,3]$, ς is taken as 3. When ω_1 and ω_2 are unknown, Λ^T is replaced by

$$\hat{\Lambda}^T = \begin{bmatrix} 3.75 & \hat{\Lambda}_2 & 28.125 & \hat{\Lambda}_4 & 7.5 \end{bmatrix}.$$

It can be verified that

$$16.125 \le \Lambda_2 \le 17.1 \quad \text{and} \quad 19.25 \le \Lambda_4 \le 21.$$

The sets Υ and δ are taken as

$$\Upsilon = \{(\Lambda_2, \Lambda_4) \mid 16.1 \le \Lambda_2 \le 17.2,\ 19.1 \le \Lambda_4 \le 21\}, \quad \delta = 0.1.$$

With $\hat{\lambda}_1 = \hat{\Lambda}_2$ and $\hat{\lambda}_2 = \hat{\Lambda}_4$, the matrix E of (5.62) is given by

$$E = \begin{bmatrix} 0 & 1 & 0 & 0 & 0 \\ 0 & 0 & 0 & 1 & 0 \end{bmatrix}.$$

Therefore, $\nu_1 = \sigma_2$ and $\nu_2 = \sigma_4$. The adaptive law is given by

$$\dot{\hat{\lambda}}_i = (\gamma/\mu^2)\Pi \mathscr{P}_i \hat{s}_1 \nu_i \quad \text{for } i = 1, 2,$$

where $\hat{s}_1 = \hat{\xi}_1 + \hat{\xi}_2$,

$$\mathscr{P}_1 = \begin{cases} 1 + (17.2 - \hat{\lambda}_1)/0.1 & \text{if } \hat{\lambda}_1 > 17.2 \text{ and } \hat{s}_1\nu_1 > 0, \\ 1 + (\hat{\lambda}_1 - 16.1)/0.1 & \text{if } \hat{\lambda}_1 < 16.1 \text{ and } \hat{s}_1\nu_1 < 0, \\ 1 & \text{otherwise}, \end{cases}$$

$$\mathscr{P}_2 = \begin{cases} 1 + (21 - \hat{\lambda}_2)/0.1 & \text{if } \hat{\lambda}_2 > 21 \text{ and } \hat{s}_1\nu_2 > 0, \\ 1 + (\hat{\lambda}_2 - 19.1)/0.1 & \text{if } \hat{\lambda}_2 < 19.1 \text{ and } \hat{s}_1\nu_2 < 0, \\ 1 & \text{otherwise}, \end{cases}$$

and Π is defined by (5.68). The simulation is carried out with the parameters $\theta_1 = 3$, $\theta_2 = 4$, $\omega_1 = 1$, $\omega_2 = 2$, $\mu = 0.1$, $\varepsilon = 10^{-4}$, and $\gamma = 10^5$ and the initial conditions $x(0) = \text{col}(1,2,0)$, $\hat{\xi}(0) = \text{col}(0,0)$, and $\hat{\lambda}(0) = \text{col}(16.5, 20)$. The simulation results of Figures 5.12 and 5.13 deal with two cases, depending on whether the persistence of excitation condition is satisfied. Figures 5.12(a) and (b) show the regulation error e and the parameter errors $\hat{\Lambda}_2 - \Lambda_2$ (solid) and $\hat{\Lambda}_4 - \Lambda_4$ (dashed) when $w(0) = \text{col}(1,2,0,0,1)$. In

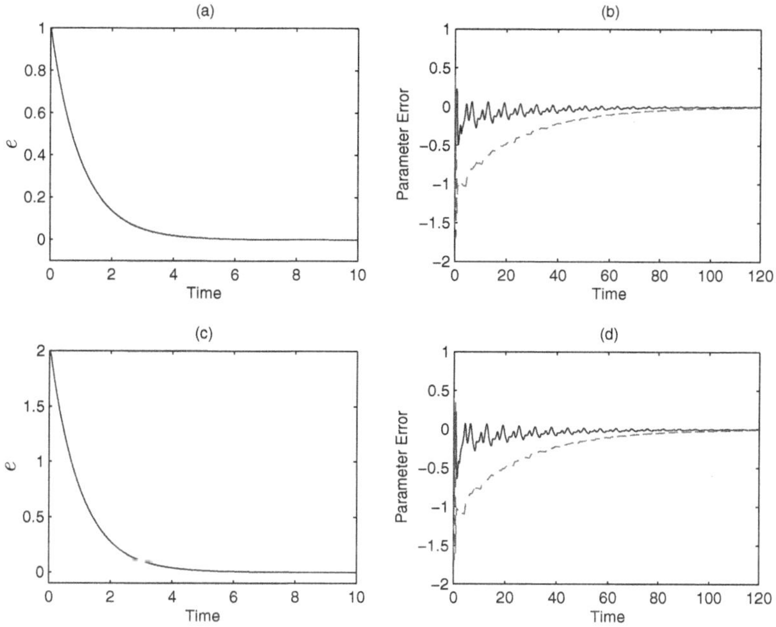

Figure 5.12. *Simulation of Example* 5.8. (a) *and* (b) *show results when all modes are excited, while in* (c) *and* (d) *the constant mode is not excited. The parameter errors are* $\tilde{\lambda}_1$ *(solid) and* $\tilde{\lambda}_2$ *(dashed).*

this case, all modes of the internal model are excited, and the parameter errors converge to zero. Figures 5.12(c) and (d) show the same variables when $w(0) = \text{col}(0,2,0,0,1)$. In this case the constant mode is not excited, but, as noted in Remark 5.14, the vector $\bar{\nu}$ is still persistently exciting. Therefore, the parameter errors converge to zero as shown in Figure 5.12(d). Figure 5.13 shows results when one of the two sinusoidal modes is absent. In Figures 5.13(a) and (b), $w(0) = \text{col}(1,0,0,0,1)$ so that the frequency ω_1 is not excited, while in Figures 5.13(c) and (d), $w(0) = \text{col}(1,2,0,0,0)$ so that the frequency ω_2 is not excited. In both cases the parameter errors do not converge to zero. However, as seen in the proof of Theorem 5.5, partial parameter convergence is achieved as $\tilde{\lambda}_a$ converges to zero. From the proof of Lemma 5.1, it can be verified that

$$L^{-T} = \begin{bmatrix} 0.0663 & -0.0074 \\ 0.0435 & -0.0194 \end{bmatrix}.$$

When ω_2 is not excited,

$$\tilde{\lambda}_a = 0.0663\tilde{\lambda}_1 - 0.0074\tilde{\lambda}_2 = 0.0663\tilde{\Lambda}_2 - 0.0074\tilde{\Lambda}_4,$$

and when ω_1 is not excited,

$$\tilde{\lambda}_a = 0.0435\tilde{\lambda}_1 - 0.0194\tilde{\lambda}_2 = 0.0435\tilde{\Lambda}_2 - 0.0194\tilde{\Lambda}_4.$$

Figures 5.13(b) and (d) confirm that $\tilde{\lambda}_a$ converges to zero in both cases. It is worthwhile to note that in all cases, the transient behavior of the regulation error is almost the same. △

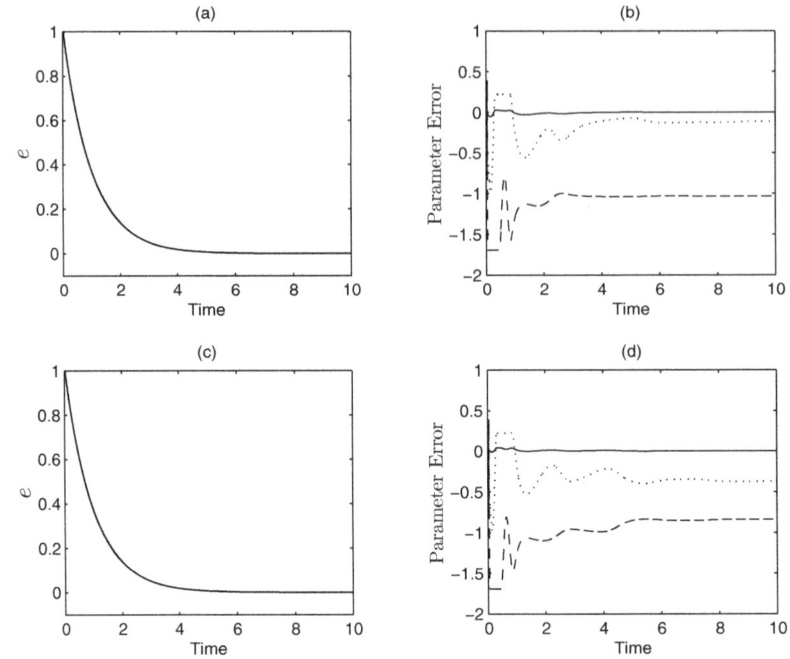

Figure 5.13. *Simulation of Example* 5.8. (a) *and* (b) *show results when the sinusoidal mode* ω_1 *is not excited, while in* (c) *and* (d) *the sinusoidal mode* ω_2 *is not excited. The parameter errors are* $\tilde{\lambda}_1$ *(dotted),* $\tilde{\lambda}_2$ *(dashed), and* $\tilde{\lambda}_a$ *(solid).*

5.7 ▪ Notes and References

The internal model principle for linear systems was developed by Francis and Wonham [44] and Davidson [33]. A key step in its extension to nonlinear systems is the work of Isidori and Byrnes [68, 27]. Different approaches to the nonlinear regulation problem are described in the books by Byrnes, Priscoli, and Isidori [28], Isidori, Marconi, and Serrani [70], Huang [60], Pavolov, van de Wouw, and Nijmeijer [115], and Chen and Huang [30]. The integral control of Section 5.2 is based on Khalil [73, 77]. The conditional integrator of Section 5.3 is based on Seshagiri and Khalil [135], while the conditional servocompensator is based on Seshagiri and Khalil [136]. A key element in the servocompensator design is the observation that the internal model should generate not only the modes of the exosystem but also higher-order harmonics induced by nonlinearities. This finding was reported, independently, by Khalil [73], Huang and Lin [61], and Priscoli [122]. The internal model perturbation result of Section 5.5 is based on Li and Khalil [96], which improves over an earlier result by Khalil [76]. The adaptive internal model of Section 5.6 is based on Li and Khalil [95], which builds on earlier work by Serrani, Isidori, and Marconi [133]. Another approach to deal with internal model uncertainty without adaptation is given in Isidori, Marconi, and Praly [69].

The results of this chapter assume that the exogenous signals are generated by the exosystem. It is intuitively clear that it is enough to assume that the exogenous signals asymptotically approach signals that are generated by the exosystem since the error is shown to converge to zero asymptotically. This relaxed assumption is used in [76, 135, 136]. The robust control technique used in the chapter is sliding mode control. Results that use Lyapunov redesign are given in [104, 105, 108, 140].

Chapter 6
Extended Observer

For the normal form

$$\begin{aligned}
\dot{\eta} &= f_0(\eta,\xi),\\
\dot{\xi}_i &= \xi_{i+1} \qquad \text{for } 1 \le i \le \rho-1,\\
\dot{\xi}_\rho &= a(\eta,\xi)+u,\\
y &= \xi_1,
\end{aligned}$$

the ρ-dimensional high-gain observer estimates x_1 and its derivatives $x_1^{(1)}$ to $x_1^{(\rho-1)}$. If an estimate of the ρth derivative $x_1^{(\rho)}$ is available and u is known, it will be possible to estimate the function a, which could play an important role in control design. An observer that estimates a is called *extended high-gain observer*. If a is a disturbance term, the observer acts as a disturbance estimator.[53] Alternatively, an estimate of a could be used to build an observer that estimates the state η of the internal dynamics $\dot{\eta} = f_0(\eta,\xi)$. In this case, the extended high-gain observer acts as a soft sensor of the internal dynamics.

The chapter starts in Section 6.1 with two motivating examples that illustrate the use of the extended observer as a disturbance estimator and a soft sensor of the internal dynamics, respectively. Section 6.2 shows how the disturbance estimator can be used to achieve feedback linearization in the presence of uncertainty. The section starts with single-input–single-output systems, followed by multi-input–multi-output systems. It ends with a dynamic inversion algorithm that can handle systems with uncertain control coefficient or systems that depend nonlinearly on the control input. Section 6.3 shows how the use of the extended observer as a soft sensor of the internal dynamics allows for relaxation of the minimum-phase assumption, which has been a common factor of the output feedback control designs in the earlier chapters. Two ideas are presented. The first one uses the soft sensor to implement a control designed for an auxiliary system that includes the internal dynamics. The second idea uses the soft sensor to build an observer for the internal dynamics to estimate η, which together with the estimate of ξ from the extended high-gain observer provide a full-order observer for the system. Notes and references are given in Section 6.4.

[53] Disturbance estimators are widely used in feedback control; see [97] for a comprehensive coverage of the topic; see also [48] and [58] for the use of disturbance estimators in Active Disturbance Rejection Control (ADRC).

6.1 ▪ Motivating Examples

Example 6.1. Consider the system

$$\dot{x}_1 = x_2, \quad \dot{x}_2 = \phi(t,x) + u, \quad y = x_1,$$

where $\phi(t,x)$ is unknown. Suppose $\phi(t,x)$ is continuously differentiable and ϕ and its partial derivatives are bounded on compact sets of x for all $t \geq 0$. The goal is to design an output feedback controller to stabilize the origin using the measured output y. Set $\sigma = \phi(t,x)$ and rewrite the system as

$$\dot{x}_1 = x_2, \quad \dot{x}_2 = \sigma + u, \quad y = x_1.$$

Had x and σ been measured, the control could have been taken as

$$u = -\sigma - x_1 - x_2,$$

which would result in the linear system

$$\dot{x} = \begin{bmatrix} 0 & 1 \\ -1 & -1 \end{bmatrix} x \stackrel{\text{def}}{=} Ax.$$

We refer to this system as the target system and to its solution as the target response. But x and σ are not measured as the only measured signal is y. To design an observer that estimates x and σ, we extend the dynamics of the system by treating σ as an additional state. The extended system is given by

$$\dot{x}_1 = x_2, \quad \dot{x}_2 = \sigma + u, \quad \dot{\sigma} = \psi(t,x,u), \quad y = x_1,$$

where

$$\psi(t,x,u) = \frac{\partial \phi}{\partial t} + \frac{\partial \phi}{\partial x_1} x_2 + \frac{\partial \phi}{\partial x_2}[\phi(t,x) + u].$$

The function ψ is unknown, but, due to the robustness of high-gain observers, we can build a third-order observer to estimate x and σ. The observer is taken as

$$\begin{aligned}
\dot{\hat{x}}_1 &= \hat{x}_2 + (\alpha_1/\varepsilon)(y - \hat{x}_1), \\
\dot{\hat{x}}_2 &= \hat{\sigma} + u + (\alpha_2/\varepsilon^2)(y - \hat{x}_1), \\
\dot{\hat{\sigma}} &= (\alpha_3/\varepsilon^3)(y - \hat{x}_1),
\end{aligned}$$

where α_1 to α_3 are chosen such that the polynomial

$$s^3 + \alpha_1 s^2 + \alpha_2 s + \alpha_3$$

is Hurwitz. We take $\alpha_1 = \alpha_2 = 3$ and $\alpha_3 = 1$ to assign the three roots of the polynomial at -1. With the estimates $\hat{x}$ and $\hat{\sigma}$ at hand, the output feedback controller is taken as

$$u = M \text{ sat}\left(\frac{-\hat{\sigma} - \hat{x}_1 - \hat{x}_2}{M}\right).$$

The control is saturated to deal with the peaking phenomenon. The saturation level M satisfies $M > \max_{x \in \Omega, t \geq 0} |-\phi(t,x) - x_1 - x_2|$, where Ω is a compact positively invariant set of the target system. To analyze the closed-loop system, let

$$\eta_1 = \frac{x_1 - \hat{x}_1}{\varepsilon^2}, \ \eta_2 = \frac{x_2 - \hat{x}_2}{\varepsilon}, \ \eta_3 = \sigma - \hat{\sigma},$$

to obtain

$$\varepsilon\dot{\eta}_1 = -\alpha_1\eta_1 + \eta_2,$$
$$\varepsilon\dot{\eta}_2 = -\alpha_2\eta_1 + \eta_3,$$
$$\varepsilon\dot{\eta}_3 = -\alpha_3\eta_1 + \varepsilon\psi(t,x,u),$$

where $\psi(t,x,u)$ is bounded on compact sets of x. Let $x(0) \in \Omega_0$, a compact set in the interior of Ω; then there exists $T(\varepsilon) > 0$, with $\lim_{\varepsilon\to 0} T(\varepsilon) = 0$, such that $x(t) \in \Omega$ for all $t \geq 0$ and $\eta(t) = O(\varepsilon)$ for all $t \geq T(\varepsilon)$. Therefore, for $t \geq T(\varepsilon)$,

$$\dot{x}_1 = x_2, \qquad \dot{x}_2 = -x_1 - x_2 + O(\varepsilon),$$

which shows that the closed-loop system under output feedback recovers the performance of the target system for sufficiently small ε. To illustrate the performance recovery, reconsider the system

$$\dot{x}_1 = x_2, \quad \dot{x}_2 = x_2^3 + u, \quad y = x_1,$$

from Example 1.2 but suppose that the term x_2^3 is unknown to the designer. The target system $\dot{x} = Ax$ has a Lyapunov function $V(x) = x^T Px$, where P is the solution of the Laypunov equation $PA + A^T P = -I$. Recall from Example 1.2 that $\Omega = \{V(x) \leq 0.3\}$ is a positively invariant set of the target system and $\max_{x\in\Omega} |-x_2^3 - x_1 - x_2| \leq 0.816$. Taking $M = 1$, the output feedback controller is given by

$$u = \text{sat}\,(-\hat{\sigma} - \hat{x}_1 - \hat{x}_2).$$

Simulation results of the target system and the system under output feedback for three values of ε are shown in Figure 6.1 for the initial conditions $x_1(0) = 0.1$ and $x_2(0) = \hat{x}_1(0) = \hat{x}_2(0) = \hat{\sigma}(0) = 0$. The figure shows that the response of the closed-loop system under output feedback approaches the target response as ε decreases. As a variation on the theme of this simulation, let the system be

$$\dot{x}_1 = x_2, \quad \dot{x}_2 = x_2^3 + \sin t + u, \quad y = x_1.$$

The observer and output feedback controller remain the same, except that the saturation level M is chosen to satisfy $M > \max_{x\in\Omega} |-x_2^3 - x_1 - x_2| + 1 = 1.816$. The choice $M = 2$ yields the controller

$$u = 2\,\text{sat}\left(\frac{-\hat{\sigma} - \hat{x}_1 - \hat{x}_2}{2}\right).$$

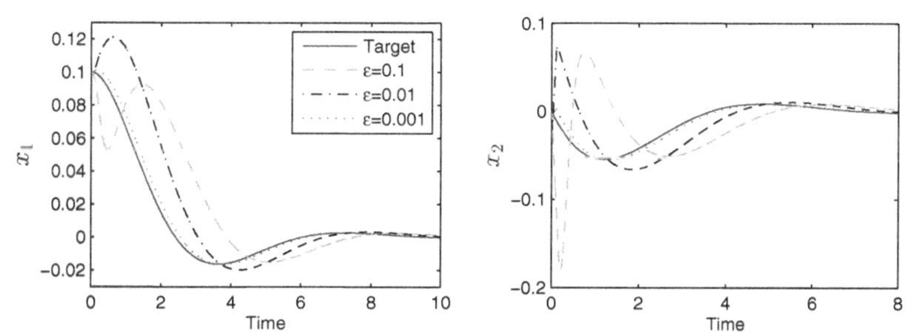

Figure 6.1. *Simulation of the output feedback controller of Example 6.1 when $\phi(t,x) = x_2^3$.*

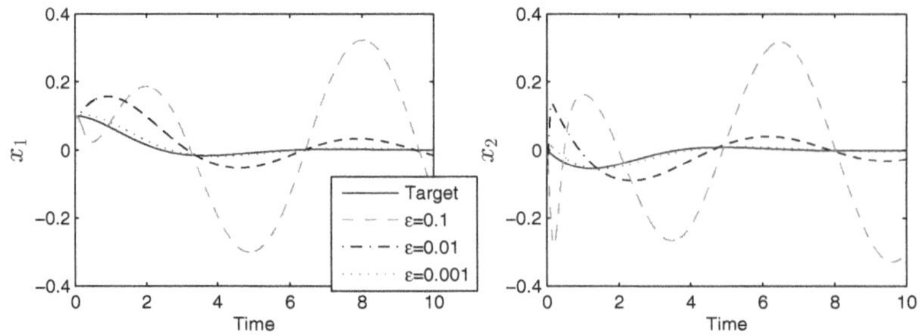

Figure 6.2. *Simulation of the output feedback controller of Example* 6.1 *when* $\phi(t,x) = x_2^3 + \sin t$.

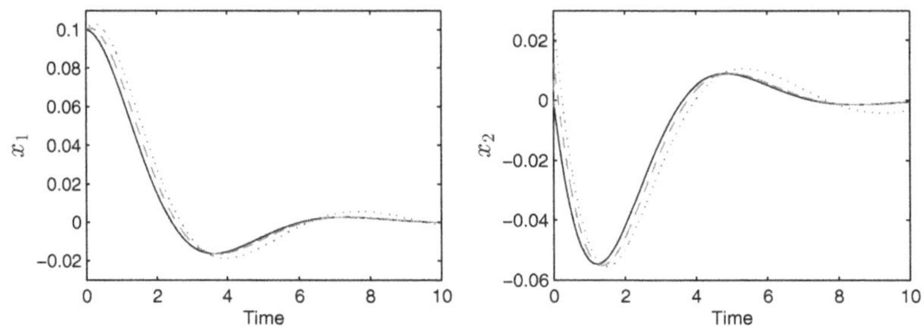

Figure 6.3. *Simulation of the output feedback controller of Example* 6.1 *when* $\phi(t,x) = x_2^3$ *(dashed) and* $\phi(t,x) = x_2^3 + \sin t$ *(dotted), with* $\varepsilon = 0.001$. *The solid line is the target response.*

The simulation results of Figure 6.2 are carried out under the same initial conditions. They confirm the same trend in Figure 6.1. Figure 6.3 shows that when $\varepsilon = 0.001$ the response is almost the same whether the unknown function is x_2^3 or $x_2^3 + \sin t$. △

Example 6.2. Consider the system

$$\dot{\eta} = \theta_1 \eta \left(1 + \frac{\eta^2}{1+\eta^2}\right) + \xi, \quad \dot{\xi} = \theta_2 \eta + u, \quad y = \xi,$$

where $\theta_1 \in [0.1, 0.2]$ and $\theta_2 \in [1, 2]$ are unknown parameters. The goal is to design an output feedback controller to stabilize the origin. This is a relative-degree-one system in the normal form, and the zero dynamics $\dot{\eta} = \theta_1 \eta[1 + \eta^2/(1+\eta^2)]$ have unstable origin; hence the system is nonminimum phase. A design that uses the extended high-gain observer is based on a procedure introduced by Isidori that aims at stabilizing the zero dynamics.[54] Had η been measured, in addition to ξ, the system could have been stabilized by a backstepping procedure that treats ξ as a virtual input to the $\dot{\eta}$-equation. After designing ξ as a feedback function of η to stabilize the zero dynamics, it would be backstepped to the $\dot{\xi}$-equation to find the actual control u that stabilizes the overall system.[55] While η is not measured, the term $\theta_2 \eta$ on the right-hand side of the $\dot{\xi}$-equation is proportional to η. This term can be estimated by an extended high-gain observer. Treat $\sigma = \theta_2 \eta$ as an additional state variable, to obtain

$$\dot{\xi} = \sigma + u, \quad \dot{\sigma} = \psi(\eta, \xi), \quad y = \xi,$$

[54] See [67] or [66, Section 12.6].

[55] See [90] or [80] for an introduction to backstepping.

where $\psi(\eta,\xi) = \theta_2\theta_1\eta[1+\eta^2/(1+\eta^2)] + \theta_2\xi$. The extended high-gain observer

$$\dot{\hat{\xi}} = \hat{\sigma} + u + (\alpha_1/\varepsilon)(y - \hat{\xi}), \quad \dot{\hat{\sigma}} = (\alpha_2/\varepsilon^2)(y - \hat{\xi})$$

with positive α_1, α_2, and ε will robustly estimate σ for sufficiently small ε. Based on the properties of high-gain observers, it is expected that a feedback controller that uses $\hat{\sigma}$ can recover the performance of a controller that uses σ. Therefore, we proceed to design a state feedback controller that assumes that both ξ and σ are available for feedback. The design starts by considering the auxiliary problem

$$\dot{\eta} = \theta_1\eta\left(1 + \frac{\eta^2}{1+\eta^2}\right) + u_a, \quad y_a = \theta_2\eta,$$

where $u_a = \xi$ is a virtual control and $y_a = \sigma$ is a virtual output. The feedback control $u_a = -ky_a$ results in

$$\dot{\eta} = \theta_1\eta\left(1 + \frac{\eta^2}{1+\eta^2}\right) - k\theta_2\eta.$$

The derivative of $V_0 = \frac{1}{2}\eta^2$ is

$$\dot{V}_0 = \theta_1\eta^2\left(1 + \frac{\eta^2}{1+\eta^2}\right) - k\theta_2\eta^2 \le 0.4\eta^2 - k\eta^2.$$

Hence, the origin $\eta = 0$ is stabilized by taking $k > 0.4$. Let $k = 0.5$. Next, we use backstepping to compute u. Let

$$s = u_a + ky_a = \xi + k\sigma = \xi + k\theta_2\eta$$

and $V = \frac{1}{2}\eta^2 + \frac{1}{2}s^2$. The derivative of V is

$$\begin{aligned}\dot{V} &= \theta_1\eta^2\left(1 + \frac{\eta^2}{1+\eta^2}\right) - k\theta_2\eta^2 + \eta s \\ &\quad + s\left\{\theta_2[1 + k(\theta_1 - k\theta_2)]\eta + k\theta_1\theta_2\frac{\eta^3}{1+\eta^2}\right\} + k\theta_2 s^2 + su \\ &\le -0.1\eta^2 + 2.9|\eta|\,|s| + s^2 + su.\end{aligned}$$

Take $u = -\beta\ \text{sat}(s/\mu)$, where β and μ are positive constants, and let $\Omega = \{V \le \frac{1}{2}c^2\}$. For $(\eta,\xi) \in \Omega \times \{|s| \ge \mu\}$

$$\dot{V} \le -0.1\eta^2 + 2.9c|s| + c|s| - \beta|s|.$$

Taking $\beta \ge 0.1 + 3.9c$ yields $\dot{V} \le -0.1(\eta^2 + |s|)$. On the other hand,

$$\begin{aligned}s\dot{s} &= s\left\{\theta_2[1 + k(\theta_1 - k\theta_2)]\eta + k\theta_1\theta_2\frac{\eta^3}{1+\eta^2}\right\} + k\theta_2 s^2 + su \\ &\le 1.9|\eta|\,|s| + s^2 - \beta|s| \le 2.9c|s| - \beta|s| = -(\beta - 2.9c)|s|.\end{aligned}$$

The choice $\beta \ge 0.1 + 3.9c$ ensures that $s\dot{s} \le -(c + 0.1)|s|$, which shows that $|s|$ decreases monotonically reaching the positively invariant set $\{|s| \le \mu\}$ in finite time. For $(\eta,\xi) \in \Omega \times \{|s| \le \mu\}$,

$$\begin{aligned}\dot{V} &\le -0.1\eta^2 + 2.9|\eta|\,|s| + s^2 - (\beta/\mu)s^2 \\ &= -\begin{bmatrix}|\eta| \\ |\xi|\end{bmatrix}^T \begin{bmatrix}0.1 & -1.45 \\ -1.45 & (\beta/\mu - 1)\end{bmatrix}\begin{bmatrix}|\eta| \\ |\xi|\end{bmatrix} < 0 \quad \text{for } \mu < \beta/22.025.\end{aligned}$$

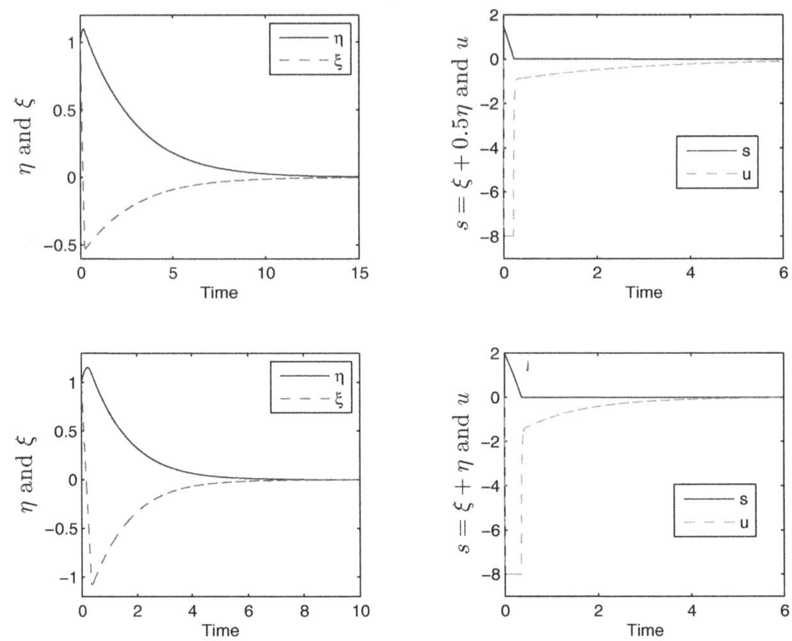

Figure 6.4. *Simulation of the output feedback controller of Example* 6.2. *The first row for* $\theta_1 = 0.1$ *and* $\theta_2 = 1$, *the second row for* $\theta_1 = 0.2$ *and* $\theta_2 = 2$.

Thus, the origin (η, s) is exponentially stable, and Ω is a subset of the region of attraction. With $c = 2$, $\beta = 8$, $\mu = 0.1$, $\alpha_1 = 2$, $\alpha_2 = 1$, and $\varepsilon = 0.01$, the output feedback controller is given by

$$\dot{\hat{\xi}} = \hat{\sigma} + u + 200(y - \hat{\xi}),$$
$$\dot{\hat{\sigma}} = 10^4(y - \hat{\xi}),$$
$$u = -8 \text{ sat}\left(\frac{\hat{\xi} + 0.5\hat{\sigma}}{0.1}\right).$$

Simulation results with the initial conditions $\eta(0) = \xi(0) = 1$ and $\hat{\eta}(0) = \hat{\xi}(0) = 0$ are shown in Figure 6.4 for two different cases of the parameters θ_1 and θ_2. △

6.2 ▪ Feedback Control via Disturbance Compensation

6.2.1 ▪ Single-Input–Single-Output Systems

Consider a single-input–single-output nonlinear system in the normal form:

$$\dot{\eta} = f_0(\eta, \xi, w), \tag{6.1}$$
$$\dot{\xi}_i = \xi_{i+1} \quad \text{for } 1 \le i \le \rho - 1, \tag{6.2}$$
$$\dot{\xi}_\rho = a(\eta, \xi, w) + b(\eta, \xi, w)u, \tag{6.3}$$
$$e = \xi_1, \tag{6.4}$$

where $\eta \in R^{n-\rho}$ and $\xi = \text{col}(\xi_1,\ldots,\xi_\rho) \in R^\rho$ are the state variables, u is the control input, e is the measured regulation error, and $w \in R^\ell$ is an exogenous input. The normal form is defined for $\eta \in D_\eta \subset R^{n-\rho}$ and $\xi \in D_\xi \subset R^\rho$ for some domains D_η and D_ξ. The functions a, b, and f_0 and the signal w could be unknown. The system is required to satisfy the following assumptions.

Assumption 6.1. *$w(t)$ and $\dot{w}(t)$ are bounded, and $w(t)$ belongs to a known compact set $\mathcal{W} \subset R^\ell$.*

Assumption 6.2. *The functions a and b are continuously differentiable with locally Lipschitz derivatives, f_0 is locally Lipschitz, and $b(\eta,\xi,w) \geq b_0 > 0$ for all $(\eta,\xi,w) \in D_\eta \times D_\xi \times \mathcal{W}$.*

The goal is to design an output feedback controller to asymptotically regulate the output $e(t)$ to zero while meeting certain requirements on the transient response. The design is pursued for systems where η is bounded for bounded ξ and w, as will be stated precisely in Assumption 6.3. Had (η,ξ,w) been available for feedback and the functions a and b been known, we could have used the feedback linearization control

$$u = \frac{-a(\eta,\xi,w)+v}{b(\eta,\xi,w)}$$

to reduce the input-output model to

$$\dot{\xi} = A\xi + Bv, \qquad e = C\xi, \tag{6.5}$$

where the triple (A,B,C) represents a chain of ρ integrators, that is,

$$A = \begin{bmatrix} 0 & 1 & 0 & \cdots & 0 \\ 0 & 0 & 1 & \cdots & 0 \\ \vdots & \vdots & \ddots & \ddots & \vdots \\ 0 & 0 & \cdots & 0 & 1 \\ 0 & 0 & \cdots & \cdots & 0 \end{bmatrix}, \qquad B = \begin{bmatrix} 0 \\ 0 \\ \vdots \\ 0 \\ 1 \end{bmatrix}, \qquad C = \begin{bmatrix} 1 & 0 & \cdots & \cdots & 0 \end{bmatrix}.$$

The auxiliary control v would then be designed as a feedback function of ξ to stabilize the origin $\xi = 0$ and meet the transient response specifications. Let $v = -K\xi$, where K is chosen using a linear control design method, such as pole placement or LQR, to make $(A-BK)$ Hurwitz and shape the transient response. Hence, the origin of

$$\dot{\xi} = (A-BK)\xi \tag{6.6}$$

is exponentially stable, and $V_s(\xi) = \xi^T P_s \xi$ is a Lyapunov function, where $P_s = P_s^T > 0$ is the solution of the Lyapunov equation $P_s(A-BK)+(A-BK)^T P_s = -Q_s$ for some $Q_s = Q_s^T > 0$. Let c be a positive constant such that $\{V_s(\xi) \leq c\} \subset D_\xi$.

While such design is not realizable, it will be shown that its performance can be recovered by using an extended high-gain observer. Toward that end, we call the system (6.6) the target system and its response the target response.

Assumption 6.3. *There is a continuously differentiable function $V_0(\eta)$, class $\mathcal{K}$ functions ϱ_1 and ϱ_2, and a nonnegative continuous nondecreasing function χ such that*

$$\varrho_1(\|\eta\|) \leq V_0(\eta) \leq \varrho_2(\|\eta\|),$$

$$\frac{\partial V_0}{\partial \eta} f_0(\eta, \xi, w) \le 0 \qquad \forall\ \|\eta\| \ge \chi(\|\xi\| + \|w\|)$$

for all $(\eta, \xi, w) \in D_\eta \times D_\xi \times \mathcal{W}$*. Moreover, there is* $c_0 \ge \varrho_2(\chi(\kappa_1(c) + \kappa_2))$ *such that the set* $\{V_0(\eta) \le c_0\}$ *is compact and contained in* D_η*, where* $\kappa_1(c) = \max_{\xi \in \{V_s(\xi) \le c\}} \|\xi\|$ *and* $\kappa_2 = \max_{w \in \mathcal{W}} \|w\|$.

Assumption 6.3 ensures that the compact set $\Omega = \{V_0(\eta) \le c_0\} \times \{V_s(\xi) \le c\}$ is positively invariant with respect to the system

$$\dot{\eta} = f_0(\eta, \xi, w), \qquad \dot{\xi} = (A - BK)\xi,$$

because on the boundary $\{V_0 = c_0\}$

$$\begin{aligned} \varrho_2(\|\eta\|) \ge c_0 \ge \varrho_2(\chi(\kappa_1(c) + \kappa_2)) &\Rightarrow \|\eta\| \ge \chi(\kappa_1(c) + \kappa_2) \\ &\Rightarrow \|\eta\| \ge \chi(\|\xi\| + \|w\|) \ \Rightarrow\ \dot{V}_0 \le 0 \end{aligned}$$

and on the boundary $\{V_s(\xi) = c\}$, $\dot{V}_s < 0$. If $D_\eta \times D_\xi = R^n$ and V_0 is radially unbounded, the constants c and c_0 can be chosen arbitrarily large, and any compact set of R^n can be put in the interior of Ω.

Let $\hat{a}(\xi)$ and $\hat{b}(\xi) \ge b_1 > 0$ be twice continuously differentiable, globally bounded functions that model $a(\eta, \xi, w)$ and $b(\eta, \xi, w)$, respectively. It is allowed to take $\hat{a} = 0$ and $\hat{b} > 0$ as a constant. The ξ_ρ-equation can be written as

$$\dot{\xi}_\rho = \sigma + \hat{a}(\xi) + \hat{b}(\xi)u,$$

where

$$\sigma = a(\eta, \xi, w) - \hat{a}(\xi) + [b(\eta, \xi, w) - \hat{b}(\xi)]u.$$

Augmenting σ as an additional state to the chain of integrators (6.2)–(6.4), a high-gain observer for the extended system is taken as

$$\dot{\hat{\xi}}_i = \hat{\xi}_{i+1} + (\alpha_i/\varepsilon^i)(e - \hat{\xi}_1) \quad \text{for } 1 \le i \le \rho - 1, \tag{6.7}$$

$$\dot{\hat{\xi}}_\rho = \hat{\sigma} + \hat{a}(\hat{\xi}) + \hat{b}(\hat{\xi})u + (\alpha_\rho/\varepsilon^\rho)(e - \hat{\xi}_1), \tag{6.8}$$

$$\dot{\hat{\sigma}} = (\alpha_{\rho+1}/\varepsilon^{\rho+1})(e - \hat{\xi}_1), \tag{6.9}$$

where α_1 to $\alpha_{\rho+1}$ are chosen such that the polynomial

$$s^{\rho+1} + \alpha_1 s^\rho + \cdots + \alpha_{\rho+1}$$

is Hurwitz and $\varepsilon > 0$ is a small parameter. Global boundedness of $\hat{a}$ and $\hat{b}$ is required to overcome the peaking phenomenon of high-gain observers. It does not exclude linear or unbounded functions because global boundedness can be always achieved by saturation outside a compact set.

From the results of Chapter 2, it is anticipated that the terms $(\alpha_i/\varepsilon^i)(e - \hat{\xi}_1)$ for $1 \le i \le \rho$ will be $O(\varepsilon)$ after a short transient period. Then equations (6.7)–(6.8) can be viewed as a perturbation of the system

$$\dot{\hat{\xi}} = A\hat{\xi} + B\left[\hat{\sigma} + \hat{a}(\hat{\xi}) + \hat{b}(\hat{\xi})u\right], \tag{6.10}$$

which can be made to coincide with the target system (6.6) by taking

$$u = \frac{-\hat{\sigma} - \hat{a}(\hat{\xi}) - K\hat{\xi}}{\hat{b}(\hat{\xi})} \stackrel{\text{def}}{=} \psi(\hat{\xi}, \hat{\sigma}). \tag{6.11}$$

To protect the system from peaking in the observer's transient response, the control is saturated outside the compact set Ω. Let

$$M > \max_{(\eta,\xi)\in\Omega,\; w\in\mathcal{W}} \left| \frac{-a(\eta,\xi,w) - K\xi}{b(\eta,\xi,w)} \right|. \tag{6.12}$$

Saturating the foregoing expression of u at $\pm M$ using the saturation function sat$(\cdot)$ yields the output feedback controller

$$u = M \,\text{sat}\left(\frac{\psi(\hat{\xi}, \hat{\sigma})}{M} \right), \tag{6.13}$$

where ψ is defined by (6.11). Saturating the control at $\pm M$ does not restrict the functions a and b because we can find M for any a, b, Ω, and $\mathcal{W}$. However, the calculation of M might not be straightforward, and one might end up with a conservative bound.

Let

$$k_u = \max_{(\eta,\xi)\in\Omega, w\in\mathcal{W}} \left| \frac{b(\eta,\xi,w) - \hat{b}(\xi)}{\hat{b}(\xi)} \right|,$$

$$\Gamma(s) = \frac{\alpha_{\rho+1}}{s^{\rho+1} + \alpha_1 s^{\rho} + \cdots + \alpha_{\rho+1}},$$

and $||\Gamma||_\infty = \max_\omega |\Gamma(j\omega)|$. Because $\Gamma(0) = 1$, $||\Gamma||_\infty \geq 1$. It can be verified that $||\Gamma||_\infty = 1$ if all the poles of $\Gamma(s)$ are real.

Theorem 6.1. *Consider the closed-loop system formed of the plant* (6.1)–(6.4), *the observer* (6.7)–(6.9), *and the controller* (6.13). *Suppose Assumptions* 6.1 to 6.3 *are satisfied,*

$$k_u < \frac{1}{||\Gamma||_\infty}, \tag{6.14}$$

the initial states of the observer belong to a compact subset of $R^{\rho+1}$, and the initial states of the plant belong to a compact set in the interior of Ω. Then

- *there exists $\varepsilon_1^* > 0$ such that for every $0 < \varepsilon \leq \varepsilon_1^*$, the trajectories of the closed-loop system are bounded for all $t \geq 0$;*
- *given any $\mu > 0$, there exist $\varepsilon_2^* > 0$, dependent on μ, such that for every $0 < \varepsilon \leq \varepsilon_2^*$,*

$$||\xi(t) - \xi_r(t)|| \leq \mu \qquad \forall\, t \geq 0, \tag{6.15}$$

where ξ_r is the solution of the target system (6.6) *with $\xi_r(0) = \xi(0)$;*

- *given any $\mu > 0$, there exist $\varepsilon_3^* > 0$ and $T_1 > 0$, both dependent on μ, such that for every $0 < \varepsilon \leq \varepsilon_3^*$,*

$$||\xi(t)|| \leq \mu \qquad \forall\, t \geq T_1. \tag{6.16}$$

◇

Remark 6.1. *The second bullet shows performance recovery because the trajectories of ξ under output feedback can be made arbitrarily close to the trajectories of the target system by choosing ε small enough. The third bullet shows that the controller achieves practical regulation because the ultimate bound on $\|\xi\|$ can be made arbitrarily small by choosing ε small enough.* ◇

Remark 6.2. *Condition* (6.14) *can always be satisfied by appropriate choice of $\hat{b}$ and the observer eigenvalues if an upper bound on $b(x,z,w)$ is known. Let $b_m \geq \max_{\{(\eta,\xi)\in\Omega, w\in\mathcal{W}\}} b(\eta,\xi,w)$. Taking $\hat{b} > b_m$ ensures that $k_u < 1$. Assigning the observer eigenvalues to be real results in $\|\Gamma\|_\infty = 1$; hence $k_u\|\Gamma\|_\infty < 1$.* ◇

Remark 6.3. *There is interplay between ε and the constants c and c_0 that define Ω. The larger c and c_0 are, the larger the saturation level M, which, in general, requires a smaller ε to achieve enough separation of time scales between the observer and the plant. This interplay can be interpreted the other way around by saying that a smaller ε would lead to a larger region of attraction.* ◇

Proof of Theorem 6.1: Similar to the analysis of high-gain observers in the previous chapters, the closed-loop system is represented as a singularly perturbed one in which the slow variables are (η,ξ) and the fast variables are the scaled estimation errors

$$\varphi_i = (\xi_i - \hat{\xi}_i)/\varepsilon^{\rho+1-i} \quad \text{for} \quad 1 \leq i \leq \rho, \tag{6.17}$$

$$\varphi_{\rho+1} = \Delta_a(\eta,\xi,w) + \Delta_b(\eta,\xi,w) M\, g_\varepsilon\left(\frac{\psi(\xi,\hat{\sigma})}{M}\right) - \hat{\sigma}, \tag{6.18}$$

where

$$\Delta_a(\eta,\xi,w) = a(\eta,\xi,w) - \hat{a}(\xi), \qquad \Delta_b(\eta,\xi,w) = b(\eta,\xi,w) - \hat{b}(\xi),$$

and g_ε is an odd function defined by

$$g_\varepsilon(y) = \begin{cases} y & \text{for } 0 \leq y \leq 1, \\ y + (y-1)/\varepsilon - 0.5(y^2-1)/\varepsilon & \text{for } 1 \leq y \leq 1+\varepsilon, \\ 1 + 0.5\,\varepsilon & \text{for } y \geq 1+\varepsilon. \end{cases}$$

The function g_ε is nondecreasing, continuously differentiable with a locally Lipschitz derivative, and bounded uniformly in ε on any bounded interval of ε and satisfies $|g_\varepsilon'(y)| \leq 1$ and $|g_\varepsilon(y) - \text{sat}(y)| \leq \varepsilon/2$ for all $y \in R$. The new feature here over the previous chapters is the additional fast variable $\varphi_{\rho+1}$, which is due to the extended dimension of the observer. The change of variables (6.18) is well defined as long as $(\eta,\xi) \in \Omega$ because

$$\frac{\partial \varphi_{\rho+1}}{\partial \hat{\sigma}} = -1 - \frac{\Delta_b(\eta,\xi,w)}{\hat{b}(\xi)} g_\varepsilon'\left(\frac{\psi(\xi,\hat{\sigma})}{M}\right) \leq -1 + k_u$$

and $k_u < 1$ in view of (6.14). The state equations of the fast variables are

$$\begin{aligned}
\varepsilon\dot{\varphi}_i &= -\alpha_i\varphi_1 + \varphi_{i+1} \quad \text{for} \quad 1 \le i \le \rho - 1,\\
\varepsilon\dot{\varphi}_\rho &= -\alpha_\rho\varphi_1 + \varphi_{\rho+1} + \Delta_0(\eta, \xi, \hat{\xi}, \hat{\sigma}, w, \varepsilon),\\
\varepsilon\dot{\varphi}_{\rho+1} &= -\alpha_{\rho+1}\varphi_1 - \frac{\Delta_b(\eta,\xi,w)}{\hat{b}(\xi)} g_\varepsilon'\left(\frac{\psi(\xi,\hat{\sigma})}{M}\right)\alpha_{\rho+1}\varphi_1\\
&\quad + \varepsilon\left[\dot{\Delta}_a + \dot{\Delta}_b M g_\varepsilon\left(\frac{\psi(\xi,\hat{\sigma})}{M}\right) + \Delta_b g_\varepsilon'\left(\frac{\psi(\xi,\hat{\sigma})}{M}\right)\frac{\partial\psi(\xi,\hat{\sigma})}{\partial\xi}\dot{\xi}\right],
\end{aligned}$$

where

$$\begin{aligned}
\Delta_0(\eta,\xi,\hat{\xi},\hat{\sigma},w,\varepsilon) &= \hat{a}(\xi) - \hat{a}(\hat{\xi})\\
&\quad + b(\eta,\xi,w)M\left[g_\varepsilon\left(\frac{\psi(\hat{\xi},\hat{\sigma})}{M}\right) - g_\varepsilon\left(\frac{\psi(\xi,\hat{\sigma})}{M}\right)\right]\\
&\quad + \hat{b}(\xi)M g_\varepsilon\left(\frac{\psi(\xi,\hat{\sigma})}{M}\right) - \hat{b}(\hat{\xi})M g_\varepsilon\left(\frac{\psi(\hat{\xi},\hat{\sigma})}{M}\right)\\
&\quad + \left[b(\eta,\xi,w) - \hat{b}(\hat{\xi})\right]M\left[\operatorname{sat}\left(\frac{\psi(\hat{\xi},\hat{\sigma})}{M}\right) - g_\varepsilon\left(\frac{\psi(\hat{\xi},\hat{\sigma})}{M}\right)\right].
\end{aligned}$$

Using (6.17), it is not hard to see that the bracketed term on the right-hand side of $\varepsilon\dot{\varphi}_{\rho+1}$ is a continuous function of $(\eta, \xi, \varphi, w, \dot{w}, \varepsilon)$. Because $\hat{a}$, $\hat{b}$, and g_ε are continuously differentiable with locally Lipschitz derivatives and globally bounded, (6.17) and the definition of g_ε can be used to show that Δ_0/ε is a locally Lipschitz function. For example,

$$\begin{aligned}
\frac{1}{\varepsilon}\left[\hat{a}(\xi) - \hat{a}(\hat{\xi})\right] &= \frac{1}{\varepsilon}\int_0^1 \frac{\partial\hat{a}}{\partial\xi}(\hat{\xi} + \lambda(\xi - \hat{\xi}))\,d\lambda\,(\xi - \hat{\xi})\\
&= \int_0^1 \frac{\partial\hat{a}}{\partial\xi}(\hat{\xi} + \lambda(\xi - \hat{\xi}))\,d\lambda\,\left[\varepsilon^{\rho-1}\varphi_1,\ \ldots,\ \varepsilon\varphi_{\rho-1},\ \varphi_\rho\right]^T.
\end{aligned}$$

Using the foregoing expressions it can be shown that, as long as $(\eta, \xi) \in \Omega$, the fast subsystem is given by

$$\varepsilon\dot{\varphi} = \Lambda\varphi - \bar{B}_1\Delta_1(\cdot)\alpha_{\rho+1}\varphi_1 + \varepsilon[\bar{B}_1\,\Delta_2(\cdot) + \bar{B}_2\Delta_3(\cdot)], \tag{6.19}$$

where

$$\Lambda = \begin{bmatrix} -\alpha_1 & 1 & 0 & \cdots & 0\\ -\alpha_2 & 0 & 1 & \ddots & 0\\ \vdots & \vdots & \ddots & \ddots & \vdots\\ -\alpha_\rho & 0 & \cdots & 0 & 1\\ -\alpha_{\rho+1} & 0 & \cdots & \cdots & 0 \end{bmatrix}, \quad \bar{B}_1 = \begin{bmatrix} 0\\ B \end{bmatrix}, \quad \bar{B}_2 = \begin{bmatrix} B\\ 0 \end{bmatrix},$$

$$\Delta_1(\eta,\xi,\varphi,w,\varepsilon) = \frac{\Delta_b(\eta,\xi,w)}{\hat{b}(\xi)} g_\varepsilon'\left(\frac{\psi(\xi,\hat{\sigma})}{M}\right),$$
$$\Delta_2(\eta,\xi,\varphi,w,\dot{w},\varepsilon) = \dot{\Delta}_a + \dot{\Delta}_b M g_\varepsilon\left(\frac{\psi(\xi,\hat{\sigma})}{M}\right) + \Delta_b g_\varepsilon'\left(\frac{\psi(\xi,\hat{\sigma})}{M}\right)\frac{\partial \psi(\xi,\hat{\sigma})}{\partial \xi}\dot{\xi},$$
$$\Delta_3(\eta,\xi,\varphi,w,\varepsilon) = \frac{\Delta_0(\eta,\xi,\hat{\xi},\hat{\sigma},w,\varepsilon)}{\varepsilon}.$$

The functions Δ_1, Δ_2, and Δ_3 are locally Lipschitz in their arguments and bounded from above by $k_a + k_b\|\varphi\|$, where k_a and k_b are positive constants independent of ε. The matrix Λ is Hurwitz by design. If it were not for the term $\bar{B}_1\Delta_1(\cdot)\alpha_{\rho+1}\varphi_1$ on the right-hand side of (6.19), it would have been the usual fast dynamics equation as in the previous chapters, and it could have been easily concluded that φ would be $O(\varepsilon)$ after a short transient period. The new term is dealt with by application of the circle criterion [80, Section 7.3]. Consider (6.19) without the $O(\varepsilon)$ terms on the right-hand side, that is,

$$\varepsilon\dot{\varphi} = \Lambda\varphi - \bar{B}_1\Delta_1(\cdot)\alpha_{\rho+1}\varphi_1. \tag{6.20}$$

Equation (6.20) can be represented as a negative feedback connection of the transfer function

$$\Gamma(\varepsilon s) = \frac{\alpha_{\rho+1}}{(\varepsilon s)^{\rho+1} + \alpha_1(\varepsilon s)^\rho + \cdots + \alpha_{\rho+1}}$$

and the time-varying gain $\Delta_1(\cdot)$, where $|\Delta_1(\cdot)| \le k_u$. The transfer function $\Gamma(\varepsilon s)$ has $\max_\omega |\Gamma(j\varepsilon\omega)| = \max_\omega |\Gamma(j\omega)| = \|\Gamma\|_\infty$. The circle criterion shows that the origin of (6.20) will be globally exponentially stable if $k_u\|\Gamma\|_\infty < 1$. By applying a loop transformation to (6.20) and using the Kalman–Yakubovich–Popov lemma [80, Lemma 5.3], we obtain a quadratic Lyapunov function $W(\varphi) = \varphi^T P\varphi$ whose derivative with respect to (6.20) is bounded from above by $-(\lambda/\varepsilon)W(\varphi)$ for some positive constant λ, independent of ε. Using $W(\varphi)$ as a Lyapunov function candidate for (6.19), arguments from the previous chapters can be repeated to show that all state variables are bounded. This is done in two steps; first, it is shown that $c_1 > 0$ can be chosen such that the set $\Omega\times\{W(\varphi)\le\varepsilon^2 c_1\}$ is positively invariant, then it is shown that φ enters $\{W(\varphi)\le\varepsilon^2 c_1\}$ in a finite time $T(\varepsilon)$, where $\lim_{\varepsilon\to 0} T(\varepsilon) = 0$.

While $(\eta,\xi,\varphi)\in\Omega\times\{W(\varphi)\le\varepsilon^2 c_1\}$, we can use $\varphi_{\rho+1} = O(\varepsilon)$ and $\xi - \hat{\xi} = O(\varepsilon)$ to show that

$$\psi(\hat{\xi},\hat{\sigma}) = \psi(\xi,\hat{\sigma}) + O(\varepsilon),$$
$$\hat{\sigma} = \Delta_a(\eta,\xi,w) + \Delta_b(\eta,\xi,w)M g_\varepsilon\left(\frac{\psi(\xi,\hat{\sigma})}{M}\right) + O(\varepsilon).$$

Thus, up to an $O(\varepsilon)$ error, $\psi(\xi,\hat{\sigma})$ satisfies the equation

$$\psi + \frac{\Delta_b(\eta,\xi,w)}{\hat{b}(\xi)} M g_\varepsilon\left(\frac{\psi}{M}\right) = \frac{-a(\eta,\xi,w) - K\xi}{\hat{b}(\xi)}.$$

This equation has a unique solution because $|\Delta_b/\hat{b}| \le k_u < 1$ and $|g_\varepsilon'(\cdot)| \le 1$. By direct substitution it can be seen that the unique solution is

$$\psi = \frac{-a(\eta,\xi,w) - K\xi}{b(\eta,\xi,w)}.$$

Hence,

$$\psi(\hat{\xi},\hat{\sigma}) = \frac{-a(\eta,\xi,w) - K\xi}{b(\eta,\xi,w)} + O(\varepsilon).$$

Recalling from (6.12) that the saturation level M is chosen greater than the maximum of $|-a-K\xi|/b$, we conclude that, for sufficiently small ε, $\psi(\hat{\xi},\hat{\sigma})$ will be in the linear region of the saturation function. Hence, $u = \psi(\hat{\xi},\hat{\sigma})$, and the closed-loop system can be represented by

$$\dot{\eta} = f_0(\eta,\xi,w), \tag{6.21}$$

$$\dot{\xi} = (A-BK)\xi + O(\varepsilon), \tag{6.22}$$

$$\varepsilon\dot{\varphi} = \Lambda\varphi - \bar{B}_1\Delta_1(\cdot)\alpha_{\rho+1}\varphi_1 + O(\varepsilon). \tag{6.23}$$

From these equations it is straightforward to argue that $\Omega\times\{W(\varphi)\le\varepsilon^2 c_1\}$ is positively invariant for sufficiently small ε.

At the initial time, $\varphi(0)$ could be outside the set $\{W(\varphi)\le\varepsilon^2 c_1\}$ but would move rapidly toward the set and will reach it within an interval $T(\varepsilon)$, where $\lim_{\varepsilon\to 0} T(\varepsilon)=0$. Because the initial state of (η,ξ) is in the interior of Ω, choosing ε small enough ensures that (η,ξ) does not leave Ω during the interval $[0,T(\varepsilon)]$, and by the end of this interval (η,ξ,φ) would be in the positively invariant set $\Omega\times\{W(\varphi)\le\varepsilon^2 c_1\}$. This establishes the boundedness of all state variables.

Since $\xi(t)$ satisfies (6.22) for $t\ge T(\varepsilon)$, $\dot{\xi}$ and $\dot{\xi}_r$ are bounded uniformly in ε, and $\xi(0)=\xi_r(0)$, it follows that

$$\xi(t)-\xi_r(t) = O(T(\varepsilon)) \qquad \text{for } 0\le t\le T(\varepsilon). \tag{6.24}$$

Hence, $\xi(T(\varepsilon))-\xi_r(T(\varepsilon)) = O(T(\varepsilon))$. Using this expression, (6.6), (6.22), continuous dependence of the solutions of differential equations on initial conditions and parameters [78, Theorem 9.1], and exponential stability of the origin of $\dot{\xi}=(A-BK)\xi$, we conclude that

$$\xi(t)-\xi_r(t) = O(\varepsilon)+O(T(\varepsilon)) \qquad \forall\, t\ge T(\varepsilon). \tag{6.25}$$

From (6.24) and (6.25), we see that

$$\xi(t)-\xi_r(t) = O(\varepsilon)+O(T(\varepsilon)) \qquad \forall\, t\ge 0, \tag{6.26}$$

which proves (6.15).

Because $\lim_{t\to\infty}\xi_r(t) = 0$, given any $\mu>0$ there is a finite time T_1 such that $\|\xi_r(t)\|\le\mu/2$ for all $t\ge T_1$. On the other hand, (6.26) shows that for sufficiently small ε, $\|\xi(t)-\xi_r(t)\|\le\mu/2$ for all $t\ge T_1$. Hence,

$$\|\xi(t)\| = \|\xi(t)-\xi_r(t)+\xi_r(t)\| \le \|\xi_r(t)\| + \|\xi(t)-\xi_r(t)\| \le \mu \qquad \forall\, t\ge T_1,$$

which proves (6.16). □

Remark 6.4. *The control is saturated by using the saturation function* sat(·), *which is locally Lipschitz but not continuously differentiable. Because the proof uses a change of variables that requires differentiation of the control variable, in* (6.18) *the function* sat(·) *is replaced by the continuously differentiable function* $g_\varepsilon(\cdot)$, *which is* $O(\varepsilon)$ *close to* sat(·). *The proof could have been simplified by saturating the control using a smooth saturation function. However, the saturation function* sat(·) *is easier to implement.* ◇

Theorem 6.1 ensures only practical regulation. The next theorem shows that if w is constant, then, under additional conditions on the zero dynamics, the controller provides integral action that ensures regulation of $e(t)$ (in fact the whole vector $\xi(t)$) to zero even when the model uncertainty does not vanish at $\xi = 0$. The following assumption states the additional requirements on $\dot{\eta} = f_0(\eta, \xi, w)$ when w is constant.

Assumption 6.4.

- *$f_0(\eta, 0, w)$ is continuously differentiable in η.*
- *For each $w \in \mathcal{W}$, the system $\dot{\eta} = f_0(\eta, 0, w)$ has a unique equilibrium point $\eta_{ss} = \eta_{ss}(w) \in \{V_0(\eta) \le c_0\}$.*
- *With $\zeta = \eta - \eta_{ss}$, there is a continuously differentiable Lyapunov function $V_a(\zeta)$, possibly dependent on w, and class $\mathcal{K}$ functions γ_1 to γ_4, independent of w, such that*

$$\gamma_1(\|\zeta\|) \le V_a(\zeta) \le \gamma_2(\|\zeta\|),$$

$$\frac{\partial V_a}{\partial \zeta} f_0(\zeta + \eta_{ss}, \xi, w) \le -\gamma_3(\|\zeta\|) \qquad \forall\ \|\zeta\| \ge \gamma_4(\|\xi\|)$$

for all $(\eta, \xi, w) \in \Omega \times \mathcal{W}$.
- *The origin of $\dot{\zeta} = f_0(\zeta + \eta_{ss}, \xi, w)$ is exponentially stable.*

The third bullet requires the system $\dot{\zeta} = f_0(\zeta + \eta_{ss}, \xi, w)$ to be regionally input-to-state stable with respect to the input ξ, uniformly in w.

Theorem 6.2. *Consider the closed-loop system formed of the plant* (6.1)–(6.4), *the observer* (6.7)–(6.9), *and the controller* (6.13), *under the assumptions of Theorem* 6.1. *Suppose w is constant and Assumption* 6.4 *is satisfied. Then there exists $\bar{\varepsilon} > 0$ such that for $\varepsilon \in (0, \bar{\varepsilon})$, $\xi(t)$ converges to zero as $t \to \infty$.* ◇

Proof: Shift the equilibrium point $\eta = \eta_{ss}$ of $\dot{\eta} = f_0(\eta, 0, w)$ to the origin by the change of variables $\zeta = \eta - \eta_{ss}$. It can be verified that the closed-loop system has an isolated equilibrium point at the origin $(\zeta, \xi, \varphi) = (0, 0, 0)$ and the corresponding equilibrium values of $\hat{\sigma}$ and u are

$$\bar{\hat{\sigma}} = \frac{\hat{b}(0)}{b(\eta_{ss}, 0, w)} a(\eta_{ss}, 0, w) - \hat{a}(0), \qquad \bar{u} = -\frac{a(\eta_{ss}, 0, w)}{b(\eta_{ss}, 0, w)}.$$

It has been already shown in the proof of Theorem 6.1 that, in finite time, $\varphi = O(\varepsilon)$ and $\|\xi\| \le \mu$ for any given $\mu > 0$. Using regional input-to-state stability of $\dot{\zeta} = f_0(\zeta + \eta_{ss}, \xi, w)$ it can be shown that (ζ, ξ, φ) will reach a neighborhood $\mathcal{N}$ of $(0, 0, 0)$, where $\mathcal{N}$ can be made arbitrarily small by choosing ε small enough. It can be also verified that ζ, ξ, and φ satisfy the equations

$$\dot{\zeta} = f_0(\zeta + \eta_{ss}, 0, w) + \delta_1(\zeta, \xi, w), \tag{6.27}$$

$$\dot{\xi} = (A - BK)\xi + \delta_2(\eta, \xi, \varphi, w, \varepsilon), \tag{6.28}$$

$$\varepsilon \dot{\varphi} = \Lambda \varphi - \bar{B}_1 \Delta_1(\eta, \xi, \varphi, w, \varepsilon) \alpha_{\rho+1} \varphi_1 + \varepsilon \delta_3(\eta, \xi, \varphi, w, \varepsilon) \tag{6.29}$$

in $\mathcal{N}$, where

$$\delta_1 = f_0(\zeta + \eta_{ss}, \xi, w) - f_0(\zeta + \eta_{ss}, 0, w),$$
$$\delta_2 = B\left\{\varphi_{\rho+1} + b(\eta, \xi, w)[\psi(\hat{\xi}, \hat{\sigma}) - \psi(\xi, \hat{\sigma})]\right\},$$
$$\delta_3 = \bar{B}_1\, \Delta_2(\eta, \xi, \varphi, w, \varepsilon) + \bar{B}_2\, \Delta_3(\eta, \xi, \varphi, w, \varepsilon).$$

The functions δ_1 to δ_3 satisfy the inequalities

$$||\delta_1|| \le k_1||\xi||, \quad ||\delta_2|| \le k_3||\varphi||, \quad ||\delta_3|| \le k_3||\xi|| + k_4||\zeta|| + k_5||\varphi||$$

in some neighborhood of the origin with some nonnegative constants k_1 to k_5 independent of ε. From the converse Lyapunov theorem of exponential stability [80, Theorem 3.8], let $V_1(\zeta)$ be a Lyapunov function for $\dot{\zeta} = f_0(\zeta + \eta_{ss}, 0, w)$, which satisfies

$$\frac{\partial V_1}{\partial \zeta} f_0(\zeta + \eta_{ss}, 0, w) \le -k_6||\zeta||^2, \quad \left\|\frac{\partial V_1}{\partial \zeta}\right\| \le k_7||\zeta||.$$

in some neighborhood of $\zeta = 0$. The Lyapunov function $V_s = \xi^T P_s \xi$ of the target system (6.6) satisfies

$$\frac{\partial V_s}{\partial \xi}(A - BK)\xi \le -k_8||\xi||^2, \qquad \left\|\frac{\partial V_s}{\partial \xi}\right\| \le k_9||\xi||$$

Using $V = k_0 V_1(\zeta) + V_s(\xi) + \varphi^T P \varphi$, with $k_0 > 0$, as a Lyapunov function candidate for (6.27)–(6.29), it can be verified that in some neighborhood of the origin[56]

$$\dot{V} \le -\mathcal{Y}^T Q \mathcal{Y},$$

where

$$\mathcal{Y} = \begin{bmatrix} ||\zeta|| \\ ||\xi|| \\ ||\varphi|| \end{bmatrix}, \qquad Q = \begin{bmatrix} k_0 k_6 & -\frac{1}{2}k_0 k_1 k_7 & \star \\ -\frac{1}{2}k_0 k_1 k_7 & k_8 & \star \\ \star & \star & \left(\frac{\lambda\, \lambda_{min}(P)}{\varepsilon}\right) - \star \end{bmatrix},$$

and each $\star$ indicates a constant independent of k_0 and ε. The matrix Q can be made positive definite by choosing $k_0 < 4k_6k_8/(k_1k_7)^2$ and then choosing ε small enough. Hence, $(\zeta(t), \xi(t), \varphi(t))$ converges to zero as t tends to infinity. □

From the proof of Theorem 6.2 it is clear that the theorem can be specialized to the stabilization case, as follows.

Corollary 6.1. *Consider the closed-loop system formed of* (6.1)–(6.4), (6.7)–(6.9), *and* (6.13), *under the assumptions of Theorem* 6.2. *Suppose* $f_0(0,0,w) = 0$ *so that* $(\eta, \xi, \varphi) = (0,0,0)$ *is an equilibrium point. Let* $(\eta(0), \xi(0)) \in \Omega_0$ *(a compact set in the interior of* Ω*) and* $(\hat{\xi}(0), \hat{\sigma}(0)) \in S$ *(a compact subset of* $R^{\rho+1}$*). Then there exists* $\bar{\varepsilon} > 0$ *such that for* $\varepsilon \in (0, \bar{\varepsilon})$*, the origin* $(\eta, \xi, \varphi) = (0,0,0)$ *is an exponentially stable equilibrium point of the closed-loop system and* $\Omega_0 \times S$ *is a subset of the region of attraction.* ◇

[56] Recall from the proof of Theorem 6.1 that the derivative of $W = \varphi^T P \varphi$ with respect to $\varepsilon\dot{\varphi} = \Lambda\varphi - \bar{B}_1\Delta_1(\cdot)\alpha_{\rho+1}\varphi_1$ satisfies the inequality $\varepsilon \dot{W} \le -\lambda W$.

Example 6.3. Consider the system

$$\dot{x}_1 = x_2 + \theta\, x_1^2, \quad \dot{x}_2 = bu + x_3 + d, \quad \dot{x}_3 = -x_3 + x_1, \quad y = x_1,$$

where b and θ are unknown parameters that satisfy $b \in [0.75, 1.75]$, $\theta \in [-1, 1]$, and $d(t)$ is a bounded disturbance with bounded derivative. The control task is to have y asymptotically track a constant reference r. With the change of variables,

$$\eta = x_3, \qquad \xi_1 = x_1 - r, \qquad \xi_2 = x_2 + \theta x_1^2, \qquad e = \xi_1.$$

The system is represented by

$$\dot{\eta} = -\eta + \xi_1 + r, \quad \dot{\xi}_1 = \xi_2, \qquad \dot{\xi}_2 = a(\eta, \xi, w) + bu, \qquad e = \xi_1,$$

where $a(x, z, w) = \eta + d + 2\theta\,(\xi_1 + r)\xi_2$ and $w = \text{col}(r, d)$. It is assumed that $|r| \leq 1$ and $|d| \leq 1$ so that $\mathscr{W} = \{w_1 \in [-1, 1],\ w_2 \in [-1, 1]\}$. The nominal values of a and b are taken as $\hat{a} = 0$ and $\hat{b} = 1.25$. Hence, $|(b - \hat{b})/\hat{b}| \leq 0.4$. The condition (6.14) is satisfied by assigning all the observer eigenvalues to be real so that $||\Gamma||_\infty = 1$; see Remark 6.2. With $K = \begin{bmatrix} 2 & 2 \end{bmatrix}$, the target system (6.6) is given by

$$\dot{\xi}_1 = \xi_2, \qquad \dot{\xi}_2 = -2\xi_1 - 2\xi_2.$$

The Lyapunov functions $V_0(\eta)$ and $V_s(\xi)$ are taken as

$$V_0(\eta) = \tfrac{1}{2}\eta^2, \qquad V_s = \xi^T P_s \xi, \quad \text{where} \quad P_s = \begin{bmatrix} 1 & \frac{0.1}{\sqrt{2}} \\ \frac{0.1}{\sqrt{2}} & \frac{1}{2} \end{bmatrix}.$$

Using the fact that [80, Equation B.4]

$$\max_{x^T P x \leq c} ||Lx|| \leq \sqrt{c}||LP^{-1/2}|| \qquad \text{for } P = P^T > 0,$$

it can be verified that $|\xi_1| \leq 1.005\sqrt{c}$, $|\xi_2| \leq 1.4213\sqrt{c}$, and $|2(\xi_1 + \xi_2)| \leq 3.3134\sqrt{c}$ in the set $\{\xi^T P_s \xi \leq c\}$. Since $\dot{V}_0 \leq 0$ for $|\eta| \geq |\xi_1| + |r|$, the positively invariant set Ω is defined by $\Omega = \{|\eta| \leq 1 + 1.005\sqrt{c}\} \times \{\xi^T P_s \xi \leq c\}$. Taking $c = 1$, it can be verified that

$$\max_{(\eta,\xi)\in\Omega,\ w\in\mathscr{W}} \left| \frac{-a(\eta, \xi, w) - 2\xi_1 - 2\xi_2}{b} \right| \leq 16.0238.$$

The control is saturated at ± 16.1. Thus, the output feedback controller is given by

$$u = 16.1\ \text{sat}\left(\frac{0.8(-\hat{\sigma} - 2\hat{\xi}_1 - 2\hat{\xi}_2)}{16.1} \right),$$

where $\hat{\xi}_1$, $\hat{\xi}_2$, and $\hat{\sigma}$ are provided by the extended observer

$$\dot{\hat{\xi}}_1 = \hat{\xi}_2 + (3/\varepsilon)(e - \hat{\xi}_1), \quad \dot{\hat{\xi}}_2 = \hat{\sigma} + 1.25u + (3/\varepsilon^2)(e - \hat{\xi}_1), \quad \dot{\hat{\sigma}} = (1/\varepsilon^3)(e - \hat{\xi}_1).$$

The observer gains are chosen to assign all three roots of $s^3 + 3s^2 + 3s + 1$ at -1. Simulation results are shown in Figures 6.5 to 6.7 with $x_i(0) = \hat{\xi}_i(0) = 0$, for $i = 1, 2, 3$,

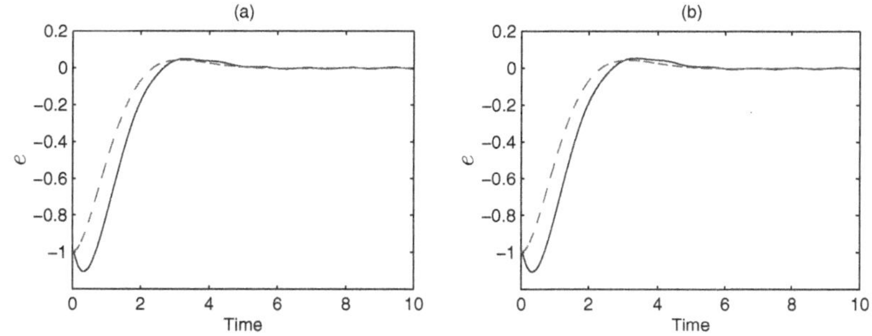

Figure 6.5. *Simulation of Example* 6.3. (a) *shows the regulation error when* $\theta = 1$, *while* (b) *shows it when* $\theta = -1$. *The target response is shown in dashed lines. In both cases* $d = \sin 5t$ *and* $\varepsilon = 0.005$.

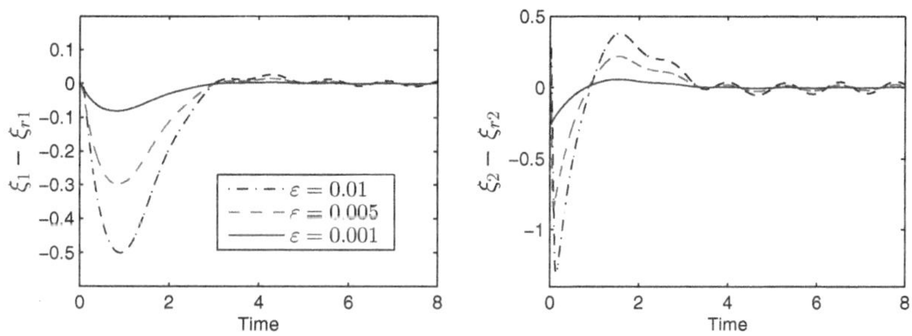

Figure 6.6. *Simulation of Example* 6.3 *with* $\theta = 1$ *and* $d = \sin 5t$ *for three different values of* ε. *The figure shows the deviation of* ξ_1 *and* ξ_2 *from the corresponding trajectories of the target system.*

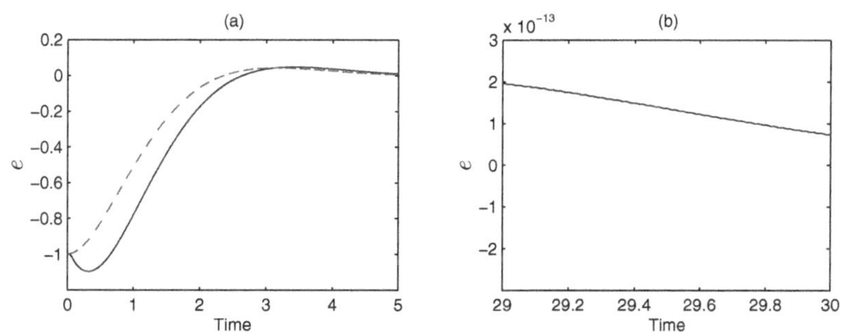

Figure 6.7. *The regulation error in Example* 6.3 *when* r *and* d *are constant. The target error is shown in dashed line.*

and $r = 1$. Figure 6.5 shows the regulation error $e = \xi_1$ for $d = \sin 5t$, $\varepsilon = 0.005$, and two different values of θ: 1 and -1. The target response is shown in dashed lines. The figure demonstrates how the closed-loop response of the uncertain nonlinear system can be shaped by shaping the response of the linear target system. It is significant to notice that changing θ from 1 to -1 has little effect on the response. Figure 6.6 demonstrates the performance recovery property by showing the difference between the two responses for three different values of ε when $\theta = 1$. Figure 6.7 demonstrates the integral action property by showing the response for $d = 1$; all other parameters and initial conditions are the same as in Figure 6.5. The regulation error is shown during the transient period and at a later time to show the convergence of the error

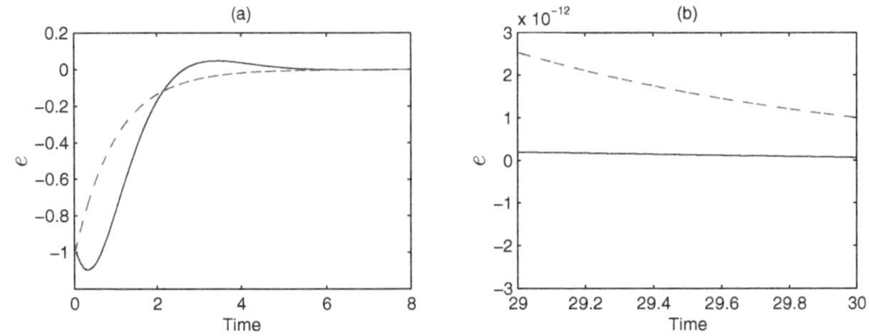

Figure 6.8. *Comparison of the regulation error in Example 6.3 between the extended-high-observer controller (solid) and the conditional-integrator controller (dashed).*

to zero. Comparison of Figures 6.5(a) and 6.7(a) shows that changing the disturbance from $d = \sin 5t$ to $d = 1$ has little effect on the transient response.

Since integral action is a key property of the controller of this section, it is useful to compare it with the conditional integrator of Section 5.3. Without getting into the details, it can be verified, following the design procedure of Section 5.3, that over the set $\{|\sigma| \le \mu\} \times \{|z - r| \le 1.32\} \times \{|\xi_1| \le 1.21\} \times \{|s| \le 1\}$, where $s = \sigma + \xi_1 + \xi_2$, the conditional-integrator, output-feedback controller can be taken as

$$\dot{\sigma} = -\sigma + \mu \,\text{sat}\left(\frac{\sigma + \hat{\xi}_1 + \hat{\xi}_2}{\mu}\right), \qquad u = -22\,\text{sat}\left(\frac{\sigma + \hat{\xi}_1 + \hat{\xi}_2}{\mu}\right),$$

where $\hat{\xi}_1$ and $\hat{\xi}_2$ are provided by the high-gain observer

$$\dot{\hat{\xi}}_1 = \hat{\xi}_2 + (2/\varepsilon)(e - \hat{\xi}_1), \qquad \dot{\hat{\xi}}_2 = (1/\varepsilon^2)(e - \hat{\xi}_1).$$

Figure 6.8 compares the regulation error under the two controllers, the extended-observer controller in solid lines and the conditional-integrator controller in dashed lines. The simulation of the two controllers is carried out for the same parameters and initial conditions, in particular, $b = \theta = d = r = 1$ and $\varepsilon = 0.005$; $\mu = 0.1$ for the conditional integrator. The transient response is comparable but not identical because the conditional integrator approaches the response under sliding mode control, while the extended-observer controller approaches a target system obtained by feedback linearization. Figure 6.8(b) shows that the convergence of the error to zero is slightly faster for the extended-observer controller. Of course, the extended-observer controller achieves performance recovery for more general classes of references and disturbances, not just constant ones, but it uses a third-order observer compared with a second-order one for the conditional integrator. △

6.2.2 ▪ Multi-Input–Multi-Output Systems

The normal form

$$\dot{\eta} = f_0(\eta, \xi, w), \tag{6.30}$$

$$\dot{\xi}^i_j = \xi^i_{j+1} \quad \text{for } 1 \le j \le \rho_i - 1,\ 1 \le i \le m, \tag{6.31}$$

$$\dot{\xi}^i_{\rho_i} = f_i(\eta, \xi, w) + G_i(\eta, \xi, w)u \quad \text{for } 1 \le i \le m, \tag{6.32}$$

$$e_i = \xi^i_1 \quad \text{for } 1 \le i \le m \tag{6.33}$$

represents a nonlinear system with vector relative degree $\{\rho_1, \dots \rho_m\}$, where

$$\xi^i = \text{col}(\xi_1^i, \xi_2^i, \dots, \xi_{\rho_i}^i), \quad \xi = \text{col}(\xi^1, \xi^2, \dots, \xi^m), \quad \rho = \rho_1 + \dots + \rho_m,$$

$\eta \in R^{n-\rho}$ and $\xi \in R^\rho$ are the state variables, $u \in R^m$ is the control input, e_1 to e_m are the measured regulation errors, and $w \in R^\ell$ is an exogenous input. The normal form is defined for $\eta \in D_\eta \subset R^{n-\rho}$ and $\xi \in D_\xi \subset R^\rho$ for some domains D_η and D_ξ. Equations (6.31)–(6.33) represents m chains of integrators with ρ_i integrators in the ith chain.[57] The system can be equivalently represented by the equations

$$\dot{\eta} = f_0(\eta, \xi, w), \tag{6.34}$$

$$\dot{\xi} = A\xi + B[f(\eta, \xi, w) + G(\eta, \xi, w)u], \tag{6.35}$$

$$e = Cx, \tag{6.36}$$

where

$$A = \text{block diag}[A_1, \dots, A_m], \ A_i = \begin{bmatrix} 0 & 1 & \cdots & \cdots & 0 \\ 0 & 0 & 1 & \cdots & 0 \\ \vdots & & & & \vdots \\ 0 & \cdots & \cdots & 0 & 1 \\ 0 & \cdots & \cdots & \cdots & 0 \end{bmatrix}_{\rho_i \times \rho_i},$$

$$B = \text{block diag}[B_1, \dots, B_m], \ B_i = \begin{bmatrix} 0 \\ \vdots \\ 0 \\ 1 \end{bmatrix}_{\rho_i \times 1},$$

$$C = \text{block diag}[C_1, \dots, C_m], \ C_i = \begin{bmatrix} 1 & 0 & \cdots & 0 \end{bmatrix}_{1 \times \rho_i},$$

$$u = \begin{bmatrix} u_1 \\ \vdots \\ u_m \end{bmatrix}, \ e = \begin{bmatrix} e_1 \\ \vdots \\ e_m \end{bmatrix}, \ f = \begin{bmatrix} f_1 \\ \vdots \\ f_m \end{bmatrix}, \ G = \begin{bmatrix} G_1 \\ \vdots \\ G_m \end{bmatrix}_{m \times m}.$$

The functions f_0, f, and G and the signal w could be unknown. Suppose w satisfies Assumption 6.1.

Assumption 6.5. *The functions f and G are continuously differentiable with locally Lipschitz derivatives, f_0 is locally Lipschitz, and the matrix $G(\eta, \xi, w)$ is nonsingular with bounded inverse for all $(\eta, \xi, w) \in D_\eta \times D_\xi \times \mathcal{W}$.*

Similar to the previous section, a target system is taken as

$$\dot{\xi} = (A - BK)\xi, \tag{6.37}$$

where K is designed to make $(A - BK)$ Hurwitz and shape the transient response of (6.37). A Lyapunov function for (6.37) can be taken as $V_s(\xi) = \xi^T P_s \xi$, where $P_s = P_s^T > 0$ is the solution of the Lyapunov equation $P_s(A - BK) + (A - BK)^T P_s = -Q_s$, for some $Q_s = Q_s^T > 0$. Let c be a positive constant such that $\{V_s(\xi) \le c\} \subset D_\xi$.

[57]See Figure 2.2.

Suppose Assumption 6.3 is satisfied. Then, the compact set $\Omega = \{V_0(\eta) \le c_0\} \times \{V_s(\xi) \le c\}$ is positively invariant with respect to the system

$$\dot{\eta} = f_0(\eta, \xi, w), \qquad \dot{\xi} = (A - BK)\xi.$$

Let the m-dimensional function $\hat{f}(\xi)$ and the $m \times m$ matrix $\hat{G}(\xi)$ be twice continuously differentiable, globally bounded models of $f(\eta, \xi, w)$ and $G(\eta, \xi, w)$, respectively, where $\hat{G}(\xi)$ is nonsingular with bounded inverse. It is allowed to take $\hat{f} = 0$ and $\hat{G}$ as a constant matrix. The ξ-equation can be written as

$$\dot{\xi} = A\xi + B[\sigma + \hat{f}(\xi) + \hat{G}(\xi)u], \qquad e = C\xi,$$

where

$$\sigma = f(\eta, \xi, w) - \hat{f}(\xi) + [G(\eta, \xi, w) - \hat{G}(\xi)]u.$$

Equivalently, for each $1 \le i \le m$,

$$\begin{aligned}
\dot{\xi}^i_j &= \xi^i_{j+1} \quad \text{for } 1 \le j \le \rho_i - 1 \\
\dot{\xi}^i_{\rho_i} &= \sigma_i + \hat{f}_i(\xi) + \hat{G}_i(\xi)u, \\
e_i &= \xi^i_1,
\end{aligned}$$

where $\hat{G}_i$ is the ith row of $\hat{G}$. Augmenting σ_i as an additional state to the ith chain of integrators, a high-gain observer of the extended system is taken as

$$\dot{\hat{\xi}}^i_j = \hat{\xi}^i_{j+1} + (\alpha^i_j/\varepsilon^j)(e_i - \hat{\xi}^i_1) \quad \text{for } 1 \le j \le \rho_i - 1, \tag{6.38}$$

$$\dot{\hat{\xi}}^i_{\rho_i} = \hat{\sigma}_i + \hat{f}_i(\hat{\xi}) + \hat{G}_i(\hat{\xi})u + (\alpha^i_{\rho_i}/\varepsilon^{\rho_i})(e_i - \hat{\xi}^i_1), \tag{6.39}$$

$$\dot{\hat{\sigma}}_i = (\alpha^i_{\rho_i+1}/\varepsilon^{\rho_i+1})(e_i - \hat{\xi}^i_1) \tag{6.40}$$

for $1 \le i \le m$, where α^i_1 to $\alpha^i_{\rho_i+1}$ are chosen such that the polynomial

$$s^{\rho_i+1} + \alpha^i_1 s^{\rho_i} + \cdots + \alpha_{\rho_i+1}$$

is Hurwitz, and $\varepsilon > 0$ is a small parameter. Let

$$\psi(\xi, \sigma) = \hat{G}^{-1}(\xi)[-\hat{f}(\xi) - \sigma - K\xi]. \tag{6.41}$$

Denote the ith component of $G^{-1}(\eta, \xi, w)[-f(\eta, \xi, w) - K\xi]$ by $h_i(\eta, \xi, w)$ and let

$$M_i > \max_{(\eta,\xi)\in\Omega,\ w\in\mathcal{W}} |h_i(\eta, \xi, w)|. \tag{6.42}$$

The output feedback controller is taken as

$$u_i = M_i \operatorname{sat}\left(\frac{\psi_i(\hat{\xi}, \hat{\sigma})}{M_i}\right) \quad \text{for } 1 \le i \le m. \tag{6.43}$$

Let

$$k_u = \max_{|d_i|\le 1} \left\{ \max_{(\eta,\xi)\in\Omega, w\in\mathcal{W}} \left\| [G(\eta, \xi, w) - \hat{G}(\xi)] D \hat{G}^{-1}(\xi) \right\| \right\}, \tag{6.44}$$

where $D = \mathrm{diag}(d_1, \ldots, d_m)$. Let

$$\Gamma(s) = \mathrm{diag}[\Gamma_1(s), \ldots \Gamma_m(s)], \quad \Gamma_i(s) = \frac{\alpha^i_{\rho_i+1}}{s^{\rho_i+1} + \alpha^i_1 s^{\rho_i} + \cdots + \alpha^i_{\rho_i+1}},$$

and $||\Gamma||_\infty = \max_{1\le i\le m} \max_\omega |\Gamma_i(j\omega)|$. Because $\Gamma_i(0) = 1$, $||\Gamma||_\infty \ge 1$. It can be verified that $||\Gamma||_\infty = 1$ if for each $1 \le i \le m$, all the poles of $\Gamma_i(s)$ are real.

The following theorem reads almost the same as Theorem 6.1.

Theorem 6.3. *Consider the closed-loop system formed of the plant* (6.30)–(6.33), *the observer* (6.38)–(6.40), *and the controller* (6.43). *Suppose Assumptions* 6.1, 6.3, *and* 6.5 *are satisfied,*

$$k_u < \frac{1}{||\Gamma||_\infty}, \tag{6.45}$$

the initial states of the observer belong to a compact subset of $R^{\rho+m}$, and the initial states of the plant belong to a compact set in the interior of Ω. Then

- *there exists $\varepsilon_1^* > 0$ such that for every $0 < \varepsilon \le \varepsilon_1^*$, the trajectories of the closed-loop system are bounded for all $t \ge 0$;*
- *given any $\mu > 0$, there exist $\varepsilon_2^* > 0$, dependent on μ, such that for every $0 < \varepsilon \le \varepsilon_2^*$,*

$$||\xi(t) - \xi_r(t)|| \le \mu \qquad \forall\, t \ge 0, \tag{6.46}$$

where ξ_r is the solution of the target system (6.37) *with $\xi_r(0) = \xi(0)$;*
- *given any $\mu > 0$, there exist $\varepsilon_3^* > 0$ and $T_1 > 0$, both dependent on μ, such that for every $0 < \varepsilon \le \varepsilon_3^*$,*

$$||\xi(t)|| \le \mu \qquad \forall\, t \ge T_1. \tag{6.47}$$

◇

Proof: For each $1 \le i \le m$, apply the change of variables

$$\varphi^i_j = (\xi^i_j - \hat{\xi}^i_j)/\varepsilon^{\rho_i+1-j} \quad \text{for} \quad 1 \le j \le \rho_i,$$
$$\varphi^i_{\rho_i+1} = \Delta_{f_i}(\eta, \xi, w) + \Delta_{G_i}(\eta, \xi, w)\,\psi_s(\xi, \hat{\sigma}) - \hat{\sigma}_i,$$

where

$$\Delta_{f_i}(\eta, \xi, w) = f_i(\eta, \xi, w) - \hat{f}_i(\xi), \qquad \Delta_{G_i}(\eta, \xi, w) = G_i(\eta, \xi, w) - \hat{G}_i(\xi),$$
$$\psi_s(\xi, \hat{\sigma}) = \mathrm{col}\left(M_1\, g_\varepsilon\left(\frac{\psi_1(\xi, \hat{\sigma})}{M_1}\right), \ldots, M_m\, g_\varepsilon\left(\frac{\psi_m(\xi, \hat{\sigma})}{M_m}\right)\right),$$

and g_ε is defined in the proof of Theorem 6.1. With $\varphi^i = \mathrm{col}(\varphi^i_1, \ldots, \varphi^i_{\rho_i})$ and $\varphi = \mathrm{col}(\varphi^1, \ldots, \varphi^m)$, it can be verified that φ satisfies the equation[58]

$$\dot{\varphi} = \Lambda\varphi - \bar{B}\Delta_\varphi(\eta, \xi, \varphi, w, \varepsilon)\bar{C}\varphi + \varepsilon\delta_\varphi(\eta, \xi, \varphi, w, \dot{w}, \varepsilon), \tag{6.48}$$

[58] The details of arriving at this equation can be found in [155].

where

$$\Lambda = \text{block diag}(\Lambda_1, \ldots, \Lambda_m), \quad \Lambda_i = \begin{bmatrix} -\alpha_1^i & 1 & 0 & \cdots & 0 \\ -\alpha_2^i & 0 & 1 & \ddots & 0 \\ \vdots & \vdots & \ddots & \ddots & \vdots \\ -\alpha_{\rho_i}^i & 0 & \cdots & 0 & 1 \\ -\alpha_{\rho_i+1}^i & 0 & \cdots & \cdots & 0 \end{bmatrix},$$

$$\bar{B} = \text{block diag}(\bar{B}_1, \ldots, \bar{B}_m), \quad \bar{B}_i = \begin{bmatrix} 0 \\ \vdots \\ 0 \\ 1 \end{bmatrix}_{(\rho_i+1)\times 1},$$

$$\bar{C} = \text{block diag}(\bar{C}_1, \ldots, \bar{C}_m), \quad \bar{C}_i = \begin{bmatrix} \alpha_{\rho_i+1}^i & 0 & \cdots & 0 \end{bmatrix}_{1\times(\rho_i+1)},$$

$$\Delta_\varphi = [G(\eta, \xi, w) - \hat{G}(\xi)] E \hat{G}^{-1}(\xi),$$

$$E = \text{diagonal}\left[g_\varepsilon'\left(\frac{\psi_1(\xi, \hat{\sigma})}{M_1}\right), \ldots, g_\varepsilon'\left(\frac{\psi_m(\xi, \hat{\sigma})}{M_m}\right)\right].$$

The function δ_φ is locally Lipschitz in its argument and $||\delta_\varphi|| \leq k||\varphi||$ for some $k > 0$, independent of ε. Equation (6.48) is an $O(\varepsilon)$ perturbation of the system

$$\dot{\varphi} = \Lambda\varphi - \bar{B}\Delta_\varphi(\eta, \xi, \varphi, w, \varepsilon)\bar{C}\varphi, \tag{6.49}$$

which can be represented as a negative feedback connection of the transfer function $\Gamma(\varepsilon s)$ and the time-varying gain $\Delta_\varphi(\cdot)$, where $||\Delta_\varphi(\cdot)|| \leq k_u$ because $|g_\varepsilon'(\cdot)| \leq 1$. The transfer function $\Gamma(\varepsilon s)$ has $\max_{1\leq i\leq m} \max_\omega |\Gamma_i(j\varepsilon\omega)| = ||\Gamma||_\infty$. By the multivariable circle criterion [80, Theorem 7.8], the origin of (6.49) will be globally exponential stable if $k_u||\Gamma||_\infty < 1$. From this point on, the proof proceeds exactly as in the proof of Theorem 6.1. □

Remark 6.5. *Because the observer can be designed to have $||\Gamma||_\infty = 1$, it is sufficient to require that $k_u < 1$.* ◇

Remark 6.6. *If G and $\hat{G}$ are diagonal matrices with diagonal elements $g_1, \ldots, g_m$ and $\hat{g}_1, \ldots, \hat{g}_m$, respectively, the constant k_u of (6.44) reduces to*

$$k_u = \max_{1\leq i\leq m} \left\{ \max_{(\eta,\xi)\in\Omega, w\in\mathcal{W}} \left| \frac{g_i(\eta, \xi, w) - \hat{g}_i(\xi)}{\hat{g}_i(\xi)} \right| \right\}.$$

If for each i the sign of g_i and an upper bound g_{im} on $|g_i|$ are known, taking $\hat{g}_i =$ $\text{sign}(g_i)g_{iM}$ with $g_{iM} > g_{im}$, for $1 \leq i \leq m$, ensures that $k_u < 1$. ◇

Remark 6.7. *If G is a symmetric positive definite matrix and there are positive constants g_1 and $g_2 > g_1$ such that*

$$g_1 I \leq G(\eta, \xi, w) \leq g_2 I$$

for all $(\eta, \xi, w) \in \Omega \times \mathcal{W}$ *and* $\hat{G} = g_m I$ *with* $g_m > g_2$, *then* k_u *of* (6.44) *satisfies* $k_u \leq 1 - g_1/g_m < 1$. *This is so because*

$$(G - \hat{G})D\hat{G}^{-1} = (G - g_m I)D\frac{1}{g_m} = \left(\frac{1}{g_m}G - I\right)D.$$

Therefore,

$$\|(G - \hat{G})D\hat{G}^{-1}\| = \left\|\left(\frac{1}{g_m}G - I\right)D\right\| \leq \left\|I - \frac{1}{g_m}G\right\| \; \|D\|$$

and

$$k_u \leq \max_{(\eta,\xi,w)\in\Omega\times\mathcal{W}} \left\|I - \frac{1}{g_m}G\right\|.$$

The matrix $[I - (1/g_m)G]$ *is positive definite because*

$$x^T\left(I - \frac{1}{g_m}G\right)x \geq \left(1 - \frac{g_2}{g_m}\right)x^T x.$$

Then

$$\left\|I - \frac{1}{g_m}G\right\| = \lambda_{\max}\left(I - \frac{1}{g_m}G\right) \leq 1 - \frac{1}{g_m}\lambda_{\min}(G) \leq 1 - \frac{g_1}{g_m}.$$

◇

In view of the properties of (6.48), it is straightforward to see that Theorem 6.2 and Corollary 6.1 hold in the multi-input–multi-output case. For completeness, they are stated without proof.

Theorem 6.4. *Consider the closed-loop system formed of the plant* (6.30)–(6.33), *the observer* (6.38)–(6.40), *and the controller* (6.43), *under the assumptions of Theorem* 6.3. *Suppose* w *is constant and Assumption* 6.4 *is satisfied. Then, there exists* $\bar{\varepsilon} > 0$ *such that for* $\varepsilon \in (0, \bar{\varepsilon})$, $\xi(t)$ *converges to zero as* $t \to \infty$. ◇

Corollary 6.2. *Consider the closed-loop system formed of the plant* (6.30)–(6.33), *the observer* (6.38)–(6.40), *and the controller* (6.43), *under the assumptions of Theorem* 6.4. *Suppose* $f_0(0, 0, w) = 0$ *so that* $(\eta, \xi, \varphi) = (0, 0, 0)$ *is an equilibrium point. Let* $(\eta(0), \xi(0)) \in \Omega_0$ *(a compact set in the interior of* Ω*) and* $(\hat{\xi}(0), \hat{\sigma}(0)) \in S$ *(a compact subset of* $R^{\rho+m}$*). Then there exists* $\bar{\varepsilon} > 0$ *such that for* $\varepsilon \in (0, \bar{\varepsilon})$, *the origin* $(\eta, \xi, \varphi) = (0, 0, 0)$ *is an exponentially stable equilibrium point of the closed-loop system and* $\Omega_0 \times S$ *is a subset of the region of attraction.* ◇

Example 6.4. A two-link robot manipulator, with neglected damping, is modeled by [59]

$$M(q)\ddot{q} + \bar{C}(q, \dot{q})\dot{q} + g(q) = u,$$

where $q \in R^2$ is a vector of generalized coordinates representing joint positions, $u \in R^2$ is the control input, and $M(q)$ is a symmetric positive definite inertia matrix. The

[59] See [146], [132], or [80].

terms $\bar{C}(q,\dot{q})\dot{q}$ and $g(q)$ account for centrifugal/Coriolis forces and gravity, respectively:

$$M = \begin{bmatrix} a_1 + 2a_4 \cos q_2 & a_2 + a_4 \cos q_2 \\ a_2 + a_4 \cos q_2 & a_3 \end{bmatrix},$$
$$\bar{C} = a_4 \sin q_2 \begin{bmatrix} -\dot{q}_2 & -(\dot{q}_1 + \dot{q}_2) \\ \dot{q}_1 & 0 \end{bmatrix}, \quad g = \begin{bmatrix} b_1 \cos q_1 + b_2 \cos(q_1 + q_2) \\ b_2 \cos(q_1 + q_2) \end{bmatrix},$$

where a_1 through a_4, b_1, and b_2 are positive constants. The goal is to regulate q_1 and q_2 to equilibrium positions q_{d1} and q_{d2}, respectively. Taking

$$e_1 = \xi_1^1 = q_1 - q_{d1}, \quad \xi_2^1 = \dot{q}_1, \quad e_2 = \xi_1^2 = q_2 - q_{d2}, \quad \xi_2^2 = \dot{q}_2,$$

the system takes the form (6.31)–(6.33) or (6.35)–(6.36) with $\rho_1 = \rho_2 = m = 2$, $f = M^{-1}(-\bar{C}\dot{q} - g)$, and $G = M^{-1}$. The matrix K that defines the target system (6.37) is designed by solving an LQR optimal control problem with identity matrices for Q and R. It is given by

$$K = \text{block diag}\left[\begin{pmatrix}1 & \sqrt{3}\end{pmatrix}, \begin{pmatrix}1 & \sqrt{3}\end{pmatrix}\right].$$

The vector $w = \text{col}(q_{d1}, q_{d2})$ is constant, and Assumption 6.5 is satisfied. Assumptions 6.3 and 6.4 are not applicable because there are no η dynamics. We need to choose $\hat{G}$ that satisfies (6.45). Because M is symmetric and positive definite, its eigenvalues are real and positive. They satisfy the inequalities

$$k_1 \le \lambda(M(q)) \le k_2 \qquad \text{for } \xi \in \{V_s(\xi) \le c\}$$

for some positive constants k_1 and $k_2 > k_1$. Consequently,

$$\frac{1}{k_2} \le \lambda(M^{-1}(q)) \le \frac{1}{k_1}.$$

Following Remark 6.5, let $\hat{G} = k_m I$, where $k_m > 1/k_1$. Then $k_u \le 1 - 1/(k_2 k_m) < 1$. The condition (6.45) is satisfied by designing the observer with real eigenvalues. We take $\hat{f} = 0$. All the conditions of Theorem 6.4 are satisfied. For $i = 1, 2$, the observer is given by

$$\begin{aligned}
\dot{\hat{\xi}}_1^i &= \hat{\xi}_2^i + (\alpha_1/\varepsilon)(e_i - \hat{\xi}_1^i), \\
\dot{\hat{\xi}}_2^i &= \hat{\sigma}_i + k_m u_i + (\alpha_2/\varepsilon^2)(e_i - \hat{\xi}_1^i), \\
\dot{\hat{\sigma}}_i &= (\alpha_3/\varepsilon^2)(e_i - \hat{\xi}_1^i),
\end{aligned}$$

where the choice $\alpha_1 = \alpha_2 = 3$ and $\alpha_3 = 1$ assigns the three roots of $s^3 + 3s^2 + 3s + 1$ at -1 and the control is

$$u_i = M_i \,\text{sat}\left(\frac{-\hat{\sigma}_i - \hat{\xi}_1^i - \sqrt{3}\,\hat{\xi}_2^i}{k_m M_i} \right).$$

For the data used in the simulation, k_m is taken as $k_m = 0.01$, and the saturation levels $M_1 = 2500$ and $M_2 = 1000$ are determined by simulation. The simulation is carried out with $q_{d1} = 0$, $q_{d2} = \pi/2$ and the initial conditions $q_1(0) = -\pi/2$ and $\dot{q}_1(0) = q_2(0) =$

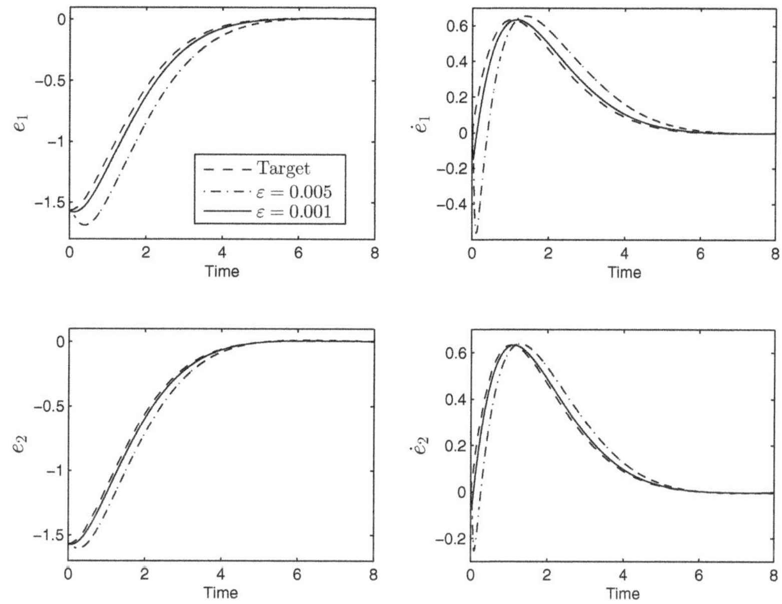

Figure 6.9. *The state trajectories of Example 6.4, compared with the target trajectories.*

$\dot{q}_2(0) = 0$. To demonstrate the robustness with respect to the system parameters, the simulation is carried out with two sets of parameters corresponding to a change in the load. The first set, referred to as Data A, is given by

$$a_1 = 200.01, \;\; a_2 = 23.5, \;\; a_3 = 122.5, \;\; a_4 = 25, \;\; b_1 = 784.8, \;\; b_2 = 245.25,$$

and the second set, referred to as Data B, is given by

$$a_1 = 259.7, a_2 = 58.19, a_3 = 157.19, a_4 = 56.25, b_1 = 1030.1, b_2 = 551.8125.$$

The simulation results in Figures 6.9 and 6.10 are for Data A. Figure 6.9 shows how the trajectories of the closed-loop system approach the trajectories of the target system as ε decreases. Figure 6.10 shows the control variables for $\varepsilon = 0.001$. To illustrate the control saturation during the peaking period of the observer, the right-side figures zoom into that period. Finally, Figure 6.11 plots the trajectory of e_1 for the two sets of parameters when $\varepsilon = 0.001$. The trajectories are almost indistinguishable. Zooming into a small area shows a slight difference between the two responses. △

6.2.3 ▪ Dynamic Inversion

Dynamic inversion is a technique that can be used with systems that depend nonlinearly on the control input.[60] Consider the system

$$\dot{x} = Ax + Bf(x,u),$$

where (A,B) is controllable and f is sufficiently smooth. Suppose it is desired to design the control u so that the response of the system matches the response of the target system

$$\dot{x} = (A - BK)x,$$

[60] See, for example, [59].

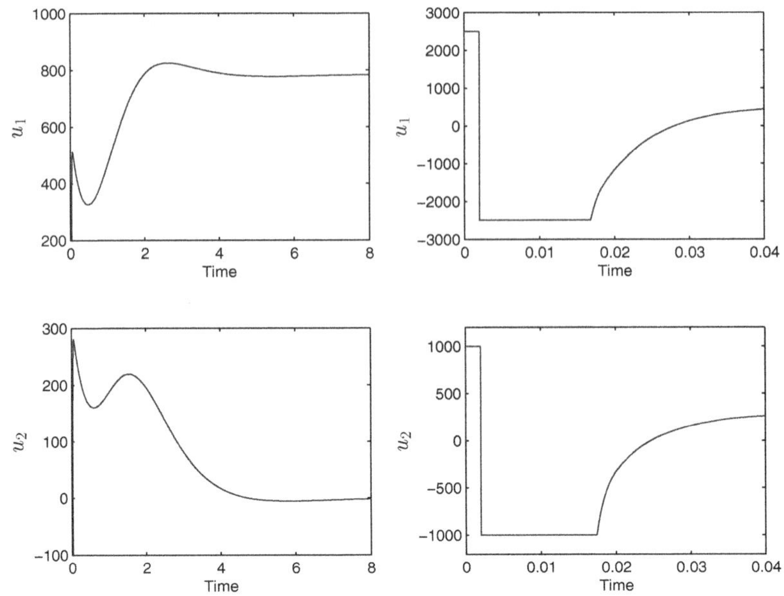

Figure 6.10. *The control variables of Example* 6.4.

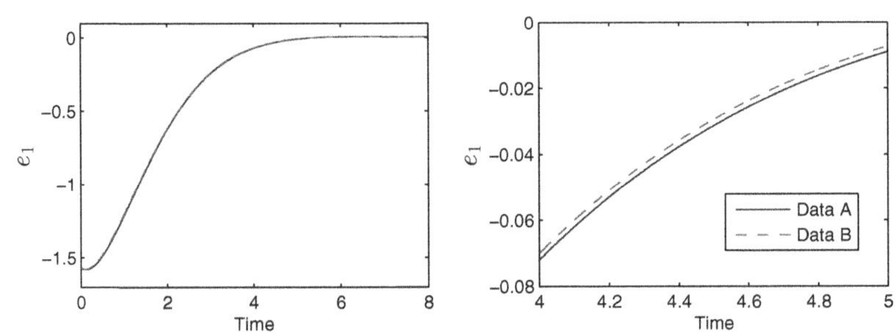

Figure 6.11. *Simulation results of Example* 6.4 *for two sets of system parameters.*

where $(A-BK)$ is Hurwitz. The control u must satisfy the equation

$$f(x,u) = -Kx.$$

Suppose this equation has a unique solution $u = \phi(x)$. The idea of dynamic inversion is to move u towards $\phi(x)$ in a fast time scale by taking u as the solution of the equation

$$\mu \dot{u} = -P(x)[f(x,u) + Kx],$$

where μ is a small positive constant and $P(x)$ is a nonsingular matrix that satisfies the condition

$$z^T P(x)[f(x,\phi(x)+z) - f(x,\phi(x))] \geq \beta ||z||^2$$

for some $\beta > 0$. The closed-loop system takes the singularly perturbed form

$$\begin{aligned} \dot{x} &= Ax + Bf(x,u), \\ \mu \dot{u} &= -P(x)[f(x,u) + Kx]. \end{aligned}$$

Setting $\mu = 0$, the equation $P(x)[f(x,u)+Kx] = 0$ has a unique solution $u = \phi(x)$ because $P(x)$ is nonsingular. The change of variables $z = u - \phi(x)$ transforms the system into the form

$$\begin{aligned}\dot{x} &= (A-BK)x + B[f(x,\phi(x)+z) - f(x,\phi(x))],\\ \mu\dot{z} &= -P(x)[f(x,\phi(x)+z) - f(x,\phi(x))] - \mu\dot{\phi},\end{aligned}$$

$$\mu = 0 \quad \Rightarrow \quad z = 0 \quad \Rightarrow \quad \dot{x} = (A-BK)x.$$

By singular perturbation theory,[61] it can be shown that the solution of the closed-loop system will be $O(\mu)$ close to the solution of the target system when both systems start from the same initial states. Exponential stability of the origin of the closed-loop system can be shown by the composite Lyapunov function $V = x^T P_0 x + \frac{1}{2} z^T z$, where P_0 is the solution of the Lyapunov equation $P_0(A-BK) + (A-BK)^T P_0 = -I$. The derivative of V is

$$\begin{aligned}\dot{V} = &-x^T x + 2x^T P_0 B[f(x,\phi(x)+z) - f(x,\phi(x))]\\ &- \frac{1}{\mu} z^T P(x)[f(x,\phi(x)+z) - f(x,\phi(x))] - z^T \frac{\partial \phi}{\partial x}\dot{x}.\end{aligned}$$

Over a compact set of x, $\dot{V}$ satisfies the inequality

$$\begin{aligned}\dot{V} &\le -\|x\|^2 + 2k_2\|x\|\,\|z\| - \frac{\beta}{\mu}\|z\|^2 + k_3\|z\|^2\\ &= -\begin{bmatrix}\|x\|\\ \|z\|\end{bmatrix}^T \begin{bmatrix}1 & -k_2\\ -k_2 & \beta/\mu - k_3\end{bmatrix} \begin{bmatrix}\|x\|\\ \|z\|\end{bmatrix}\end{aligned}$$

for some positive constants k_2 and k_3, independent of μ. Hence, the origin is exponentially stable for sufficiently small μ. Implementation of the dynamic inversion algorithm requires knowledge of $f(x,u)$. This can be avoided by replacing f by an estimate $\hat{f}$ provided by an extended high-gain observer, as we shall see in the rest of this section.

Consider a multi-input–multi-output system represented by the equations

$$\dot{\eta} = f_0(\eta, \xi, w), \tag{6.50}$$

$$\dot{\xi}^i_j = \xi^i_{j+1} \quad \text{for } 1 \le j \le \rho_i - 1,\ 1 \le i \le m, \tag{6.51}$$

$$\dot{\xi}^i_{\rho_i} = f_i(\eta, \xi, w, u) \quad \text{for } 1 \le i \le m, \tag{6.52}$$

$$e_i = \xi^i_1 \quad \text{for } 1 \le i \le m, \tag{6.53}$$

where

$$\xi^i = \text{col}(\xi^i_1, \xi^i_2, \ldots, \xi^i_{\rho_i}), \quad \xi = \text{col}(\xi^1, \xi^2, \ldots, \xi^m), \quad \rho = \rho_1 + \cdots + \rho_m,$$

$\eta \in R^{n-\rho}$ and $\xi \in R^{\rho}$ are the state variables, $u \in R^m$ is the control input, e_1 to e_m are the measured regulation errors, and $w \in R^{\ell}$ is an exogenous input. The system

[61] See [87] or [78, Chapter 11].

is defined for $\eta \in D_\eta \subset R^{n-\rho}$ and $\xi \in D_\xi \subset R^\rho$ for some domains D_η and D_ξ. The system can be equivalently represented by the equations

$$\dot{\eta} = f_0(\eta, \xi, w), \tag{6.54}$$

$$\dot{\xi} = A\xi + Bf(\eta, \xi, w, u), \tag{6.55}$$

$$e = Cx, \tag{6.56}$$

where the block diagonal matrices $\{A, B, C\}$ are defined in the previous section. Suppose w satisfies Assumption 6.1 and f and f_0 satisfy the following assumption.

Assumption 6.6. *The function f is continuously differentiable with locally Lipschitz derivatives and f_0 is locally Lipschitz for all $(\eta, \xi, w) \in D_\eta \times D_\xi \times \mathcal{W}$.*

Consider the target system

$$\dot{\xi} = (A - BK)\xi, \tag{6.57}$$

where $(A - BK)$ is Hurwitz. The goal is to design an output feedback controller to asymptotically regulate the regulation error e to zero while bringing $\xi(t)$ of the closed-loop system arbitrarily close to the response of the target system when both systems start from the same initial states.

Assumption 6.7. *For $(\eta, \xi, w) \in D_\eta \times D_\xi \times \mathcal{W}$, the equation*

$$f(\eta, \xi, w, u) = -K\xi \tag{6.58}$$

has a unique solution $u = \phi(\eta, \xi, w)$ and ϕ is continuously differentiable.

It follows that $\dot{\phi}$ is bounded on compact sets of (η, ξ).

Assumption 6.8. *There is a known, locally Lipschitz, $m \times m$ nonsingular matrix $P(\xi)$ such that*

$$z^T P(\xi)[f(\eta, \xi, w, z + \phi) - f(\eta, \xi, w, \phi)] \geq \beta \|z\|^2 \tag{6.59}$$

for all $(\eta, \xi, w, z) \in D_\eta \times D_\xi \times \mathcal{W} \times R^m$, where β is a positive constant.

Remark 6.8. *When $f(\eta, \xi, w, u) = h(\eta, \xi, w) + G(\eta, \xi, w)u$,* (6.59) *is satisfied if*

$$P(\xi)G(\eta, \xi, w) + (P(\xi)G(\eta, \xi, w))^T \geq 2\beta I. \tag{6.60}$$

This condition encompasses the ones used in the previous sections. For single-input–single-output systems where $G = b \geq b_0$, (6.60) *is satisfied with $P = 1$ and $\beta = b_0$. For multi-input–multi-output systems where*

$$\|\hat{G}^{-1}(\xi)[G(\eta, \xi, w) - \hat{G}(\xi)]\| \leq k_u < 1 \tag{6.61}$$

for a known nonsingular matrix $\hat{G}(\xi)$, it can be verified that

$$\hat{G}^{-1}(\xi)G(\eta, \xi, w) + (\hat{G}^{-1}(\xi)G(\eta, \xi, w))^T \geq 2(1 - k_u)I.$$

The condition (6.61) *is not the same as* (6.44)*, but it is fairly close.* ◇

Let $V_s(\xi)=\xi^T P_s \xi$, where $P_s=P_s^T>0$ is the solution of the Lyapunov equation $P_s(A-BK)+(A-BK)^T P_s=-Q_s$ for some $Q_s=Q_s^T>0$. Let c be a positive constant such that $\{V_s(\xi)\le c\}\subset D_\xi$. Suppose Assumption 6.3 is satisfied. Then the compact set $\Omega=\{V_0(\eta)\le c_0\}\times\{V_s(\xi)\le c\}$ is positively invariant with respect to the system

$$\dot{\eta}=f_0(\eta,\xi,w), \qquad \dot{\xi}=(A-BK)\xi. \tag{6.62}$$

Let $\hat{f}(\xi,u)$ be a twice continuously differentiable, globally bounded function that models $f(\eta,\xi,w,u)$. The choice $\hat{f}=0$ is allowed. Set $\sigma=f(\eta,\xi,w,u)-\hat{f}(\xi,u)$. If ξ and σ were available for feedback, the dynamic inversion algorithm could have been taken as

$$\mu\dot{u}=-P(\xi)[\hat{f}(\xi,u)+\sigma+K\xi],$$

and $z=u-\phi(\eta,\xi,w)$ would satisfy the equation

$$\mu\dot{z}=-P(\xi)[f(\eta,\xi,w,z+\phi)-f(\eta,\xi,w,\phi)]-\mu\dot{\phi}.$$

Then, from Assumption 6.8,

$$\mu z^T\dot{z}\le-\beta\|z\|^2+\mu\|z\|\,\|\dot{\phi}\|.$$

For all $(\eta,\xi,w)\in\Omega\times\mathcal{W}$ and $\|u\|\le r_0$, $\dot{\phi}$ is bounded. Therefore, the set $\{\|z\|\le r\}$, for any $r>0$, is positively invariant for sufficiently small μ. Taking $r\ge r_0+\max_{(\eta,\xi,w)\in\Omega\times\mathcal{W}}\|\phi(\eta,\xi,w)\|$ ensures that $\|z(t)\|\le r$ for all $\|u(t)\|\le r_0$.

An estimate $\hat{\sigma}$ of σ is provided by the extended high-gain observer

$$\dot{\hat{\xi}}^i_j=\hat{\xi}^i_{j+1}+(\alpha^i_j/\varepsilon^j)(e_i-\hat{\xi}^i_1)\quad \text{for } 1\le j\le\rho_i-1, \tag{6.63}$$

$$\dot{\hat{\xi}}^i_{\rho_i}=\hat{\sigma}_i+\hat{f}_i(\hat{\xi},u)+(\alpha^i_{\rho_i}/\varepsilon^{\rho_i})(e_i-\hat{\xi}^i_1), \tag{6.64}$$

$$\dot{\hat{\sigma}}_i=(\alpha^i_{\rho_i+1}/\varepsilon^{\rho_i+1})(e_i-\hat{\xi}^i_1) \tag{6.65}$$

for $1\le i\le m$, where α^i_1 to $\alpha^i_{\rho_i+1}$ are chosen such that the polynomial

$$s^{\rho_i+1}+\alpha^i_1 s^{\rho_i}+\cdots+\alpha_{\rho_i+1}$$

is Hurwitz, and $\varepsilon>0$ is a small parameter. Let

$$M_i>\max_{\xi^T P_s\xi\le c}|K_i\xi|,$$

where K_i is the ith row of K,

$$N_i>\max_{(\eta,\xi)\in\Omega,w\in\mathcal{W},\|z\|\le r}\left|f_i(\eta,\xi,w,z+\phi)-\hat{f}_i(\xi,z+\phi)\right|,$$

and define the saturated estimates $(K\hat{\xi})_s$ and $\hat{\sigma}_s$ by

$$(K\hat{\xi})_s=\text{col}\left(M_1\,\text{sat}\left(\frac{K_1\hat{\xi}}{M_1}\right),\ldots,M_m\,\text{sat}\left(\frac{K_m\hat{\xi}}{M_m}\right)\right),$$

$$\hat{\sigma}_s=\text{col}\left(N_1\,\text{sat}\left(\frac{\hat{\sigma}_1}{N_1}\right),\ldots,N_m\,\text{sat}\left(\frac{\hat{\sigma}_m}{N_m}\right)\right).$$

Similarly, $\hat{\xi}_s$ is defined by saturating each component of $\hat{\xi}$ outside the compact set $\{\xi^T P_s \xi \le c\}$.
The dynamic inversion algorithm is defined by

$$\mu \dot{u} = -P(\hat{\xi}_s)[\hat{f}(\hat{\xi}, u) + \hat{\sigma}_s + (K\hat{\xi})_s], \tag{6.66}$$

where μ is a small positive constant. The estimates on the right-hand side of (6.66) are saturated to protect against peaking of the high-gain observer. The arguments of $\hat{f}$ are not saturated because $\hat{f}$ is globally bounded. The output feedback controller is defined by the extended high-gain observer (6.63)–(6.65) and the dynamic inversion algorithm (6.66).

Remark 6.9. *The output feedback controller has two small parameters μ and ε, which determine the time scales of the dynamic inversion and the observer, respectively. The observer dynamics should be faster than the dynamic inversion so that the estimates $\hat{\xi}$ and $\hat{\sigma}$ approach ξ and σ rapidly. The dynamic inversion should be faster than the dynamics of the system* (6.62). *Thus, $0 < \varepsilon \ll \mu \ll 1$. In tuning these parameters, one would consider first the closed-loop system under state feedback, that is,*

$$\dot{\eta} = f_0(\eta, \xi, w), \quad \dot{\xi} = f(\eta, \xi, w, u), \quad \mu \dot{u} = -P(\xi)[f(\eta, \xi, w, u) + K\xi],$$

and reduce μ until its trajectories are brought close enough to the trajectories of (6.62). *Then the closed-loop system under output feedback is considered, and ε is reduced until its trajectories are brought close enough to the trajectories of* (6.62). *The tuning procedure would use simulation while allowing the uncertain quantities to vary over their respective ranges.* ◇

Theorem 6.5. *Consider the closed-loop system formed of the plant* (6.50)–(6.53), *the observer* (6.63)–(6.65), *and the dynamic inversion algorithm* (6.66). *Suppose Assumptions* 6.1, 6.3, 6.6, 6.7, *and* 6.8 *are satisfied, $(\eta(0), \xi(0))$ belongs to a compact subset in the interior of Ω, $||z(0)|| < r$, and the initial states of the observer belong to a compact subset of $R^{\rho+m}$. Then*

- *there exists $\lambda_1^* > 0$ such that for ε and μ with $\max\{\mu, \varepsilon/\mu\} \le \lambda_1^*$, the trajectories of the closed-loop system are bounded for all $t \ge 0$;*
- *given any $\varsigma > 0$, there exits $\lambda_2^* > 0$ such that for ε and μ with $\max\{\mu, \varepsilon/\mu\} \le \lambda_2^*$,*

$$||\xi(t) - \xi_r(t)|| \le \zeta \qquad \forall\, t \ge 0, \tag{6.67}$$

where ξ_r is the solution of the target system (6.57) *with $\xi_r(0) = \xi(0)$.* ◇

Proof: Set $\lambda = \max\{\mu, \varepsilon/\mu\}$. With the change of variables

$$\varphi_j^i = (\xi_j^i - \hat{\xi}_j^i)/\varepsilon^{\rho_i+1-j} \quad \text{for} \quad 1 \le j \le \rho_i,\ 1 \le i \le m,$$
$$\varphi_{\rho_i+1}^i = \sigma_i - \hat{\sigma}_i = f_i(\eta, \xi, w, u) - \hat{f}_i(\xi, u) - \hat{\sigma}_i \quad \text{for} \quad 1 \le i \le m,$$
$$z = u - \phi(\eta, \xi, w),$$

the closed-loop system is represented by

$$\dot{\eta} = f_0(\eta, \xi, w), \tag{6.68}$$
$$\dot{\xi} = (A - BK)\xi + B[f(\eta, \xi, w, z + \phi) - f(\eta, \xi, w, \phi)], \tag{6.69}$$
$$\mu \dot{z} = -P(\xi)[f(\eta, \xi, w, z + \phi) - f(\eta, \xi, w, \phi)] - \mu \dot{\phi} + \psi_1, \tag{6.70}$$
$$\varepsilon \dot{\varphi} = \Lambda \varphi + \varepsilon(\bar{E}\psi_2 + \bar{B}\psi_3) + (\varepsilon/\mu)\bar{B}\psi_4, \tag{6.71}$$

where

$$\begin{aligned}
\psi_1 &= -[P(\hat{\xi}_s) - P(\xi)][f(\eta, \xi, w, z + \phi) - f(\eta, \xi, w, \phi)] \\
&\quad - P(\hat{\xi}_s)[(K\hat{\xi})_s - K\xi + \hat{\sigma}_s - \sigma + \hat{f}(\hat{\xi}, z + \phi) - \hat{f}(\xi, z + \phi)], \\
\psi_2 &= (1/\varepsilon)[\hat{f}(\xi, z + \phi) - \hat{f}(\hat{\xi}, z + \phi)], \\
\psi_3 &= \frac{\partial \sigma}{\partial \eta} f_0 + \frac{\partial \sigma}{\partial \xi}(A\xi + Bf) + \frac{\partial \sigma}{\partial w}\dot{w}, \\
\psi_4 &= -\frac{\partial \sigma}{\partial u} P(\hat{\xi}_s)[f(\eta, \xi, w, z + \phi) - f(\eta, \xi, w, \phi) + (K\hat{\xi})_s - K\xi \\
&\quad + \hat{\sigma}_s - \sigma + \hat{f}(\hat{\xi}, z + \phi) - \hat{f}(\xi, z + \phi)],
\end{aligned}$$

$$\Lambda = \text{block diag}(\Lambda_1, \ldots, \Lambda_m), \quad \Lambda_i = \begin{bmatrix} -\alpha_1^i & 1 & 0 & \cdots & 0 \\ -\alpha_2^i & 0 & 1 & \ddots & 0 \\ \vdots & \vdots & \ddots & \ddots & \vdots \\ -\alpha_{\rho_i}^i & 0 & \cdots & 0 & 1 \\ -\alpha_{\rho_i+1}^i & 0 & \cdots & \cdots & 0 \end{bmatrix},$$

$$\bar{B} = \text{block diag}(\bar{B}_1, \ldots, \bar{B}_m), \quad \bar{B}_i = \begin{bmatrix} 0 \\ \vdots \\ 0 \\ 1 \end{bmatrix}_{(\rho_i+1)\times 1},$$

$$\bar{E} = \text{block diag}(\bar{E}_1, \ldots, \bar{E}_m), \quad \bar{E}_i = \begin{bmatrix} 0 \\ \vdots \\ 0 \\ 1 \\ 0 \end{bmatrix}_{(\rho_i+1)\times 1}.$$

The functions f_0, f, and ϕ are written without their arguments. Their complete forms are $f_0 = f_0(\eta, \xi, w)$, $f = f(\eta, \xi, w, z + \phi)$, and $\phi = \phi(\eta, \xi, w)$. Let $P_0 = P_0^T > 0$ be the solution of the Lyapunov equation $P_0\Lambda + \Lambda^T P_0 = -I$, which exists since Λ is Hurwitz. Due to the continuous differentiability of f and $\hat{f}$, the local Lipschitz property of f_0 and P, and the global boundedness of $\hat{f}$, $\hat{\xi}_s$ and $\hat{\sigma}_s$, the functions ψ_2 to ψ_4 satisfy inequalities of the form $|\psi_1| \le k_1\|\varphi\|$, $|\psi_2| \le k_2\|\varphi\|$, $|\psi_3| \le k_3$, and $|\psi_4| \le k_4 + k_5\|\varphi\|$, for all $\hat{\xi} \in R^\rho$ and $(\eta, \xi, z) \in \Omega \times \{\|z\| \le r\}$, where k_2 to k_5 are positive constants independent of ε and μ. With $\lambda \le 1$, $\varepsilon \le \varepsilon/\mu$. The proof proceeds in four steps:

Step 1: Show that there exist positive constants κ_1, κ_2, and $\bar{\lambda}_1$ such that for $\lambda \le \bar{\lambda}_1$ the set $\Omega \times \{\|z\| \le \kappa_1 \lambda\} \times \{\varphi^T P_0 \varphi \le \kappa_2 (\varepsilon/\mu)^2\}$ is positively invariant.

Step 2: Show that for any bounded $\hat{\xi}(0)$ and any $(\eta(0), \xi(0), z(0))$ in the interior of $\Omega \times \{\|z\| \le r\}$, there exists $\bar{\lambda}_2 > 0$ such that for $\lambda \le \bar{\lambda}_2$ the trajectory enters the set $\Omega \times \{\|z\| \le r\} \times \{\varphi^T P_0 \varphi \le \kappa_2 (\varepsilon/\mu)^2\}$ in finite time $T_1(\varepsilon)$ with $\lim_{\varepsilon\to 0} T_1(\varepsilon) = 0$.

Step 3: Show that there exists $\bar{\lambda}_3 > 0$ such that for $\lambda \le \bar{\lambda}_3$ every trajectory in $\Omega \times \{\|z\| \le r\} \times \{\varphi^T P_0 \varphi \le \kappa_2 (\varepsilon/\mu)^2\}$ enters $\Omega \times \{\|z\| \le \kappa_1 \lambda\} \times \{\varphi^T P_0 \varphi \le \kappa_2 (\varepsilon/\mu)^2\}$ in finite time $T_2(\mu)$, where $\lim_{\mu\to 0} T_2(\mu) = 0$.

Step 4: Show (6.67).

The first step is shown by calculating the derivatives of $V_s = \xi^T P_s \xi$, $V_1 = \frac{1}{2} z^T z$, and $V_2 = \varphi^T P_0 \varphi$ on the boundaries $\{V_s = c\}$, $\{\|z\| = \kappa_1 \lambda\}$, and $\{V_2 = \kappa_2 (\varepsilon/\mu)^2\}$, respectively. The derivative $\dot{V}_2$ satisfies the inequality

$$\begin{aligned}
\varepsilon \dot{V}_2 &\le -\|\varphi\|^2 + 2\varepsilon\|P_0\| \left(k_2\|\varphi\|^2 + k_3\|\varphi\| + \frac{k_4}{\mu}\|\varphi\| + \frac{k_5}{\mu}\|\varphi\|^2 \right) \\
&= -\left(1 - 2\varepsilon k_2\|P_0\| - 2k_5\|P_0\|\frac{\varepsilon}{\mu} \right)\|\varphi\|^2 + 2\varepsilon\|P_0\|\left(k_3 + \frac{k_4}{\mu} \right)\|\varphi\| \\
&\le \frac{1}{2}\left(-\|\varphi\|^2 + \frac{4\varepsilon}{\mu}\|P_0\|(k_3 + k_4)\|\varphi\| \right)
\end{aligned}$$

for $\varepsilon/\mu \le 1/(4\|P_0\|(k_2 + k_5))$. Thus,

$$\varepsilon \dot{V}_2 \le -\frac{1}{4}\|\varphi\|^2 \qquad \text{for } \|\varphi\| \ge \frac{8\varepsilon}{\mu}\|P_0\|(k_3 + k_4).$$

Taking $\kappa_2 = 64\|P_0\|^3 (k_3 + k_4)^2$ ensures that

$$\varepsilon \dot{V}_2 \le -\frac{1}{4}\|\varphi\|^2 \qquad \forall\ V_2 \ge \kappa_2 (\varepsilon/\mu)^2. \tag{6.72}$$

The derivative of V_1 satisfies

$$\begin{aligned}
\mu \dot{V}_1 = \mu z^T \dot{z} &\le -\beta\|z\|^2 + \mu\|z\|\,\|\dot{\phi}\| + k_1\|z\|\,\|\varphi\| \\
&\le -\beta\|z\|^2 + \lambda k_6\|z\|,
\end{aligned}$$

where $k_6 > 0$ is independent of λ. Thus

$$\mu \dot{V}_1 \le -\frac{\beta}{2}\|z\|^2 \qquad \text{for } \|z\| \ge \frac{2k_6\lambda}{\beta}.$$

Taking $\kappa_1 = 2k_6/\beta$ ensures that

$$\mu \dot{V}_1 \le -\frac{\beta}{2}\|z\|^2 \qquad \forall\ \|z\| \ge \kappa_1 \lambda. \tag{6.73}$$

Choosing λ small enough ensures that $\{\|z\| \le \kappa_1 \lambda\}$ is in the interior of $\{\|z\| \le r\}$. The derivative of V_s satisfies

$$\begin{aligned}
\dot{V}_s &= -\xi^T Q_s \xi + 2\xi^T P_s B[f(\eta,\xi,w,z+\phi) - f(\eta,\xi,w,\phi)] \\
&\le -\lambda_{\min}(Q_s)\|\xi\|^2 + k_7\|z\|\,\|\xi\| \le -\lambda_{\min}(Q_s)\|\xi\|^2 + \kappa_1 k_7 \lambda\|\xi\|,
\end{aligned}$$

where $k_7 > 0$ is independent of λ. For sufficiently small λ, $\dot{V}_s < 0$ on the boundary $V_s = c$. The positive invariance of $\{V_0(\eta) \le c_0\}$ follows from Assumption 6.3.

In the second step, it is seen from (6.72) that when $\varphi(0) \notin \{\varphi^T P_0 \varphi \leq \kappa_2 (\varepsilon/\mu)^2\}$,

$$V_2(t) \leq e^{-k_8 t/\varepsilon} V_2(0) \leq k_9 e^{-k_8 t/\varepsilon} \|\varphi(0)\|^2,$$

where $k_8 = 1/(4\lambda_{\max}(P_s))$ and $k_9 = \lambda_{\max}(P_s)$. Taking into consideration that $\varphi(0) = O(1/\varepsilon^{\rho_{\max}})$, where $\rho_{\max} = \max\{\rho_1, \ldots, \rho_m\}$, it can been seen that

$$V_2(t) \leq \frac{k_{10}}{\varepsilon^{2\rho_{\max}}} e^{-k_8 t/\varepsilon}$$

for some $k_{10} > 0$. Since $\varepsilon \leq \varepsilon/\mu$ for $\lambda \leq 1$, the time it takes $\varphi(t)$ to enter $\{V_2 \leq \kappa_2 (\varepsilon/\mu)^2\}$ can be estimated by the more conservative time $T_1(\varepsilon)$ when $V_2 = \kappa_2 \varepsilon^2$, which is given by

$$\frac{k_{10}}{\varepsilon^{2\rho_{\max}}} e^{-k_8 T_1/\varepsilon} = \kappa_2 \varepsilon^2 \quad \Leftrightarrow \quad T_1 = \frac{\varepsilon}{k_8} \ln\left(\frac{k_{10}}{\kappa_2 \varepsilon^{2(\rho_{\max}+1)}}\right).$$

By l'Hôpital's rule, it can be shown that $\lim_{\varepsilon\to 0} T_1(\varepsilon) = 0$. Because the initial state $(\eta(0), \xi(0), z(0))$ is in the interior of $\Omega \times \{\|z\| \leq r\}$ and the right-hand-side functions of (6.68) to (6.70) are bounded uniformly in ε, there is time T_0 such that $(\eta(t), \xi(t), z(t)) \in \Omega \times \{\|z\| \leq r\}$ for all $t \in [0, T_0]$. For sufficiently small ε, $T_1(\varepsilon) \leq (1/2)T_0$. Thus, $\varphi(t)$ enters $\{V_1 \leq \kappa_2 (\varepsilon/\mu)^2\}$, while $(\eta(t), \xi(t), z(t))$ remain in the interior of $\Omega \times \{\|z\| \leq r\}$. The second step is complete.

The third step is shown by repeating the argument of the second step to show that z enters $\{\|z\| \leq \kappa_1 \lambda\}$. It follows from (6.73) that when $z(0) \notin \{\|z\| \leq \kappa_1 \lambda\}$,

$$V_1(t) \leq e^{-\beta t/\mu} V_1(0) \leq \tfrac{1}{2} e^{-\beta t/\mu} r^2.$$

The time it takes $z(t)$ to enter $\{\|z\| \leq \kappa_1 \lambda\}$ can be estimated by $T_2(\mu)$, which is given by

$$e^{-\beta T_2/(2\mu)} r = \kappa_1 \mu \quad \Leftrightarrow \quad T_2 = \frac{2\mu}{\beta} \ln\left(\frac{r}{\kappa_1 \mu}\right).$$

By l'Hôpital's rule, it can be shown that $\lim_{\mu\to 0} T_2(\mu) = 0$. Once again, by choosing μ small enough, (η, ξ) remains in Ω. Thus, within the time $T_1(\varepsilon) + T_2(\mu)$, the trajectory enters the positively invariant set $\Omega \times \{\|z\| \leq \kappa_1 \lambda\} \times \{\varphi^T P_0 \varphi \leq \kappa_2 (\varepsilon/\mu)^2\}$.

Next, we show (6.67). Since the right-hand-side functions of (6.57) and (6.69) are bounded, uniformly in ε and μ, and $\xi(0) = \xi_r(0)$, it follows that

$$\xi(t) - \xi_r(t) = O(T_1(\varepsilon) + T_2(\mu)) \qquad \text{for } 0 \leq t \leq T_1(\varepsilon) + T_2(\mu).$$

Hence, $\xi(T_1(\varepsilon) + T_2(\mu)) - \xi_r(T_1(\varepsilon) + T_2(\mu)) = O(T_1(\varepsilon) + T_2(\mu))$. For $t \geq T_1(\varepsilon) + T_2(\mu)$, ξ satisfies the equation

$$\dot{\xi} = (A - BK)\xi + O(\lambda).$$

By continuous dependence of the solutions of differential equations on initial conditions and parameters [78, Theorem 9.1] and exponential stability of the origin of $\dot{\xi} = (A - BK)\xi$, we conclude that

$$\xi(t) - \xi_r(t) = O(\lambda) + O(T_1(\varepsilon) + T_2(\mu)) \qquad \forall\, t \geq 0.$$

Given any $\varsigma > 0$, $\|\xi(t) - \xi_r(t)\|$ can be made smaller than ς by choosing λ small enough, which proves (6.67). □

Remark 6.10. *Since* $\lim_{t\to\infty} \xi_r(t) = 0$, (6.67) *implies that for every* $\varsigma > 0$ *there is* $T > 0$ *such that for sufficiently small* μ *and* ε/μ, $\|\xi(t)\| \le \varsigma$ *for all* $t \ge T$. ◇

As in Theorem 6.2 and Corollary 6.1, when w is constant the dynamic inversion controller provides integral action and stabilization under appropriate conditions. These results are stated in Theorem 6.6 and Corollary 6.3.

Theorem 6.6. *Consider the closed-loop system formed of the plant* (6.50)–(6.53), *the observer* (6.63)–(6.65), *and the dynamic inversion algorithm* (6.66), *under the assumptions of Theorem* 6.5. *Suppose* w *is constant and Assumption* 6.4 *is satisfied. Then there exists* $\bar{\lambda} > 0$ *such that for* ε *and* μ *with* $\max\{\mu, \varepsilon/\mu\} \le \bar{\lambda}$, $\xi(t)$ *converges to zero as* $t \to \infty$. ◇

Corollary 6.3. *Consider the closed-loop system formed of the plant* (6.50)–(6.53), *the observer* (6.63)–(6.65), *and the dynamic inversion algorithm* (6.66), *under the assumptions of Theorem* 6.6. *Suppose* $f(0,0,w,\phi(0,0,w)) = 0$, $\hat{f}(0,\phi(0,0,w)) = 0$, *and* $f_0(0,0,w) = 0$ *so that* $(\eta,\xi,z,\varphi) = (0,0,0,0)$ *is an equilibrium point, where* $z = u - \phi(\eta,\xi,w)$. *Let* $(\eta(0),\xi(0)) \in \Omega_0$ *(a compact set in the interior of* Ω*),* $u(0) \in U$ *(a compact subset of* R^m*), and* $(\hat{\xi}(0),\hat{\sigma}(0)) \in S$ *(a compact subset of* $R^{\rho+m}$*). Then there exists* $\bar{\lambda} > 0$ *such that for* ε *and* μ *with* $\max\{\mu,\varepsilon/\mu\} \le \bar{\lambda}$, *the origin* $(\eta,\xi,z,\varphi) = (0,0,0,0)$ *is an exponentially stable equilibrium point of the closed-loop system and* $\Omega_0 \times U \times S$ *is a subset of the region of attraction.* ◇

Proof of Theorem 6.6: Shift the equilibrium point $\eta = \eta_{ss}$ of $\dot{\eta} = f_0(\eta,0,w)$ to the origin by the change of variables $\zeta = \eta - \eta_{ss}$, and let $z = u - \phi(\eta,\xi,w)$. It can be verified that the closed-loop system has an isolated equilibrium point at the origin $(\zeta,\xi,z,\varphi) = (0,0,0,0)$. It has been already shown in the proof of Theorem 6.5 that, in finite time, $\varphi = O(\varepsilon/\mu)$, $z = O(\lambda)$, and $\|\xi\| \le \varsigma$ for any given $\varsigma > 0$, where $\lambda = \max\{\mu,\varepsilon/\mu\}$. Using regional input-to-state stability of $\dot{\zeta} = f_0(\zeta + \eta_{ss},\xi,w)$, it can be shown that (ζ,ξ,z,φ) reaches a neighborhood $\mathcal{N}$ of $(0,0,0,0)$, where $\mathcal{N}$ can be made arbitrarily small by choosing μ and ε/μ small enough. It can be also verified that ζ, ξ, z, and φ satisfy the equations

$$\dot{\zeta} = f_0(\zeta + \eta_{ss},0,w) + \delta_1(\zeta,\xi,w), \tag{6.74}$$

$$\dot{\xi} = (A-BK)\xi + \delta_2(\zeta,\xi,z,w), \tag{6.75}$$

$$\begin{aligned} \mu\dot{z} = &-P(\xi)[f(\eta,\xi,w,z+\phi) - f(\eta,\xi,w,\phi)] \\ &+ \mu\delta_3(\zeta,\xi,z,w) + \delta_4(\zeta,\xi,z,\varphi,w,\varepsilon), \end{aligned} \tag{6.76}$$

$$\varepsilon\dot{\varphi} = \Lambda\varphi + \varepsilon\delta_5(\zeta,\xi,z,\varphi,w,\varepsilon) + (\varepsilon/\mu)\delta_6(\zeta,\xi,z,\varphi,w,\varepsilon) \tag{6.77}$$

in $\mathcal{N}$, where

$$\begin{aligned} \delta_1 &= f_0(\zeta + \eta_{ss},\xi,w) - f_0(\zeta + \eta_{ss},0,w), \\ \delta_2 &= B[f(\eta,\xi,w,z+\phi) - f(\eta,\xi,w,\phi)], \\ \delta_3 &= -\frac{\partial\phi}{\partial\eta} f_0(\eta,\xi,w) - \frac{\partial\phi}{\partial\xi}[(A-BK)\xi + \delta_2], \\ \delta_4 &= \psi_1, \quad \delta_5 = \bar{E}\psi_2 + \bar{B}\psi_3, \quad \delta_6 = \bar{B}\psi_4; \end{aligned}$$

$\bar{B}$, $\bar{E}$, and ψ_1 to ψ_4 are defined in the proof of Theorem 6.5.

The functions δ_1 to δ_6 satisfy the inequalities

$$\|\delta_1\| \le k_1\|\xi\|, \quad \|\delta_2\| \le k_2\|z\|, \quad \|\delta_3\| \le k_3\|\zeta\| + k_4\|\xi\| + k_5\|z\|$$

$$\|\delta_4\| \le k_6\|\varphi\|, \; \|\delta_5\| \le k_7\|\zeta\| + k_8\|\xi\| + k_9\|z\| + k_{10}\|\varphi\|, \; \|\delta_6\| \le k_{11}\|z\| + k_{12}\|\varphi\|,$$

in some neighborhood of the origin with some nonnegative constants k_1 to k_{12} independent of μ and ε. From the converse Lyapunov theorem of exponential stability, [80, Theorem 3.8], let $V_3(\zeta)$ be a Lyapunov function for $\dot{\zeta} = f_0(\zeta + \eta_{ss}, 0, w)$, which satisfies

$$\frac{\partial V_3}{\partial \zeta} f_0(\zeta + \eta_{ss}, 0, w) \le -c_1\|\zeta\|^2, \quad \left\|\frac{\partial V_3}{\partial \zeta}\right\| \le c_2\|\zeta\|$$

in some neighborhood of $\zeta = 0$. The Lyapunov function $V_s = \xi^T P_s \xi$ satisfies

$$\frac{\partial V_s}{\partial \xi}(A - BK)\xi \le -c_3\|\xi\|^2, \qquad \left\|\frac{\partial V_s}{\partial \xi}\right\| \le c_4\|\xi\|.$$

Using $V = k_0 V_3(\zeta) + V_s(\xi) + \frac{1}{2}z^T z + \varphi^T P_0 \varphi$, with $k_0 > 0$, as a Lyapunov function candidate for (6.74)–(6.77), it can be verified that in some neighborhood $\mathcal{N}_0$ of the origin

$$\dot{V} \le -\mathcal{Y}^T Q \mathcal{Y},$$

where $\mathcal{Y} = \mathrm{col}(\|\zeta\|, \|\xi\|, \|z\|, \|\varphi\|)$ and

$$Q = \begin{bmatrix} k_0 c_1 & -\frac{1}{2}k_0 c_2 k_1 & \star & \star \\ -\frac{1}{2}k_0 c_2 k_1 & c_3 & \star & \star \\ \star & \star & \left(\frac{\beta}{\mu} - \star\right) & -\left(\frac{k_6 + 2k_{11}\|P_0\|}{2\mu} + \star\right) \\ \star & \star & -\left(\frac{k_6 + 2k_{11}\|P_0\|}{2\mu} + \star\right) & \left(\frac{1}{\varepsilon} - \frac{2k_{12}\|P_0\|}{\mu} - \star\right) \end{bmatrix}$$

and each $\star$ indicates a constant independent of k_0, μ, and ε. The matrix Q can be made positive definite by choosing $k_0 < 4c_1c_3/(k_1c_2)^2$, then choosing μ small enough to make the 3×3 principal minor positive, then choosing ε/μ small enough to make $\det(Q)$ positive. Hence, $(\zeta(t), \xi(t), z(t), \varphi(t))$ converges to zero as t tends to infinity. □

Example 6.5. It is required to stabilize the origin of the system

$$\dot{x}_1 = x_2, \quad \dot{x}_2 = \theta_1 \sin(x_1) + \theta_2 \tanh(\theta_3 u), \quad y = x_1,$$

using feedback from the output y while recovering the response of the target system

$$\dot{x}_1 = x_2, \quad \dot{x}_2 = -x_1 - 2x_2.$$

The unknown parameters θ_1 to θ_3 satisfy $\theta_1 \in [0.5, 1]$, $\theta_2 \in [10, 12]$, and $\theta_3 \in [0.5, 2]$. Assumption 6.7 requires the equation

$$\theta_1 \sin(x_1) + \theta_2 \tanh(\theta_3 u) = -x_1 - 2x_2$$

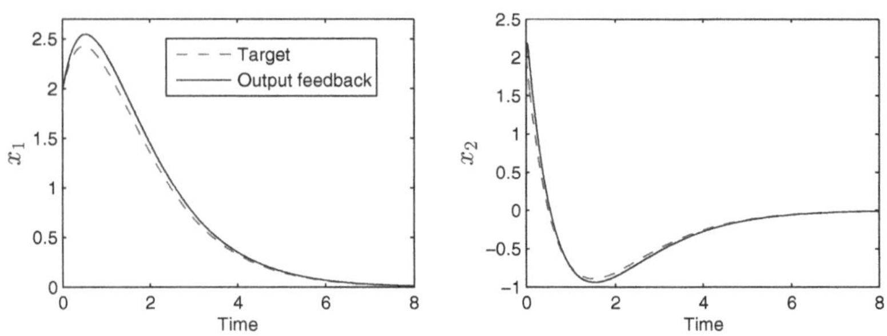

Figure 6.12. *Simulation results of Example* 6.5.

to have a unique solution, which will be the case if x is restricted to a set where $|x_1 + 2x_2| < 9$. A quadratic Lyapunov function for the target system can be taken as $V_s = x^T P_s x$, where

$$P_s = \begin{bmatrix} 1 & 0.05 \\ 0.05 & 0.95 \end{bmatrix} \text{ and } \max_{x^T P_s x \le c} |x_1 + 2x_2] = \sqrt{c}\|\begin{bmatrix} 1 & 2 \end{bmatrix} P^{-1/2}\| = 2.239\sqrt{c}.$$

Taking $c = 16$ ensures that for all $x \in \{x^T P_s x \le 16\}$, $|x_1 + 2x_2| < 9$. Assumption 6.8 is satisfied with $P = 1$. With $\hat{f} = 0$, the extended high-gain observer is taken as

$$\dot{\hat{x}}_1 = \hat{x}_2 + (3/\varepsilon)(y - \hat{x}_1), \quad \dot{\hat{x}}_2 = \hat{\sigma} + (3/\varepsilon^2)(y - \hat{x}_1), \quad \dot{\hat{\sigma}} = (1/\varepsilon^3)(y - \hat{x}_1),$$

and the dynamic inversion algorithm is

$$\mu \dot{u} = -N \text{ sat}\left(\frac{\hat{\sigma}}{N}\right) - M \text{ sat}\left(\frac{\hat{x}_1 + 2\hat{x}_2}{M}\right),$$

where the saturation levels are $N = 13$ and $M = 9$. Simulation is carried out with $\mu = 0.1$, $\varepsilon = 0.001$, $\theta_1 = 1$, $\theta_2 = 10$, $\theta_3 = 1$, $x_1(0) = x_2(0) = 2$, and $u(0) = \hat{x}_1(0) = \hat{x}_2(0) = \hat{\sigma}(0) = 0$. The simulation results in Figure 6.12 demonstrate the recovery of the target response. △

6.3 ▪ Nonminimum Phase Systems

6.3.1 ▪ Stabilization

Consider a single-input–single-output nonlinear system in the normal form

$$\dot{\eta} = f_0(\eta, \xi_1, \ldots, \xi_{\rho-1}, \xi_\rho, \theta), \tag{6.78}$$

$$\dot{\xi}_i = \xi_{i+1} \qquad \text{for } 1 \le i \le \rho - 1, \tag{6.79}$$

$$\dot{\xi}_\rho = a(\eta, \xi_1, \ldots, \xi_{\rho-1}, \xi_\rho, \theta) + a_0(\xi) + b(\xi)u, \tag{6.80}$$

$$y = \xi_1, \tag{6.81}$$

where $\eta \in R^{n-\rho}$ and $\xi = \text{col}(\xi_1, \ldots, \xi_\rho) \in R^\rho$ are the state variables, u is the control input, y is the measured output, $\theta \in \Theta \subset R^\ell$ is a vector of unknown constant parameters, and Θ is a compact set. The normal form is defined for $\eta \in D_\eta \subset R^{n-\rho}$ and $\xi \in D_\xi \subset R^\rho$ for some domains D_η and D_ξ that contain $\eta = 0$ and $\xi = 0$, respectively. The functions f_0 and a could be unknown, while a_0 and b are known.

Assumption 6.9. *The functions f_0 and a are continuous in θ and continuously differentiable in (η, ξ) with locally Lipschitz derivatives; a_0, and b are continuously differentiable with locally Lipschitz derivatives; $f_0(0,0,\theta) = 0$, $a(0,0,\theta) = 0$, $a_0(0) = 0$, and $b(\xi) \geq k_0 > 0$ for all $\xi \in D_\xi$.*

The goal is to design an output feedback controller to stabilize the origin. Toward that end, consider the auxiliary system

$$\dot{\eta} = f_0(\eta, \xi_1, \ldots, \xi_{\rho-1}, u_a, \theta), \tag{6.82}$$

$$\dot{\xi}_i = \xi_{i+1} \qquad \text{for } 1 \leq i \leq \rho - 2, \tag{6.83}$$

$$\dot{\xi}_{\rho-1} = u_a, \tag{6.84}$$

$$y_a = a(\eta, \xi_1, \ldots, \xi_{\rho-1}, u_a, \theta), \tag{6.85}$$

which is constructed from (6.78)–(6.80) with ξ_ρ as the input u_a and $a(\eta, \xi_1, \ldots, \xi_{\rho-1}, \xi_\rho, \theta)$ as the output y_a. Once an output feedback controller is designed for the auxiliary system, the output y_a will be estimated using an extended high-gain observer and the control u will be obtained from u_a using sliding mode control. The output feedback controller for the auxiliary system takes the form

$$\dot{z} = f_c(z, \xi_1, \ldots, \xi_{\rho-1}, y_a), \tag{6.86}$$

$$u_a = h_c(z, \xi_1, \ldots, \xi_{\rho-1}, y_a), \tag{6.87}$$

where $f_c(0) = 0$, $h_c(0) = 0$, and f_c and h_c are sufficiently smooth. The controller is defined for $z \in D_z \subset R^r$, where D_z is a domain that contains $z = 0$. If $a(\cdot)$ depends on ξ_ρ, h_c is restricted to be independent of y_a, which ensures that the closed-loop auxiliary system will be well-posed. With this restriction,

$$h_c(z, \xi_1, \ldots, \xi_{\rho-1}, a) = \begin{cases} h_c(z, \xi_1, \ldots, \xi_{\rho-1}) & \text{if } a \text{ depends on } \xi_\rho, \\ h_c(z, \xi_1, \ldots, \xi_{\rho-1}, a(\eta, \xi_1, \ldots, \xi_{\rho-1}, \theta)) & \text{otherwise.} \end{cases}$$

Thus, $h_c(z, \xi_1, \ldots, \xi_{\rho-1}, a)$ is independent of ξ_ρ. The closed-loop auxiliary system is given by

$$\dot{\eta} = f_0(\eta, \xi_1, \ldots, \xi_{\rho-1}, h_c(\cdot)), \tag{6.88}$$

$$\dot{\xi}_i = \xi_{i+1} \qquad \text{for } 1 \leq i \leq \rho - 2, \tag{6.89}$$

$$\dot{\xi}_{\rho-1} = h_c(\cdot), \tag{6.90}$$

$$\dot{z} = f_c(z, \xi_1, \ldots, \xi_{\rho-1}, a(\eta, \xi_1, \ldots, \xi_{\rho-1}, h_c(\cdot)), \tag{6.91}$$

where $h_c(\cdot) = h_c(z, \xi_1, \ldots, \xi_{\rho-1}, a)$.

Assumption 6.10. *There is an output feedback controller of the form* (6.86)–(6.87) *such that the origin of the closed-loop system* (6.88)–(6.91) *is exponentially stable, uniformly in θ.*[62]

[62]There is no general result that guarantees the existence of a stabilizing controller that satisfies Assumption 6.10 for a class of nonlinear systems, but it is shown in [67] that for linear systems the assumption is satisfied if the original system (6.78)–(6.81) is controllable and observable.

Taking $y_a = a$ in (6.86)–(6.87), augmenting the equations with the system (6.78)–(6.80), and applying the change of variables $s = \xi_\rho - h_c(z, \xi_1, \ldots, \xi_{\rho-1}, a)$, the augmented system can be written as

$$\dot{\zeta} = f_a(\zeta, s, \theta), \tag{6.92}$$
$$\dot{s} = f_b(\zeta, s, \theta) + a_0(\xi) + b(\xi)u, \tag{6.93}$$

where

$$\zeta = \begin{bmatrix} \eta \\ \xi_1 \\ \vdots \\ \vdots \\ \xi_{\rho-1} \\ z \end{bmatrix}, \quad f_a = \begin{bmatrix} f_0(\eta, \xi_1, \ldots, \xi_{\rho-1}, s + h_c(\cdot), \theta) \\ \xi_2 \\ \vdots \\ \xi_{\rho-1} \\ s + h_c(\cdot) \\ f_c(z, \xi_1, \ldots, \xi_{\rho-1}, a(\eta, \xi_1, \ldots, \xi_{\rho-1}, s + h_c(\cdot), \theta)) \end{bmatrix},$$

and

$$\begin{aligned} f_b = {} & a(\eta, \xi_1, \ldots, \xi_{\rho-1}, s + h_c(\cdot), \theta) \\ & - \frac{\partial h_c}{\partial \eta} f_0(\eta, \xi_1, \ldots, \xi_{\rho-1}, s + h_c(\cdot), \theta) \\ & - \sum_{i=1}^{\rho-2} \frac{\partial h_c}{\partial \xi_i} \xi_{i+1} - \frac{\partial h_c}{\partial \xi_{\rho-1}} [s + h_c(\cdot)] \\ & - \frac{\partial h_c}{\partial z} f_c(z, \xi_1, \ldots, \xi_{\rho-1}, a(\eta, \xi_1, \ldots, \xi_{\rho-1}, s + h_c(\cdot), \theta)). \end{aligned}$$

With s as the output, the system (6.92)–(6.93) has relative degree one and its zero dynamics are given by the equation $\dot{\zeta} = f_a(\zeta, 0, \theta)$, which is the closed-loop auxiliary system (6.88)–(6.91). Thus, the origin $\zeta = 0$ of $\dot{\zeta} = f_a(\zeta, 0, \theta)$ is exponentially stable uniformly in θ. Let $D \subset R^{n+r}$ be a domain containing the origin such that

$$(\zeta, s) \in D \quad \Rightarrow \quad (\eta, \xi, z) \in D_\eta \times D_\xi \times D_z.$$

Assumption 6.11. *There is a continuously differentiable Lyapunov function $V(\zeta)$, possibly dependent on θ, such that*

$$\gamma_1(\|\zeta\|) \le V(\zeta) \le \gamma_2(\|\zeta\|), \tag{6.94}$$

$$\frac{\partial V}{\partial \zeta} f_a(\zeta, s, \theta) \le -\gamma_3(\|\zeta\|) \quad \forall \; \|\zeta\| \ge \gamma_4(|s|) \tag{6.95}$$

for all $(\zeta, s) \in D$, where γ_1 to γ_4 are class $\mathcal{K}$ functions independent of θ.

This assumption implies that the system $\dot{\zeta} = f_a(\zeta, s, \theta)$ is regionally input-to-state stable with s viewed as the input [80, Section 4.4]. Had ξ and a been available for feedback, a continuously implemented sliding mode controller could have been taken as

$$u = \frac{1}{b(\xi)} \left[-a_0(\xi) - \beta(\xi, z) \, \text{sat}\left(\frac{s}{\mu} \right) \right], \tag{6.96}$$

where μ is a small positive constant and β satisfies

$$\beta(\xi,z) \geq |f_b(\zeta,s,\theta)| + \beta_0 \quad \forall\, (\zeta,s) \in \Omega \text{ and } \theta \in \Theta, \tag{6.97}$$

where $\beta_0 > 0$ and $\Omega = \{V(\zeta) \leq c_0\} \times \{|s| \leq c\}$, in which $c > \mu$ and $c_0 \geq \gamma_2(\gamma_4(c))$ are chosen such that Ω is a compact subset of D. This choice of β ensures that $s\dot{s} \leq -\beta_0|s|$ for $|s| \geq \mu$, which can be used to show that Ω is positively invariant and every trajectory starting in Ω reaches the positively invariant set $\Omega_\mu = \{V(\zeta) \leq \gamma_2(\gamma_4(\mu))\} \times \{|s| \leq \mu\}$ in finite time. Analysis inside Ω_μ shows that the origin of the closed-loop system is exponentially stable and every trajectory in Ω_μ converges to the origin as time tends to infinity. Thus, Ω is a subset of the region of attraction.[63]

For output feedback control, an extended high-gain observer is used to estimate ξ and a. The observer is given by

$$\dot{\hat{\xi}}_i = \hat{\xi}_{i+1} + (\alpha_i/\varepsilon^i)(y - \hat{\xi}_1) \quad \text{for } 1 \leq i \leq \rho - 1, \tag{6.98}$$

$$\dot{\hat{\xi}}_\rho = \hat{\sigma} + a_0^s(\hat{\xi}) + b^s(\hat{\xi})u + (\alpha_\rho/\varepsilon^\rho)(y - \hat{\xi}_1), \tag{6.99}$$

$$\dot{\hat{\sigma}} = (\alpha_{\rho+1}/\varepsilon^{\rho+1})(y - \hat{\xi}_1), \tag{6.100}$$

where ε is a small positive constant, α_1 to $\alpha_{\rho+1}$ are chosen such that the polynomial

$$\lambda^{\rho+1} + \alpha_1\lambda^\rho + \cdots + \alpha_\rho\lambda + \alpha_{\rho+1}$$

is Hurwitz, and $a_0^s(\xi)$ and $b^s(\xi)$ are globally bounded functions, which are obtained by saturating the ξ_1 to ξ_ρ arguments of a_0 and b outside Ω. In particular, with $M_i > \max_{(\zeta,s)\in\Omega} |\xi_i|$, a_0^s is defined by

$$a_0^s(\xi_1, \ldots, \xi_\rho) = a_0(M_1 \operatorname{sat}(\xi_1/M_1), \ldots, M_\rho \operatorname{sat}(\xi_\rho/M_\rho)).$$

The function b^s is defined similarly. The variables $\hat{\xi}_1$ to $\hat{\xi}_\rho$ are estimates of ξ_1 to ξ_ρ, while $\hat{\sigma}$ is an estimate of a. The output feedback controller is given by

$$\dot{z} = f_c(z, M_1 \operatorname{sat}(\hat{\xi}_1/M_1), \ldots, M_\rho \operatorname{sat}(\hat{\xi}_\rho/M_\rho), M_{\rho+1} \operatorname{sat}(\hat{\sigma}/M_{\rho+1})), \tag{6.101}$$

$$u = -\psi_1(\hat{\xi}) - \psi_2(\hat{\xi}) \operatorname{sat}\left(\frac{\hat{\xi}_\rho - h_c(z, \hat{\xi}_1, \ldots, \hat{\xi}_{\rho-1}, \hat{\sigma})}{\mu}\right), \tag{6.102}$$

where $M_{\rho+1} > \max_{\{(\zeta,s)\in\Omega, \theta\in\Theta\}} |a(\eta,\xi,\theta)|$ and $\psi_1(\xi)$ and $\psi_2(\xi)$ are globally bounded functions, which are obtained by saturating the ξ_1 to ξ_ρ arguments of $a_0(\xi)/b(\xi)$ and $\beta(\xi,z)/b(\xi)$, respectively, outside Ω. The functions a_0^s and b^s in (6.99), f_0 in (6.101), and ψ_1 and ψ_2 in (6.102) are chosen to be globally bounded in $\hat{\xi}$ and $\hat{\sigma}$ to deal with the peaking phenomenon of the observer.

Theorem 6.7. *Consider the closed-loop system formed of the plant* (6.78)–(6.81), *the observer* (6.98)–(6.100), *and the controller* (6.101)–(6.102). *Suppose Assumptions* 6.9–6.11 *are satisfied. Let Ψ be a compact set in the interior of Ω and $\mathcal{Q}$ be a compact subset of*

[63] The details of this argument are standard for continuously implemented sliding mode control, as shown in [80, Section 10.1].

$R^{\rho+1}$. Suppose $(\zeta(0), s(0)) \in \Psi$ and $(\hat{\xi}(0), \hat{\sigma}(0)) \in \mathcal{Q}$. Then there is $\mu^* > 0$ and for every $\mu \in (0, \mu^*]$ there is $\varepsilon^* > 0$, dependent on μ, such that for every $0 < \varepsilon \leq \varepsilon^*$, the origin of the closed-loop system is exponentially stable and $\Psi \times \mathcal{Q}$ is a subset of the region of attraction. ◇

Proof: The scaled estimation errors

$$\varphi_i = (\xi_i - \hat{\xi}_i)/\varepsilon^{\rho+1-i} \qquad \text{for} \quad 1 \leq i \leq \rho,$$
$$\varphi_{\rho+1} = a(\eta, \xi, \theta) - \hat{\sigma}$$

satisfy the equation

$$\varepsilon\, \dot{\varphi} = \Lambda \varphi + \varepsilon[\bar{B}_1\, \Delta_1(\cdot) + \bar{B}_2\, \Delta_2(\cdot)],$$

where

$$\Lambda = \begin{bmatrix} -\alpha_1 & 1 & 0 & \cdots & 0 \\ -\alpha_2 & 0 & 1 & \ddots & 0 \\ \vdots & \vdots & \ddots & \ddots & \vdots \\ -\alpha_\rho & 0 & \cdots & 0 & 1 \\ -\alpha_{\rho+1} & 0 & \cdots & \cdots & 0 \end{bmatrix}, \quad \bar{B}_1 = \begin{bmatrix} 0 \\ \vdots \\ 0 \\ 0 \\ 1 \end{bmatrix}, \quad \bar{B}_2 = \begin{bmatrix} 0 \\ \vdots \\ 0 \\ 1 \\ 0 \end{bmatrix},$$

$$\Delta_1(\eta, \xi, z, \varphi, \theta, \varepsilon) = \frac{\partial a}{\partial \eta} f_0(\eta, \xi, \theta) + \sum_{i=1}^{\rho-1} \frac{\partial a}{\partial \xi_i} \xi_{i+1}$$
$$+ \frac{\partial a}{\partial \xi_\rho}[a(\eta, \xi, \theta) + a_0(\xi) + b(\xi)u],$$

$$\Delta_2(\eta, \xi, z, \varphi, \theta, \varepsilon) = (1/\varepsilon)\{a_0(\xi) - a_0^s(\hat{\xi}) + [b(\xi) - b^s(\hat{\xi})]u\}.$$

The matrix Λ is Hurwitz, and the functions Δ_1 and Δ_2 satisfy $|\Delta_1| \leq k_1$ and $|\Delta_2| \leq k_2\|\varphi\|$ for all $(\zeta, s) \in \Omega$ and $\varphi \in R^{\rho+1}$. Let $W(\varphi) = \varphi^T P \varphi$, where $P = P^T > 0$ is the solution of the Lyapunov equation $P\Lambda + \Lambda^T P = -I$. It can be verified that

$$\varepsilon \dot{W} \leq -\|\varphi\|^2 + 2\varepsilon k_1 \|P\bar{B}_1\|\,\|\varphi\| + 2\varepsilon k_2 \|P\bar{B}_2\|\,\|\varphi\|^2 \leq -\tfrac{1}{2}\|\varphi\|^2 + 2\varepsilon k_1 \|P\bar{B}_1\|\,\|\varphi\|$$

for $2\varepsilon k_2 \|P\bar{B}_2\| \leq \frac{1}{2}$. Thus, there is $\varrho_1 > 0$ such that, as long as $(\zeta, s) \in \Omega$, $\{W(\varphi) \leq \varepsilon^2 \varrho_1\}$ is positively invariant and φ enters $\{W(\varphi) \leq \varepsilon^2 \varrho_1\}$ in finite time $T(\varepsilon)$, where $\lim_{\varepsilon \to 0} T(\varepsilon) = 0$. On the other hand, since $(\zeta(0), s(0))$ is in the interior of Ω and the right-hand sides of (6.101) and (6.102) are globally bounded functions in $\hat{\xi}$ and $\hat{\sigma}$, there is time T_1, independent of ε, such that $(\zeta(t), s(t))$ remain in Ω for $t \in [0, T_1]$. By choosing ε small enough, it can be ensured that $T(\varepsilon) < T_1$. While $(\zeta, s, \varphi) \in \Omega \times \{W(\varphi) \leq \varepsilon^2 \varrho_1\}$, $\|\varphi\| = O(\varepsilon)$ can be used to show that $\Omega \times \{W(\varphi) \leq \varepsilon^2 \varrho_1\}$ is positively invariant and the trajectories enter the set $\Omega_\mu \times \{W(\varphi) \leq \varepsilon^2 \varrho_1\}$ in finite time. Inside this set, the closed-loop system can be represented by the equations

$$\dot{\zeta} = f_a(\zeta, 0, \theta) + \delta_1(\cdot),$$
$$\dot{s} = -\beta(\xi, z)s/\mu + \delta_2(\cdot),$$
$$\varepsilon \dot{\varphi} = \Lambda \varphi + \varepsilon[\bar{B}_1 \Delta_1(\cdot) + \bar{B}_2 \Delta_2(\cdot)],$$

where, for $\mu \leq 1$, δ_1, δ_2, Δ_1, and Δ_2 satisfy the inequalities

$$\|\delta_1\| \leq L_1|s| + L_2\|\varphi\|, \quad |\delta_2| \leq L_3\|\zeta\| + L_4|s| + (L_5/\mu)\|\varphi\|,$$

$$|\Delta_1| \le L_6\|\zeta\| + (L_7/\mu)|s| + (L_8/\mu)\|\varphi\|, \quad |\Delta_2| \le L_9\|\varphi\|$$

for some positive constants L_3 to L_9. The factor $1/\mu$ arises from $\text{sat}(s/\mu)$. Because the origin of $\dot{\zeta} = f_a(\zeta, 0, \theta)$ is exponentially stable uniformly in θ, by the converse Lyapunov theorem [80, Lemma 9.8], there is a Lyapunov function $V_1(\zeta)$, possibly dependent on θ, and positive constants c_1 to c_4, independent of θ, such that

$$c_1\|\zeta\|^2 \le V_1 \le c_2\|\zeta\|^2, \quad \frac{\partial V_1}{\partial \zeta} f_a(\zeta, 0, \theta) \le -c_3\|\zeta\|^2, \quad \left\|\frac{\partial V_1}{\partial \zeta}\right\| \le c_4\|\zeta\|$$

in some neighborhood of the origin. Using $V = V_1(\zeta) + \frac{1}{2}s^2 + \varphi^T P \varphi$ as a Lyapunov function candidate for the closed-loop system, it can be shown that $\dot{V} \le -\mathcal{Y}^T Q \mathcal{Y}$, where

$$\mathcal{Y} = \begin{bmatrix} \|\zeta\| \\ |s| \\ \|\varphi\| \end{bmatrix}, \qquad Q = \begin{bmatrix} c_1 & -\ell_1 & -\ell_2 \\ -\ell_1 & (\beta_0/\mu - \ell_3) & -\ell_4/\mu \\ -\ell_2 & -\ell_4/\mu & (1/\varepsilon - \ell_5/\mu) \end{bmatrix},$$

and ℓ_1 to ℓ_5 are positive constants independent of μ and ε for $\mu \le 1$. Choosing μ small enough to make the 2×2 principal minor positive, followed by choosing ε small enough to make the determinant positive, makes the matrix Q positive definite, which shows that the origin is exponentially stable. Finally, choosing μ and ε small enough brings $\Omega_\mu \times \{W(\varphi) \le \varepsilon^2 \varrho_1\}$ to an arbitrarily small neighborhood of the origin, which shows that every trajectory starting in $\Psi \times \mathcal{Q}$ converges to the origin. □

Remark 6.11. *If Assumptions* 6.9 *and* 6.11 *hold globally and* γ_1 *belongs to class* $\mathcal{K}_\infty$, *the controller achieves semiglobal stabilization because* Ψ *can be chosen to include any compact subset of* R^{n+r}. ◇

Example 6.6. The Translational Oscillator with Rotating Actuator (TORA) system is modeled by the equations [64]

$$\begin{aligned}
\dot{x}_1 &= x_2, \\
\dot{x}_2 &= \frac{1}{\Delta(x_1)}\left[(m+M)\tau - mL\cos x_1(mLx_2^2 \sin x_1 - kx_3)\right], \\
\dot{x}_3 &= x_4, \\
\dot{x}_4 &= \frac{1}{\Delta(x_1)}\left[-mL\tau\cos x_1 + (J + mL^2)(mLx_2^2 \sin x_1 - kx_3)\right],
\end{aligned}$$

where x_1 is the angular position of the proof mass, x_3 is the translational position of the platform, τ is the toque input, and $\Delta(x_1) = (J + mL^2)(m+M) - m^2L^2\cos^2 x_1$. The masses m and M, moment of inertia J, distance L, and spring constant k are given by

$$M = 1.3608, \quad m = 0.096, \quad J = 0.0002175, \quad L = 0.0592, \quad k = 186.3.$$

It is required to design an output feedback controller using only measurement of x_1 to stabilize the system at the origin $x = 0$. The change of variables

$$\eta_1 = x_3 + \frac{mL}{m+M}\sin x_1, \quad \eta_2 = x_4 + \frac{mLx_2}{m+M}\cos x_1, \quad \xi_1 = x_1, \quad \xi_2 = x_2$$

[64] See [24] or [80, Appendix A].

transforms the system into the normal form

$$
\begin{aligned}
\dot{\eta}_1 &= \eta_2, \\
\dot{\eta}_2 &= \frac{k}{m+M}\left(\frac{mL}{m+M}\sin\xi_1 - \eta_1\right), \\
\dot{\xi}_1 &= \xi_2, \\
\dot{\xi}_2 &= \frac{1}{\Delta(\xi_1)}\left[(m+M)\tau + kmL\eta_1\cos\xi_1 - m^2L^2\sin\xi_1\cos\xi_1\left(\xi_2^2 + \frac{k}{m+M}\right)\right], \\
y &= \xi_1.
\end{aligned}
$$

With $y(t) \equiv 0$, the zero dynamics are

$$
\dot{\eta}_1 = \eta_2, \quad \dot{\eta}_2 = -\frac{k}{m+M}\eta_1,
$$

whose origin is stable but not asymptotically stable. Hence, the system is not minimum phase. Taking

$$
\tau = \frac{m^2L^2}{m+M}\sin\xi_1\cos\xi_1\left(\xi_2^2 + \frac{k}{m+M}\right) + \frac{\Delta(\xi_1)}{m+M}u
$$

results in the system

$$
\begin{aligned}
\dot{\eta}_1 &= \eta_2, \\
\dot{\eta}_2 &= \frac{k}{m+M}\left(\frac{mL}{m+M}\sin\xi_1 - \eta_1\right), \\
\dot{\xi}_1 &= \xi_2, \\
\dot{\xi}_2 &= \frac{kmL\cos\xi_1}{\Delta(\xi_1)}\eta_1 + u, \\
y &= \xi_1,
\end{aligned}
$$

which takes the form (6.78)–(6.81) with $a = (kmL\,\eta_1\cos\xi_1)/\Delta(\xi_1)$, $a_0 = 0$, and $b = 1$. The design is carried out for nominal parameters, and the effect of perturbations in m, M, and J will be studied by simulation. The auxiliary system (6.82)–(6.85) is given by

$$
\begin{aligned}
\dot{\eta}_1 &= \eta_2, \\
\dot{\eta}_2 &= \frac{k}{m+M}\left(\frac{mL}{m+M}\sin\xi_1 - \eta_1\right), \\
\dot{\xi}_1 &= u_a, \\
y_a &= \frac{kmL\cos\xi_1}{\Delta(\xi_1)}\eta_1.
\end{aligned}
$$

With the change of variables,

$$
\tilde{y}_a = \frac{\Delta(\xi_1)}{kmL\cos\xi_1}y_a,
$$

the output of the auxiliary system is taken as $\tilde{y}_a = \eta_1$. The change of variables is valid when $\cos\xi_1 \neq 0$. Therefore, ξ_1 is restricted to $|\xi_1| < \pi/2$. The design of a stabilizing controller for the auxiliary system is pursued by a separation approach. First a state feedback controller that uses η_1, η_2, and ξ_1 is designed; then an observer is designed to estimate η_2. Recall that $\eta_1 = \tilde{y}_a$ and ξ_1 are available. The observer is taken as

$$\dot{\hat{\eta}}_1 = \hat{\eta}_2 + h_1(\tilde{y}_a - \hat{\eta}_1),$$
$$\dot{\hat{\eta}}_2 = \frac{k}{m+M}\left(\frac{mL}{m+M}\sin\xi_1 - \hat{\eta}_1\right) + h_2(\tilde{y}_a - \hat{\eta}_1).$$

The estimation error $\tilde{\eta} = \eta - \hat{\eta}$ satisfies the linear equation

$$\dot{\tilde{\eta}} = \begin{bmatrix} -h_1 & 1 \\ -(h_2 + k/(m+M)) & 0 \end{bmatrix} \tilde{\eta},$$

which is exponentially stable for any $h_1 > 0$ and $h_2 > -k/(m+M)$. The design of the state feedback controller uses high-gain feedback to reduce the system to a linear one. Taking

$$u_a = \frac{1}{\nu}(v_a - \sin\xi_1),$$

where ν is s small positive constant, results in the singularly perturbed equation

$$\nu\dot{\xi}_1 = -\sin\xi_1 + v_a,$$

which has a well-defined quasi–steady-state solution $v_a = \sin\xi_1$ provided $\cos\xi_1 > 0$, which is the case for $|\xi_1| < \pi/2$. Substitution of the quasi–steady-state solution in the η-equation results in the linear system

$$\dot{\eta}_1 = \eta_2, \quad \dot{\eta}_2 = \frac{k}{m+M}\left(\frac{mL}{m+M}v_a - \eta_1\right).$$

The linear feedback control $v_a = -k_1\eta_1 - k_2\eta_2$ is designed to stabilize the system subject to the constraint $|v_a| \le 1$. Thus, the stabilizing controller of the auxiliary system is given by

$$\dot{\hat{\eta}}_1 = \hat{\eta}_2 + h_1\left(\frac{\Delta(\xi_1)}{kmL\cos\xi_1}y_a - \hat{\eta}_1\right),$$
$$\dot{\hat{\eta}}_2 = \frac{k}{m+M}\left(\frac{mL}{m+M}\sin\xi_1 - \hat{\eta}_1\right) + h_2\left(\frac{\Delta(\xi_1)}{kmL\cos\xi_1}y_a - \hat{\eta}_1\right),$$
$$u_a = \frac{1}{\nu}\left[\text{sat}(-k_1\hat{\eta}_1 - k_2\hat{\eta}_2) - \sin\xi_1\right],$$

where the term $-k_1\hat{\eta}_1 - k_2\hat{\eta}_2$ is saturated at ± 1 to ensure that the $\dot{\xi}_1$-equation will have a well-defined quasi–steady-state solution. Using simulation with the initial conditions $|\eta_1(0)| \le 0.025$ and $\eta_2(0) = 0$, the parameters k_1, k_2, ν, h_1, and h_2 are chosen to be 1, 8, 0.01, 2, and 1, respectively. The state feedback controller for the original system is taken as

$$u = -\beta\,\text{sat}\left(\frac{s}{\mu}\right),$$

where $s = \xi_2 + [\text{sat}(k_1\hat{\eta}_1 + k_2\hat{\eta}_2) + \sin\xi_1]/\nu$. Using simulation with $|x_1(0)| \le 1.3$, $|x_3(0)| \le 0.025$, $x_2(0) = x_4(0) = 0$, the parameters β and μ are chosen as 500 and 0.1, respectively. For later use with output feedback, it is determined from simulation that the maximum values of $|\xi_2|$ and $|a|$ are less than 40.

Using an extended high-gain observer, the output feedback controller is given by

$$\begin{aligned}
\dot{\hat{\xi}}_1 &= \hat{\xi}_2 + (\alpha_1/\varepsilon)(y - \hat{\xi}_1), \\
\dot{\hat{\xi}}_2 &= \hat{\sigma} + u + (\alpha_2/\varepsilon^2)(y - \hat{\xi}_1), \\
\dot{\hat{\sigma}} &= (\alpha_3/\varepsilon^3)(y - \hat{\xi}_1), \\
\dot{\hat{\eta}}_1 &= \hat{\eta}_2 + h_1\left(\frac{\Delta(\hat{\xi}_1)}{kmL\cos\hat{\xi}_1}\hat{\sigma} - \hat{\eta}_1\right), \\
\dot{\hat{\eta}}_2 &= \frac{k}{m+M}\left(\frac{mL}{m+M}\sin\hat{\xi}_1 - \hat{\eta}_1\right) + h_2\left(\frac{\Delta(\hat{\xi}_1)}{kmL\cos\hat{\xi}_1}\hat{\sigma} - \hat{\eta}_1\right), \\
\hat{s} &= \hat{\xi}_2 + [\text{sat}(k_1\hat{\eta}_1 + k_2\hat{\eta}_2) + \sin\hat{\xi}_1]/\nu, \\
u &= -\beta\,\text{sat}(\hat{s}/\mu), \\
\tau &= \frac{m^2L^2}{m+M}\sin\hat{\xi}_1\cos\hat{\xi}_1\left(\hat{\xi}_2^2 + \frac{k}{m+M}\right) + \frac{\Delta(\hat{\xi}_1)}{m+M}u,
\end{aligned}$$

where $\hat{\sigma}$ on the right-hand side of the $\dot{\hat{\eta}}$-equations and $\hat{\xi}_2$ on the right-hand side of the τ-equation are saturated at ± 40. The parameters α_1, α_2, α_3, and ε are taken as 3, 3, 1, and 0.001, respectively.

Figure 6.13 shows simulation results for two sets of initial conditions. To examine robustness to parameter perturbations, Figure 6.14 shows results when m, M, and J are increased by 75% and decreased by 25%. The simulation is carried out for $x(0) = \text{col}(1, 0, 0.02, 0)$. The perturbed response deteriorates, compared with the nominal one, but the system is still stabilized. For larger perturbations, the controller fails to stabilize the system. △

6.3.2 ▪ Full-Order Observer

Consider a single-input–single-output nonlinear system modeled by

$$\dot{\eta} = f(\eta, \xi), \tag{6.103}$$
$$\dot{\xi}_i = \xi_{i+1} \quad \text{for } 1 \le i \le \rho - 1, \tag{6.104}$$
$$\dot{\xi}_\rho = g(\xi, u) + h(\eta, \xi), \tag{6.105}$$
$$y = \xi_1, \tag{6.106}$$

where $\eta \in R^{n-\rho}$ and $\xi = \text{col}(\xi_1, \dots, \xi_\rho) \in R^\rho$ are the state variables, u is the control input, and y is the measured output. The known functions f, g, and h are continuously differentiable with locally Lipschitz derivatives, and $\eta(t)$, $\xi(t)$, and $u(t)$ are bounded for all $t \ge 0$. In particular, $\eta(t) \in \Pi \subset R^{n-\rho}$, $\xi(t) \in \Xi \subset R^\rho$, and $u(t) \in U \subset R$ for some compact sets Π, Ξ, and U. It is required to design an observer to estimate (η, ξ) such that the estimation error converges asymptotically to zero. A high-gain observer can be designed to estimate ξ. A full-order observer that estimates both η and ξ can be built if the origin of $\dot{\eta} = f(\eta, 0)$ is asymptotically stable.[65] Such an observer will

[65] See, for example, [39] and [129].

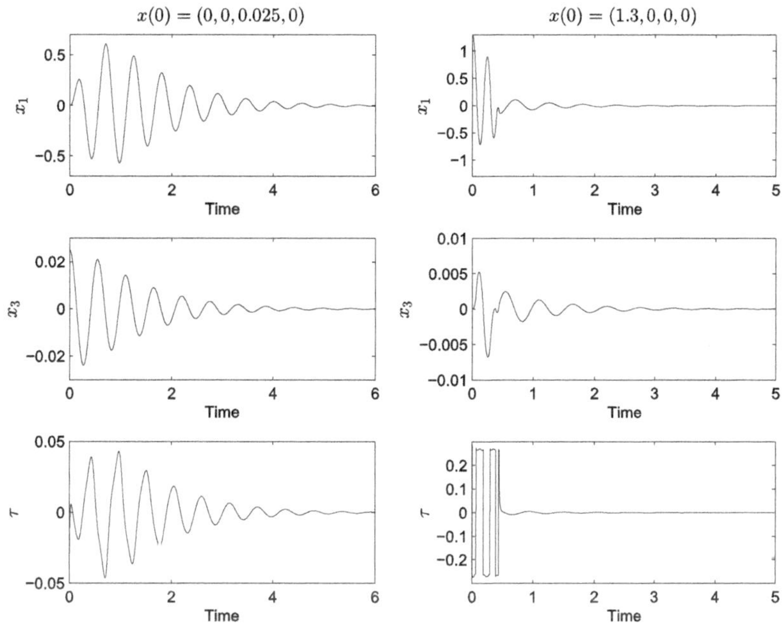

Figure 6.13. *Simulation results of Example* 6.6.

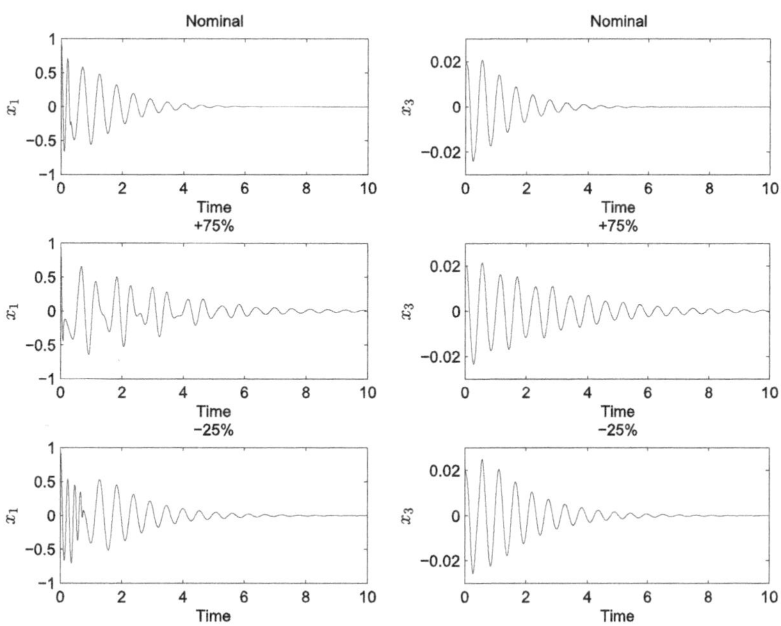

Figure 6.14. *Simulation results of Example* 6.6 *when* m, M, *and* J *are increased by* 75% *and decreased by* 25%. *The initial state is* $x(0) = (1, 0, 0.02, 0)$.

estimate $\hat{\eta}$ by the open-loop observer $\dot{\hat{\eta}} = f(\hat{\eta}, \hat{\xi})$. To estimate η without this stability assumption, an extended high-gain observer is designed to estimate $h(\eta, \xi)$, in addition to estimating ξ. Anticipating that the estimates of ξ and h can be made arbitrarily close to their true values in arbitrarily small time, we consider the auxiliary system

$$\dot{\eta} = f(\eta, \xi), \tag{6.107}$$
$$\sigma = h(\eta, \xi), \tag{6.108}$$

where ξ is a known input and σ is the measured output. The task now is to design an observer for the auxiliary system. This is a nonlinear observer problem for the lower-dimensional system (6.107)–(6.108), and the design of the observer will depend on the properties of the functions f and h. For example, if the system happens to take the form that is treated in Chapter 2, a high-gain observer can be designed to estimate η. Instead of assuming a special structure for f and h, we design an extended Kalman filter [80, Section 11.2], which does not require a special structure for the system but yields only a local result that requires the initial estimation error to be sufficiently small. The extended Kalman filter is given by

$$\dot{\hat{\eta}} = f(\hat{\eta}, \xi) + P(t)C^T(\hat{\eta}, \xi)R^{-1}[\sigma - h(\hat{\eta}, \xi)],$$

where $P(t)$ is the solution of the Riccati equation

$$\dot{P} = A(\hat{\eta}, \xi)P(\hat{\eta}, \xi) + PA^T(\hat{\eta}, \xi) + Q - PC^T(\hat{\eta}, \xi)R^{-1}C(\hat{\eta}, \xi)P, \quad P(t_0) = P_0,$$

in which Q, R, and P_0 are symmetric positive definite constant matrices, while A and C are defined by

$$A(\eta, \xi) = \frac{\partial f}{\partial \eta}(\eta, \xi), \qquad C(\eta, \xi) = \frac{\partial h}{\partial \eta}(\eta, \xi).$$

With the use of the extended high-gain observer to estimate ξ and σ, the full-order observer is given by

$$\dot{\hat{\xi}}_i = \hat{\xi}_{i+1} + (\alpha_i/\varepsilon^i)(y - \hat{\xi}_1) \quad \text{for } 1 \le i \le \rho - 1, \tag{6.109}$$
$$\dot{\hat{\xi}}_\rho = \hat{\sigma} + g(\hat{\xi}, u) + (\alpha_\rho/\varepsilon^\rho)(y - \hat{\xi}_1), \tag{6.110}$$
$$\dot{\hat{\sigma}} = \dot{h}(\hat{\eta}, \hat{\xi}, u) + (\alpha_{\rho+1}/\varepsilon^{\rho+1})(y - \hat{\xi}_1), \tag{6.111}$$
$$\dot{\hat{\eta}} = f(\hat{\eta}, \hat{\xi}) + PC^T(\hat{\eta}, \hat{\xi})R^{-1}[\hat{\sigma} - h(\hat{\eta}, \hat{\xi})], \tag{6.112}$$
$$\dot{P} = A(\hat{\eta}, \hat{\xi})P + PA^T(\hat{\eta}, \hat{\xi}) + Q - PC^T(\hat{\eta}, \hat{\xi})R^{-1}C(\hat{\eta}, \hat{\xi})P, \tag{6.113}$$

with $P(t_0) = P_0$, where ε is a small positive constant, α_1 to $\alpha_{\rho+1}$ are chosen such that the polynomial

$$\lambda^{\rho+1} + \alpha_1\lambda^\rho + \cdots + \alpha_\rho\lambda + \alpha_{\rho+1}$$

is Hurwitz, and

$$\dot{h}(\eta, \xi, u) = \frac{\partial h}{\partial \eta} f(\eta, \xi) + \sum_{i=1}^{\rho-1} \frac{\partial h}{\partial \xi_i}\xi_{i+1} + \frac{\partial h}{\partial \xi_\rho}[g(\xi, u) + h(\eta, \xi)].$$

The $\hat{\xi}$ argument of the functions $g(\hat{\xi}, u)$, $\dot{h}(\hat{\eta}, \hat{\xi}, u)$, $f(\hat{\eta}, \hat{\xi})$, $A(\hat{\eta}, \hat{\xi})$, and $C(\hat{\eta}, \hat{\xi})$ in the preceding equations is saturated outside the compact set Ξ to overcome the peaking of the high-gain observer. In particular, $\hat{\xi}_i$ is replaced by $M_i \operatorname{sat}(\hat{\xi}_i/M_i)$, where $M_i > \max_{\xi\in\Xi}|\xi_i|$. Moreover, $\hat{\sigma}$ in equation (6.112) is replaced by $M_{\rho+1}\operatorname{sat}(\hat{\sigma}/M_{\rho+1})$, where $M_{\rho+1} > \max_{\eta\in\Pi, \xi\in\Xi}|h(\eta, \xi)|$.

Remark 6.12. *The effect of $\dot{h}$ in* (6.111) *is asymptotically attenuated as $\varepsilon \to 0$, but it is included to show that the estimation error converges to zero as $t \to \infty$. Without its inclusion, the error would eventually be of the order $O(\varepsilon)$.* ◇

Assumption 6.12. *The Riccati equation* (6.113) *has a positive definite solution that satisfies the inequality*

$$c_1 I \le P(t) \le c_2 I \tag{6.114}$$

for some positive constants c_1 and c_2.

This assumption is crucial for the validity of the extended Kalman filter, yet it is hard to verify. Although it is known from the properties of the Riccati equation that the assumption is satisfied if $A(t) = (\hat{\xi}(t), \hat{\eta}(t))$ and $C(t) = C(\hat{\xi}(t), \hat{\eta}(t))$ are bounded and the pair $(A(t), C(t))$ is uniformly observable [91], the matrices $A(t)$ and $C(t)$ are generated in real time, so observability cannot be checked offline. For certain classes of nonlinear systems it can be shown that the assumption holds under certain observability conditions.[66]

Theorem 6.8. *Consider the system* (6.103)–(6.106) *and the observer* (6.109)–(6.113). *Suppose Assumption* 6.12 *is satisfied. Let $\mathcal{Q}$ be a compact subset of $R^{\rho+1}$. Then there exist positive constants r and ε^* such that for all $\|\eta(0)-\hat{\eta}(0)\| \le r$ and $(\hat{\xi}(0), \hat{\sigma}(0)) \in \mathcal{Q}$, and for every $\varepsilon \in (0, \varepsilon^*]$, the estimation errors $\eta(t)-\hat{\eta}(t)$ and $\xi(t)-\hat{\xi}(t)$ converge to zero as $t \to \infty$.* ◇

Proof: The scaled estimation errors

$$\begin{aligned}
\tilde{\eta} &= \eta - \hat{\eta}, \\
\varphi_i &= (\xi_i - \hat{\xi}_i)/\varepsilon^{\rho+1-i} \qquad \text{for} \quad 1 \le i \le \rho, \\
\varphi_{\rho+1} &= h(\eta, \xi) - \hat{\sigma}
\end{aligned}$$

satisfy the equations

$$\begin{aligned}
\dot{\tilde{\eta}} &= [A(t) - H(t)C(t)]\tilde{\eta} + \Delta(\cdot), \\
\varepsilon\dot{\varphi} &= \Lambda\varphi + \varepsilon[\bar{B}_1 \Delta_1(\cdot) + \bar{B}_2 \Delta_2(\cdot)],
\end{aligned}$$

where $A(t) = A(\hat{\eta}(t), \hat{\xi}(t))$, $C(t) = C(\hat{\eta}(t), \hat{\xi}(t))$, $H(t) = P(t)C^T(t)R^{-1}$,

$$\Lambda = \begin{bmatrix} -\alpha_1 & 1 & 0 & \cdots & 0 \\ -\alpha_2 & 0 & 1 & \ddots & 0 \\ \vdots & \vdots & \ddots & \ddots & \vdots \\ -\alpha_\rho & 0 & \cdots & 0 & 1 \\ -\alpha_{\rho+1} & 0 & \cdots & \cdots & 0 \end{bmatrix}, \quad \bar{B}_1 = \begin{bmatrix} 0 \\ \vdots \\ 0 \\ 0 \\ 1 \end{bmatrix}, \quad \bar{B}_2 = \begin{bmatrix} 0 \\ \vdots \\ 0 \\ 1 \\ 0 \end{bmatrix},$$

[66] It is shown in [20] that the assumption is satisfied if the auxiliary system (6.107)–(6.108) has a uniform detectability property. It is shown in [88] that the assumption holds if the auxiliary system is uniformly observable for any input ξ; uniform observability is defined in [49].

$$\begin{aligned}\Delta &= f(\eta,\hat{\xi})-f(\hat{\eta},\hat{\xi})-A(t)\tilde{\eta}-H(t)[h(\eta,\hat{\xi})-h(\hat{\eta},\hat{\xi})-C(t)\tilde{\eta}]\\ &\quad +f(\eta,\xi)-f(\eta,\hat{\xi})+H(t)\varphi_{\rho+1}-H(t)[h(\eta,\xi)-h(\eta,\hat{\xi})],\\ \Delta_1 &= \dot{h}(\eta,\xi,u)-\dot{h}(\hat{\eta},\hat{\xi},u),\\ \Delta_2 &= (1/\varepsilon)[g(\xi,u)-g(\hat{\xi},u)].\end{aligned}$$

The matrix Λ is Hurwitz by design. Let $V_1(t,\tilde{\eta}) = \tilde{\eta}^T P^{-1}\tilde{\eta}$. By Assumption 6.12, there are positive constants ϱ_1 and ϱ_2 such that

$$\varrho_1\|\tilde{\eta}\|^2 \le V_1 \le \varrho_2\|\tilde{\eta}\|^2.$$

Let $\Omega = \{\tilde{\eta} \mid V_1(t,\tilde{\eta}) \le c\}$, where $c > 0$ is to be determined. The forthcoming analysis is carried out for $\tilde{\eta} \in \Omega$, for which $\hat{\eta}$ is bounded, and for $\varepsilon \le 1$. Consequently, $A(t)$ and $C(t)$ are bounded since they are globally bounded functions of $\hat{\xi}$. Since

$$\left\|f(\eta,\hat{\xi})-f(\hat{\eta},\hat{\xi})-A(t)\tilde{\eta}\right\| = \left\|\int_0^1\left[\frac{\partial f}{\partial \eta}(z\tilde{\eta}+\hat{\eta},\hat{\xi})-\frac{\partial f}{\partial \eta}(\hat{\eta},\hat{\xi})\right]dz\ \tilde{\eta}\right\| \le \tfrac{1}{2}L_1\|\tilde{\eta}\|^2$$

and

$$\left\|h(\eta,\hat{\xi})-h(\hat{\eta},\hat{\xi})-C(t)\tilde{\eta}\right\| = \left\|\int_0^1\left[\frac{\partial h}{\partial \eta}(z\tilde{\eta}+\hat{\eta},\hat{\xi})-\frac{\partial h}{\partial \eta}(\hat{\eta},\hat{\xi})\right]dz\ \tilde{\eta}\right\| \le \tfrac{1}{2}L_2\|\tilde{\eta}\|^2,$$

where L_1 and L_2 are Lipschitz constants of $[\partial f/\partial \eta]$ and $[\partial h/\partial \eta]$, respectively, there are positive constants k_a and k_b such that

$$\|\Delta\| \le k_a\|\tilde{\eta}\|^2 + k_b\|\varphi\|.$$

By the Lipschitz properties of $\dot{h}(\eta,\xi,u)$ and $g(\xi,u)$,

$$|\Delta_1| = \left|\dot{h}(\eta,\xi,u)-\dot{h}(\hat{\eta},\hat{\xi},u)\right| \le L_3\|\tilde{\eta}\| + L_4\|\xi-\hat{\xi}\| \le L_3\|\tilde{\eta}\| + \varepsilon L_4\|\varphi\|,$$

$$|\Delta_2| = \frac{1}{\varepsilon}\left|g(\xi,u)-g(\hat{\xi},u)\right| \le \frac{L_5}{\varepsilon}\|\xi-\hat{\xi}\| \le L_5\|\varphi\|.$$

Let $V_2(\varphi) = \varphi^T P_2\varphi$, where $P_2 = P_2^T > 0$ is the solution of the Lyapunov equation $P_2\Lambda + \Lambda^T P_2 = -I$, and define the set $\Sigma = \{\varphi^T P_2\varphi \le \varrho\varepsilon^2\}$, where $\varrho > 0$ is to be determined. The derivative of V_2 satisfies

$$\begin{aligned}\varepsilon\dot{V}_2 &= -\varphi^T\varphi + 2\varepsilon\varphi^T P_2(\bar{B}_1\Delta_1 + \bar{B}_2\Delta_2)\\ &\le -\|\varphi\|^2 + 2\varepsilon\|\varphi\|\left[\ \|P_2\bar{B}_1\|(L_3\|\tilde{\eta}\| + L_4\|\varphi\|) + \|P_2\bar{B}_2\|L_5\|\varphi\|\ \right]\\ &\le -\|\varphi\|^2 + \varepsilon k_c\|\varphi\|^2 + \varepsilon k_d\|\varphi\|,\end{aligned}$$

where k_c and k_d are positive constants. For $2\varepsilon k_c < 1$,

$$\varepsilon\dot{V}_2 \le -\tfrac{1}{2}\|\varphi\|^2 + \varepsilon k_d\|\varphi\| \le -\tfrac{1}{4}\|\varphi\|^2 \quad \text{for } \|\varphi\| \ge 4\varepsilon k_d.$$

Taking $\varrho = 16k_d^2\lambda_{\max}(P_2)$ shows that

$$V_2(\varphi) \geq \varrho\varepsilon^2 \quad \Rightarrow \quad \|\varphi\| \geq 4\varepsilon k_d \quad \Rightarrow \quad \varepsilon \dot{V}_2 \leq -\frac{1}{4\lambda_{\max}(P_2)} V_2.$$

On the other hand,

$$\begin{aligned}
\dot{V}_1 &= \frac{\partial V_1}{\partial t} + \frac{\partial V_1}{\partial \tilde{\eta}}(A - HC)\tilde{\eta} + \frac{\partial V_1}{\partial \tilde{\eta}}\Delta \\
&= \tilde{\eta}^T \left[\frac{d}{dt}P^{-1} + P^{-1}(A - PC^TR^{-1}C) + (A - PC^TR^{-1}C)^TP^{-1}\right]\tilde{\eta} + 2\tilde{\eta}P\Delta \\
&= \tilde{\eta}^T P^{-1}(AP + PA - 2PC^TR^{-1}CP - \dot{P})P^{-1}\tilde{\eta} + 2\tilde{\eta}P\Delta \\
&= -\tilde{\eta}^T(P^{-1}QP^{-1} + C^TR^{-1}C)\tilde{\eta} + 2\tilde{\eta}P\Delta \\
&\leq -k_1\|\tilde{\eta}\|^2 + k_e\|\tilde{\eta}\|^3 + k_f\|\tilde{\eta}\|\,\|\varphi\| \\
&\leq -(k_1 - k_2c)\|\tilde{\eta}\|^2 + k_f\|\tilde{\eta}\|\,\|\varphi\|
\end{aligned}$$

for some positive constants k_1, k_2, and k_f. Choosing c small enough to make $k_1 - k_2c > 0$, we conclude that the set $\Omega \times \Sigma$ is positively invariant for sufficiently small ε. For all $\tilde{\eta}(0) \in \{\tilde{\eta} \mid V_1(t,\tilde{\eta}) \leq c_0\}$, with $c_0 < c$, there is finite time $T_1 > 0$ such that $\tilde{\eta}(t) \in \{\tilde{\eta} \mid V_1(t,\tilde{\eta}) \leq c\}$ for all $t \in [0, T_1]$. During this time, $\varepsilon\dot{V}_2 \leq -\frac{1}{4\lambda_{\max}(P_2)}V_2$; hence, φ enters the set Σ in finite time $T(\varepsilon)$ with $\lim_{\varepsilon\to 0} T(\varepsilon) = 0$. Choosing ε small enough shows that $(\tilde{\eta}(t), \varphi(t)) \in \Omega \times \Sigma$ for all $t \geq T(\varepsilon)$. Inside $\Omega \times \Sigma$, the derivative of the composite Lyapunov function $V(t,\tilde{\eta},\varphi) = V_1(t,\tilde{\eta}) + V_2(\varphi)$ satisfies

$$\dot{V} \leq -\mathcal{Y}^T Q \mathcal{Y},$$

where

$$\mathcal{Y} = \begin{bmatrix} \|\tilde{\eta}\| \\ \|\varphi\| \end{bmatrix} \quad \text{and} \quad Q = \begin{bmatrix} (k_1 - k_2c) & -k_3 \\ -k_3 & 1/\varepsilon - k_4 \end{bmatrix},$$

where k_1 to k_4 are positive constants independent of ε and $k_1 - k_2c > 0$. For sufficiently small ε, Q is positive definite, which shows that $\lim_{t\to\infty} \tilde{\eta}(t) = 0$ and $\lim_{t\to\infty} \varphi(t) = 0$. □

Example 6.7. The one-axis model of a synchronous generator connected to an infinite bus is given by [67]

$$\begin{aligned}
\dot{\eta}_1 &= \eta_2, \\
M\dot{\eta}_2 &= P - D\eta_2 - \beta_1\xi\sin\eta_1, \\
\tau\dot{\xi} &= -\beta_2\xi + \beta_3\cos\eta_1 + u,
\end{aligned}$$

where ξ is a voltage proportional to the field-flux linkage, η_1 is the angle between the voltage of the infinite bus and ξ, P is the mechanical input power (assumed constant), M is the moment of inertia, D is the damping coefficient, τ is a time constant, u is the field voltage (assumed constant), and β_1 to β_3 are positive parameters. Assuming that the only measured signal is ξ, that is, $y = \xi$, it is desired to build an observer to estimate η_1 and η_2. With $y \equiv 0$, the zero dynamics

$$\dot{\eta}_1 = \eta_2, \qquad M\dot{\eta}_2 = P - D\eta_2$$

[67] See [7] or [114].

do not have an asymptotically stable equilibrium point; hence, the system is not minimum phase. The system takes the form (6.103)–(6.106) with

$$f=\begin{bmatrix} \eta_2 \\ (P-D\eta_2-\beta_1\xi\sin\eta_1)/M \end{bmatrix}, \quad g=(-\beta_2\xi+u)/\tau, \quad \text{and} \quad h=(\beta_3/\tau)\cos\eta_1.$$

The matrices A and C are given by

$$A(\eta,\xi)=\begin{bmatrix} 0 & 1 \\ -(\beta_1/M)\xi\cos\eta_1 & -D/M \end{bmatrix}, \quad C=-(\beta_3/\tau)\sin\eta_1\begin{bmatrix} 1 & 0 \end{bmatrix},$$

and $\dot{h}=-(\beta_3/\tau)\eta_2\sin\eta_1$. Taking $Q=I$, $R=1$, and $P=\left[\begin{smallmatrix} p_{11} & p_{12} \\ p_{12} & p_{22} \end{smallmatrix}\right]$, the full-order observer is given by

$$\begin{aligned}
\dot{p}_{11} &= 2p_{12}+1-[(\beta_3/\tau)\sin\hat{\eta}_1]^2 p_{11}^2, \\
\dot{p}_{12} &= p_{22}-(\beta_1/M)\xi\, p_{11}\cos\hat{\eta}_1-(D/M)p_{12}-[(\beta_3/\tau)\sin\hat{\eta}_1]^2 p_{11}p_{12}, \\
\dot{p}_{22} &= -(2\beta_1/M)\xi\, p_{12}\cos\hat{\eta}_1-2(D/M)p_{22}+1-[(\beta_3/\tau)\sin\hat{\eta}_1]^2 p_{12}^2, \\
\dot{\hat{\eta}}_1 &= \hat{\eta}_2-(\beta_3/\tau)p_{11}\sin\hat{\eta}_1\,[M_1\ \mathrm{sat}(\hat{\sigma}/M_1)-(\beta_3/\tau)\cos\hat{\eta}_1], \\
\dot{\hat{\eta}}_2 &= P/M-(D/M)\hat{\eta}_2-(\beta_1/M)\xi\sin\hat{\eta}_1 \\
&\quad -(\beta_3/\tau)p_{12}\sin\hat{\eta}_1\,[M_1\ \mathrm{sat}(\hat{\sigma}/M_1)-(\beta_3/\tau)\cos\hat{\eta}_1], \\
\dot{\hat{\xi}} &= -(\beta_2/\tau)\hat{\xi}+\hat{\sigma}+u/\tau+(\alpha_1/\varepsilon)(y-\hat{\xi}), \\
\dot{\hat{\sigma}} &= -(\beta_3/\tau)\hat{\eta}_2\sin\hat{\eta}_1+(\alpha_2/\varepsilon^2)(y-\hat{\xi}),
\end{aligned}$$

where M_1 is the saturation level of $\hat{\sigma}$. Since ξ is measured, it is used in the extended Kalman filter instead of $\hat{\xi}$. With $P=0.815$, $\beta_1=2$, $\beta_2=2.7$, $\beta_3=1.7$, $M=0.0147$, $D/M=4$, $\tau=6.6$, and $u=1.22$, the system has an asymptotically stable equilibrium point at ($\eta_1=0.4067$, $\eta_2=0$, $\xi=1.0301$). Using simulation, it is found that, with $\eta_2(0)=0$, the set $\{-0.6\le\eta_1\le 1.4, 0.6\le\xi\le 2\}$ is a subset of the region of attraction. Since $|\sigma|=|(\beta_3/\tau)\cos\eta_1|\le\beta_3/\tau=0.2576$, $\hat{\sigma}$ is saturated at ± 0.3. Figures 6.15 and 6.16 show simulation results for $\alpha_1=2$, $\alpha_2=1$, and $\varepsilon=0.01$ and the initial conditions $\eta_1(0)=1$, $\eta_2(0)=0$, $\xi(0)=2$, $p_{11}(0)=p_{22}(0)=1$, $p_{12}(0)=0$, $\hat{\eta}_1(0)=\hat{\eta}_2(0)=\hat{\xi}(0)=\hat{\sigma}(0)=0$. Figure 6.15 shows the convergence of the estimation errors to zero. As expected, the estimation errors $\xi-\hat{\xi}$ and $\sigma-\hat{\sigma}$ of the extended high-gain observer converge much faster than the estimation errors $\eta_1-\hat{\eta}_1$ and $\eta_2-\hat{\eta}_2$ of the extended Kalman filter. Figure 6.16 shows the solution of the Riccati equation and the determinant of the matrix P. The figure confirms that Assumption 6.12 is satisfied since p_{11} and $p_{11}p_{22}-p_{12}^2$ are bounded from below by positive constants, while all three elements p_{11}, p_{12} and p_{22} are bounded from above. △

Theorem 6.8 requires the initial estimation error $\eta(0)-\hat{\eta}(0)$ to be sufficiently small. A special case where this restriction can be removed is when the system (6.103)–(6.106) takes the special form

$$\dot{\eta}=A(\xi)\eta+\phi(\xi), \tag{6.115}$$

$$\dot{\xi}_i=\xi_{i+1} \quad \text{for } 1\le i\le\rho-1, \tag{6.116}$$

$$\dot{\xi}_\rho=C(\xi)\eta+g(\xi,u), \tag{6.117}$$

$$y=\xi_1. \tag{6.118}$$

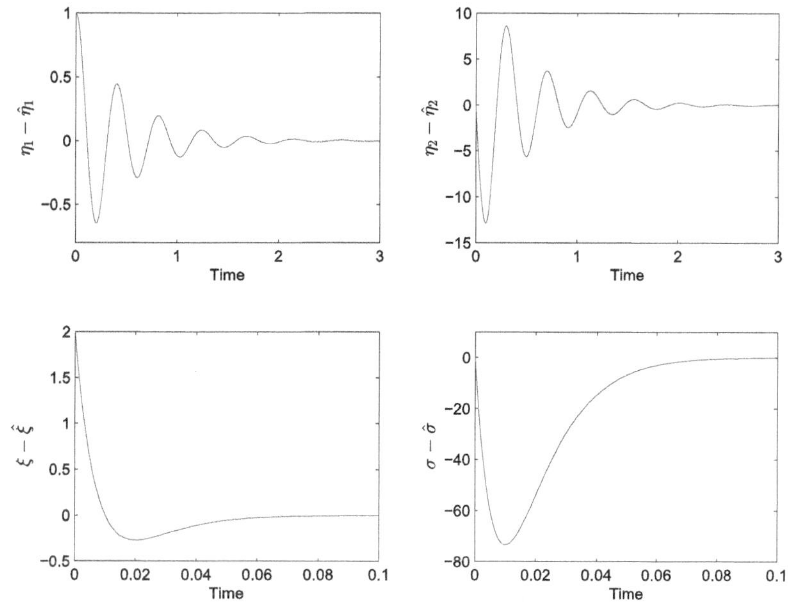

Figure 6.15. *The estimation errors of Example* 6.7.

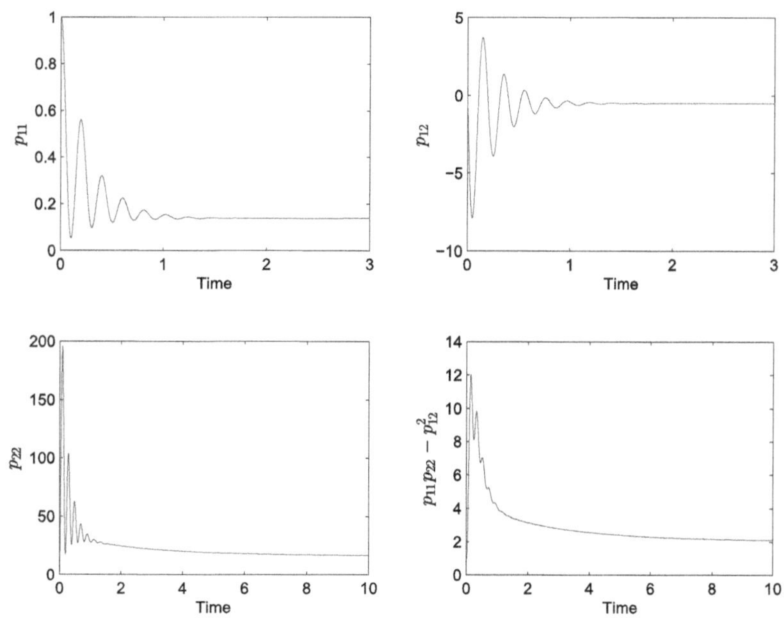

Figure 6.16. *The solution of the Riccati equation of Example* 6.7.

In this case, the full-order observer specializes to

$$\dot{\hat{\xi}}_i = \hat{\xi}_{i+1} + (\alpha_i/\varepsilon^i)(y - \hat{\xi}_1) \quad \text{for } 1 \le i \le \rho - 1, \tag{6.119}$$

$$\dot{\hat{\xi}}_\rho = \hat{\sigma} + g(\hat{\xi}, u) + (\alpha_\rho/\varepsilon^\rho)(y - \hat{\xi}_1), \tag{6.120}$$

$$\dot{\hat{\sigma}} = \dot{h}(\hat{\eta}, \hat{\xi}, u) + (\alpha_{\rho+1}/\varepsilon^{\rho+1})(y - \hat{\xi}_1), \tag{6.121}$$

$$\dot{\hat{\eta}} = A(\hat{\xi})\hat{\eta} + P\hat{C}^T(\hat{\xi})R^{-1}[\hat{\sigma} - C(\hat{\xi})\hat{\eta}], \tag{6.122}$$

$$\dot{P} = A(\hat{\xi})P + PA^T(\hat{\xi}) + Q - PC^T(\hat{\xi})R^{-1}C(\hat{\xi})P, \tag{6.123}$$

with $P(t_0) = P_0$, where

$$\dot{h}(\eta,\xi,u) = C(\xi)[A(\xi)\eta + \phi(\xi)] + \sum_{i=1}^{\rho-1} \frac{\partial}{\partial \xi_i}[C(\xi)\eta]\xi_{i+1} + \frac{\partial}{\partial \xi_\rho}[C(\xi)\eta][C(\xi)\eta + g(\xi,u)].$$

The choice of parameters and the saturation of $\hat{\xi}$ and $\hat{\sigma}$, which are described after (6.109)–(6.113), carry over to the current case.

Theorem 6.9. *Consider the system* (6.115)–(6.119) *and the observer* (6.119)–(6.123). *Suppose Assumption* 6.12 *is satisfied. Let $\mathcal{Q}$ be a compact subset of $R^{\rho+1}$, and let r be any positive constant. Then, there is $\varepsilon^* > 0$ such that for all $\|\eta(0) - \hat{\eta}(0)\| \le r$ and $(\hat{\xi}(0), \hat{\sigma}(0)) \in \mathcal{Q}$, and for every $\varepsilon \in (0, \varepsilon^*]$, the estimation errors $\eta(t) - \hat{\eta}(t)$ and $\xi(t) - \hat{\xi}(t)$ converge to zero as $t \to \infty$.* ◇

The proof of this theorem proceeds exactly as the proof of Theorem 6.8, with one difference.[68] The derivative of V_1 satisfies the inequality

$$\dot{V}_1 \le -k_1\|\tilde{\eta}\|^2 + k_f\|\tilde{\eta}\|\,\|\varphi\|$$

instead of

$$\dot{V}_1 \le -(k_1 - k_2 c)\|\tilde{\eta}\|^2 + k_f\|\tilde{\eta}\|\,\|\varphi\|.$$

Therefore, c can be chosen arbitrarily large.

6.4 ▪ Notes and References

The performance recovery result of Section 6.2.1 is based on [47], and its extension to multivariable systems in Section 6.2.2 is based on [155]. The dynamic inversion algorithm of Section 6.2.3 is based on [93]. An experimental application of the algorithm to the inverted pendulum on a cart is given in [92]. The stabilization result of Section 6.3.1 is based on [111], which deals with the more general case when the control coefficient b is allowed to be uncertain. The auxiliary system tool, which is the core of this stabilization technique, was introduced by Isidori in [67]; see also [66, Section 12.6]. The full-order observer of Section 6.3.2 is based on [22]. Its use in feedback stabilization of nonminimum phase systems is discussed in [23].

[68]The complete proof is given in [22].

Chapter 7

Low-Power Observers

The standard high-gain observer (2.17)–(2.18) faces a challenge in numerical implementation when its dimension ρ is high because the observer gain is $O(1/\varepsilon^{\rho})$ and its internal states could be $O(1/\varepsilon^{\rho-1})$ during the peaking period. The reduced-order observer (2.48)–(2.50) is not of much help because its internal states could still be $O(1/\varepsilon^{\rho-1})$, although its gain will be $O(1/\varepsilon^{\rho-1})$ instead of $O(1/\varepsilon^{\rho})$. This chapter presents modified high-gain observers where the observer gain is $O(1/\varepsilon)$ and the observer internal states peak to $O(1/\varepsilon)$. The modified observers are developed for the special case when the functions ψ_1 to $\psi_{\rho-1}$ of (2.2) are zero. In this case, the state variables x_2 to x_ρ are the derivatives of the output $y = x_1$.

The cascade observer of Section 7.1 is formed by cascading a second-order linear observer and $\rho - 1$ first-order linear observers, with saturation functions in between them. It is shown that the observer is robust to model uncertainty and the estimation error decays to $O(\varepsilon)$ values within a time period $[0, T(\varepsilon)]$, where $\lim_{\varepsilon\to 0} T(\varepsilon) = 0$. Section 7.2 proves a nonlinear separation theorem when the observer is used in feedback control. The performance of this observer in feedback control is comparable to the performance of the standard observer, but as a state estimator it has some drawbacks when compared with the standard observer. These drawbacks are removed in the cascade observer with feedback injection of Section 7.3. This observer is formed of the cascade connection of $\rho - 1$ second-order observers with injected feedback signals in each stage and saturation functions between the different stages. It is shown that the estimation error of this observer has properties similar to the standard observer.

In preparation for the cascade observers, some features of first- and second-order high-gain observers are recalled. When $\rho = 2$ and $\psi_1 = 0$, the observer (2.17)–(2.18) specializes to

$$\dot{\hat{x}}_1 = \hat{x}_2 + (\alpha_1/\varepsilon)(y - \hat{x}_1), \quad \dot{\hat{x}}_2 = \phi_0 + (\alpha_2/\varepsilon^2)(y - \hat{x}_1),$$

where α_1, α_2, and ε are positive constants. The observer gain is $O(1/\varepsilon^2)$. By scaling the observer variables as $z_1 = \hat{x}_1$ and $z_2 = \varepsilon\hat{x}_2$, the observer is implemented by the equations

$$\dot{z}_1 = (1/\varepsilon)[z_2 + \alpha_1(y - z_1)], \qquad \dot{z}_2 = \varepsilon\phi_0 + (\alpha_2/\varepsilon)(y - z_1),$$
$$\hat{x}_1 = z_1, \qquad \hat{x}_2 = z_2/\varepsilon,$$

where now the highest gain is $O(1/\varepsilon)$. When $\phi_0 = 0$, the observer is linear, and its transfer function from y to $\hat{x}_2$ is

$$\frac{\alpha_2 s}{(\varepsilon s)^2 + \alpha_1(\varepsilon s) + \alpha_2}.$$

The state $x_2 = \dot{y}$ can also be estimated using the reduced-order linear observer

$$\dot{z} = -(\beta/\varepsilon)(z + y), \qquad \hat{x}_2 = (\beta/\varepsilon)(z + y),$$

where β and ε are positive constants, which is a special case of the observer (2.48)–(2.50) with a change of variables that replaces z by $\varepsilon z/\beta$. The observer's transfer function from y to $\hat{x}_2$ is

$$\frac{\beta s}{\varepsilon s + \beta}.$$

7.1 ▪ Cascade Observer

Consider the system

$$\dot{w} = f_0(w, x, u), \tag{7.1}$$
$$\dot{x}_i = x_{i+1} \qquad \text{for } 1 \le i \le \rho - 1, \tag{7.2}$$
$$\dot{x}_\rho = \phi(w, x, u), \tag{7.3}$$
$$y = x_1, \tag{7.4}$$

where f_0 and ϕ are locally Lipschitz in their arguments and $w(t)$, $x(t)$, and $u(t)$ are bounded for all $t \ge 0$. The states x_2 to x_ρ are derivatives of x_1. The derivative of a signal can be estimated by a first-order or second-order linear observer. By cascading such low-order observers we can estimate higher derivatives of the signal. We choose to estimate x_2 using a second-order observer because of its low-pass filtering characteristics but use first-order observers to estimate x_3 to x_ρ so that the dimension of the cascade connection is ρ, as in the standard observer. The cascade connection is given by

$$\dot{z}_1 = (1/\varepsilon)[z_2 + \beta_1(y - z_1)], \tag{7.5}$$
$$\dot{z}_2 = (\beta_2/\varepsilon)(y - z_1), \tag{7.6}$$
$$\hat{x}_1 = z_1, \qquad \hat{x}_2 = z_2/\varepsilon, \tag{7.7}$$
$$\dot{z}_i = -(\beta_i/\varepsilon)(z_i + \hat{x}_{i-1}) \quad \text{for } 3 \le i \le \rho, \tag{7.8}$$
$$\hat{x}_i = (\beta_i/\varepsilon)(z_i + \hat{x}_{i-1}) \quad \text{for } 3 \le i \le \rho, \tag{7.9}$$

where β_1 to β_ρ are positive constants. A block diagram representation of the cascade connection, in terms of the transfer functions of its stages, is shown in Figure 7.1. The highest gain in this observer is $O(1/\varepsilon)$, but the estimate of the highest derivative could

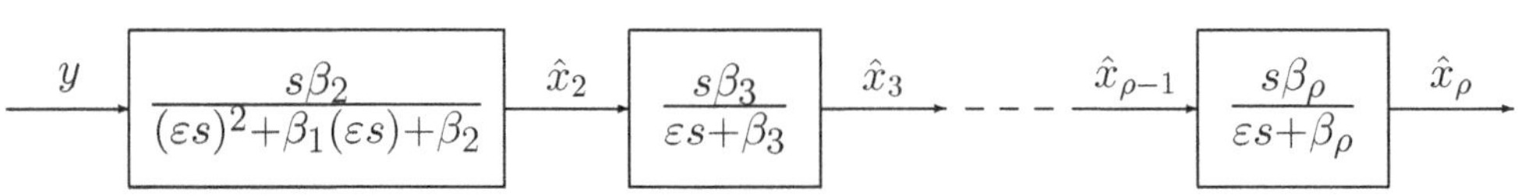

Figure 7.1. *Cascade realization of high-gain observer.*

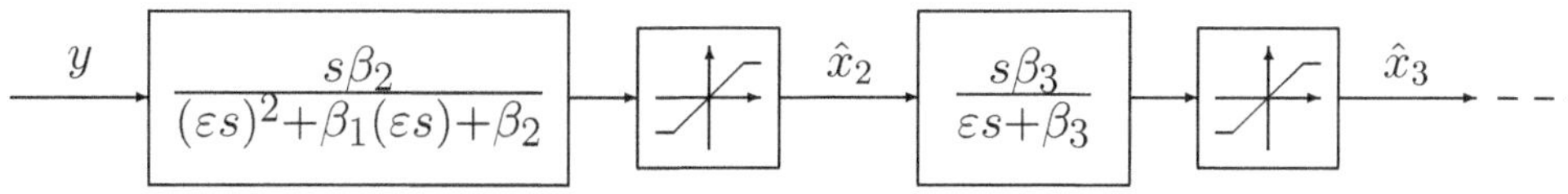

Figure 7.2. *Cascade realization of high-gain observer with saturation.*

be $O(1/\varepsilon^{\rho-1})$ during the peaking period. This peaking can be reduced by saturating the estimates. Suppose $x(t)$ belongs to a compact set X and take $M_i > \max_{x\in X}|x_i|$ for $2 \le i \le \rho$. The estimate of x_i is saturated at $\pm M_i$. The cascade observer with saturation is given by the following equations, and its block diagram representation is shown in Figure 7.2:

$$\dot{z}_1 = (1/\varepsilon)[z_2 + \beta_1(y - z_1)], \tag{7.10}$$
$$\dot{z}_2 = (\beta_2/\varepsilon)(y - z_1), \tag{7.11}$$
$$\hat{x}_1 = z_1, \qquad \hat{x}_2 = M_2 \operatorname{sat}\left(\frac{z_2}{\varepsilon M_2}\right), \tag{7.12}$$
$$\dot{z}_i = -(\beta_i/\varepsilon)(z_i + \hat{x}_{i-1}), \tag{7.13}$$
$$\hat{x}_i = M_i \operatorname{sat}\left(\frac{\beta_i(z_i + \hat{x}_{i-1})}{\varepsilon M_i}\right), \tag{7.14}$$

where $\operatorname{sat}(y) = \min\{|y|, 1\}\operatorname{sign}(y)$ and (7.13) and (7.14) hold for $3 \le i \le \rho$. In the first stage, the signal z_2/ε could be $O(1/\varepsilon)$, but the estimate $\hat{x}_2$ is bounded uniformly in ε. For all the other stages, the deriving input $\hat{x}_{i-1}$ is bounded uniformly in ε, and so is z_i. Therefore, $(\beta_i/\varepsilon)(z_i + \hat{x}_{i-1})$ is limited to be $O(1/\varepsilon)$.

Theorem 7.1. *Under the stated assumptions, there is $\varepsilon^* > 0$ such that for $0 < \varepsilon \le \varepsilon^*$, the estimation errors $\tilde{x}_i = x_i - \hat{x}_i$, for $1 \le i \le \rho$, of the cascade high-gain observer* (7.10)–(7.14) *are bounded uniformly in ε and there is time $T(\varepsilon)$, with $\lim_{\varepsilon\to 0} T(\varepsilon) = 0$, such that*

$$|\tilde{x}_1(t)| \le a\varepsilon^2, \tag{7.15}$$
$$|\tilde{x}_i(t)| \le a\varepsilon \quad \textit{for } 2 \le i \le \rho \tag{7.16}$$

for all $t \ge T(\varepsilon)$, for some positive constant a. ◇

Proof: Since y is bounded by assumption and each stage of the observer is an exponentially stable system, the observer variables z_1 to z_ρ are bounded. Let

$$\xi_1 = z_1, \qquad \xi_2 = z_2/\varepsilon, \qquad \xi_i = (\beta_i/\varepsilon)(z_i + \xi_{i-1}) \quad \text{for } 3 \le i \le \rho \tag{7.17}$$

and

$$\eta_i = \varepsilon^{i-1}(x_i - \xi_i) \qquad \text{for } 1 \le i \le \rho. \tag{7.18}$$

When $\hat{x}_2, \dots, \hat{x}_i$ are not saturated, they are equal to $\xi_2, \dots, \xi_i$, respectively. In this case, $\eta_2, \dots, \eta_i$ are scaled estimation errors and $\dot{z}_k = -\xi_k$ for $3 \le k \le \rho$. Our analysis of the estimation error will proceed stage by stage starting from the second-order observer. In each step it is shown that after a short time period the estimation error will be small and the estimate will not be saturated. For the first stage, it can be shown that $\psi_2 = \operatorname{col}(\eta_1, \eta_2)$ satisfies the equation

$$\varepsilon\dot{\psi}_2 = F_2\psi_2 + \varepsilon^2 G_{23} x_3, \tag{7.19}$$

where

$$F_2 = \begin{bmatrix} -\beta_1 & 1 \\ -\beta_2 & 0 \end{bmatrix}, \quad G_{23} = \begin{bmatrix} 0 \\ 1 \end{bmatrix}.$$

The matrix F_2 is Hurwitz. Since $x_3(t)$ is bounded uniformly in ε, it can be shown that there is $T_2(\varepsilon) > 0$, with $\lim_{\varepsilon \to 0} T_2(\varepsilon) = 0$, such that $\|\psi_2(t)\| \le k_1 \varepsilon^2$ for all $t \ge T_2$, for some constant k_1.[69] In view of (7.18), we have

$$|x_1 - \xi_1| \le k_1 \varepsilon^2 \quad \text{and} \quad |x_2 - \xi_2| \le k_1 \varepsilon \qquad \forall\, t \ge T_2.$$

We conclude that, for sufficiently small ε, $|\xi_2| \le M_2$. Therefore, $\hat{x}_i = \xi_i$ for $i = 1, 2$, which proves (7.15) and the case $i = 2$ of (7.16).

We proceed to the next stage where, for $t \ge T_2$, η_3 satisfies the equation

$$\begin{aligned} \varepsilon \dot{\eta}_3 = \varepsilon^3(\dot{x}_3 - \dot{\xi}_3) &= \varepsilon^3 \left[\dot{x}_3 - (\beta_3/\varepsilon)\dot{z}_3 - (\beta_3/\varepsilon^2)\dot{z}_2 \right] \\ &= \varepsilon^3 \left[x_4 + (\beta_3/\varepsilon)\xi_3 - (\beta_2\beta_3/\varepsilon^3)(x_1 - z_1) \right] \\ &= -\beta_2\beta_3\eta_1 - \beta_3\eta_3 + \varepsilon^2\beta_3 x_3 + \varepsilon^3 x_4. \end{aligned}$$

This equation can be used to show an ultimate bound on $|\eta_3|$ of the form $k_2\varepsilon^2$. This bound and (7.18) show that $|x_3 - \xi_3| \le k_2$, which does not yield (7.16). We need a sharper bound on $|\eta_3|$ that takes the form $k_3\varepsilon^3$. Toward that end we note that η_3 is the output of the system

$$\varepsilon \dot{\psi}_3 = F_3\psi_3 + \varepsilon^2 G_{33} x_3 + \varepsilon^3 G_{34} x_4, \qquad \eta_3 = H_3 \psi_3,$$

where

$$\psi_3 = \begin{bmatrix} \eta_1 \\ \eta_2 \\ \eta_3 \end{bmatrix}, \quad F_3 = \begin{bmatrix} -\beta_1 & 1 & 0 \\ -\beta_2 & 0 & 0 \\ -\beta_2\beta_3 & 0 & -\beta_3 \end{bmatrix}, \quad G_{33} = \begin{bmatrix} 0 \\ 1 \\ \beta_3 \end{bmatrix}, \quad G_{34} = \begin{bmatrix} 0 \\ 0 \\ 1 \end{bmatrix},$$

and $H_3 = \begin{bmatrix} 0 & 0 & 1 \end{bmatrix}$. The matrix F_3 is Hurwitz. To eliminate the $\varepsilon^2 x_3$ term on the right-hand side of the $\dot{\psi}_3$-equation, we apply the transformation

$$\tilde{\psi}_3 = \psi_3 + \varepsilon^2 F_3^{-1} G_{33} x_3,$$

which yields

$$\varepsilon \dot{\tilde{\psi}}_3 = F_3 \tilde{\psi}_3 + \varepsilon^3 (F_3^{-1} G_{33} + G_{34}) x_4,$$

and η_3 is given by

$$\eta_3 = H_3 \tilde{\psi}_3 - \varepsilon^2 H_3 F_3^{-1} G_{33} x_3.$$

The second term of the foregoing expression vanishes because

$$H_3 F_3^{-1} G_{33} = \begin{bmatrix} 0 & 0 & 1 \end{bmatrix} \begin{bmatrix} 0 & -1/\beta_2 & 0 \\ 1 & -\beta_1/\beta_2 & 0 \\ 0 & 1 & -1/\beta_3 \end{bmatrix} \begin{bmatrix} 0 \\ 1 \\ \beta_3 \end{bmatrix} = 0.$$

Thus, η_3 is the output of the system

$$\varepsilon \dot{\tilde{\psi}}_3 = F_3 \tilde{\psi}_3 + \varepsilon^3 (F_3^{-1} G_{33} + G_{34}) x_4, \qquad \eta_3 = H_3 \tilde{\psi}_3,$$

[69]Throughout the proof, k_i denotes a positive constant independent of ε.

which shows that there exists $T_3(\varepsilon) > T_2(\varepsilon)$, with $\lim_{\varepsilon\to 0} T_3(\varepsilon) = 0$, such that $|\eta_3(t)| \le k_3\varepsilon^3$ for all $t \ge T_3$, for some constant k_3. In view of (7.18), we have $|x_3 - \xi_3| \le k_3\varepsilon$ for all $t \ge T_3$. We conclude that, for sufficiently small ε, $|\xi_3| \le M_3$. Therefore, $\hat{x}_3 = \xi_3$, which proves the case $i = 3$ of (7.16).

Proceeding to the next stage, we note that for $t \ge T_3$, η_4 satisfies the equation

$$\begin{aligned}
\varepsilon\dot{\eta}_4 = \varepsilon^4(\dot{x}_4 - \dot{\xi}_4) &= \varepsilon^4\left[\dot{x}_4 - (\beta_4/\varepsilon)\dot{z}_4 - (\beta_4/\varepsilon)\dot{\xi}_3\right] \\
&= \varepsilon^3\beta_4\xi_4 - \varepsilon^3\beta_4\dot{\xi}_3 + \varepsilon^4 x_5 \\
&= -\beta_4(\eta_4 - \varepsilon^3 x_4) + \beta_4(\varepsilon\dot{\eta}_3 - \varepsilon^3\dot{x}_3) + \varepsilon^4 x_5 \\
&= \beta_4\varepsilon\dot{\eta}_3 - \beta_4\eta_4 + \varepsilon^4 x_5 \\
&= \beta_4 H_3(F_3\psi_3 + \varepsilon^2 G_{33}x_3 + \varepsilon^3 G_{34}x_4) - \beta_4\eta_4 + \varepsilon^4 x_5 \\
&= -\beta_2\beta_3\beta_4\eta_1 - \beta_3\beta_4\eta_3 - \beta_4\eta_4 + \varepsilon^2\beta_3\beta_4 x_3 + \varepsilon^3\beta_4 x_4 + \varepsilon^4 x_5.
\end{aligned}$$

Thus, η_4 is the output of the system

$$\varepsilon\dot{\psi}_4 = F_4\psi_4 + \varepsilon^2 G_{43}x_3 + \varepsilon^3 G_{44}x_4 + \varepsilon^4 G_{45}x_5, \qquad \eta_4 = H_4\psi_4,$$

where

$$\psi_4 = \begin{bmatrix} \eta_1 \\ \eta_2 \\ \eta_3 \\ \eta_4 \end{bmatrix}, \quad F_4 = \begin{bmatrix} -\beta_1 & 1 & 0 & 0 \\ -\beta_2 & 0 & 0 & 0 \\ -\beta_2\beta_3 & 0 & -\beta_3 & 0 \\ -\beta_2\beta_3\beta_4 & 0 & -\beta_3\beta_4 & -\beta_4 \end{bmatrix},$$

$$G_{43} = \begin{bmatrix} 0 \\ 1 \\ \beta_3 \\ \beta_3\beta_4 \end{bmatrix}, \quad G_{44} = \begin{bmatrix} 0 \\ 0 \\ 1 \\ \beta_4 \end{bmatrix}, \quad G_{45} = \begin{bmatrix} 0 \\ 0 \\ 0 \\ 1 \end{bmatrix},$$

and $H_4 = \begin{bmatrix} 0 & 0 & 0 & 1 \end{bmatrix}$. The matrix F_4 is Hurwitz. We pay attention to the $\varepsilon^2 x_3$ and $\varepsilon^3 x_4$ terms on the right-hand side of the $\dot{\psi}_4$-equation because the power of ε in these two terms is less than 4. We apply two transformations to eliminate these terms. The first transformation eliminates the ε^2 term, and the second one eliminates the ε^3 term. Then we combine the two transformations into one. The transformation

$$\hat{\psi}_4 = \psi_4 + \varepsilon^2 F_4^{-1} G_{43} x_3$$

results in

$$\varepsilon\dot{\hat{\psi}}_4 = F_4\hat{\psi}_4 + \varepsilon^3(F_4^{-1}G_{43} + G_{44})x_4 + \varepsilon^4 G_{45}x_5.$$

Then the transformation

$$\tilde{\psi}_4 = \hat{\psi}_4 + \varepsilon^3 F_4^{-1}(F_4^{-1}G_{43} + G_{44})x_4$$

yields

$$\varepsilon\dot{\tilde{\psi}}_4 = F_4\tilde{\psi}_4 + \varepsilon^4(F_4^{-2}G_{43} + F_4^{-1}G_{44} + G_{45})x_5.$$

The overall transformation is

$$\tilde{\psi}_4 = \psi_4 + \varepsilon^2 F_4^{-1} G_{43} x_3 + \varepsilon^3 F_4^{-1}(F_4^{-1}G_{43} + G_{44})x_4$$

and

$$\eta_4 = H_4\tilde{\psi}_4 - \varepsilon^2 H_4F_4^{-1}G_{43}x_3 - \varepsilon^3 H_4F_4^{-2}G_{43}x_4 - \varepsilon^3 H_4F_4^{-1}G_{44}x_4.$$

It can be verified that

$$H_4F_4^{-1}G_{43} = 0, \qquad H_4F_4^{-2}G_{43} = 0, \qquad \text{and} \qquad H_4F_4^{-1}G_{44} = 0. \tag{7.20}$$

Thus, η_4 is the output of the system

$$\varepsilon\dot{\tilde{\psi}}_4 = F_4\tilde{\psi}_4 + \varepsilon^4(F_4^{-2}G_{43} + F_4^{-1}G_{44} + G_{45})x_5, \qquad \eta_4 = H_4\tilde{\psi}_4,$$

and there exists $T_4(\varepsilon) > T_3(\varepsilon)$, with $\lim_{\varepsilon\to 0} T_4(\varepsilon) = 0$, such that $|\eta_4(t)| \le k_4\varepsilon^4$ for all $t \ge T_4$, for some constant k_4. In view of (7.18), we have $|x_4 - \xi_4| \le k_4\varepsilon$ for all $t \ge T_4$. We conclude that, for sufficiently small ε, $|\xi_4| \le M_4$. Therefore, $\hat{x}_4 = \xi_4$, which proves the case $i = 4$ of (7.16). Proceeding in the same way to the next stages, consider the stage $i > 4$, where, for $t \ge T_{i-1}(\varepsilon)$, we have

$$\begin{aligned}
\varepsilon\dot{\eta}_i &= \varepsilon^i(\dot{x}_i - \dot{\xi}_i) = \varepsilon^i\left[\dot{x}_i - (\beta_i/\varepsilon)\dot{z}_i - (\beta_i/\varepsilon)\dot{\xi}_{i-1}\right]\\
&= \varepsilon^{i-1}\beta_i\xi_i - \varepsilon^{i-1}\beta_i\dot{\xi}_{i-1} + \varepsilon^i x_{i+1}\\
&= -\beta_i(\eta_i - \varepsilon^{i-1}x_i) + \beta_i(\varepsilon\dot{\eta}_{i-1} - \varepsilon^{i-1}\dot{x}_{i-1}) + \varepsilon^i x_{i+1}\\
&= \beta_i\varepsilon\dot{\eta}_{i-1} - \beta_i\eta_i + \varepsilon^i x_{i+1}.
\end{aligned}$$

Thus, η_i is the output of the system

$$\varepsilon\dot{\psi}_i = F_i\psi_i + \sum_{k=3}^{i} G_{ik}\varepsilon^{k-1}x_k + G_{i,i+1}\varepsilon^i x_{i+1}, \qquad \eta_i = H_i\psi_i, \tag{7.21}$$

where

$$\psi_i = \begin{bmatrix}\psi_{i-1}\\ \eta_i\end{bmatrix}, \quad F_i = \begin{bmatrix}F_{i-1} & 0\\ \beta_i H_{i-1}F_{i-1} & -\beta_i\end{bmatrix}, \quad G_{ik} = \begin{bmatrix}G_{i-1,k}\\ \beta_i H_{i-1}G_{i-1,k}\end{bmatrix},$$

$$H_i = \begin{bmatrix}0 & \dots & 0 & 1\end{bmatrix}_{1\times i}, \quad G_{i,i+1} = H_i^T.$$

When $i = \rho$, the x_{i+1} term in the $\dot{\psi}_\rho$-equation is replaced by ϕ. From the block triangular structure of F_i and the fact that F_4 is Hurwitz, it is clear that the matrices $F_5, F_6, \dots, F_\rho$ are all Hurwitz.

Claim. *For* $3 \le i \le \rho$,

$$H_iF_i^{-j}G_{ik} = 0 \quad \text{for } 1 \le j \le i-k+1,\ 3 \le k \le i. \tag{7.22}$$

Proof of the claim: We have already shown (7.22) for $i = 3$. We prove it for $i = 4, 5, 6, \dots, \rho$ by mathematical induction. Suppose it is true for $i-1$ and show it for i:

$$F_i^{-1}G_{ik} = \begin{bmatrix}F_{i-1}^{-1} & 0\\ H_{i-1} & -1/\beta_i\end{bmatrix}\begin{bmatrix}G_{i-1,k}\\ \beta_i H_{i-1}G_{i-1,k}\end{bmatrix} = \begin{bmatrix}F_{i-1}^{-1}G_{i-1,k}\\ 0\end{bmatrix},$$

$$F_i^{-2}G_{ik} = \begin{bmatrix}F_{i-1}^{-1} & 0\\ H_{i-1} & -1/\beta_i\end{bmatrix}\begin{bmatrix}F_{i-1}^{-1}G_{i-1,k}\\ 0\end{bmatrix} = \begin{bmatrix}F_{i-1}^{-2}G_{i-1,k}\\ 0\end{bmatrix},$$

$$\vdots \quad \vdots$$

$$F_i^{-(i-k+1)}G_{ik} = \begin{bmatrix}F_{i-1}^{-(i-k+1)}G_{i-1,k}\\ 0\end{bmatrix}.$$

Premultiplication by H_i completes the proof. □

Applying the transformation

$$\tilde{\psi}_i = \psi_i + \sum_{k=3}^{i} \sum_{j=1}^{i-k+1} F_i^{-j} G_{ik} \varepsilon^{k+j-2} x_{k+j-1}$$

to (7.21) and using (7.22) shows, after some lengthy but straightforward manipulation, that η_i is the output of the system

$$\varepsilon \dot{\tilde{\psi}}_i = F_i \tilde{\psi}_i + \varepsilon^i \sum_{k=3}^{i+1} F_i^{-(i-k+1)} G_{ik} x_{i+1}, \qquad \eta_i = H_i \tilde{\psi}_i. \tag{7.23}$$

Therefore, there exists $T_i(\varepsilon)$, with $\lim_{\varepsilon \to 0} T_i(\varepsilon) = 0$, such that $|\eta_i(t)| \le k_i \varepsilon^i$ for all $t \ge T_i$, for some constant k_i. In view of (7.18), we have $|x_i - \xi_i| \le k_i \varepsilon$ for all $t \ge T_i$, which completes the proof of the theorem. □

While the cascade high-gain observer (7.10)–(7.14) has a numerical advantage over the standard high-gain observer (2.17)–(2.18) when ρ is high, it has some drawbacks as well. First, if the standard high-gain observer is implemented with $\phi_0 = \phi$, the estimation error will converge to zero as time tends to infinity. This can be seen from (2.24) because in this case $M = 0$. Such convergence will not take place in the cascade observer, which does not use the function ϕ. Second, the ultimate bound on the estimation error in the case of the standard high-gain observer is seen from (2.24) to be $|\tilde{x}_i| \le \varepsilon^{\rho+1-i} cM$. When contrasted with the ultimate bound in the cascade high-gain observer, given by (7.15) and (7.16), we see that, except for the bound on $|x_\rho - \hat{x}_\rho|$, the ultimate bound is orders of magnitude smaller in the case of the standard observer. This point is illustrated further by simulation in Example 7.1. Table 7.1 compares the features of the standard and cascade observers.

Example 7.1. We use the standard and cascade high-gain observers to estimate the states of the system

$$\dot{x}_1 = x_2, \quad \dot{x}_2 = x_3, \quad \dot{x}_3 = x_4, \quad \dot{x}_4 = \phi(x), \quad y = x_1,$$

where

$$\begin{aligned}\phi(x) = &-(x_1 + 2x_2 + 2x_3 + 2x_4) \\ &+ [1 - (x_1 + 2x_2 + x_3)^2](x_2 + 2x_3 + x_4).\end{aligned}$$

Table 7.1. *Comparison of the features of the standard and cascade observers.*

Feature	Standard	Cascade
Dimension	ρ	ρ
Observer gain	$O(1/\varepsilon^\rho)$	$O(1/\varepsilon)$
Peaking signal	$O(1/\varepsilon^{\rho-1})$	$O(1/\varepsilon)$
Ultimate bound on $\|\tilde{x}_1\|$	$O(\varepsilon^\rho)$	$O(\varepsilon^2)$
Ultimate bound on $\|\tilde{x}_i\|$ for $2 \le i \le \rho$	$O(\varepsilon^{\rho+1-i})$	$O(\varepsilon)$
Can the observer be designed so that the error converges to zero?	Yes	No

The system is constructed by cascading a van der Pol equation with a second-order exponentially stable linear system, then transforming the whole system into the normal form. Therefore, the solution $x(t)$ is periodic. It can be seen by simulation that for all initial states in the set $\{|x_i| \leq 1, i = 1,2,3,4\}$, the solution satisfies $|x_2(t)| \leq 1.7$, $|x_3(t)| \leq 1.2$, and $|x_4(t)| \leq 2.7$. Therefore, we choose the saturation levels for the estimates $\hat{x}_2$, $\hat{x}_3$, and $\hat{x}_4$ as ± 2, ± 1.5, and ± 3, respectively. For the two observers, we assign all eigenvalues at $-1/\varepsilon$. The standard observer is designed as a linear one by taking the function ϕ_0 of (2.18) to be zero. It is given by

$$\begin{aligned}
\dot{z}_1 &= z_2 + (4/\varepsilon)(y - z_1), & \hat{x}_1 &= z_1, \\
\dot{z}_2 &= z_3 + (6/\varepsilon^2)(y - z_1), & \hat{x}_2 &= z_2, \\
\dot{z}_3 &= z_4 + (4/\varepsilon^3)(y - z_1), & \hat{x}_3 &= z_3, \\
\dot{z}_4 &= (1/\varepsilon^4)(y - z_1), & \hat{x}_4 &= z_4.
\end{aligned}$$

The cascade observer is given by

$$\begin{aligned}
\dot{z}_1 &= (1/\varepsilon)[z_2 + 2(y - z_1)], & \hat{x}_1 &= z_1, \\
\dot{z}_2 &= (1/\varepsilon)(y - z_1), & \hat{x}_2 &= 2\,\mathrm{sat}(z_2/(2\varepsilon)), \\
\dot{z}_3 &= -(1/\varepsilon)(z_3 + \hat{x}_2), & \hat{x}_3 &= 1.5\,\mathrm{sat}((z_3 + \hat{x}_2)/(1.5\varepsilon)), \\
\dot{z}_4 &= -(1/\varepsilon)(z_4 + \hat{x}_3), & \hat{x}_4 &= 3\,\mathrm{sat}((z_4 + \hat{x}_3)/(3\varepsilon)).
\end{aligned}$$

Simulation results are shown in Figures 7.3 to 7.5 for $\varepsilon = 0.001$, $x_i(0) = 1$ for $i = 1,2,3,4$, and zero initial conditions for the observers. Figure 7.3 shows the states x_3

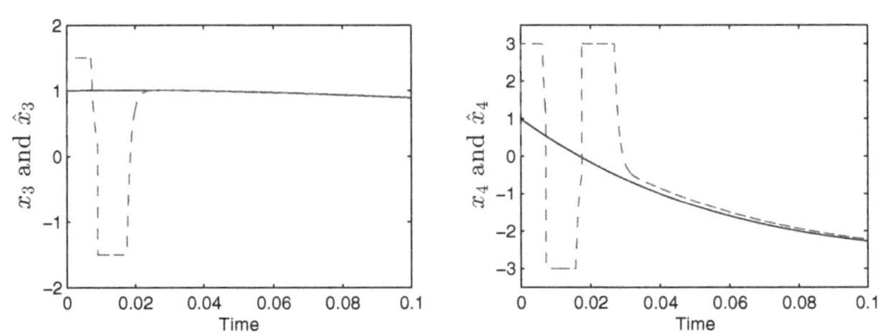

Figure 7.3. *Simulation of Example 7.1. The states x_3 and x_4 (solid) and their estimates $\hat{x}_3$ and $\hat{x}_4$ (dashed) for the cascade observer.*

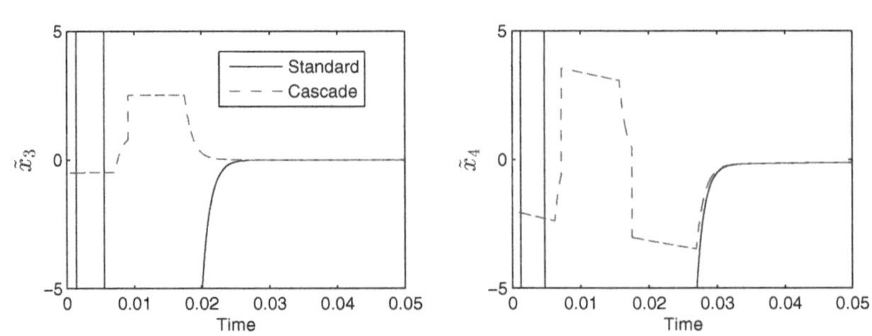

Figure 7.4. *Simulation of Example 7.1. The estimation errors $\tilde{x}_3$ and $\tilde{x}_4$ for the standard observer (solid) and the cascade observer (dashed).*

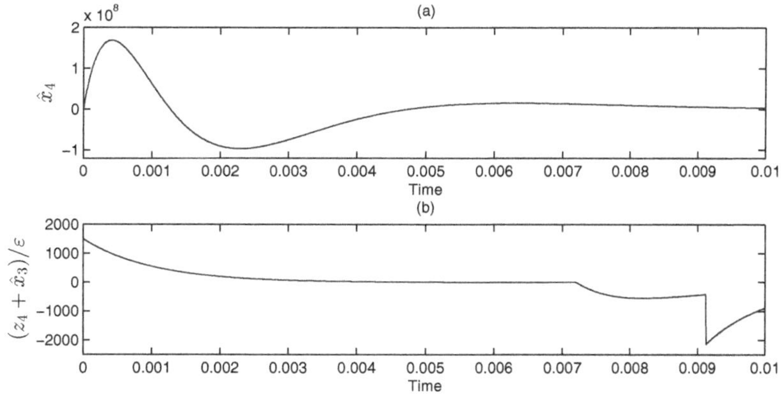

Figure 7.5. *Simulation of Example 7.1. The most peaking signal when $\varepsilon = 0.001$: (a) the standard observer; (b) the cascade observer.*

Table 7.2. *Simulation of Example 7.1. The steady state estimation errors for the standard and cascade observers when $\varepsilon = 0.001$. The maximum is taken over the time interval $[25, 30]$.*

Observer	$\max\|\tilde{x}_1\|$	$\max\|\tilde{x}_2\|$	$\max\|\tilde{x}_3\|$	$\max\|\tilde{x}_4\|$
Standard	4.0709×10^{-12}	1.6284×10^{-8}	2.4425×10^{-5}	0.0163
Cascade	1.0042×10^{-6}	0.002	0.0048	0.0163

and x_4 and their estimates provided by the cascade observer. It illustrates the saturation of the estimates during the peaking period and the closeness of the estimates to the states beyond the peaking period. Figure 7.4 compares the estimation errors $\tilde{x}_3$ and $\tilde{x}_4$ for the standard and cascade observers. It illustrates that the cascade observer has convergence properties comparable to the standard one. Note that by choosing the y-axis to be between ± 5, the peaking behavior of the standard observer is truncated. Figure 7.5 shows the largest peaking signals in the two observers. For the standard observer $\hat{x}_4$ peaks to $O(1/\varepsilon^3)$ value while for the cascade observer $z_4 + \hat{x}_3/\varepsilon$ peaks to $O(1/\varepsilon)$ value. Finally, Table 7.2 compares the steady-state estimation errors of the two observers. It illustrates the fact that $\tilde{x}_1$, $\tilde{x}_2$, $\tilde{x}_3$, and $\tilde{x}_4$ are $O(\varepsilon^4)$, $O(\varepsilon^3)$, $O(\varepsilon^2)$, and $O(\varepsilon)$, respectively, in the standard observer, compared with $O(\varepsilon^2)$, $O(\varepsilon)$, $O(\varepsilon)$, and $O(\varepsilon)$ in the cascade observer. △

7.2 ▪ Stabilization

Consider the system

$$\dot{w} = f_0(w, x, u), \tag{7.24}$$
$$\dot{x}_i = x_{i+1} \quad \text{for } 1 \le i \le \rho - 1, \tag{7.25}$$
$$\dot{x}_\rho = \phi(w, x, u), \tag{7.26}$$
$$y = x_1, \tag{7.27}$$
$$z = f_a(w, x), \tag{7.28}$$

where $u \in R^m$ is the control input, $y \in R$ and $z \in R^s$ are the measured outputs, and $w \in R^\ell$ and $x \in R^\rho$ constitute the state vector. The functions f_0, f_a, and ϕ are locally

Lipschitz in their arguments for $(w,x,u) \in D_w \times D_x \times R^m$, where $D_w \subset R^\ell$ and $D_x \subset R^\rho$ are domains that contain their respective origins. Moreover, $f_0(0,0,0)=0$, $f_a(0,0)=0$, and $\phi(0,0,0)=0$. Our goal is to design an output feedback controller to stabilize the origin.

We use a two-step approach. First, a partial state feedback controller that uses measurements of x and z is designed to asymptotically stabilize the origin. Then a cascade high-gain observer is used to estimate x from y. The state feedback controller takes the form[70]

$$u=\gamma(x,z), \qquad \gamma(0,0)=0, \tag{7.29}$$

where γ is locally Lipschitz. For convenience, we write the closed-loop system under state feedback as

$$\dot{\chi}=f(\chi), \tag{7.30}$$

where $\chi=\text{col}(w,x)$. By design, the origin of (7.30) is asymptotically stable. Let $V(\chi)$ be a continuously differentiable, positive definite Lyapunov function that satisfies the inequality

$$\frac{\partial V}{\partial \chi} f(\chi) \le -W(\chi) \tag{7.31}$$

for all $\chi \in D_w \times D_x$, where $W(\chi)$ is a continuous positive definite function. Choose a positive constant c such that $\Omega_c = \{V(\chi) \le c\}$ is a compact subset of $D_w \times D_x$. Then Ω_c is a positively invariant set of (7.30); that is, $\chi(0) \in \Omega_c$ implies that $\chi(t) \in \Omega_c$ for all $t \ge 0$. The existence of a Lyapunov function satisfying (7.31) follows from the converse Lyapunov theorem [78, Theorem 4.17], but here we need to know $V(\chi)$. In output feedback control we saturate the state estimates and the feedback control outside Ω_c. Toward that end, let $M_i > \max_{x \in \Omega_c} |x_i|$ for $i=1,\dots,\rho$. The cascade observer (7.5)–(7.9) is implemented with $\hat{x}_2$ to $\hat{x}_\rho$ saturated at $\pm M_2$ to $\pm M_\rho$, respectively. The output feedback control is taken as

$$u=\gamma(\hat{x},z). \tag{7.32}$$

Theorem 7.2. *Consider the closed-loop system of the plant (7.24)–(7.28), the observer (7.10)–(7.14), and the output feedback controller (7.32). Let $\Omega_a = \{V(x) \le a\}$, with $a<c$, and let $\mathcal{Q}$ be any compact subset of R^ρ. Then, given any $\mu>0$, there exist $\varepsilon^*>0$ and $T^*>0$, dependent on μ, such that for every $0<\varepsilon\le\varepsilon^*$, the solutions $(\chi(t),\hat{x}(t))$ of the closed-loop system, starting in $\Omega_a \times \mathcal{Q}$, are bounded for all $t\ge 0$ and satisfy*

$$\|\chi(t)\| \le \mu \quad and \quad \|\hat{x}(t)\| \le \mu \quad \forall\, t \ge T^*, \tag{7.33}$$

$$\|\chi(t)-\chi_r(t)\| \le \mu \quad \forall\, t \ge 0, \tag{7.34}$$

where χ_r is the solution of (7.30) with $\chi_r(0)=\chi(0)$. Moreover, if the origin of (7.30) is exponentially stable, then the origin of the closed-loop system is exponentially stable and $\Omega_a \times \mathcal{Q}$ is a subset of its region of attraction. ◇

Proof: Since $\hat{x}$ is bounded uniformly in ε, $\gamma(\hat{x},z)$ is bounded for all $\chi \in \Omega_c$. Because $\chi(0)$ belongs to Ω_a, where $a<c$, there exist positive constants T_0 and $b \in (a,c)$, both independent of ε, such that $\chi(t) \in \Omega_b$ for all $t \in [0,T_0]$. During this time interval, the conditions of Theorem 7.1 hold, and we conclude that the estimation errors satisfy

[70] In the separation result of Chapter 3, the state feedback control is allowed to be a dynamic controller. This extension can be allowed here as well, but it is omitted for conciseness of the presentation.

(7.15) and (7.16) for $t \geq T(\varepsilon)$, where $\lim_{\varepsilon\to 0} T(\varepsilon) = 0$. Thus, there is $\varepsilon_1^* > 0$ such that $T(\varepsilon) < T_0$ for all $\varepsilon \in (0, \varepsilon_1^*]$. The upper bounds on the estimation error imply that there is $\varepsilon_2^* > 0$ such that for all $\varepsilon \in (0, \varepsilon_2^*]$, $\hat{x}(t) \in \Omega_c$, and since $\hat{x}(t)$ is saturated outside Ω_c, we see that for $t \in [T(\varepsilon), T_0]$, $\hat{x}(t)$ is not saturated. Then the $\dot{\chi}$-equation under output feedback can be written as

$$\dot{\chi} = f_c(\chi, x - \hat{x}), \tag{7.35}$$

where $f_c(\chi, 0) = f(\chi)$. The derivative of $V(\chi)$ along the trajectories of (7.35) is given by

$$\begin{aligned}\dot{V} &= \frac{\partial V}{\partial \chi} f(\chi) + \frac{\partial V}{\partial \chi}[f_c(\chi, x - \hat{x}) - f_c(\chi, 0)] \\ &\leq -W(\chi) + k_a\|x - \hat{x}\| \leq -W(\chi) + k_b\varepsilon \end{aligned} \tag{7.36}$$

for some positive constants k_a and k_b. Thus, there is ε_3^* such that for all $\varepsilon \in (0, \varepsilon_3^*]$, $\dot{V} < 0$ on the boundary of Ω_b. Hence, $\chi(t)$ never leaves Ω_b, and the estimation errors satisfy (7.15) and (7.16) for all $t \geq 0$. Inequality (7.36) shows also that $\|\chi\|$ is ultimately bounded by a class $\mathcal{K}$ function of ε, and from (7.15) and (7.16) we see that $\|\hat{x}\|$ is also ultimately bounded by a class $\mathcal{K}$ function of ε. Therefore, given any $\mu > 0$ we can find $\varepsilon_4^* > 0$ and $T^* > 0$ such that (7.33) is satisfied for $\varepsilon \in (0, \varepsilon_4^*]$.

To prove (7.34) we note that, since $\lim_{t\to\infty} \chi_r(t) = 0$, given any $\mu > 0$, there is $T_1 > 0$ such that $\|\chi_r(t)\| \leq \mu/2$ for all $t \geq T_1$. On the other hand, in view of (7.33), there is $T_2 > 0$ such that $\|\chi(t)\| \leq \mu/2$ for all $t \geq T_2$. Taking $T_3 = \max\{T_1, T_2\}$, we see that $\|\chi(t) - \chi_r(t)\| \leq \mu$ for all $t \geq T_3$. So we are left with the task of showing (7.34) on the compact interval $[0, T_3]$. We divide $[0, T_3]$ into $[0, T(\varepsilon)]$ and $[T(\varepsilon), T_3]$ and show (7.34) on each interval. Consider first $[0, T(\varepsilon)]$. Since $\gamma(\hat{x}, z)$ is bounded uniformly in ε, there is a positive constant k_c such that $\|\chi(t) - \chi(0)\| \leq k_c t$. Similarly, there is a positive constant k_d such that $\|\chi_r(t) - \chi_r(0)\| \leq k_d t$. Because $\chi(0) = \chi_r(0)$, $\|\chi(t) - \chi_r(t)\| \leq 2k_e T(\varepsilon)$ for all $t \in [0, T(\varepsilon)]$, where $k_e = \max\{k_c, k_d\}$. Since $T(\varepsilon) \to 0$ as $\varepsilon \to 0$, there exists $\varepsilon_5^* > 0$ such that (7.34) holds on $[0, T(\varepsilon)]$ for every $\varepsilon \in (0, \varepsilon_5^*]$. Next, consider the interval $[T(\varepsilon), T_3]$. Inequality (7.34) follows from continuous dependence of the solutions of differential equations on initial states and right-hand-side functions [78, Theorem 3.5] because $f_c(\chi, x - \hat{x})$ is $O(\varepsilon)$ close to $f_c(\chi, 0) = f(\chi)$ and $\|\chi(T(\varepsilon)) - \chi_r(T(\varepsilon))\| \leq 2k_e T(\varepsilon)$.

To prove exponential stability, we construct a nonminimal realization of the closed-loop system under output feedback for $t \geq T(\varepsilon)$. We have seen in the proof of Theorem 7.1 that $\psi_2 = \text{col}(\eta_1, \eta_2)$ satisfies (7.19) and, for $i = 3, \ldots, \rho$, η_i is the output of the system (7.23), where for $i = \rho$, $x_{\rho+1}$ is replaced by ϕ. The change of variables

$$\varphi_2 = \frac{1}{\varepsilon}\psi_2, \qquad \varphi_i = \frac{1}{\varepsilon^{i-1}}\tilde{\psi}_i \quad \text{for } 3 \leq i \leq \rho$$

results in

$$\begin{aligned}\varepsilon\dot{\varphi}_2 &= F_2\varphi_2 + \varepsilon G_{23}x_3, \\ \varepsilon\dot{\varphi}_i &= F_i\varphi_i + \varepsilon G_i x_{i+1} \quad \text{for } 3 \leq i \leq \rho,\end{aligned}$$

where $G_i = \sum_{k=3}^{i+1} F_i^{-(i-k+1)} G_{ik}$. Using (7.18), we conclude that $x - \hat{x}$ is the output of the system

$$\varepsilon\dot{\varphi} = F\varphi + \varepsilon[Gx + E\phi(w, x, u)], \qquad x - \hat{x} = H\varphi,$$

where

$$\varphi = \text{col}(\varphi_2, \varphi_3, \ldots, \varphi_\rho), \quad F = \text{block diag}[F_2, F_3, \ldots, F_\rho],$$

$$G = \begin{bmatrix} 0 & G_{23} & 0 & 0 & \cdots & 0 \\ 0 & 0 & G_3 & 0 & \cdots & 0 \\ \vdots & & & & & \vdots \\ 0 & & \cdots & & 0 & G_{\rho-1} \\ 0 & & \cdots & & 0 & 0 \end{bmatrix}, \quad E = \begin{bmatrix} 0 \\ \vdots \\ \vdots \\ 0 \\ G_\rho \end{bmatrix},$$

$$H = \text{block diag}[H_2, H_3, \ldots, H_\rho], \quad \text{in which } H_2 = \begin{bmatrix} \varepsilon & 0 \\ 0 & 1 \end{bmatrix}.$$

The matrix F is Hurwitz because all the matrices F_2 to F_ρ are Hurwitz. This realization is nonminimal because the dimension of φ is $2+3+4+\cdots+\rho$. The closed-loop system under output feedback can be represented by

$$\dot{\chi} = f_c(\chi, H\varphi), \tag{7.37}$$

$$\varepsilon\dot{\varphi} = F\varphi + \varepsilon[Gx + E\phi(w, x, \gamma(x - H\varphi, f_a(w, x)))]. \tag{7.38}$$

Assuming that the origin of $\dot{\chi} = f(\chi)$ is exponentially stable, it follows from the converse Lyapunov theorem [78, Theorem 4.14] that there is a continuously differentiable Lyapunov function $V_1(\chi)$, which satisfies the inequalities

$$b_1\|\chi\|^2 \le V_1(\chi) \le b_2\|\chi\|^2, \quad \frac{\partial V_1}{\partial \chi} f(\chi) \le -b_3\|\chi\|^2, \quad \left\|\frac{\partial V_1}{\partial \chi}\right\| \le b_4\|\chi\|$$

over a ball $B_r \subset R^{\ell+\rho}$ for some positive constants r and b_1 to b_4. Consider the composite Lyapunov function $V_2(\chi, \varphi) = V_1(\chi) + \varphi^T P_0 \varphi$, where P_0 is the positive definite solution of the Lyapunov equation $P_0 F + F^T P_0 = -I$. Using the local Lipschitz properties of f, f_a, ϕ, and γ and the fact that ϕ, f_a, and γ vanish at the origin, it can be shown that the derivative of V_2 along the trajectories of the system (7.37)–(7.38) satisfies the inequality

$$\dot{V}_2 \le -\mathcal{Y}^T Q_0 \mathcal{Y},$$

where

$$Q_0 = \begin{bmatrix} b_3 & -b_5 \\ -b_5 & (1/\varepsilon) - b_6 \end{bmatrix}, \quad \mathcal{Y} = \begin{bmatrix} \|\chi\| \\ \|\varphi\| \end{bmatrix}$$

for some nonnegative constants b_5 and b_6. The matrix Q_0 is positive definite for sufficiently small ε. Hence, there is a neighborhood $\mathcal{N}$ of the origin ($\chi = 0, \varphi = 0$), independent of ε, and $\varepsilon_6^* > 0$ such that for every $\varepsilon \in (0, \varepsilon_6^*]$, the origin is exponentially stable and every trajectory in $\mathcal{N}$ converges to the origin as $t \to \infty$. By (7.33), there is $\varepsilon_7^* > 0$ such that for every $\varepsilon \in (0, \varepsilon_7^*]$ and every $(\chi(0), \hat{x}(0)) \in \Omega_a \times \mathcal{Q}$, the solution of (7.37)–(7.38) enters $\mathcal{N}$ in finite time. Hence, for every $\varepsilon \in (0, \min\{\varepsilon_6^*, \varepsilon_7^*\}]$, the origin of the closed-loop system is exponentially stable and $\Omega_a \times \mathcal{Q}$ is a subset of the region of attraction. □

Theorem 7.2 shows a number of properties of the output feedback controller when ε is sufficiently small. First, it shows that the trajectories are bounded. Second, (7.33) shows practical stabilization because the trajectories can be brought to an arbitrarily small neighborhood of the origin by choosing ε small enough. Third, (7.34) shows that the solution $\chi(t)$ under output feedback approaches the solution $\chi_r(t)$ under state feedback as ε tends to zero. Fourth, the output feedback controller recovers exponential stability when the origin under state feedback is exponentially stable. Finally, if the assumptions hold globally, the origin under state feedback is globally asymptotically stable and locally exponentially stable, and the function $V(x)$ is radially unbounded, then for sufficiently small ε, the output feedback controller achieves semiglobal stabilization because the set $\Omega_a = \{V(\chi) \le a\}$ can be chosen to contain any given compact subset of $R^{\ell+\rho}$.

Theorem 7.2 reads almost the same as Theorem 3.1 for the standard high-gain observer, but there is a difference between the two cases. In the case of the standard observer we require the function $\gamma(x,z)$ to be globally bounded in x, but we do not require knowledge of the Lyapunov function $V(\chi)$. Global boundedness of γ is typically achieved by saturation outside a compact set, but we do not have to explicitly describe such set. Once we know that the globally bounded γ stabilizes the origin, we know that there is a region of attraction $\mathscr{R}$, and the initial conditions can belong to any compact subset of $\mathscr{R}$. The Lyapunov function $V(\chi)$ is established in the proof from a converse Lyapunov theorem. In the current case, the situation is different because we have to saturate the estimates $\hat{x}_2$ to $\hat{x}_\rho$ in the cascade observer, so we need to explicitly specify the compact set over which saturation is carried out. This is achieved here by explicitly requiring knowledge of the Lyapunov function $V(\chi)$, and saturation is performed outside the compact set $\Omega_c = \{V(\chi) \le c\}$.

Example 7.2. Consider the system

$$\dot{x}_1 = x_2, \quad \dot{x}_2 = x_3, \quad \dot{x}_3 = x_4, \quad \dot{x}_4 = \phi(x) + u, \quad y = x_1,$$

where $\phi(x)$ is the same as in Example 7.1. The goal is to stabilize the origin without the knowledge of ϕ. The state feedback is designed using Lyapunov redesign [80, Section 10.2]. Rewrite the state equation as $\dot{x} = Ax + B[\phi(x) + u]$, where the pair (A,B) represents a chain of four integrators. The matrix $K = \begin{bmatrix} 1 & 4 & 6 & 4 \end{bmatrix}$ assigns all the eigenvalues of $A - BK$ at -1. By adding and subtracting Kx, we rewrite the state equation as

$$\dot{x} = (A - BK)x + B[u + \delta(x)],$$

where

$$\delta(x) = \phi(x) + Kx = [3 - (x_1 + 2x_2 + x_3)^2](x_2 + 2x_3 + x_4).$$

Let $P = P^T > 0$ be the solution of the Lyapunov equation

$$P(A - BK) + (A - BK)^T P = -Q,$$

with $Q = \text{diag}[0.1, 0.1, 1, 1]$; $V(x) = x^T P x$ is a Lyapunov function for the linear system $\dot{x} = (A - BK)x$. According to the Lyapunov redesign procedure, a stabilizing state feedback control can be taken as

$$u = -\beta \ \text{sat}\left(\frac{2\beta B^T P x}{\mu}\right),$$

where β is an upper bound on $|\delta(x)|$ over the compact set of interest and μ is a positive constant. Suppose all initial states of interest are in the interior of $\Omega = \{V(x) \le 1.2\}$ and calculate an upper bound $|\delta(x)|$. Using the fact that $\max_{x^T P x \le c} |b^T x| = \sqrt{c}\|b^T P^{-1/2}\|$, it can be shown that $|x_1| \le 2.7397$, $|x_2| \le 1.4881$, $|x_3| \le 1.0747$, $|x_4| \le$

2.7337, $|x_1+2x_2+x_3|^2 \le 3.6085$, $|3-(x_1+2x_2+x_3)^2| \le 3$, and $|x_2+2x_3+x_4| \le 2.5595$ over Ω. From the last two bounds, it can be shown that $|\delta(x)| \le \varrho$ and $|\delta(x)| \le \varrho_1\|x\|$ for all $x \in \Omega$, where $\varrho = 7.6785$ and $\varrho_1 = 7.35$. The derivative of $V(x)$ satisfies

$$\dot{V} = -x^T Q x + w\ \delta(x) - \beta w \ \text{sat}\left(\frac{\beta w}{\mu}\right),$$

where $w = 2x^T PB$. For $\beta|w| \ge \mu$,

$$\dot{V} \le -x^T Q x + \varrho|w| - \beta|w| \le -x^T Q x$$

when $\beta \ge \rho$. For $\beta|w| < \mu$,

$$\dot{V} \le -x^T Q x + \varrho_1\|x\|\,|w| - \frac{\beta^2|w|^2}{\mu}.$$

With $0 < \theta < 1$,

$$\begin{aligned}
\dot{V} &\le -(1-\theta)x^T Q x - \theta x^T Q x + \varrho_1\|x\|\,|w| - \frac{\beta^2|w|^2}{\mu} \\
&= -(1-\theta)x^T Q x - \begin{bmatrix} \|x\| \\ |w| \end{bmatrix}^T \begin{bmatrix} \theta\lambda_{\min}(Q) & -\varrho_1/2 \\ -\varrho_1/2 & \beta^2/\mu \end{bmatrix} \\
&\le -(1-\theta)x^T Q x
\end{aligned}$$

for $\mu \le 4\theta\lambda_{\min}(Q)(\beta/\varrho_1)^2$. Thus, choosing

$$\beta \ge \varrho \quad \text{and} \quad \mu \le 4\theta\lambda_{\min}(Q)(\beta/\varrho_1)^2$$

ensures that $\dot{V} \le -(1-\theta)x^T Q x$ for all $x \in \Omega$, which shows that the origin is exponentially stable and Ω is a subset of the region of attraction. The foregoing inequalities are satisfied with the choices $\beta = 12$, $\theta = 0.95$, and $\mu = 1$. Thus, the state feedback control is given by

$$u = -12 \ \text{sat}\,(1.2x_1 + 4.05x_2 + 7.8x_3 + 4.95x_4) \stackrel{\text{def}}{=} \gamma(x),$$

and the output feedback control is taken as $u = \gamma(\hat{x})$. Simulation is used to compare the performance of the standard and cascade observers. Both observers are taken from Example 7.1. The estimates $\hat{x}_2$, $\hat{x}_3$, and $\hat{x}_4$ of the cascade observer are saturated at ± 1.6, ± 1.2, and ± 3, respectively. The simulations of Figures 7.6 and 7.7 are carried out with $x(0) = \text{col}(1, 0, 0, -2) \in \Omega$ and zero initial conditions for the observers. Figure 7.6 shows the state trajectories under state and output feedback when the cascade observer is used. It demonstrates that the trajectories under output feedback approach the ones under state feedback as ε decreases. Figure 7.7 compares the performance of the two observers for $\varepsilon = 0.001$. It shows that the performance of the cascade observer is comparable to the performance of the standard observer. △

7.3 ▪ Cascade Observer with Feedback Injection

The performance of the cascade observer of Section 7.1 in feedback control is comparable to the performance of the standard observer. However, as a state estimator it has

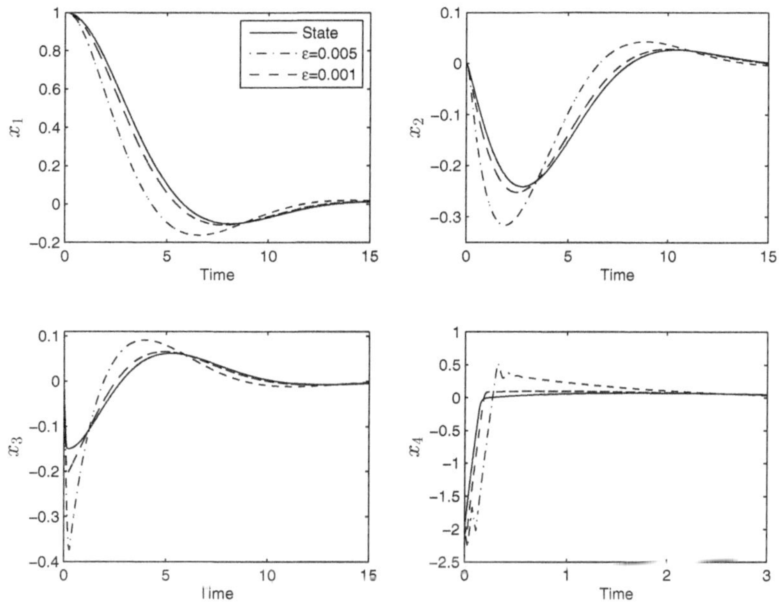

Figure 7.6. *Simulation of Example 7.2. The state trajectories under state feedback (solid) and output feedback with the cascade observer for* $\varepsilon = 0.005$ *(dash-dotted) and* $\varepsilon = 0.001$ *(dashed).*

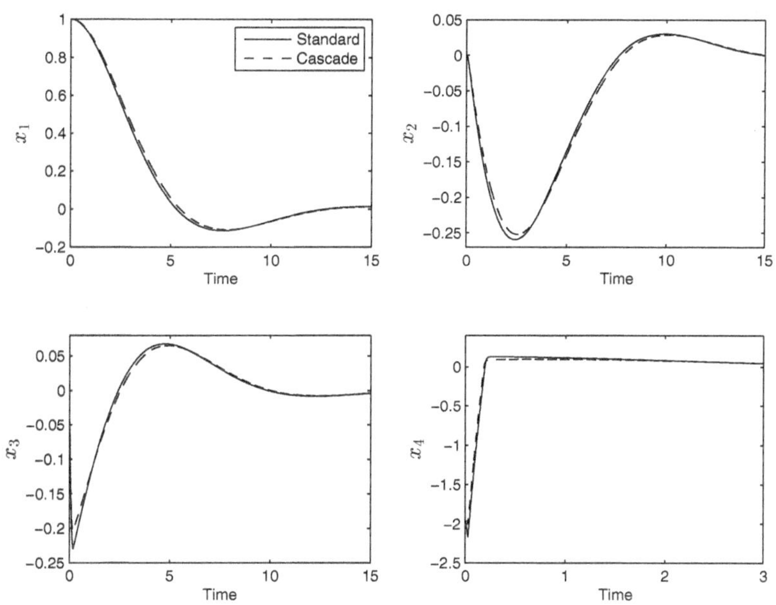

Figure 7.7. *Simulation of Example 7.2. The state trajectories under output feedback with the standard (solid) and cascade (dashed) observers for* $\varepsilon = 0.001$.

two drawbacks. First, the ultimate bound on the estimation error is not as small as with the standard observer, except for the estimate of x_ρ. Second, even when the function ϕ is known, the estimation error does not converge to zero as t tends to infinity. Both of these drawbacks are removed by the observer of this section.

Reconsider the system (7.1)–(7.4) with $\phi = \phi(x)$ and rewrite (7.2)–(7.4) for the case $\rho = 4$:

$$\dot{x}_1 = x_2, \quad \dot{x}_2 = x_3, \quad \dot{x}_3 = x_4, \quad \dot{x}_4 = \phi(x), \quad y = x_1. \tag{7.39}$$

Suppose $\phi(x)$ is locally Lipschitz. To avoid $O(1/\varepsilon^4)$ gain, as in the standard observer, a sequence of second-order observers are designed to estimate the states. In the first stage, a second-order observer is designed for system

$$\dot{x}_1 = x_2, \quad \dot{x}_2 = x_3, \quad y = x_1,$$

to estimate x_1 and x_2 by $\hat{x}_1$ and $\hat{x}_2$. In designing this observer the x_3 term in the second equation is not neglected as in Section 7.1. Instead, the observer uses the estimate of x_3, $\hat{x}_3$, which is provided by next stage. In the second stage, a second-order observer is designed for the system

$$\dot{x}_2 = x_3, \quad \dot{x}_3 = x_4,$$

to estimate x_3 by $\hat{x}_3$, using $\hat{x}_2$ as a virtual measurement of x_2 and feeding $\hat{x}_4$ back from the next stage. Finally, $\hat{x}_4$ is estimated by designing a second-order observer for the system

$$\dot{x}_3 = x_4, \quad \dot{x}_4 = \phi,$$

by using $\hat{x}_3$ as a virtual measurement of x_3 and including a model ϕ_0 of ϕ, which is calculated in terms of the state estimates. A block diagram representation of this observer is shown in Figure 7.8. By scaling the states of the second-order observer, as in Section 7.1, the observer gain in each stage will be $O(1/\varepsilon)$. The reduction of the gain power comes at the expense of increasing the dimension of the observer from four in the standard observer to six in the proposed one. In general, the dimension increases from ρ to $2(\rho - 1)$. This cascade structure with feedback injection avoids the $O(1/\varepsilon^\rho)$ gain, but the observer states may still peak to $O(1/\varepsilon^{\rho-1})$. As in Section 7.1, this peaking can be eliminated by saturating the signals in between the different stages. For example, for the second-stage observer, both the driving signal $\hat{x}_2$ and the feedback signal $\hat{x}_4$ are saturated. A block diagram representation of the observer with saturation is shown in Figure 7.9.

The gains of each second-order observer need to be chosen to ensure that the estimation errors decay to $O(\varepsilon)$ values within a time interval $T(\varepsilon)$, where $\lim_{\varepsilon \to 0} T(\varepsilon) = 0$. Moreover, beyond this period, the estimates should be within the linear regions of the saturation functions. A procedure for choosing the gains and analysis of the observer will be illustrated shortly, but first some preliminary results are stated.

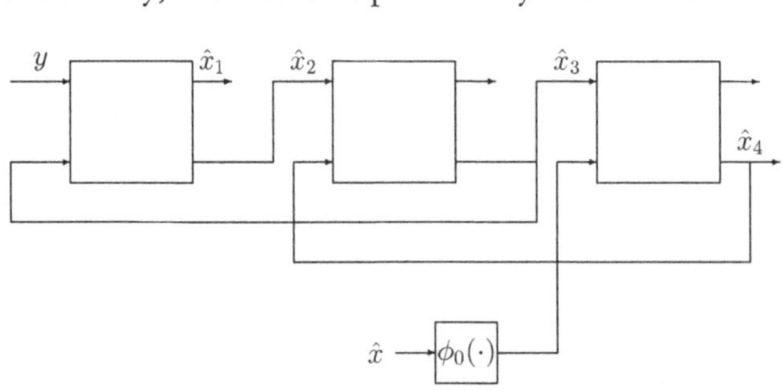

Figure 7.8. *Cascade realization with feedback injection.*

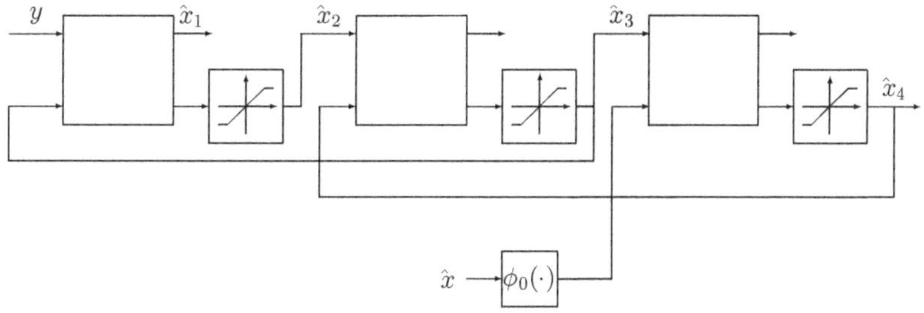

Figure 7.9. *Cascade realization with feedback injection and saturation.*

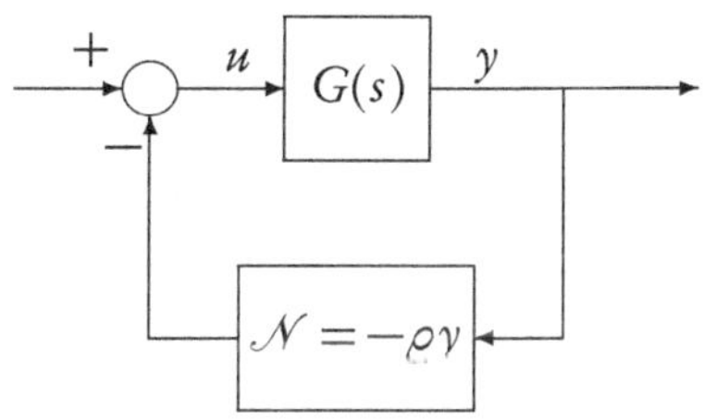

Figure 7.10. *Feedback connection.*

Lemma 7.1. *Consider the linear time-varying system*

$$\dot{x} = [A + \varrho(t)BC]x,$$

where (A, B) is controllable, (A, C) is observable, and $\varrho(t)$ is a continuous function such that $\varrho(t) \in [0, 1]$ for all $t \geq 0$. Let $G(s) = C(sI - A)^{-1}B$. If

$$\max_{\omega \in R} |G(j\omega)| < 1,$$

then the origin of the system is exponentially stable and a matrix $P = P^T > 0$ and a positive constant μ exist such that

$$P[A + \varrho(t)BC] + [A + \varrho(t)BC]^T P \leq -\mu P \tag{7.40}$$

for all $t \geq 0$.

Proof: The system can be represented by the feedback connection of Figure 7.10, where $\mathcal{N}$ belongs to the sector $[-1, 0]$. By the circle criterion [78, Example 7.1], the origin is exponential stable if $\max_{\omega \in R} |G(j\omega)| < 1$. To find P, apply a loop transformation and the Kalman–Yakubovich–Popov lemma [78, Lemma 6.3]. A loop transformation that transforms $\mathcal{N}$ into a passive function yields the strictly positive real system[71]

$$\dot{x} = (A + BC)x + B\tilde{u}, \quad \tilde{y} = Cx + \tilde{u}.$$

[71]See Figure 7.3 of [78].

By the Kalman–Yakubovich–Popov lemma, there exist matrices $P = P^T > 0$, L, W, and a positive constant μ such that

$$\begin{aligned} P(A+BC)+(A+BC)^T P &= -L^T L - \mu P, \\ PB &= C^T - LW, \\ W^T W &= 2I. \end{aligned}$$

Using $V(x) = x^T P x$ as a Lyapunov function for the system $\dot{x} = [A + \varrho(t)BC]x$, we obtain

$$\begin{aligned} \dot{V} &= x^T\left[P(A+\varrho BC)+(A+\varrho BC)^T P\right]x \\ &= x^T\left[P(A+BC)+(A+BC)^T P\right]x - 2(1-\varrho)x^T PBCx \\ &= -\mu x^T Px - x^T L^T Lx - 2(1-\varrho)[x^T C^T Cx - x^T L^T WCx] \\ &= -\mu x^T Px - x^T L^T Lx - (1-\varrho)x^T C^T W^T WCx + 2(1-\varrho)x^T L^T WCx \\ &\leq -\mu x^T Px - \|Lx\|^2 - (1-\varrho)\|WCx\|^2 + 2\sqrt{1-\varrho}\,\|Lx\|\,\|WCx\| \\ &= -\mu x^T Px - \left(\|Lx\| - \sqrt{1-\varrho}\,\|WCx\|\right)^2 \leq -\mu x^T Px, \end{aligned}$$

where $1-\varrho \leq \sqrt{1-\varrho}$ since $1-\varrho \in [0,1]$. Thus, $\dot{V} \leq -\mu V$, and (7.40) is satisfied. □

Lemma 7.2. *Let x and y be real numbers with $|x| \leq K < M$. Then*

$$x - M\,\text{sat}\left(\frac{y}{M}\right) = \varrho(x,y)(x-y),$$

where $\varrho \in [0,1]$ is a continuous function.

Proof: ϱ is given by

$$\varrho(x,y) = \begin{cases} (M-x)/(y-x) & \text{if } y > M, \\ 1 & \text{if } |y| \leq M, \\ (x+M)/(x-y) & \text{if } y < -M. \end{cases}$$

It is defined and continuous since

$$y > M \Rightarrow y - x > M - K \quad \text{and} \quad y < -M \Rightarrow x - y > M - K.$$

□

The observer design and its analysis are illustrated for the system (7.39). Suppose $x(t)$ is bounded for all $t \geq 0$ and positive constants M_2 to M_4 are known such that $M_i \geq |x_i(t)| + k$ for all $t \geq 0$, for some $k > 0$. In the first stage, the observer

$$\begin{aligned} \dot{z}_1 &= \frac{1}{\varepsilon}[\zeta_1 + \alpha_1(y - z_1)], & \hat{x}_1 &= z_1, \\ \dot{\zeta}_1 &= \varepsilon\hat{x}_3 + \frac{\beta_1}{\varepsilon}(y - z_1), & \hat{x}_2 &= M_2\,\text{sat}\left(\frac{\zeta_1}{\varepsilon M_2}\right) \end{aligned}$$

is used to estimate x_1 and x_2, where α_1 and β_1 are positive constants and $\hat{x}_3$ is the saturated estimate of x_3 from the next stage. Taking

$$\eta_1 = (x_1 - z_1) \quad \text{and} \quad \xi_1 = \varepsilon\left(x_2 - \frac{\zeta_1}{\varepsilon}\right),$$

it can be shown that

$$\varepsilon\dot{\psi}_1 = F_1\psi_1 + \varepsilon^2 G_2(x_3 - \hat{x}_3),$$

where

$$\psi_1 = \begin{bmatrix} \eta_1 \\ \xi_1 \end{bmatrix}, \quad F_1 = \begin{bmatrix} -\alpha_1 & 1 \\ -\beta_1 & 0 \end{bmatrix}, \quad \text{and} \quad G_2 = \begin{bmatrix} 0 \\ 1 \end{bmatrix}.$$

The matrix F_1 is Hurwitz. In the forthcoming analysis, G_i for $i \geq 2$ is defined by

$$G_i = \text{col}(0, 0, \ldots, 0, 1)_{i\times 1}. \tag{7.41}$$

Since $(x_3 - \hat{x}_3)$ is bounded, $\psi_1(t)$ is bounded for all $t \geq 0$, and there is time $T_1(\varepsilon)$, with $\lim_{\varepsilon\to 0} T_1(\varepsilon) = 0$, such that $\psi_1(t) = O(\varepsilon^2)$ for all $t \geq T_1$. Therefore, $x_1(t) - \hat{x}_1(t) = O(\varepsilon^2)$ and $x_2(t) - \zeta_1(t)/\varepsilon = O(\varepsilon)$ for all $t \geq T_1$. Since $M_2 \geq |x_2(t)| + k$, for sufficiently small ε, $|\zeta_1(t)|/\varepsilon \leq M_2$ and $\hat{x}_2(t) = \zeta_1(t)/\varepsilon$ for all $t \geq T_1$.

In the second stage, the observer

$$\dot{z}_2 = \frac{1}{\varepsilon}[\zeta_2 + \alpha_2(\hat{x}_2 - z_2)],$$
$$\dot{\zeta}_2 = \varepsilon\hat{x}_4 + \frac{\beta_2}{\varepsilon}(\hat{x}_2 - z_2), \qquad \hat{x}_3 = M_3 \,\text{sat}\left(\frac{\zeta_2}{\varepsilon M_3}\right)$$

is used to estimate x_3, where α_2 and β_2 are positive constants and $\hat{x}_2$ and $\hat{x}_4$ are the saturated estimates from the previous and next stages, respectively. Taking

$$\eta_2 = \varepsilon(x_2 - z_2) \quad \text{and} \quad \xi_2 = \varepsilon^2\left(x_3 - \frac{\zeta_2}{\varepsilon}\right),$$

it can be shown that

$$\varepsilon\dot{\psi}_2 = F_2\psi_2 + \varepsilon E_2(x_2 - \hat{x}_2) + \varepsilon^3 G_2(x_4 - \hat{x}_4),$$

where

$$\psi_2 = \begin{bmatrix} \eta_2 \\ \xi_2 \end{bmatrix}, \quad F_2 = \begin{bmatrix} -\alpha_2 & 1 \\ -\beta_2 & 0 \end{bmatrix}, \quad \text{and} \quad E_2 = \begin{bmatrix} \alpha_2 \\ \beta_2 \end{bmatrix}.$$

The matrix F_2 is Hurwitz. Since $(x_2 - \hat{x}_2)$ and $(x_4 - \hat{x}_4)$ are bounded, $\psi_2(t)$ is bounded for all $t \geq 0$. For $t \geq T_1$,

$$\hat{x}_2 = \frac{\zeta_1}{\varepsilon} \quad \Rightarrow \quad x_2 - \hat{x}_2 = x_2 - \frac{\zeta_1}{\varepsilon} = \frac{\xi_1}{\varepsilon}.$$

Therefore, the equations for ψ_1 and ψ_2 can be augmented together to obtain

$$\varepsilon\dot{\Psi}_2 = A_2\Psi_2 + \varepsilon^2 B_2(x_3 - \hat{x}_3) + \varepsilon^3 G_4(x_4 - \hat{x}_4),$$

where

$$\Psi_2 = \begin{bmatrix} \psi_1 \\ \psi_2 \end{bmatrix}, \quad A_2 = \begin{bmatrix} F_1 & 0 \\ E_2 G_2^T & F_2 \end{bmatrix}, \quad \text{and} \quad B_2 = \begin{bmatrix} G_2 \\ 0 \end{bmatrix}.$$

Using Lemma 7.2 we can write $x_3 - \hat{x}_3$ as

$$x_3 - \hat{x}_3 = x_3 - M_3 \,\text{sat}\left(\frac{\zeta_2}{\varepsilon M_2}\right) = \varrho_2(\cdot)\left(x_3 - \frac{\zeta_2}{\varepsilon}\right),$$

where $\varrho_2 \in [0,1]$ is continuous. Therefore,

$$\varepsilon^2(x_3 - \hat{x}_3) = \varrho_2 \varepsilon^2 \left(x_3 - \frac{\zeta_2}{\varepsilon} \right) = \varrho_2 \xi_2.$$

Substituting this expression in the $\dot{\Psi}_2$-equation yields

$$\varepsilon \dot{\Psi}_2 = (A_2 + \varrho_2 B_2 C_2)\Psi_2 + \varepsilon^3 G_4(x_4 - \hat{x}_4),$$

where $C_2 = G_4^T$. The transfer function $H_2(s) = C_2(sI - A_2)^{-1}B_2$ can be written as the cascade connection

$$H_2(s) = H_{2a}(s)\, H_{2b}(s),$$

where H_{2a} and H_{2b} are Hurwitz transfer functions given by

$$H_{2a}(s) = G_2^T(sI - F_2)^{-1}E_2 = \frac{\beta_2 s}{s^2 + \alpha_2 s + \beta_2} \quad \text{and} \quad H_{2b}(s) = G_2^T(sI - F_1)^{-1}G_2.$$

Choose α_2 and β_2 such that $\{A_2, B_2, C_2\}$ is a minimal realization. Since $\{F_1, G_2, G_2^T\}$ and $\{F_2, E_2, G_2^T\}$ are minimal realizations, this is achieved by choosing α_2 and β_2 such that the poles of $H_{2a}(s)$ are not located at the zeros of $H_{2b}(s)$. It can be verified that $\max_{\omega \in R} |H_{2a}(j\omega)| = \beta_2/\alpha_2$.[72] Choosing $\beta_2/\alpha_2 < 1/\gamma_2$, where $\gamma_2 = \max_{\omega \in R} |H_{2b}(j\omega)|$, ensures that $\max_{\omega \in R} |H_2(j\omega)| < 1$. By Lemma 7.1, there exist a matrix $P_2 = P_2^T > 0$ and a positive constant μ_2 such that

$$P_2(A_2 + \varrho_2 B_2 C_2) + (A_2 + \varrho_2 B_2 C_2)^T P_2 \le -\mu_2 P_2.$$

Using $V_2 = \sqrt{\Psi_2^T P_2 \Psi_2}$ as a Lyapunov function for the $\dot{\Psi}_2$-equation, it can be shown that

$$\begin{aligned}
\varepsilon \dot{V}_2 &= \frac{1}{2V_2}[\Psi_2^T P_2(A_2 + \varrho_2 B_2 C_2)\Psi_2 + \Psi_2^T (A_2 + \varrho_2 B_2 C_2)P_2\Psi_2 \\
&\quad + 2\varepsilon^3 \Psi_2^T P_2 G_4(x_4 - \hat{x}_4)] \\
&\le -\frac{\mu_2}{2} V_2 + \varepsilon^3 k_2 \|x_4 - \hat{x}_4\|
\end{aligned}$$

for some $k_2 > 0$. Thus, there is $T_2(\varepsilon) \ge T_1(\varepsilon)$, with $\lim_{\varepsilon \to 0} T_2(\varepsilon) = 0$, such that for $t \ge T_2$,

$$V(\Psi_2(t)) = O(\varepsilon^3) \Rightarrow \Psi_2(t) = O(\varepsilon^3) \Rightarrow \xi_2(t) = O(\varepsilon^3) \Rightarrow x_3(t) - \frac{\zeta_2(t)}{\varepsilon} = O(\varepsilon).$$

For sufficiently small ε,

$$\left| \frac{\zeta_2(t)}{\varepsilon} \right| \le M_3 \Rightarrow \hat{x}_3(t) = \frac{\zeta_2(t)}{\varepsilon}.$$

Hence,

$$\varepsilon \dot{\Psi}_2 = (A_2 + B_2 C_2)\Psi_2 + \varepsilon^3 G_4(x_4 - \hat{x}_4).$$

The matrix $A_2 + B_2 C_2$ is Hurwitz because it is equal to $A_2 + \varrho_2 B_2 C_2$ at $\varrho_2 = 1$.

[72] $\max_{\omega \in R} |H(j\omega)|$ can be calculated with the MATLAB command getPeakGain.

In the third stage, the observer

$$\dot{z}_3 = \frac{1}{\varepsilon}[\zeta_3 + \alpha_3(\hat{x}_3 - z_3)],$$
$$\dot{\zeta}_3 = \varepsilon\phi_0(\hat{x}) + \frac{\beta_3}{\varepsilon}(\hat{x}_3 - z_3), \qquad \hat{x}_4 = M_4 \text{ sat}\left(\frac{\zeta_3}{\varepsilon M_4}\right)$$

is used to estimate x_4, where α_3 and β_3 are positive constants, $\hat{x}_3$ is the saturated estimated from the previous stage, and ϕ_0 is a locally Lipschitz model of ϕ, which is bounded since $\hat{x}_1$ to $\hat{x}_4$ are bounded. Taking

$$\eta_3 = \varepsilon^2(x_3 - z_3) \quad \text{and} \quad \xi_3 = \varepsilon^3\left(x_4 - \frac{\zeta_3}{\varepsilon}\right),$$

it can be shown that

$$\varepsilon\dot{\psi}_3 = F_3\psi_3 + \varepsilon^2 E_3(x_3 - \hat{x}_3) + \varepsilon^4 G_2[\phi(x) - \phi_0(\hat{x})],$$

where

$$\psi_3 = \begin{bmatrix} \eta_3 \\ \xi_3 \end{bmatrix}, \quad F_3 = \begin{bmatrix} -\alpha_3 & 1 \\ -\beta_3 & 0 \end{bmatrix}, \quad \text{and} \quad E_3 = \begin{bmatrix} \alpha_3 \\ \beta_3 \end{bmatrix}.$$

The matrix F_3 is Hurwitz. Because $(x_3 - \hat{x}_3)$ and $[\phi(x) - \phi_0(\hat{x})]$ are bounded, $\psi_3(t)$ is bounded for all $t \geq 0$. For $t \geq T_2$,

$$\varepsilon^2(x_3 - \hat{x}_3) = \varepsilon^2\left(x_3 - \frac{\zeta_2}{\varepsilon}\right) = \xi_2 = G_4^T \Psi_2.$$

By Lemma 7.2,

$$\varepsilon^3(x_4 - \hat{x}_4) = \varepsilon^3\left[x_4 - M_4 \text{ sat}\left(\frac{\zeta_3}{\varepsilon M_4}\right)\right] = \varepsilon^3 \varrho_3(\cdot)\left(x_4 - \frac{\zeta_3}{\varepsilon}\right) = \varrho_3(\cdot)\xi_3,$$

where $\varrho_3 \in [0, 1]$ is continuous. Augmenting the equations of Ψ_2 and ψ_3, we arrive at

$$\varepsilon\dot{\Psi}_3 = (A_3 + \varrho_3 B_3 C_3)\Psi_3 + \varepsilon^4 G_6[\phi(x) - \phi_0(\hat{x})],$$

where

$$\Psi_3 = \begin{bmatrix} \Psi_2 \\ \psi_3 \end{bmatrix}, \quad A_3 = \begin{bmatrix} (A_2 + B_2C_2) & 0 \\ E_3 G_4^T & F_3 \end{bmatrix}, \quad B_3 = \begin{bmatrix} G_4 \\ 0 \end{bmatrix}, \quad \text{and} \quad C_3 = G_6^T.$$

The matrix A_3 is Hurwitz. The transfer function $H_3(s) = C_3(sI - A_3)^{-1}B_3 = H_{3a}(s)H_{3b}(s)$, where H_{3a} and H_{3b} are Hurwitz transfer functions given by

$$H_{3a}(s) = G_2^T(sI - F_3)^{-1}E_3 = \frac{\beta_3 s}{s^2 + \alpha_3 s + \beta_3} \quad \text{and} \quad H_{3b}(s) = G_4^T[sI - (A_2 + B_2C_2)]^{-1}G_4.$$

Choosing α_3 and β_3 such that $\{A_3, B_3, C_3\}$ is a minimal realization and $\beta_3/\alpha_3 < 1/\gamma_3$, where $\gamma_3 = \max_{\omega \in R} |H_{3b}(j\omega)|$, ensures that $\max_{\omega \in R} |H_3(j\omega)| < 1$ and Lemma 7.1 can be applied to show the existence of $P_3 = P_3^T > 0$ and $\mu_3 > 0$ such that

$$P_3(A_3 + \varrho_3 B_3 C_3) + (A_3 + \varrho_3 B_3 C_3)^T P_3 \leq -\mu_3 P_3.$$

Using $V_3 = \sqrt{\Psi_3^T P_3 \Psi_3}$ as a Lyapunov function for the $\dot{\Psi}_3$-equation, we can repeat the previous derivation to show that there is $T_3(\varepsilon) \geq T_2(\varepsilon)$, with $\lim_{\varepsilon\to 0} T_3(\varepsilon) = 0$, such that for $t \geq T_3$,

$$\Psi_3(t) = O(\varepsilon^4) \Rightarrow \xi_3(t) = O(\varepsilon^4) \Rightarrow x_4(t) - \frac{\zeta_3(t)}{\varepsilon} = O(\varepsilon^3).$$

For sufficiently small ε,

$$\left|\frac{\zeta_3(t)}{\varepsilon}\right| \leq M_4 \Rightarrow \hat{x}_4(t) = \frac{\zeta_3(t)}{\varepsilon}.$$

From this time on, Ψ_3 satisfies the equation

$$\varepsilon\dot{\Psi}_3 = (A_3 + B_3C_3)\Psi_3 + \varepsilon^4 G_6[\phi(x) - \phi_0(\hat{x})],$$

where $A_3 + B_3C_3$ is Hurwitz. Recalling the definition of Ψ_3, we have

$$\begin{bmatrix} \tilde{x}_1 \\ \varepsilon\tilde{x}_2 \\ \varepsilon(x_2 - z_2) \\ \varepsilon^2\tilde{x}_3 \\ \varepsilon^2(x_3 - z_3) \\ \varepsilon^3\tilde{x}_4 \end{bmatrix} = O(\varepsilon^4) \Rightarrow \tilde{x}_i = O(\varepsilon^{4+1-i}) \quad \text{for } 1 \leq i \leq 4.$$

If $\phi_0 = \phi$, using the Lipschitz property of ϕ,

$$\varepsilon^4\|\phi(x) - \phi(\hat{x})\| \leq \varepsilon^4 L_1\|\tilde{x}\| \leq \varepsilon L_2\|\Psi_3\|.$$

Hence, for sufficiently small ε,

$$\lim_{t\to\infty} \Psi_3(t) = 0 \quad \Rightarrow \quad \lim_{t\to\infty} \tilde{x}_i(t) = 0 \text{ for } 1 \leq i \leq 4.$$

We are now ready to describe the observer design for the general case. Consider the system

$$\dot{w} = f_0(w, x, u), \tag{7.42}$$
$$\dot{x}_i = x_{i+1} \qquad \text{for } 1 \leq i \leq \rho - 1, \tag{7.43}$$
$$\dot{x}_\rho = \phi(w, x, u), \tag{7.44}$$
$$y = x_1, \tag{7.45}$$

where f_0 and ϕ are locally Lipschitz in their arguments and $w(t)$, $x(t)$, and $u(t)$ are bounded for all $t \geq 0$. Suppose $x(t)$ belongs to a compact set X and take $M_i \geq k + \max_{x\in X} |x_i|$ for $2 \leq i \leq \rho$ for some $k > 0$. The observer consists of $\rho - 1$ second-order systems. The estimates $\hat{x}_1$ and $\hat{x}_2$ are provided by the system

$$\dot{z}_1 = \frac{1}{\varepsilon}[\zeta_1 + \alpha_1(y - z_1)], \qquad \hat{x}_1 = z_1, \tag{7.46}$$
$$\dot{\zeta}_1 = \varepsilon\hat{x}_3 + \frac{\beta_1}{\varepsilon}(y - z_1), \qquad \hat{x}_2 = M_2 \operatorname{sat}\left(\frac{\zeta_1}{\varepsilon M_2}\right), \tag{7.47}$$

where ε is a small positive constant, and α_1 and β_1 are positive constants that assign the eigenvalues of

$$F_1 = \begin{bmatrix} -\alpha_1 & 1 \\ -\beta_1 & 0 \end{bmatrix}.$$

The estimate $\hat{x}_{i+1}$ for $2 \le i \le \rho - 2$ is provided by the system

$$\dot{z}_i = \frac{1}{\varepsilon}[\zeta_i + \alpha_i(\hat{x}_i - z_i)], \tag{7.48}$$

$$\dot{\zeta}_i = \varepsilon \hat{x}_{i+2} + \frac{\beta_i}{\varepsilon}(\hat{x}_i - z_i), \qquad \hat{x}_{i+1} = M_{i+1} \operatorname{sat}\left(\frac{\zeta_i}{\varepsilon M_{i+1}}\right), \tag{7.49}$$

while the last estimate $\hat{x}_\rho$ is provided by

$$\dot{z}_{\rho-1} = \frac{1}{\varepsilon}[\zeta_{\rho-1} + \alpha_{\rho-1}(\hat{x}_{\rho-1} - z_{\rho-1})], \tag{7.50}$$

$$\dot{\zeta}_{\rho-1} = \varepsilon \phi_0(\hat{x}, u) + \frac{\beta_{\rho-1}}{\varepsilon}(\hat{x}_{\rho-1} - z_{\rho-1}), \qquad \hat{x}_\rho = M_\rho \operatorname{sat}\left(\frac{\zeta_{\rho-1}}{\varepsilon M_\rho}\right), \tag{7.51}$$

where ϕ_0 is locally Lipschitz. The positive constants α_i and β_i for $2 \le i \le \rho - 1$ are chosen to satisfy the forthcoming conditions (7.52) and (7.53). For $2 \le i \le \rho - 1$, let

$$F_i = \begin{bmatrix} -\alpha_i & 1 \\ -\beta_i & 0 \end{bmatrix} \quad \text{and} \quad E_i = \begin{bmatrix} \alpha_i \\ \beta_i \end{bmatrix}.$$

Define the matrices $\{A_2, B_2, C_2\}$ and the transfer function $H_{2b}(s)$ by

$$A_2 = \begin{bmatrix} F_1 & 0 \\ E_2 G_2^T & F_2 \end{bmatrix}, \quad B_2 = \begin{bmatrix} G_2 \\ 0 \end{bmatrix}, \quad C_2 = G_4^T,$$
$$H_{2b}(s) = G_2^T (sI - F_1)^{-1} G_2.$$

For $3 \le i \le \rho - 1$, define the matrices $\{A_i, B_i, C_i\}$ and the transfer function $H_{ib}(s)$ by

$$A_i = \begin{bmatrix} (A_{i-1} + B_{i-1}C_{i-1}) & 0 \\ E_i G_{2(i-1)}^T & F_i \end{bmatrix}, \quad B_i = \begin{bmatrix} G_{2(i-1)} \\ 0 \end{bmatrix}, \quad C_i = G_{2i}^T,$$
$$H_{ib}(s) = G_{2(i-1)}^T [sI - (A_{i-1} + B_{i-1}C_{i-1})]^{-1} G_{2(i-1)}.$$

For each $2 \le i \le \rho - 1$, choose α_i and β_i such that

$$\{A_i, B_i, C_i\} \text{ is a minimal realization}, \tag{7.52}$$

$$\frac{\beta_i}{\alpha_i} < \frac{1}{\gamma_i}, \quad \text{where } \gamma_i = \max_{\omega \in R} |H_{ib}(j\omega)|. \tag{7.53}$$

Remark 7.1. *The system $\{A_i, B_i, C_i\}$ is the cascade connection of $\{F_i, E_i, G_2^T\}$ and $\{A_{i-1} + B_{i-1}C_{i-1}, G_{2(i-1)}, G_{2(i-1)}^T\}$, where the latter system's transfer function is $H_{ib}(s)$. It is clear that $\{F_i, E_i, G_2^T\}$ is a minimal realization. It can be verified that $\{A_{i-1} + B_{i-1}C_{i-1}, G_{2(i-1)}, G_{2(i-1)}^T\}$ is a minimal realization. Therefore, $\{A_i, B_i, C_i\}$ is a minimal realization if the eigenvalues of F_i are different from the zeros of $H_{ib}(s)$.* ◇

Theorem 7.3. *Under the stated assumptions, there is* $\varepsilon^* > 0$ *such that for* $0 < \varepsilon \le \varepsilon^*$ *the estimation errors* $\tilde{x}_i = x_i - \hat{x}_i$*, for* $1 \le i \le \rho$ *of the cascade high-gain observer with feedback injection* (7.46)–(7.51) *are bounded uniformly in* ε*, and there is time* $T(\varepsilon)$*, with* $\lim_{\varepsilon\to 0} T(\varepsilon) = 0$*, such that*

$$|\tilde{x}_i(t)| \le a\varepsilon^{\rho+1-i} \quad \textit{for } 1 \le i \le \rho \tag{7.54}$$

for all $t \ge T(\varepsilon)$*, for some positive constant* a*. If* $\phi_0 = \phi$*,*

$$\lim_{t\to\infty} \tilde{x}(t) = 0. \qquad \diamond$$

The proof proceeds as the foregoing analysis of the case $\rho = 4$.

Example 7.3. Reconsider the system

$$\dot{x}_1 = x_2, \quad \dot{x}_2 = x_3, \quad \dot{x}_3 = x_4, \quad \dot{x}_4 = \phi(x), \quad y = x_1,$$

from Example 7.1, where

$$\begin{aligned}\phi(x) = &-(x_1 + 2x_2 + 2x_3 + 2x_4)\\ &+[1-(x_1+2x_2+x_3)^2](x_2+2x_3+x_4).\end{aligned}$$

The observer (7.46)–(7.51) is designed in the following steps:

- $\alpha_1 = 2$ and $\beta_1 = 1$ assign the eigenvalues of F_1 at $-1,-1$;
- $\gamma_2 = 2$; $\alpha_2 = 2$ and $\beta_2 = 0.9$ satisfy (7.52) and (7.53);
- $\gamma_3 = 4.5851$; $\alpha_3 = 2$ and $\beta_3 = 0.4$ satisfy (7.52) and (7.53).

As in Example 7.1, the saturation levels are taken as $M_1 = 2$, $M_2 = 1.5$, and $M_3 = 3$. Two choices of the function ϕ_0 are considered: $\phi_0 = 0$ and $\phi_0 = \phi$. The performance of the observer (7.46)–(7.51) is compared with the standard observer from Example 7.1. Simulation is carried out with $\varepsilon = 0.001$, $x_i(0) = 1$ for $i = 1,2,3,4$, and zero initial conditions for the observers. For the case $\phi_0 = 0$, Figure 7.11 shows the states x_3 and x_4 and their estimates, and Figure 7.12 compares the estimation errors $\tilde{x}_3$ and $\tilde{x}_4$ with the standard observer. These are repeated in Figures 7.13 and 7.14 for the case $\phi_0 = \phi$. Figures 7.11 and 7.13 illustrate the saturation of the estimates during the peaking period and the closeness of the estimates to the states beyond this period. Figures 7.12

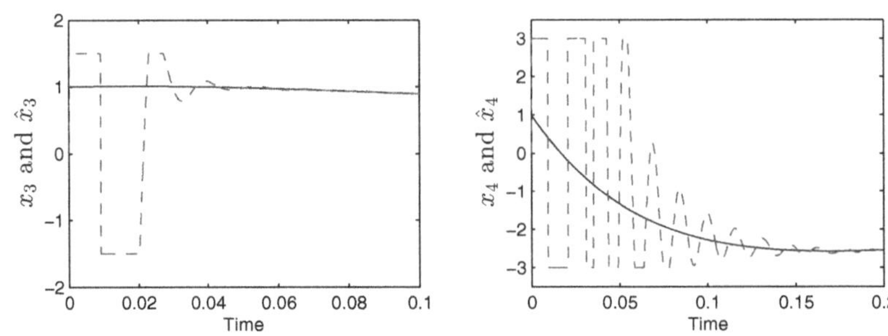

Figure 7.11. *Simulation of Example 7.3 when* $\phi_0 = 0$*. The states* x_3 *and* x_4 *(solid) and their estimates* $\hat{x}_3$ *and* $\hat{x}_4$ *(dashed) for the cascade observer with feedback injection.*

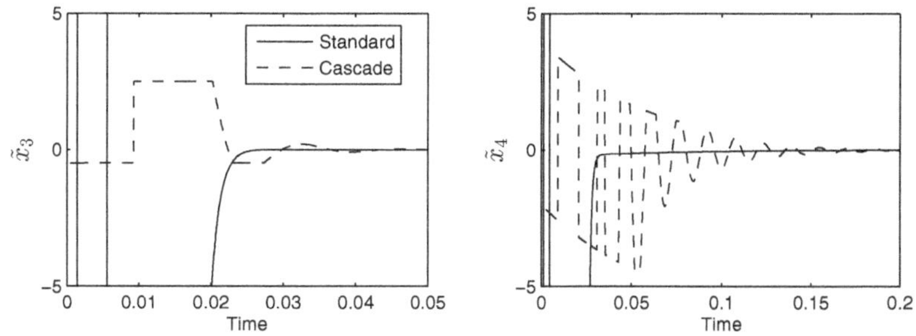

Figure 7.12. *Simulation of Example 7.3 when $\phi_0 = 0$. The estimation errors $\tilde{x}_3$ and $\tilde{x}_4$ for the standard observer (solid) and the cascade observer with feedback injection (dashed).*

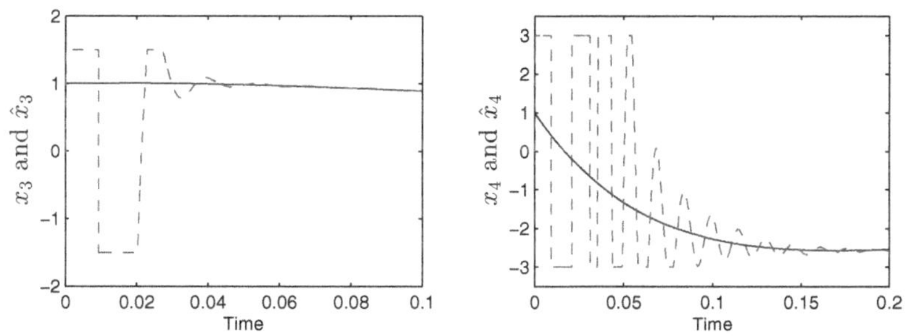

Figure 7.13. *Simulation of Example 7.3 when $\phi_0 = \phi$. The states x_3 and x_4 (solid) and their estimates $\hat{x}_3$ and $\hat{x}_4$ (dashed) for the cascade observer with feedback injection.*

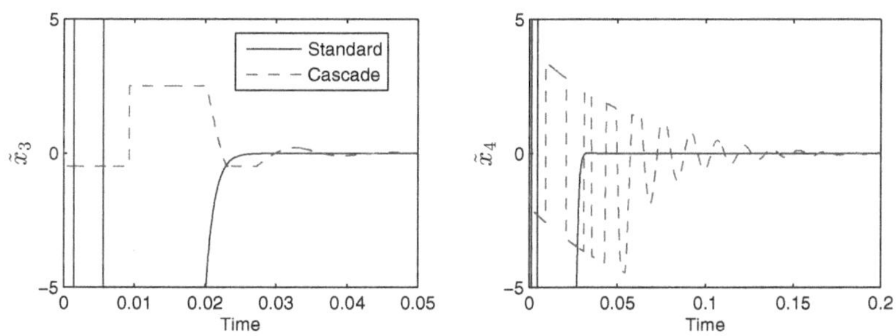

Figure 7.14. *Simulation of Example 7.3 when $\phi_0 = \phi$. The estimation errors $\tilde{x}_3$ and $\tilde{x}_4$ for the standard observer (solid) and the cascade observer with feedback injection (dashed).*

and 7.14 illustrate that the cascade observer with feedback injection has convergence properties comparable to the standard one, although the transient behavior of $\tilde{x}_4$ is more oscillatory. This is probably because the eigenvalues of the standard observer can be freely assigned, while in the current observer they are determined by the procedure for choosing α_i and β_i. The choice of ϕ_0 has no effect on the transient behavior, but it has a significant effect on the steady-state error, as shown in Figure 7.15, where the estimation error converges to zero when $\phi_0 = \phi$. Figure 7.16 shows the largest peaking signals in the two observers. Peaking in the standard observer is $O(1/\varepsilon^3)$, while in the current observer it is $O(1/\varepsilon)$. Finally, Table 7.3 compares the steady-state estimation errors of the two observers when $\phi_0 = 0$. It shows that they are of the same order of magnitude. $\triangle$

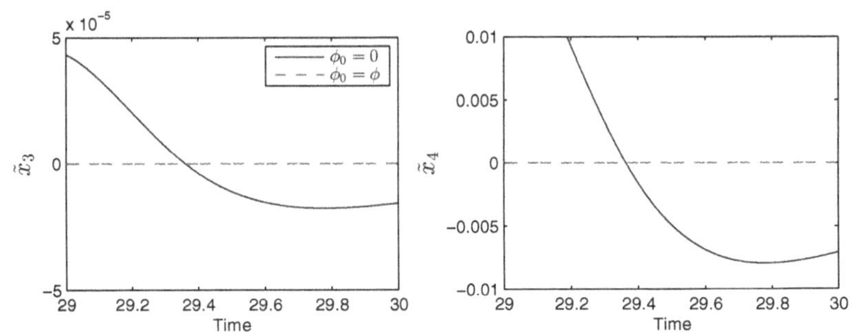

Figure 7.15. *Simulation of Example 7.3. The estimation errors* $\tilde{x}_3$ *and* $\tilde{x}_4$ *of the cascade observer with feedback injection when* $\phi_0 = 0$ *(solid) and* $\phi_0 = \phi$ *(dashed).*

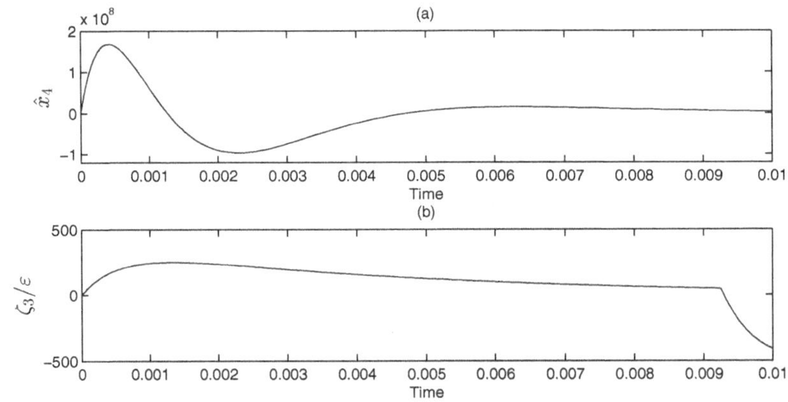

Figure 7.16. *Simulation of Example 7.3. The most peaking signal when* $\varepsilon = 0.001$*:* (a) *the standard observer;* (b) *the cascade observer with feedback injection.*

Table 7.3. *Simulation of Example 7.3. The steady state estimation errors for the standard and cascade with feedback injection (CWFI) observers when* $\varepsilon = 0.001$*. The maximum is taken over the time interval* $[25, 30]$.

Observer	$\max \lvert\tilde{x}_1\rvert$	$\max \lvert\tilde{x}_2\rvert$	$\max \lvert\tilde{x}_3\rvert$	$\max \lvert\tilde{x}_4\rvert$
Standard	4.0709×10^{-12}	1.6284×10^{-8}	2.4425×10^{-5}	0.0163
CWFI	4.5235×10^{-11}	9.047×10^{-8}	4.5236×10^{-5}	0.0204

7.4 ▪ Notes and References

Sections 7.1 and 7.2 are taken from Khalil [81, 82]. Section 7.3 is taken from Astolfi, Marconi, and Teel [14], which is based on earlier work by Astolfi and Marconi [11]. For pedagogical reasons, the cascade observer of Section 7.1 is presented first, followed by the cascade observer with feedback injection of Section 7.3. However, Asotlfi and Marconi introduced their observer first in [11] without saturation functions. Motivated by their ideas, Khalil introduced the cascade observer with saturation in [81, 82]. The more recent work [14] introduced the saturation functions in the observer of [11]. The dimension of the observer of Section 7.3 is higher than the dimension of the observer of Section 7.1. It can be shown that the additional dynamics act as a low-pass filter that reduces the effect of high-frequency noise [12].

Chapter 8
Measurement Noise

Measurement noise presents a challenge to the implementation of high-gain observers. This chapter characterizes the effect of measurement noise on the accuracy of the state estimates as well as on the performance of the observer in feedback control. It presents two ideas to reduce the effect of the noise. The chapter starts by calculating an ultimate bound on the estimation error due to noise in Section 8.1. The bound, which is proportional to a negative power of ε, is calculated for any bounded noise. While the rest of the chapter focuses attention on this case, it is shown in the same section that the bound is conservative because it does not take into consideration the frequency content of the noise. Two results are shown. If the frequencies of the noise are much higher than $1/\varepsilon$, which is the order of the bandwidth of the observer, the low-pass filtering property of the observer attenuates the noise and yields a less conservative bound. If the frequencies of the noise are much lower than $1/\varepsilon$, an ultimate bound independent of ε is derived. Section 8.2 studies a closed-loop system in which the observer is used to implement a stabilizing state feedback controller. A result similar to the separation principle of Section 3.1 is proved in the presence of noise. When the system has a chain of integrators where the estimated states are the derivatives of the output, the closed-loop effect of the noise on the state variables differs depending on the place of the variable in the chain of integrators, with an ultimate bound on the output that does not depend on a negative power ε. This is particularly significant in the tracking problem, which is studied in Section 8.3 for linear systems. Section 8.4 shows how to reduce the effect of measurement noise by using low-pass filters and nonlinear gains.

8.1 ▪ Observer

Reconsider the system

$$\dot{w} = f_0(w, x, u), \tag{8.1}$$

$$\dot{x}_i = x_{i+1} + \psi_i(x_1, \ldots, x_i, u) \qquad \text{for } 1 \le i \le \rho - 1, \tag{8.2}$$

$$\dot{x}_\rho = \phi(w, x, u), \tag{8.3}$$

$$y = x_1 + v \tag{8.4}$$

from Section 2.2, where f_0, $\psi_1, \ldots, \psi_{\rho-1}$, and ϕ are locally Lipschitz in their arguments; $u(t)$ and $v(t)$ are piecewise continuous functions of t; and $w(t)$, $x(t)$, $u(t)$, and $v(t)$ are bounded for all $t \geq 0$. In particular, let $|v(t)| \leq N$, $w(t) \in W \subset R^\ell$, $x(t) \in X \subset R^\rho$, and $u(t) \in U \subset R^m$ for all $t \geq 0$, for some compact sets W, X, and U. The new feature here is the presence of bounded measurement noise $v(t)$. It is assumed that ψ_1 to $\psi_{\rho-1}$ are known and, for any compact set $S \subset R^\rho$, there are positive constants L_1 to $L_{\rho-1}$ such that

$$|\psi_i(x_1,\dots,x_i,u) - \psi_i(z_1,\dots,z_i,u)| \leq L_i \sum_{k=1}^{i} |x_k - z_k| \tag{8.5}$$

for all $x, z \in S$ and $u \in U$. The observer is taken as

$$\dot{\hat{x}}_i = \hat{x}_{i+1} + \psi_i^s(\hat{x}_1,\dots,\hat{x}_i,u) + \frac{\alpha_i}{\varepsilon^i}(y - \hat{x}_1) \qquad \text{for } 1 \leq i \leq \rho - 1, \tag{8.6}$$

$$\dot{\hat{x}}_\rho = \phi_0(\hat{x},u) + \frac{\alpha_\rho}{\varepsilon^\rho}(y - \hat{x}_1), \tag{8.7}$$

where ϕ_0 is a nominal model of ϕ, ε is a sufficiently small positive constant, and α_1 to α_ρ are chosen such that the polynomial

$$s^\rho + \alpha_1 s^{\rho-1} + \cdots + \alpha_{\rho-1} s + \alpha_\rho \tag{8.8}$$

is Hurwitz. The functions ψ_1^s to $\psi_{\rho-1}^s$ are locally Lipschitz and satisfy the conditions

$$\psi_i^s(x_1,\dots,x_i,u) = \psi_i(x_1,\dots,x_i,u) \qquad \forall\, x \in X \tag{8.9}$$

and

$$|\psi_i^s(x_1,\dots,x_i,u) - \psi_i^s(z_1,\dots,z_i,u)| \leq L_i \sum_{k=1}^{i} |x_k - z_k| \tag{8.10}$$

for all $x, z \in R^\rho$ and $u \in U$. The function ϕ_0 is assumed to be locally Lipschitz in its arguments and

$$|\phi(w,x,u) - \phi_0(z,u)| \leq L\,\|x - z\| + M \tag{8.11}$$

for all $w \in W$, $x \in X$, $z \in R^\rho$, and $u \in U$.

Theorem 8.1. *Consider the system* (8.1)–(8.4) *and the observer* (8.6)–(8.7). *Suppose* (8.9) *to* (8.11) *are satisfied. Then there is* $\varepsilon^* \in (0,1]$ *such that for* $0 < \varepsilon \leq \varepsilon^*$, *the estimation error* $\tilde{x}_i = x_i - \hat{x}_i$, *for* $1 \leq i \leq \rho$, *satisfies the bound*

$$|\tilde{x}_i| \leq \max\left\{ \frac{b}{\varepsilon^{i-1}}\, e^{-at/\varepsilon} \|\tilde{x}(0)\|,\ \varepsilon^{\rho+1-i} cM + \frac{Nd}{\varepsilon^{i-1}} \right\} \tag{8.12}$$

for some positive constants a, b, c, d. ◇

Proof: Let $\eta = \text{col}(\eta_1, \eta_2, \dots, \eta_\rho)$, where

$$\eta_i = \frac{x_i - \hat{x}_i}{\varepsilon^{\rho-i}} \quad \text{for} \quad 1 \leq i \leq \rho.$$

It can be shown that η satisfies the equation

$$\varepsilon\dot{\eta} = F\eta + \varepsilon\delta - (1/\varepsilon^{\rho-1})Ev,$$

where

$$F = \begin{bmatrix} -\alpha_1 & 1 & 0 & \cdots & 0 \\ -\alpha_2 & 0 & 1 & \cdots & 0 \\ \vdots & & \ddots & & \vdots \\ -\alpha_{\rho-1} & & & 0 & 1 \\ -\alpha_\rho & 0 & \cdots & \cdots & 0 \end{bmatrix}, \quad E = \begin{bmatrix} \alpha_1 \\ \alpha_2 \\ \vdots \\ \alpha_{\rho-1} \\ \alpha_\rho \end{bmatrix},$$

$$\delta = \mathrm{col}(\delta_1, \delta_2, \ldots, \delta_\rho),$$

$$\delta_i = \frac{1}{\varepsilon^{\rho-i}}[\psi_i(x_1, \ldots, x_i, u) - \psi_i^s(\hat{x}_1, \ldots, \hat{x}_i, u)] \quad \text{for} \quad 1 \le i \le \rho - 1,$$

$$\delta_\rho = \phi(w, x, u) - \phi_0(\hat{x}, u).$$

The matrix F is Hurwitz by design and

$$\|\delta\| \le L_\delta \|\eta\| + M,$$

where L_δ is independent of ε for $\varepsilon \le 1$. Let $V = \eta^T P \eta$, where $P = P^T > 0$ is the solution of the Lyapunov equation $PF + F^T P = -I$. Then

$$\varepsilon \dot{V} \le -\|\eta\|^2 + 2\varepsilon L_\delta \|P\| \, \|\eta\|^2 + 2\varepsilon M \|P\| \, \|\eta\| + (2/\varepsilon^{\rho-1}) N \|PE\| \, \|\eta\|.$$

For $\varepsilon L_\delta \|P\| \le \frac{1}{4}$,

$$\varepsilon \dot{V} \le -\frac{1}{2}\|\eta\|^2 + 2\left[\varepsilon M \|P\| + \frac{N}{\varepsilon^{\rho-1}}\|PE\|\right]\|\eta\|,$$

$$\varepsilon \dot{V} \le -\frac{1}{4}\|\eta\|^2 \quad \forall \ \|\eta\| \ge 8\left[\varepsilon M \|P\| + \frac{N}{\varepsilon^{\rho-1}}\|PE\|\right].$$

Therefore [80, Theorem 4.5],

$$\|\eta(t)\| \le \max\left\{ b e^{-at/\varepsilon} \|\eta(0)\|, \varepsilon c M + \frac{Nd}{\varepsilon^{\rho-1}} \right\}$$

for some positive constants a, b, c, d. Using

$$\|\eta(0)\| \le \|\tilde{x}(0)\| / \varepsilon^{\rho-1} \quad \text{and} \quad \tilde{x}_i = \varepsilon^{\rho-i} \eta_i$$

yields (8.12). □

Similar to the bound (2.24) of Theorem 2.1, the bound (8.12) has a transient term due to the initial estimation error $\tilde{x}(0)$ and an ultimate bound term. The transient term decays to $O(\varepsilon)$ values within a time period $[0, T(\varepsilon)]$, where $\lim_{\varepsilon \to 0} T(\varepsilon) = 0$. The ultimate bound is the sum of two terms: $\varepsilon^{\rho+1-i} cM$, due to uncertainty in modeling ϕ, and Nd/ε^{i-1}, due to measurement noise. Except for $\tilde{x}_1$, the term Nd/ε^{i-1} grows as ε decreases. A sketch of the ultimate bound as function of ε is shown in Figure 1.11 for $\rho = 2$. Decreasing ε reduces the bound until a minimum point where ε is proportional to $(N/M)^{1/\rho}$. Reducing ε beyond this point increases the ultimate bound. Thus, the presence of measurement noise puts a lower bound on ε. Another trade-off exists between the speed of convergence of the observer and the ultimate bound on the estimation error.

The ultimate bound given by (8.12) is conservative because it does not take into account the low-pass filtering characteristics of the high-gain observer. To illustrate this point, consider the second-order observer

$$\dot{\hat{x}}_1 = \hat{x}_2 + (\alpha_1/\varepsilon)(y - \hat{x}_1), \quad \dot{\hat{x}}_2 = (\alpha_2/\varepsilon^2)(y - \hat{x}_1),$$

when $y = x_1 + v$. The transfer function from v to $\hat{x}_2$ is

$$G(s) = \frac{\alpha_2 s}{\varepsilon^2 s^2 + \alpha_1 \varepsilon s + \alpha_2}. \tag{8.13}$$

For the sinusoidal noise $v(t) = N \sin(\omega t + \theta)$, the amplitude of the steady-state estimation error due to noise is given by

$$N|G(j\omega)| = \frac{\alpha_2 \omega N}{\sqrt{(\alpha_2 - \varepsilon^2\omega^2)^2 + (\alpha_1 \varepsilon \omega)^2}}.$$

At $\omega = \sqrt{\alpha_2}/\varepsilon$, $N|G(j\omega)| = N\alpha_1/(\alpha_2\varepsilon)$, which agrees with the ultimate bound of (8.12). However, for $\omega \gg \sqrt{\alpha_2}/\varepsilon$ the low-pass feature of $|G(j\omega)|$ will diminish the steady-state error since $\lim_{\omega\to\infty} |G(j\omega)| = 0$. Therefore, the high-frequency components of the noise will be attenuated. The low-pass filtering characteristics of the high-gain observer are captured in the following theorem when $v(t)$ is the sum of sinusoidal signals generated by the model

$$\tau \dot{z} = Sz, \quad v = Hz, \tag{8.14}$$

where $\tau > 0$ and

$$S = \text{block diag}(S_1, \ldots, S_m), \quad S_i = \begin{bmatrix} 0 & \omega_i \\ -\omega_i & 0 \end{bmatrix},$$

and $\|z(t)\| \le N$ for all $t \ge 0$. The sinusoidal signals have frequencies ω_1/τ to ω_m/τ and for $\tau/\varepsilon \ll 1$, $\omega_i/\tau \gg \omega_i/\varepsilon$.

Theorem 8.2. *Consider the system* (8.1)–(8.4) *and the observer* (8.6)–(8.7). *Suppose* (8.9) *to* (8.11) *are satisfied and the measurement noise v is generated by* (8.14). *Then there are $\varepsilon^* \in (0,1]$, $\lambda > 0$, and $T > 0$ such that for $0 < \varepsilon \le \varepsilon^*$ and $0 < \tau/\varepsilon \le \lambda$ the estimation error $\tilde{x}_i = x_i - \hat{x}_i$, for $1 \le i \le \rho$, satisfies the bound*

$$|\tilde{x}_i(t)| \le \varepsilon^{\rho+1-i} c_1 M + \frac{\tau N d_1}{\varepsilon^i} \quad \forall\, t \ge T \tag{8.15}$$

for some positive constants c_1 and d_1. ◇

Proof: Similar to the proof of Theorem 8.1, η satisfies the equation

$$\varepsilon \dot{\eta} = F\eta + \varepsilon\delta - (1/\varepsilon^{\rho-1})Ev,$$

where $\|\delta\| \le L_\delta \|\eta\| + M$. Let p be the solution of

$$\varepsilon \dot{p} = Fp - (1/\varepsilon^{\rho-1})Ev, \quad p(0) = 0,$$

and set $q = \eta - p$. Then $q(0) = \eta(0)$ and

$$\varepsilon \dot{q} = Fq + \varepsilon\delta,$$

where $\|\delta\| \le L_\delta(\|q\|+\|p\|)+M$. Using (8.14), it can be seen that p is the solution of

$$\varepsilon\dot{p} = Fp - (1/\varepsilon^{\rho-1})EHz, \quad \tau\dot{z} = Sz.$$

Let Π be the solution of the Sylvester equation

$$\Pi S = \frac{\tau}{\varepsilon}F\Pi - \frac{\tau}{\varepsilon^\rho}EH. \tag{8.16}$$

Since F is Hurwitz and the eigenvalues of S are on the imaginary axis, the Sylvester equation has a unique solution [156, Problem 9.6]. Using (8.16), it can be shown that $\tilde{p}(t) = p(t) - \Pi z(t)$ satisfies the equation $\varepsilon\dot{\tilde{p}} = F\tilde{p}$. Since F is Hurwitz,

$$\lim_{t\to\infty}[p(t) - \Pi z(t)] = 0.$$

Seeking the solution of (8.16) as a power series in τ/ε, it can be seen that

$$\Pi = -\frac{\tau}{\varepsilon^\rho}\left[EH + \frac{\tau}{\varepsilon}\sum_{k=0}^{\infty}\left(\frac{\tau}{\varepsilon}\right)^k F^{k+1}EHS^{-(k+1)}\right]S^{-1} = -\frac{\tau}{\varepsilon^\rho}[EH + O(\tau/\varepsilon)]S^{-1},$$

where the infinite series converges for sufficiently small τ/ε. Let $W = \sqrt{q^TPq}$, where $P = P^T > 0$ is the solution of the Lyapunov equation $PF + F^TP = -I$. The derivative of W satisfies

$$\begin{aligned}
\varepsilon\dot{W} &\le \frac{1}{2W}\left[-\|q\|^2 + 2\varepsilon\|P\|\,\|q\|(L_\delta\|q\| + L_\delta\|p\| + M)\right] \\
&\le \frac{1}{2W}\left[-\|q\|^2 + 2\varepsilon\|P\|\,\|q\|(L_\delta\|q\| + L_\delta\|\Pi z\| + L_\delta\|p - \Pi z\| + M)\right] \\
&\le \frac{1}{2W}\left[-\tfrac{1}{2}\|q\|^2 + 2\varepsilon\|P\|\,\|q\|(L_\delta\|\Pi z\| + L_\delta\|p(t) - \Pi z(t)\| + M)\right]
\end{aligned}$$

for $2\varepsilon\|P\|L_\delta \le \frac{1}{2}$. Using

$$\sqrt{\lambda_{\min}(P)}\,\|q\| \le W \le \sqrt{\lambda_{\max}(P)}\,\|q\|,$$

we arrive at

$$\varepsilon\dot{W} \le -\frac{1}{4\lambda_{\max}(P)}W + \frac{\|P\|\varepsilon}{\sqrt{\lambda_{\min}(P)}}(L_\delta\|\Pi z\| + L_\delta\|p - \Pi z\| + M).$$

There is a positive constant k_1 such that

$$L_\delta\|\Pi z(t)\| \le \frac{k_1\tau N}{\varepsilon^\rho} \quad \forall\, t \ge 0.$$

Hence,

$$\varepsilon\dot{W} \le -\ell_1 W + \varepsilon\ell_2\left(\frac{k_1\tau N}{\varepsilon^\rho} + M + \gamma(t)\right),$$

where ℓ_1 and ℓ_2 are positive constants and

$$\gamma(t) = L_\delta\|p(t) - \Pi z(t)\| \to 0 \quad \text{as} \quad t \to \infty.$$

By the comparison lemma [78, Lemma 3.4], $W(t) \le U(t)$ where

$$\varepsilon \dot{U} = -\ell_1 U + \varepsilon \ell_2 \left(\frac{k_1 \tau N}{\varepsilon^\rho} + M + \gamma(t) \right), \quad U(0) = W(0).$$

Therefore,

$$\begin{aligned} U(t) &= e^{-\ell_1 t/\varepsilon} W(0) + \int_0^t e^{-\ell_1 (t-\sigma)/\varepsilon} \ell_2 \left(\frac{k_1 \tau N}{\varepsilon^\rho} + M + \gamma(\sigma) \right) d\sigma \\ &= e^{-\ell_1 t/\varepsilon} W(0) + \frac{\varepsilon \ell_2}{\ell_1} \left(\frac{k_1 \tau N}{\varepsilon^\rho} + M \right) \left(1 - e^{-\ell_1 t/\varepsilon} \right) \\ &\quad + \int_0^t e^{-\ell_1 (t-\sigma)/\varepsilon} \ell_2 \gamma(\sigma)\, d\sigma. \end{aligned}$$

Since $\lim_{t\to\infty} \gamma(t) = 0$, the integral term in the preceding expression tends to zero as $t \to \infty$. Hence, there is $T_1 > 0$ such that

$$\|q(t)\| \le k_2 \varepsilon \left(\frac{k_1 \tau N}{\varepsilon^\rho} + M \right) \quad \forall\, t \ge T_1$$

for some positive constant k_2. On the other hand, since $\lim_{t\to\infty} \|p(t) - \Pi z(t)\| = 0$ and $\|\Pi z(t)\| \le k_1 \tau N/\varepsilon^\rho$, there are positive constants k_3 and T_2 such

$$\|p(t)\| \le \frac{k_3 \tau N}{\varepsilon^\rho} \quad \forall\, t \ge T_2.$$

Thus, for all $t \ge \max\{T_1, T_2\}$,

$$\begin{aligned} \|\eta(t)\| &= \|q(t) + p(t)\| \\ &\le k_2 \varepsilon \left(\frac{k_1 \tau N}{\varepsilon^\rho} + M \right) + \frac{k_3 \tau N}{\varepsilon^\rho} \\ &= (\varepsilon k_1 k_2 + k_3) \frac{\tau N}{\varepsilon^\rho} + \varepsilon k_2 M \\ &\le 2 k_3 \frac{\tau N}{\varepsilon^\rho} + \varepsilon k_2 M \end{aligned}$$

for $\varepsilon k_1 k_2 \le k_3$. Using $\tilde{x}_i = \varepsilon^{\rho - i} \eta_i$ yields (8.15). □

Remark 8.1. *The ultimate bound on the estimation error due to measurement noise in Theorem* 8.1 *is* $Nd/\varepsilon^{\rho-1}$*, while in Theorem* 8.2 *it is* $\tau N d_1/\varepsilon^\rho$*. When* $\tau/\varepsilon \ll 1$*, the latter bound will be much smaller, which demonstrates the low-pass filtering characteristics of the high-gain observer.* ◇

The emphasis in this chapter is on measurement noise that causes the estimation error to be of the order of a negative power of ε. It is useful, however, to note that such a situation does not arise if the measurement noise is low frequency. Consider again the second-order high-gain observer with the transfer function $G(s)$ of (8.13) and suppose $v(t) = N \sin(\omega t + \theta)$, where $\omega \ll 1/\varepsilon$. The amplitude of the steady-state estimation error due to noise is given by

$$N|G(j\omega)| = \frac{\alpha_2 \omega N}{\sqrt{(\alpha_2 - \varepsilon^2 \omega^2)^2 + (\alpha_1 \varepsilon \omega)^2}} \approx \omega N.$$

To generalize this observation, consider a special case of (8.1)–(8.4) where $\psi_1 = \cdots = \psi_{\rho-1} = 0$, that is,

$$\dot{w} = f_0(w, x, u), \tag{8.17}$$
$$\dot{x}_i = x_{i+1} \qquad \text{for } 1 \le i \le \rho - 1, \tag{8.18}$$
$$\dot{x}_\rho = \phi(w, x, u), \tag{8.19}$$
$$y = x_1 + v. \tag{8.20}$$

The high-gain observer (8.6)–(8.7) specializes to

$$\dot{\hat{x}}_i = \hat{x}_{i+1} + \frac{\alpha_i}{\varepsilon^i}(y - \hat{x}_1) \qquad \text{for } 1 \le i \le \rho - 1, \tag{8.21}$$
$$\dot{\hat{x}}_\rho = \phi_0(\hat{x}, u) + \frac{\alpha_\rho}{\varepsilon^\rho}(y - \hat{x}_1). \tag{8.22}$$

Suppose $v(t)$ and its derivatives up to the ρth derivative are bounded. In particular,

$$|v^{(i)}(t)| \le N_{i+1} \quad \text{for } 0 \le i \le \rho \ \forall \ t \ge 0, \tag{8.23}$$

where $v^{(0)} \stackrel{\text{def}}{=} v$. Let $\nu = \text{col}(v, v^{(1)}, \ldots, v^{(\rho-1)})$ and $N = \sqrt{N_1^2 + \cdots N_\rho^2}$ so that $\|\nu\| \le N$.

Theorem 8.3. *Consider the system* (8.17)–(8.20) *and the observer* (8.21)–(8.22). *Suppose* (8.11) *and* (8.23) *are satisfied. Then there is $\varepsilon^* \in (0,1]$ such that for $0 < \varepsilon \le \varepsilon^*$ the estimation error $\tilde{x}_i = x_i - \hat{x}_i$, for $1 \le i \le \rho$, satisfies the bound*

$$|\tilde{x}_i(t)| \le N_i + \max\left\{ \frac{b}{\varepsilon^{i-1}} \, e^{-at/\varepsilon}(\|\tilde{x}(0)\| + N), \ \varepsilon^{\rho+1-i} c(LN + M + N_{\rho+1}) \right\} \tag{8.24}$$

for some positive constants a, b, c. ◇

Proof: Let $\eta = \text{col}(\eta_1, \eta_2, \ldots, \eta_\rho)$, where

$$\eta_i = \frac{y^{(i-1)} - \hat{x}_i}{\varepsilon^{\rho-i}} = \frac{x_i - \hat{x}_i + v^{(i-1)}}{\varepsilon^{\rho-i}} \quad \text{for} \quad 1 \le i \le \rho.$$

It can be shown that η satisfies the equation

$$\varepsilon \dot{\eta} = F\eta + \varepsilon B\left[\delta + v^{(\rho)}\right],$$

where

$$F = \begin{bmatrix} -\alpha_1 & 1 & 0 & \cdots & 0 \\ -\alpha_2 & 0 & 1 & \cdots & 0 \\ \vdots & & \ddots & & \vdots \\ -\alpha_{\rho-1} & & & 0 & 1 \\ -\alpha_\rho & 0 & \cdots & \cdots & 0 \end{bmatrix}, \quad B = \begin{bmatrix} 0 \\ 0 \\ \vdots \\ 0 \\ 1 \end{bmatrix},$$

$$\delta = \phi(w, x, u) - \phi_0(\hat{x}, u).$$

The matrix F is Hurwitz by design and

$$|\delta| \le L\|x - \hat{x}\| + M \le L(\|\eta\| + \|\nu\|) + M$$

for $\varepsilon \leq 1$. Let $V = \eta^T P \eta$, where $P = P^T > 0$ is the solution of the Lyapunov equation $PF + F^T P = -I$. Then

$$\varepsilon \dot{V} \leq -\|\eta\|^2 + 2\varepsilon L\|PB\| \, \|\eta\|^2 + 2\varepsilon\|PB\|(LN + M + N_{\rho+1}) \, \|\eta\|.$$

For $\varepsilon L\|PB\| \leq \frac{1}{4}$,

$$\varepsilon \dot{V} \leq -\frac{1}{2}\|\eta\|^2 + 2\varepsilon\|PB\|(LN + M + N_{\rho+1}) \, \|\eta\|,$$

$$\varepsilon \dot{V} \leq -\frac{1}{4}\|\eta\|^2 \quad \forall \, \|\eta\| \geq 8\varepsilon\|PB\|(LN + M + N_{\rho+1}).$$

Therefore [80, Theorem 4.5],

$$\|\eta(t)\| \leq \max\left\{be^{-at/\varepsilon}\|\eta(0)\|, \varepsilon c(LN + M + N_{\rho+1})\right\}.$$

Using

$$\|\eta(0)\| \leq \frac{1}{\varepsilon^{\rho-1}}(\|\tilde{x}(0)\| + \|\nu(0)\|) \leq \frac{1}{\varepsilon^{\rho-1}}(\|\tilde{x}(0)\| + N)$$

and

$$\tilde{x}_i = \varepsilon^{\rho-i}\eta_i - v^{(i-1)},$$

we arrive at (8.24). □

Example 8.1. Consider the van der Pol oscillator with corrupted measurement

$$\dot{x}_1 = x_2, \quad \dot{x}_2 = -x_1 + (1 - x_1^2)x_2, \quad y = x_1 + v,$$

together with the high-gain observer

$$\dot{\hat{x}}_1 = \hat{x}_2 + (2/\varepsilon)(y - \hat{x}_1), \quad \dot{\hat{x}}_2 = (1/\varepsilon^2)(y - \hat{x}_1).$$

We use simulation to study the effect of reducing ε on the steady-state estimation error for two cases of the measurement noise v. In the first case, v is wide-band random signal generated by the Simulink block "Uniform Random Number" with amplitude between ± 0.1 and sample time 0.0001. In the second case, v is a low-frequency signal, which is obtained by passing the noise of the first case through a filter with the transfer function $8000/(s^2 + 10\sqrt{2}s + 100)$, which filters out the high-frequency noise. The numerator of the transfer function is chosen such v has almost the same amplitude in both cases. The noise and filtered noise are shown in Figure 8.1. Simulation is carried out with the initial conditions $x_1(0) = 1$, $x_2(0) = \hat{x}_1(0) = \hat{x}_2(0) = 0$ and two values of ε: 0.01 and 0.001. The estimation error $\tilde{x}_2$ at steady state is shown in Figure 8.2. The upper graphs are for the wide-band noise and the lower ones for the low-frequency noise. For the wide-band noise, reducing ε from 0.01 to 0.001 increases the steady-state error by an order of magnitude. This is not the case for the low-frequency noise, where the estimation error is of the same order of magnitude for the two values of ε.

△

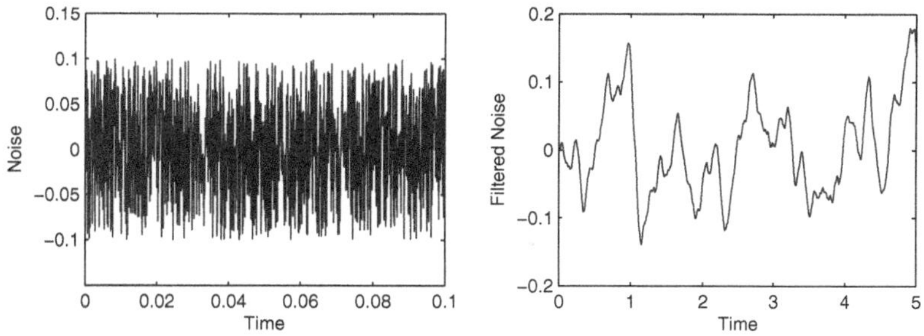

Figure 8.1. *The measurement noise of Example* 8.1. *The noise is generated by a random number generator, and the filtered noise is obtained by passing the random signal through a low-pass filter to remove its high-frequency content.*

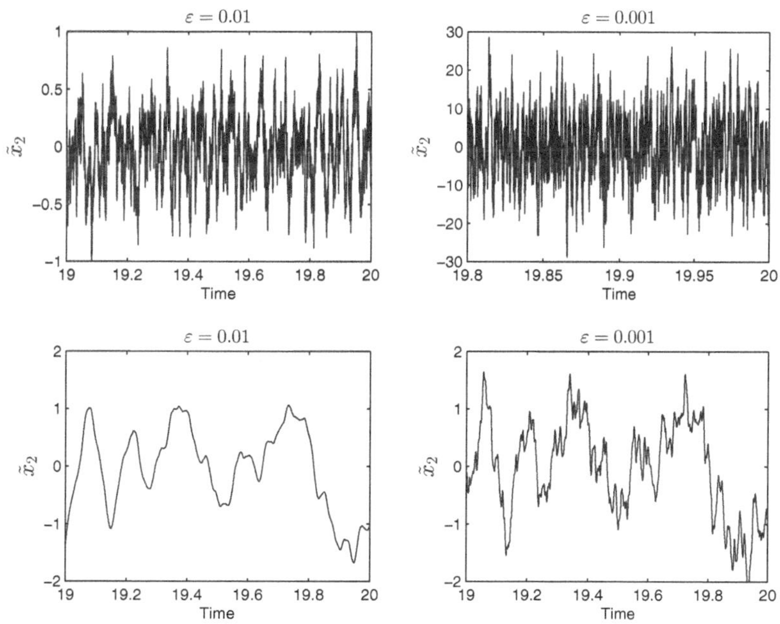

Figure 8.2. *The estimation error* $\tilde{x}_2 = x_2 - \hat{x}_2$ *of Example* 8.1. *The upper graphs are for the wide-band noise and the lower ones for the low-frequency noise.*

8.2 ▪ Closed-Loop Control

Consider the single-input–single-output nonlinear system

$$\dot{w} = f_0(w, x, u), \tag{8.25}$$
$$\dot{x} = Ax + B\phi(w, x, u), \tag{8.26}$$
$$y = Cx + v, \tag{8.27}$$
$$z = \psi(w, x), \tag{8.28}$$

where $u \in R$ is the control input, $y \in R$ and $z \in R^s$ are measured outputs, and $w \in R^\ell$ and $x \in R^\rho$ constitute the state vector. The $\rho \times \rho$ matrix A, the $\rho \times 1$ matrix B, and the $1 \times \rho$ matrix C, given by

$$A = \begin{bmatrix} 0 & 1 & \cdots & \cdots & 0 \\ 0 & 0 & 1 & \cdots & 0 \\ \vdots & & & & \vdots \\ 0 & \cdots & \cdots & 0 & 1 \\ 0 & \cdots & \cdots & \cdots & 0 \end{bmatrix}, \quad B = \begin{bmatrix} 0 \\ 0 \\ \vdots \\ 0 \\ 1 \end{bmatrix}, \quad C = \begin{bmatrix} 1 & 0 & \cdots & \cdots & 0 \end{bmatrix},$$

represent a chain of ρ integrators. The functions f_0, ϕ, and ψ are locally Lipschitz in their arguments for $(w, x, u) \in D_w \times D_x \times R$, where $D_w \subset R^\ell$ and $D_x \subset R^\rho$ are domains that contain their respective origins. Moreover, $f_0(0,0,0) = 0$, $\phi(0,0,0) = 0$, and $\psi(0,0) = 0$. The measurement noise v is a bounded, piecewise continuous function of time. Because the interest in this section is to study the effect of measurement noise on the performance of the high-gain observer, measurement noise is considered only in the measurement y, which will be the driving input of the observer. The state feedback controller is given by

$$\dot{\vartheta} = \Gamma(\vartheta, x, z), \qquad u = \gamma(\vartheta, x, z), \tag{8.29}$$

where γ and Γ are locally Lipschitz functions in their arguments over the domain of interest and globally bounded functions of x. Moreover, $\gamma(0,0,0) = 0$ and $\Gamma(0,0,0) = 0$. A static state feedback controller $u = \gamma(x, z)$ is a special case of the foregoing equation where the $\dot{\vartheta}$-equation is dropped. For convenience, the closed-loop system under state feedback is written as

$$\dot{\chi} = f_r(\chi), \tag{8.30}$$

where $\chi = \text{col}(w, x, \vartheta)$. The output feedback controller is given by

$$\dot{\vartheta} = \Gamma(\vartheta, \hat{x}, z), \qquad u = \gamma(\vartheta, \hat{x}, z), \tag{8.31}$$

where $\hat{x}$ is generated by the high-gain observer

$$\dot{\hat{x}} = A\hat{x} + B\phi_0(z, \hat{x}, u) + H(y - C\hat{x}). \tag{8.32}$$

The observer gain H is chosen as

$$H = \begin{bmatrix} \alpha_1/\varepsilon \\ \alpha_2/\varepsilon^2 \\ \vdots \\ \alpha_{\rho-1}/\varepsilon^{\rho-1} \\ \alpha_\rho/\varepsilon^\rho \end{bmatrix}, \tag{8.33}$$

where ε is a small positive constant and the positive constants α_i are chosen such that the polynomial

$$s^\rho + \alpha_1 s^{\rho-1} + \cdots + \alpha_{\rho-1}s + \alpha_\rho$$

is Hurwitz. The function $\phi_0(z, x, u)$ is a nominal model of $\phi(w, x, u)$, which is locally Lipschitz in its arguments over the domain of interest and globally bounded in x. Moreover, $\phi_0(0,0,0) = 0$.

Theorem 8.4. *Consider the closed-loop system of the plant* (8.25)–(8.28) *and the output feedback controller* (8.31)–(8.32). *Suppose the origin of* (8.30) *is asymptotically stable with*

$\mathscr{R}$ as its region of attraction and $|v(t)| \leq N$ for all $t \geq 0$. Let $\mathscr{S}$ be any compact set in the interior of $\mathscr{R}$, and let $\mathscr{Q}$ be any compact subset of R^{ρ}. Suppose $(\chi(0), \hat{x}(0)) \in \mathscr{S} \times \mathscr{Q}$. Then

- *there exist positive constants c_a and N^* such that for each $N < N^*$ there is a constant $\varepsilon_a = \varepsilon_a(N) > c_a N^{1/\rho}$, with $\lim_{N\to 0} \varepsilon_a(N) = \varepsilon_a^* > 0$, such that for each $\varepsilon \in (c_a N^{1/\rho}, \varepsilon_a]$ the solutions $(\chi(t), \hat{x}(t))$ of the closed-loop system are bounded for all $t \geq 0$;*

- *there exist $N_1^* > 0$ and class $\mathscr{K}$ function ϱ_1 such that for every $N < N_1^*$ and every $\mu_1 > \varrho_1(N)$ there are constants $T_1 = T_1(\mu_1) > 0$ and $\varepsilon_b = \varepsilon_b(N, \mu_1) > c_a N^{1/\rho}$, with $\lim_{N\to 0} \varepsilon_b(N, \mu_1) = \varepsilon_b^*(\mu_1) > 0$, such that for $\varepsilon \in (c_a N^{1/\rho}, \varepsilon_b]$, the solutions of the closed-loop system satisfy*

$$\|\chi(t)\| \leq \mu_1 \quad and \quad \|x(t) - \hat{x}(t)\| \leq \mu_1 \quad \forall\, t \geq T_1; \tag{8.34}$$

- *there exist $N_2^* > 0$ and class $\mathscr{K}$ function ϱ_2 such that for every $N < N_2^*$ and every $\mu_2 > \varrho_2(N)$ there is a constant $\varepsilon_c = \varepsilon_c(N, \mu_2) > c_a N^{1/\rho}$, with $\lim_{N\to 0} \varepsilon_c(N, \mu_2) = \varepsilon_c^*(\mu_2) > 0$, such that for each $\varepsilon \in (c_a N^{1/\rho}, \varepsilon_c]$, the solutions of the closed-loop system satisfy*

$$\|\chi(t) - \chi_r(t)\| \leq \mu_2 \quad \forall\, t \geq 0, \tag{8.35}$$

  where χ_r is the solution of (8.30) with $\chi_r(0) = \chi(0)$.

◇

Remark 8.2. *The three bullets of the theorem show, respectively, boundedness of all trajectories, ultimate boundedness where the trajectories come close to the equilibrium point $\{x = 0\} \times \{x - \hat{x} = 0\}$ as time progresses, and closeness of the trajectories under output feedback to the ones under state feedback. In the absence of measurement noise, it is shown in Theorem 3.1 that the output feedback controller recovers asymptotic stability of the origin if either $\phi_0 = \phi$ or the origin of (8.30) is exponentially stable. Comparison of Theorem 8.4 with Theorem 3.1 shows that the presence of measurement noise is manifested in three points, which are intuitively expected:*

- *The amplitude of measurement noise N is limited by the restriction $N < N^*$.*
- *There is a lower bound on ε, which is of the order $O(N^{1/\rho})$.*
- *The constants μ_1 and μ_2, which measure ultimate boundedness and closeness of trajectories, respectively, cannot be made arbitrarily small. Instead, they are bounded from below by class $\mathscr{K}$ functions of N.*

◇

Proof of Theorem 8.4: With

$$\eta_i = \varepsilon^{i-1}(x_i - \hat{x}_i) \qquad \text{for } 1 \leq i \leq \rho,$$

the closed-loop system can be represented by

$$\dot{\chi} = f(\chi, D^{-1}(\varepsilon)\eta), \tag{8.36}$$

$$\varepsilon\dot{\eta} = F\eta + \varepsilon^{\rho} B g(\chi, D^{-1}(\varepsilon)\eta) - Ev, \tag{8.37}$$

where

$$F = \begin{bmatrix} -\alpha_1 & 1 & 0 & \cdots & 0 \\ -\alpha_2 & 0 & 1 & \cdots & 0 \\ \vdots & & & \ddots & \vdots \\ -\alpha_{\rho-1} & & & 0 & 1 \\ -\alpha_\rho & 0 & \cdots & \cdots & 0 \end{bmatrix}, \quad E = \begin{bmatrix} \alpha_1 \\ \alpha_2 \\ \vdots \\ \alpha_{\rho-1} \\ \alpha_\rho \end{bmatrix},$$

$$D(\varepsilon) = \text{diag}(1,\ \varepsilon, \ldots, \varepsilon^{\rho-1}),$$

$$g(\chi, D^{-1}(\varepsilon)\eta) = \phi(w, x, \gamma(\vartheta, \hat{x}, z)) - \phi_0(z, \hat{x}, \gamma(\vartheta, \hat{x}, z)),$$

and $f(\chi, 0) = f_r(\chi)$. The matrix F is Hurwitz. The initial states are $\chi(0) = \text{col}(w(0), x(0), \vartheta(0)) \in \mathcal{S}$, $\hat{x}(0) \in \mathcal{Q}$, and $\eta(0) = D(\varepsilon)[x(0) - \hat{x}(0)]$. Setting $\eta = 0$ in (8.36) results in

$$\dot{\chi} = f_r(\chi), \tag{8.38}$$

which is the closed-loop system under state feedback. Since the origin of (8.38) is asymptotically stable and $\mathcal{R}$ is its region of attraction, by the converse Lyapunov theorem of asymptotic stability [78, Theorem 4.17], there is a smooth, positive definite function $V(\chi)$ and a continuous, positive definite function $U(\chi)$, both defined for all $\chi \in \mathcal{R}$, such that

$$V(\chi) \to \infty \quad \text{as } \chi \to \partial\mathcal{R},$$

$$\frac{\partial V}{\partial \chi} f_r(\chi) \le -U(\chi) \quad \forall\, \chi \in \mathcal{R},$$

and for any $c > 0$, $\{V(\chi) \le c\}$ is a compact subset of $\mathcal{R}$. Let $\mathcal{S}$ be any compact set in the interior of $\mathcal{R}$. Choose positive constants b and c such that $c > b > \max_{\chi \in \mathcal{S}} V(\chi)$. Then

$$\mathcal{S} \subset \Omega_b = \{V(\chi) \le b\} \subset \Omega_c = \{V(\chi) \le c\} \subset \mathcal{R}.$$

Due to the global boundedness of g in $D^{-1}(\varepsilon)\eta$, there is a constant $k_g > 0$, independent of ε, such that

$$\|g(\chi, D^{-1}(\varepsilon)\eta)\| \le k_g \tag{8.39}$$

for all $\chi \in \Omega_c$ and $\eta \in R^\rho$. We start by analyzing equation (8.37) for $\chi \in \Omega_c$. Since F is Hurwitz, the Lyapunov equation $PF + F^T P = -I$ has a positive definite symmetric solution P. Take $W(\eta) = \eta^T P \eta$ as a Lyapunov function candidate for (8.37). It can be shown that

$$\dot{W} \le -\frac{1}{\varepsilon}\|\eta\|^2 + 2\varepsilon^{\rho-1}\|\eta\|\,\|PB\|k_g + \frac{2}{\varepsilon}\|\eta\|\,\|PE\|N.$$

Hence,

$$\dot{W} \le -\frac{1}{2\varepsilon\|P\|} W \tag{8.40}$$

for $W \ge (\sigma_1 \varepsilon^\rho + \sigma_2 N)^2$, where $\sigma_1 = 4\|PB\|k_g\sqrt{\|P\|}$ and $\sigma_2 = 4\|PE\|\sqrt{\|P\|}$. Therefore, the set

$$\Sigma = \{W(\eta) \le (\sigma_1 \varepsilon^\rho + \sigma_2 N)^2\} \tag{8.41}$$

is positively invariant. When $\eta(0)$ is outside Σ, (8.40) shows that

$$W(\eta(t)) \le W(\eta(0)) \exp\left(-\frac{\sigma_3 t}{\varepsilon}\right), \tag{8.42}$$

where $\sigma_3 = 1/(2\|P\|)$. From the scaling $\eta_i = \varepsilon^{i-1}(x_i - \hat{x}_i)$, we see that whenever $x(0)$, $\hat{x}(0)$, and ε are bounded, there is a constant $k_w > 0$, independent of ε, such that $W(\eta(0)) \le k_w^2$. It follows from (8.41) and (8.42) that η reaches the set Σ within the time interval $[0, T(\varepsilon)]$, where

$$T(\varepsilon) = \frac{2\varepsilon}{\sigma_3} \ln\left(\frac{k_w}{\sigma_1 \varepsilon^\rho}\right) \to 0 \quad \text{as } \varepsilon \to 0. \tag{8.43}$$

Inside Σ, η is bounded by

$$\|\eta\| \le c_1 \varepsilon^\rho + c_2 N, \tag{8.44}$$

where $c_1 = \sigma_1/\sqrt{\lambda_{\min}(P)}$ and $c_2 = \sigma_2/\sqrt{\lambda_{\min}(P)}$. Therefore,

$$\|x(t) - \hat{x}(t)\| = \|D^{-1}(\varepsilon)\eta(t)\| \le \varepsilon c_1 + \frac{N}{\varepsilon^{\rho-1}} c_2 \stackrel{\text{def}}{=} h(\varepsilon, N). \tag{8.45}$$

Lemma 8.1. *$h(\varepsilon, N)$ has the following properties for $\varepsilon > 0$ and $N \ge 0$:*

- *$h(\varepsilon, N)$ has a global minimum at $\varepsilon = [(\rho - 1)c_2 N/c_1]^{(1/\rho)} \stackrel{\text{def}}{=} c_a N^{1/\rho}$ and $\min_{\varepsilon>0} h(\varepsilon, N) = (c_1 c_a + c_2/c_a^{\rho-1})N^{1/\rho} \stackrel{\text{def}}{=} k_a N^{1/\rho}$.*
- *For $\varepsilon > c_a N^{1/\rho}$, $h(\varepsilon, N)$ is a strictly increasing function of ε and $h(\varepsilon, N) \le k_b \varepsilon$, where $k_b = c_1 + c_2/c_a^\rho$.*
- *Given $k > 0$, for every $N \in [0, (k/k_a)^\rho)$ there exist $\varepsilon_m = \varepsilon_m(N, k) \ge 0$ and $\varepsilon_M = \varepsilon_M(N, k) > 0$, with $\varepsilon_m \le c_a N^{1/\rho} < \varepsilon_M$, $\lim_{N\to 0} \varepsilon_m = 0$, and $\lim_{N\to 0} \varepsilon_M = k/c_1$, such that $h(\varepsilon, N) \le k$ for $\varepsilon \in (\varepsilon_m, \varepsilon_M]$.* ◇

Proof of Lemma 8.1: From

$$\frac{\partial h}{\partial \varepsilon} = c_1 - \frac{(\rho - 1)N c_2}{\varepsilon^\rho},$$

we see that h is strictly decreasing for $\varepsilon < [(\rho - 1)c_2 N/c_1]^{(1/\rho)} = c_a N^{1/\rho}$, is strictly increasing for $\varepsilon > c_a N^{1/\rho}$, and has a global minimum at $\varepsilon = c_a N^{1/\rho}$. For $N < (k/k_a)^\rho$, $\min h < k$, and the equation $h(\varepsilon, N) = k$ has two solutions at $\varepsilon_m \le c_a N^{1/\rho}$ and $\varepsilon_M > c_a N^{1/\rho}$, where the equality in $\varepsilon_m \le c_a N^{1/\rho}$ happens only at $N = 0$. Since $\lim_{N\to 0} h(\varepsilon, N) = \varepsilon c_1$, it is clear that $\lim_{N\to 0} \varepsilon_M = k/c_1$. □

Due to global boundedness of f in $D^{-1}(\varepsilon)\eta$, there exist positive constants k_f and L_1, independent of ε, such that

$$\|f(\chi, D^{-1}(\varepsilon)\eta)\| \le k_f \quad \text{and} \quad \|f(\chi, D^{-1}(\varepsilon)\eta) - f(\chi, 0)\| \le L_1 \|D^{-1}(\varepsilon)\eta\|$$

for all $\chi \in \Omega_c$ and $\eta \in R^\rho$. Let L_2 be an upper bound for $\|\partial V/\partial \chi\|$ over Ω_c and set $L = L_1 L_2$. Then

$$\dot{V} \le -U(\chi) + L\|D^{-1}(\varepsilon)\eta\| \le -U(\chi) + Lh(\varepsilon, N) \tag{8.46}$$

for all $(\chi, \eta) \in \Omega_c \times \Sigma$. Let $L_3 = (1/L)\min_{\chi\in\partial\Omega_c} U(\chi)$, apply Lemma 8.1 with $k = L_3$, and set $N_3^* = (k/k_a)^\rho$ and $\bar{\varepsilon}_a = \varepsilon_M$. Then, for $N < N_3^*$ and $\varepsilon \in \left(c_a N^{1/\rho}, \bar{\varepsilon}_a\right]$, $\dot{V} \le 0$ for all $(\chi,\eta)\in\partial\Omega_c\times\Sigma$. From the foregoing analysis, $\dot{W}\le 0$ for all $(\chi,\eta)\in\Omega_c\times\partial\Sigma$. Therefore, the set $\Omega_c\times\Sigma$ is positively invariant. Since $\chi(0)$ is in the interior of Ω_c,

$$\|\chi(t)-\chi(0)\|\le k_1 t \tag{8.47}$$

for some positive constant k_1, independent of ε, as long as $\chi(t)\in\Omega_c$. Therefore, there exists a finite time T^*, independent of ε, such that $\chi(t)\in\Omega_c$ for all $t\in[0,T^*]$. We know from the foregoing analysis that η enters the set Σ during the finite time period $[0,T(\varepsilon)]$, where $T(\varepsilon)\to 0$ as $\varepsilon\to 0$. Hence, there is $\varepsilon_u>0$ such that for all $0<\varepsilon\le\varepsilon_u$, $T(\varepsilon)\le T^*$. Choosing N_4^* small enough that $c_a(N_4^*)^{1/\rho}<\varepsilon_u$ and setting $N^*=\min\{N_3^*,N_4^*\}$, $\varepsilon_a=\min\{\bar{\varepsilon}_a,\varepsilon_u\}$, we conclude that for $0\le N<N^*$ and $c_aN^{1/\rho}<\varepsilon\le\varepsilon_a$, the trajectory (χ,η) enters the set $\Omega_c\times\Sigma$ during the finite time period $[0,T(\varepsilon)]$ and remains in thereafter. Prior to entering this set, $\chi(t)$ and $\eta(t)$ are bounded by (8.47) and (8.42), respectively. Thus, the closed-loop trajectories are bounded.

Inequalities (8.45) and (8.46) are satisfied for all $t\ge T^*$. Take a positive constant $\tau\in(k_aN^{1/\rho},L_3]$, apply Lemma 8.1 with $k=\tau$, and set $\bar{\varepsilon}_b=\varepsilon_M$. Furthermore, take $\varepsilon_b=\min\{\varepsilon_a,\bar{\varepsilon}_b\}$. Then $h(\varepsilon,N)\le\tau$ for $\varepsilon\in\left(c_aN^{1/\rho},\varepsilon_b\right]$, and from (8.46) we have

$$\dot{V}\le -U(\chi)+L\tau\le-\tfrac{1}{2}U(\chi)\ \text{ for }\chi\notin\{U(\chi)\le 2L\tau\}.$$

Since $U(\chi)$ is positive definite and continuous, there is a positive constant $\tau^*<L_3$ such that the set $\{U(\chi)\le 2L\tau\}$ is compact for $\tau\le\tau^*$. Let $c_0(\tau)=\max_{U(\chi)\le 2L\tau}\{V(\chi)\}$; $c_0(\tau)$ is nondecreasing, and $\lim_{\tau\to 0}c_0(\tau)=0$. Let $\varphi(\tau)$ be a class $\mathscr{K}$ function such that $\varphi(\tau)\ge c_0(\tau)$. Then

$$\dot{V}\le-\tfrac{1}{2}U(\chi)\quad\text{for }\ \varphi(\tau)\le V(\chi)\le c.$$

Hence, there exists $T_a=T_a(\tau)\ge 0$ such that

$$V(\chi(t))\le\varphi(\tau)\qquad\forall\ t\ge T^*+T_a(\tau).$$

It can be shown that there is a class $\mathscr{K}_\infty$ function φ_1 such that $\varphi_1(\|\chi\|)\le V(\chi)$ for all $\chi\in\mathscr{R}$.[73] Therefore,

$$\|\chi(t)\|\le\varphi_1^{-1}(\varphi(\tau))\stackrel{\text{def}}{=}\varrho_a(\tau)\qquad\forall\ t\ge T^*+T_a(\tau).$$

Thus,

$$\max\{\|\chi(t)\|,\|x(t)-\hat{x}(t)\|\}\le\max\{\tau,\varrho_a(\tau)\}\stackrel{\text{def}}{=}\varrho_b(\tau).$$

The function ϱ_b is a class $\mathscr{K}_\infty$ function. Define the class $\mathscr{K}_\infty$ function ϱ_1 by $\varrho_1(N)=\varrho_b(k_aN^{1/\rho})$ and suppose $N<N_1^*=\min\{N^*,(\tau^*/k_a)^\rho\}$. Then, given $\mu_1>\varrho_1(N)$, take $\tau=\min\{\tau^*,\varrho_b^{-1}(\mu_1)\}$ and set $T_1=T^*+T_a(\tau)$, to obtain (8.34).

Finally, we show (8.35). From (8.34) and asymptotic stability of the origin of (8.38), given $\mu_2>2\varrho_1(N)$, there is $T_2=T_2(\mu_2)>0$, independent of ε, such that

$$\|\chi(t)\|\le\mu_2/2,\qquad\|\chi_r(t)\|\le\mu_2/2\ \ \forall\ t\ge T_2 \tag{8.48}$$

for $\varepsilon\in\left(c_aN^{1/\rho},\varepsilon_b\right]$. Thus,

$$\|\chi(t)-\chi_r(t)\|\le\|\chi(t)\|+\|\chi_r(t)\|\le\mu_2\ \ \forall\ t\ge T_2. \tag{8.49}$$

[73] See the proof of Theorem 4.17 of [78].

From (8.47) we have

$$\|\chi(t)-\chi(0)\| \le k_1 t \quad \forall\ t \in [0, T(\varepsilon)].$$

Similarly, it can be shown that

$$\|\chi_r(t)-\chi(0)\| \le k_1 \quad \forall\ t \in [0, T(\varepsilon)].$$

Hence,

$$\|\chi(t)-\chi_r(t)\| \le 2k_1 T(\varepsilon) \quad \forall\ t \in [0, T(\varepsilon)]. \tag{8.50}$$

View (8.36) as a perturbation of (8.38). From continuous dependence of the solutions on initial conditions and parameters [78, Theorem 3.4], it follows that

$$\|\chi(t)-\chi_r(t)\| \le 2k_1 c_3 T(\varepsilon) + c_4 h(\varepsilon, N) \quad \forall\ t \in [T(\varepsilon), T_2] \tag{8.51}$$

for some constants $c_3 \ge 1$ and $c_4 > 0$, independent of ε. From (8.50) and (8.51) we conclude that

$$\|\chi(t)-\chi_r(t)\| \le 2k_1 c_3 T(\varepsilon) + c_4 h(\varepsilon, N) \quad \forall\ t \in [0, T_2]. \tag{8.52}$$

It can be verified that $T(\varepsilon)$ is a class $\mathcal{K}$ function for $\varepsilon \le (1/e)(k_w/\sigma_1)^{1/\rho} \stackrel{\text{def}}{=} \varepsilon_g$. Choose N_5^* small enough that $c_a(N_5^*)^{1/\rho} < \varepsilon_g$. Then, for each $N < N_5^*$,

$$\min_{\varepsilon \in (c_a N^{1/\rho},\ \varepsilon_g]} \{2k_1 c_3 T(\varepsilon) + c_4 h(\varepsilon, N)\} = 2k_1 c_3 T(c_a N^{1/\rho}) + c_4 k_a N^{1/\rho} \stackrel{\text{def}}{=} \varrho_3(N).$$

It can be verified that $\varrho_3(N)$ is a class $\mathcal{K}$ function. For each $\mu_2 > \varrho_3(N)$, there exists $\bar{\varepsilon}_c = \bar{\varepsilon}_c(N, \mu_2) > c_a N^{1/\rho}$, with $\lim_{N\to 0} \bar{\varepsilon}_c(N, \mu_2) = \bar{\varepsilon}_c^*(\mu_2) > 0$, such that

$$2k_1 c_3 T(\varepsilon) + c_4 h(\varepsilon, N) \le \mu_2 \tag{8.53}$$

for all $\varepsilon \in \left(c_a N^{1/\rho},\ \bar{\varepsilon}_c\right]$. Taking $\varrho_2(N) = \max\{2\varrho_1(N), \varrho_3(N)\}$, $N_2^* = \min\{N_1^*, N_5^*\}$, $\varepsilon_c = \min\{\varepsilon_b, \bar{\varepsilon}_c, \varepsilon_g\}$, and using (8.49), (8.52), and (8.53), we arrive at (8.35). □

Example 8.2. Reconsider the pendulum equation

$$\dot{x}_1 = x_2, \qquad \dot{x}_2 = \sin x_1 - c_1 x_2 + c_2 u$$

from Example 3.4, where $0 \le c_1 \le 0.2$ and $0.5 \le c_2 \le 2$. It is shown there that, with $y = x_1$ as the measured output, the output feedback controller

$$\begin{aligned}
\dot{\hat{x}}_1 &= \hat{x}_2 + (2/\varepsilon)(y - \hat{x}_1),\\
\dot{\hat{x}}_2 &= \sin \hat{x}_1 - 0.1\hat{x}_2 + 1.25u + (1/\varepsilon^2)(y - \hat{x}_1),\\
u &= -2\left[1.25\pi \ \mathrm{sat}\left(\frac{|\hat{x}_1|}{1.25\pi}\right) + 2.25\pi \ \mathrm{sat}\left(\frac{|\hat{x}_2|}{2.25\pi}\right) + 1\right] \mathrm{sat}\left(\frac{\hat{x}_1 + \hat{x}_2}{0.1}\right)
\end{aligned}$$

stabilizes the origin ($x = 0, \hat{x} = 0$), for sufficiently small ε, and the set $\Omega = \{|x_1| \le 1.25\pi\} \times \{|x_1 + x_2| \le \pi\}$ is included in the region of attraction. Suppose $y = x_1 + v$, where $|v(t)| \le 0.001$, which is consistent with the roundoff error in typical optical encoders. Simulation is carried out with the initial conditions $x_1(0) = -\pi$, $x_2(0) =$

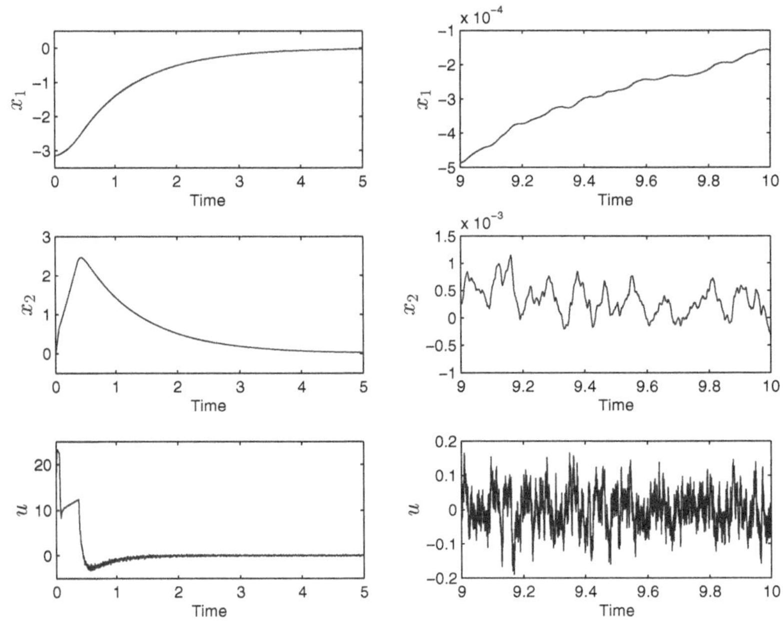

Figure 8.3. *The trajectories of Example* 8.2 *when* $\varepsilon = 0.01$.

Table 8.1. *The maximum over* $t \in [9, 10]$ *of* $|x_1|$, $|x_2|$, $|\tilde{x}_1|$, $|\tilde{x}_2|$, *and* $|u|$.

ε	x_1	x_2	$\tilde{x}_1$	$\tilde{x}_2$	u
0.05	3.614×10^{-4}	4.9417×10^{-4}	1.0958×10^{-4}	7.1861×10^{-4}	0.0147
0.01	4.8717×10^{-4}	0.0012	2.0667×10^{-4}	0.009	0.1899
0.002	6.057×10^{-4}	0.0022	5.1809×10^{-4}	0.1179	2.2367

$\hat{x}_1(0) = \hat{x}_2(0) = 0$, the parameters $c_1 = 0.01$, $c_2 = 0.5$, and with three values of ε: 0.05, 0.01, and 0.002. The noise v is random signal generated by the Simulink block "Uniform Random Number" with amplitude between ± 0.001 and sample time 0.0001. Figure 8.3 shows the trajectories of x_1, x_2, and u for $\varepsilon = 0.01$. It can be seen that the measurement noise has little effect on the transient behavior of the system. Zooming on the interval $t \in [9, 10]$ shows the effect on the steady-state behavior. To examine how the effect of measurement noise changes with ε, Table 8.1 shows the maximum over $t \in [9, 10]$ of the absolute values of x_1, x_2, $\tilde{x}_1 = x_1 - \hat{x}_1$, $\tilde{x}_2 = x_2 - \hat{x}_2$, and u. The table demonstrates that, in general, the steady state of these signals increases as ε decreases. There is, however, an important difference between the various variables. To illustrate this difference, Figure 8.4 displays the numbers for x_1, $\tilde{x}_1$, $\tilde{x}_2$, and u, which are shown as "dots." The same figure displays, in "+", points proportional to $1/\varepsilon$. It can be seen that the increase in $\tilde{x}_2$ and u is of the order of $1/\varepsilon$, while the increase in x_1 and $\tilde{x}_1$ is not of that order. The difference between $\tilde{x}_1$ and $\tilde{x}_2$ is consistent with Theorem 8.1, which shows that the ultimate bound on $\tilde{x}_2$ due to measurement noise is $O(1/\varepsilon)$ while the ultimate bound on $\tilde{x}_1$ is independent of ε. Since the control u depends on $\hat{x}_2$, it is not surprising that its change with ε will be similar to the change of $\tilde{x}_2$. The behavior of x_1 is not unique to this example. It is a general trend that is explored in the next section. △

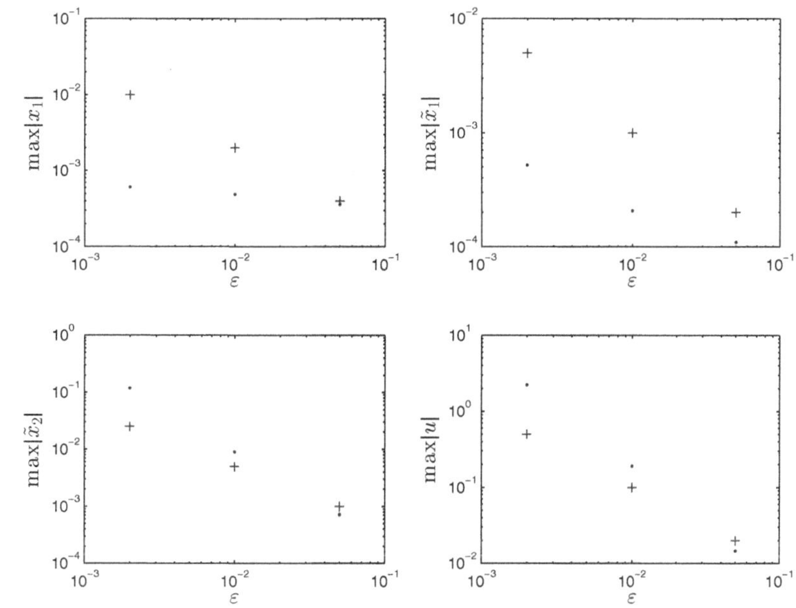

Figure 8.4. *Display of the numbers of Table 8.1 in log-log scale.*

8.3 ▪ Tracking

It is observed in Example 8.2 that the steady state of x_1 in the presence of measurement noise is not of the order of $1/\varepsilon$. This is particularly significant in tracking problems where x_1 is required to track a given reference signal. In this section, we elaborate on this observation for a special case of the tracking problem of Section 3.3 where the plant is linear and has no zero dynamics.

Consider the single-input–single-output linear system

$$\dot{x}_i = x_{i+1} \qquad \text{for } 1 \le i \le n-1, \tag{8.54}$$

$$\dot{x}_n = \sum_{i=1}^{n} a_i x_i + bu + \delta(t), \tag{8.55}$$

$$y = x_1 + v, \tag{8.56}$$

where $b > 0$, δ is a disturbance input, and v is measurement noise. Both δ and v are piecewise continuous and bounded. It is required to design an output feedback controller such that x_1 tracks a reference signal r that satisfies Assumption 3.1 with $\rho = n$. In the error coordinates

$$e_i = x_i - r^{(i-1)} \qquad \text{for } 1 \le i \le n,$$

where $r^{(0)} = r$, the system is represented by

$$\dot{e} = Ae + B(ae + bu + \theta + \delta), \qquad y - r = Ce + v, \tag{8.57}$$

where the triple (A,B,C) represents a chain of n integrators, $e = \text{col}(e_1,\ldots,e_n)$, $a = \begin{bmatrix} a_1 & a_2 & \ldots & a_n \end{bmatrix}$, and $\theta(t) = \sum_{i=1}^{n} a_i r^{(i-1)}(t) - r^{(n)}(t)$.

The state feedback controller is taken as

$$u = -\frac{1}{\mu}(k_1 e_1 + \cdots + k_{n-1} e_{n-1} + e_n),$$

where k_1 to k_{n-1} are chosen such that the polynomial

$$\lambda^{n-1} + k_{n-1}\lambda^{n-2} + \cdots + k_1$$

is Hurwitz, and μ is a small positive constant to be determined. This linear high-gain feedback control is similar to the sliding mode control of Section 3.3.2, but it maintains the linearity of the closed-loop system, which is given by

$$\dot{e} = [A + Ba - (b/\mu)BK]e + B(\theta + \delta), \tag{8.58}$$

where $K = \begin{bmatrix} k_1 & \dots & k_{n-1} & 1 \end{bmatrix}$.

Lemma 8.2. *There exist $\mu^* > 0$ and $T > 0$ such that the matrix $A + Ba - (b/\mu)BK$ is Hurwitz for all $0 < \mu \le \mu^*$ and the solution of (8.58) satisfies $e = O(\mu)$ for all $t \ge T$.* $\diamond$

Proof: Let

$$s = k_1 e_1 + \cdots + k_{n-1} e_{n-1} + e_n \quad \text{and} \quad \zeta = \text{col}(e_1, \dots, e_{n-1}).$$

The system (8.58) can be represented in the singularly perturbed form

$$\dot{\zeta} = A_1 \zeta + B_1 s, \tag{8.59}$$

$$\mu \dot{s} = \mu C_1 \zeta - (b - \mu a_n - \mu k_{n-1}) s + \mu(\theta + \delta), \tag{8.60}$$

where

$$A_1 = \begin{bmatrix} 0 & 1 & \cdots & \cdots & 0 \\ 0 & 0 & 1 & \cdots & 0 \\ \vdots & & & & \vdots \\ 0 & \cdots & \cdots & 0 & 1 \\ -k_1 & \cdots & \cdots & \cdots & -k_{n-1} \end{bmatrix}, \quad B_1 = \begin{bmatrix} 0 \\ 0 \\ \vdots \\ 0 \\ 1 \end{bmatrix},$$

and $C_1 = \begin{bmatrix} c_1 & c_2 & \dots & c_{n-2} & c_{n-1} \end{bmatrix}$, in which $c_1 = a_1 - (a_n + k_{n-1})k_1$ and $c_i = a_i + k_{i-1} - (a_n + k_{n-1})k_i$ for $2 \le i \le n-1$. The matrix A_1 is Hurwitz by design. Let L be the solution of

$$0 = C_1 + (b - \mu a_n - \mu k_{n-1})L + \mu L A_1 - \mu^2 L B_1 L. \tag{8.61}$$

Equation (8.61) has a unique solution for sufficiently small μ and $L = -(1/b)C_1 + O(\mu)$ [87, Section 2.2]. The change of variables $\psi = s + \mu L \zeta$ transforms the system (8.59)–(8.60) into the block triangular form

$$\dot{\zeta} = (A_1 - \mu B_1 L)\zeta + B_1 \psi, \tag{8.62}$$

$$\mu \dot{\psi} = -(b - \mu a_n - \mu k_{n-1} - \mu^2 L B_1)\psi + \mu(\theta + \delta), \tag{8.63}$$

which shows that the closed-loop eigenvalues are $-(b - \mu a_n - \mu k_{n-1} - \mu^2 L B_1)/\mu$ and the eigenvalues of $(A_1 - \mu B_1 L)$. Hence, the closed-loop matrix is Hurwitz for sufficiently small μ. Since $\theta + \delta$ is bounded, it is seen from (8.63) that ψ is ultimately bounded by $O(\mu)$. Then (8.62) shows that ζ is ultimately bounded by $O(\mu)$. Therefore, e is ultimately bounded by $O(\mu)$. □

In preparation for output feedback, the control is saturated outside a compact set of interest. Since $A+Ba-(b/\mu)BK$ is Hurwitz, let $P_0 = P_0^T > 0$ be the solution of the Lyapunov equation

$$P_0[A+Ba-(b/\mu)BK]+[A+Ba-(b/\mu)BK]^T P_0 = -I, \tag{8.64}$$

$V = e^T P_0 e$, and $\Omega = \{e^T P_0 e \le c\}$, where $c > 0$ is chosen large enough that the initial error $e(0)$ is in the interior of Ω and $\dot{V} < 0$ on the boundary $\{V = c\}$, which is possible because $\theta + \delta$ is bounded. Take

$$U > \max_{e\in\Omega} \left| \frac{1}{\mu}(k_1 e_1 + \cdots + k_{n-1}e_{n-1} + e_n) \right|.$$

The output feedback controller is given by

$$u = -U \text{ sat}\left(\frac{k_1\hat{e}_1 + \cdots + k_{n-1}\hat{e}_{n-1} + \hat{e}_n}{\mu U} \right), \tag{8.65}$$

where the estimates $\hat{e}_1$ to $\hat{e}_n$ are provided by the high-gain observer

$$\dot{\hat{e}}_i = \hat{e}_{i+1} + \frac{\alpha_i}{\varepsilon^i}(y - r - \hat{e}_1), \quad 1 \le i \le n-1, \tag{8.66}$$

$$\dot{\hat{e}}_n = \frac{\alpha_n}{\varepsilon^n}(y - r - \hat{e}_1), \tag{8.67}$$

where ε is a sufficiently small positive constant, and α_1 to α_n are chosen such that the polynomial

$$\lambda^n + \alpha_1\lambda^{n-1} + \cdots + \alpha_{n-1}\lambda + \alpha_n$$

is Hurwitz.

Theorem 8.5. *Consider the closed-loop system of the plant* (8.54)–(8.56) *and the output feedback controller* (8.65)–(8.67). *Let $\mathcal{S}$ be any compact set in the interior of Ω, and let $\mathcal{Q}$ be any compact subset of R^n. Suppose $(e(0), \hat{e}(0)) \in \mathcal{S} \times \mathcal{Q}$. Let $\bar{e}(t)$ be the tracking error under output feedback without noise. Then there is $\mu^* > 0$, and for each $\mu \in (0, \mu^*]$, there are $N^* > 0$ and $T^* > 0$, dependent on μ, such that for each $N < N^*$ there is $\varepsilon_a = \varepsilon_a(N) > c_a N^{1/\rho}$, with $\lim_{N\to 0} \varepsilon_a(N) = \varepsilon_a^* > 0$, such that for each $\varepsilon \in (c_a N^{1/\rho}, \varepsilon_a]$, the solutions $(e(t), \hat{e}(t))$ of the closed-loop system are bounded for all $t \ge 0$ and*

$$e_i(t) - \bar{e}_i(t) = O(N) \quad \textit{for } i = 1, 2, \tag{8.68}$$

$$e_i(t) - \bar{e}_i(t) = O\left(\frac{N}{\varepsilon^{i-2}}\right) \quad \textit{for } 3 \le i \le n \tag{8.69}$$

for all $t \ge T^*$. ◇

Proof: With

$$\varphi_i = \varepsilon^{i-1}(e_i - \hat{e}_i) \qquad \text{for } 1 \le i \le n,$$

the closed-loop system takes the singularly perturbed form

$$\dot{e} = Ae + B\left[ae + \theta + \delta - b\ U\ \text{sat}\left(\frac{K(e - D^{-1}(\varepsilon)\varphi)}{\mu U}\right)\right],$$
$$\varepsilon\dot{\varphi} = F\varphi + \varepsilon^n B\left[ae + \theta + \delta - b\ U\ \text{sat}\left(\frac{K(e - D^{-1}(\varepsilon)\varphi)}{\mu U}\right)\right] - Ev,$$

where

$$F = \begin{bmatrix} -\alpha_1 & 1 & 0 & \cdots & 0 \\ -\alpha_2 & 0 & 1 & \cdots & 0 \\ \vdots & & \ddots & & \vdots \\ -\alpha_{n-1} & & & 0 & 1 \\ -\alpha_n & 0 & \cdots & \cdots & 0 \end{bmatrix}, \quad E = \begin{bmatrix} \alpha_1 \\ \alpha_2 \\ \vdots \\ \alpha_{n-1} \\ \alpha_n \end{bmatrix},$$

and

$$D(\varepsilon) = \text{diag}(1,\ \varepsilon, \ldots, \varepsilon^{n-1}).$$

The matrix F is Hurwitz. In analyzing this system, as ε tends to zero, μ is maintained at a value small enough for the conclusions of Lemma 8.2 to hold. Similar to the proof of Theorem 8.4, it can be shown that there is $T(\varepsilon) > 0$, with $\lim_{\varepsilon \to 0} T(\varepsilon) = 0$, such that for all $t \geq T(\varepsilon)$ the trajectories of the closed-loop system are confined to the positively invariant set $\Omega \times \Sigma$, where Σ is defined by (8.41). Inside $\Omega \times \Sigma$ the control saturation is not active, and the closed-loop system is linear. The closed-loop system without noise is represented by the same linear equations with $v = 0$. Therefore $\tilde{e} = e - \bar{e}$ is determined by the linear singularly perturbed system

$$\dot{\tilde{e}} = [A + Ba - (b/\mu)BK]\tilde{e} + (b/\mu)BKD^{-1}(\varepsilon)\eta, \tag{8.70}$$
$$\varepsilon\dot{\eta} = \varepsilon^n B[a - (b/\mu)K]\tilde{e} + [F + (\varepsilon^n b/\mu)BKD^{-1}(\varepsilon)]\eta - Ev, \tag{8.71}$$

where η is the difference between φ and the corresponding variable in the absence of noise. Since $F + (\varepsilon^n b/\mu)BKD^{-1}(\varepsilon) = F + O(\varepsilon)$, it is Hurwitz for sufficiently small ε. Next, a change of variables is used to transform the system (8.70)–(8.71) into a block triangular form where the fast variable η is removed from the slow equation [87, Section 2.2]. Let M be the solution of the equation

$$\begin{aligned} 0 = {} & \varepsilon[A + Ba - (b/\mu)BK]M + (b/\mu)BK\varepsilon^{n-1}D^{-1}(\varepsilon) \\ & - \varepsilon^2 MB[a - (b/\mu)K]M - M[F + (\varepsilon^n b/\mu)BKD^{-1}(\varepsilon)]. \end{aligned} \tag{8.72}$$

Noting that $\varepsilon^n D^{-1}(\varepsilon) = O(\varepsilon)$, it can be seen that (8.72) has a unique solution. The change of variables

$$\xi = \tilde{e} - \frac{1}{\varepsilon^{n-2}}M\eta \tag{8.73}$$

transforms (8.70)–(8.71) into the form

$$\dot{\xi} = [A + (I - \varepsilon M)B(a - (b/\mu)K)]\xi + (1/\varepsilon^{n-1})MEv, \tag{8.74}$$
$$\begin{aligned} \varepsilon\dot{\eta} = {} & [F + (\varepsilon^n b/\mu)BKD^{-1}(\varepsilon) + \varepsilon^2 B(a - (b/\mu)K)M]\eta \\ & + \varepsilon^n B[a - (b/\mu)K]\xi - Ev. \end{aligned} \tag{8.75}$$

Deriving an ultimate bound on $\tilde{e}$ is carried out by deriving an ultimate bound on ξ from (8.74), then using (8.73). In both cases it is important to reveal how M depends on ε. This can be done by seeking the solution of (8.72) using successive approximation. Rewrite (8.72) as

$$\begin{aligned} M = {} & \varepsilon[A+Ba-(b/\mu)BK]MF^{-1}+(b/\mu)(I-\varepsilon M)BKS(\varepsilon)F^{-1} \\ & -\varepsilon^2 MB[a-(b/\mu)K]MF^{-1}, \end{aligned} \tag{8.76}$$

where

$$S(\varepsilon)=\varepsilon^{n-1}D^{-1}(\varepsilon)=\text{diag}(\varepsilon^{n-1},\varepsilon^{n-2},\ldots,\varepsilon,1).$$

For sufficiently small ε, the right-hand side of (8.76) is a contraction mapping. Therefore, we can solve for M using the successive approximation [101]

$$\begin{aligned} M_{k+1} = {} & \varepsilon[A+Ba-(b/\mu)BK]M_k F^{-1}+(b/\mu)(I-\varepsilon M_k)BKS(\varepsilon)F^{-1} \\ & -\varepsilon^2 M_k B[a-(b/\mu)K]M_k F^{-1}, \end{aligned} \tag{8.77}$$

starting with $M_0=(b/\mu)BKSF^{-1}$. After k iterations, the exact solution is approximated to within $O(\varepsilon^k)$ error [87, Section 2.2]. It can be seen that the successive terms $M_0, M_1, \ldots$ take the form

$$\begin{aligned} M_0 &= (\cdots)SF^{-1}, \\ M_1 &= (\cdots)SF^{-1}+\varepsilon(\cdots)SF^{-2}, \\ M_2 &= (\cdots)SF^{-1}+\varepsilon(\cdots)SF^{-2}+\varepsilon^2(\cdots)SF^{-3}, \\ M_3 &= (\cdots)SF^{-1}+\varepsilon(\cdots)SF^{-2}+\varepsilon^2(\cdots)SF^{-3}+\varepsilon^3(\cdots)SF^{-4}, \\ &\vdots \quad \vdots \\ M_{n-2} &= (\cdots)SF^{-1}+\varepsilon(\cdots)SF^{-2}+\cdots+\varepsilon^{n-2}(\cdots)SF^{-(n-1)}+\varepsilon^{n-1}(\cdots). \end{aligned}$$

It can be also verified that

$$SF^{-1}E=\begin{bmatrix} -\varepsilon^{n-1} \\ 0 \\ \vdots \\ \vdots \\ 0 \end{bmatrix}, \quad SF^{-2}E=\begin{bmatrix} 0 \\ -\varepsilon^{n-2} \\ 0 \\ \vdots \\ 0 \end{bmatrix}, \quad \ldots, \quad SF^{-(n-1)}E=\begin{bmatrix} 0 \\ \vdots \\ 0 \\ -\varepsilon \\ 0 \end{bmatrix}.$$

Therefore, $ME=O(\varepsilon^{n-1})$, and the term $(1/\varepsilon^{n-1})MEv$ on the right-hand side of (8.74) is $O(N)$, uniformly in ε for sufficiently small ε. Similar to the proof of Lemma 8.2, it can be shown that the ultimate bound of ξ is $O(N)$. Alternatively, it can be seen that

$$\begin{aligned} M_0 &= B(\cdots), \\ M_1 &= B(\cdots)+\varepsilon AB(\cdots), \\ M_2 &= B(\cdots)+\varepsilon AB(\cdots)+\varepsilon^2 A^2B(\cdots), \\ &\vdots \quad \vdots \\ M_{n-2} &= B(\cdots)+\varepsilon AB(\cdots)+\cdots+\varepsilon^{n-2}A^{n-2}B(\cdots)+\varepsilon^{n-1}(\cdots), \end{aligned}$$

and

$$B = \begin{bmatrix} 0 \\ \vdots \\ \vdots \\ 0 \\ 0 \\ 1 \end{bmatrix}, \quad AB = \begin{bmatrix} 0 \\ \vdots \\ \vdots \\ 0 \\ 1 \\ 0 \end{bmatrix}, \quad A^2B = \begin{bmatrix} 0 \\ \vdots \\ 0 \\ 1 \\ 0 \\ 0 \end{bmatrix}, \quad \ldots, \quad A^{n-2}B = \begin{bmatrix} 0 \\ 1 \\ 0 \\ \vdots \\ \vdots \\ 0 \end{bmatrix}.$$

Therefore,

$$M = M_{n-2} + O(\varepsilon^{n-1}) = \begin{bmatrix} O(\varepsilon^{n-1}) \\ O(\varepsilon^{n-2}) \\ \vdots \\ O(\varepsilon) \\ O(1) \end{bmatrix}.$$

Since the matrix $F + (\varepsilon^n b/\mu)BKD^{-1}(\varepsilon) + \varepsilon^2 B(a - (b/\mu)K)M$ is Hurwitz for sufficiently small ε, it follows from (8.75) that the ultimate bound of η is $O(N)$. Using the ultimate bounds of ξ and η, (8.73), and the foregoing orders of magnitude of the components of M, we arrive at (8.68) and (8.69). □

Remark 8.3. *The significant finding of Theorem* 8.5 *is that, unlike the estimation error, the component of the tracking error due to measurement noise does not increase as a negative power of* ε. *This is important because it is the tracking error that usually has to meet stringent accuracy requirements. This does not mean, however, that we can arbitrarily decrease* ε *without consequences. For high-dimensional systems, the effect of noise on* e_3 *to* e_n *will increase as a negative power of* ε. *More importantly, the effect of noise on the control signal, which is a function of all the state estimates, will be* $O(N/\varepsilon^{n-1})$. *Another important observation about the control signal is seen from the proof of the theorem, which requires the noise level to be small enough such that the control signal saturates only during the peaking period of the observer. The ultimate bound estimates* (8.68) *and* (8.69) *are not guaranteed to hold if the noise causes the control signal to saturate beyond the peaking period.* ◇

Example 8.3. Consider the system

$$\dot{x}_1 = x_2, \quad \dot{x}_2 = x_3, \quad \dot{x}_3 = -x_1 - x_3 + u, \quad y = x_1 + v,$$

where x_1 is required to track $r = \cos t$. The state feedback controller is taken as

$$u = -\frac{1}{\mu}(e_1 + 2e_2 + e_3)$$

with $\mu = 0.05$. The maximum of $|u|$ over $\Omega = \{e^T P_0 e \le 2\}$, where P_0 is the solution of the Lyapunov equation (8.64), is 177.1. The control is saturated at ± 200. The third-order observer is designed with $\alpha_1 = \alpha_2 = 3$ and $\alpha_3 = 1$. The simulation is carried out with $x(0) = 0$, $\hat{e}(0) = 0$, and the measurement noise is generated by the uniform random number block of Simulink with amplitude ± 0.0001 and sampling time 0.0001. Four values of ε are tried: 0.01, 0.005, 0.002, and 0.001. Figure 8.5 shows that tracking error $e_1 = x_1 - r$ under state feedback (noise free) and output feedback with two

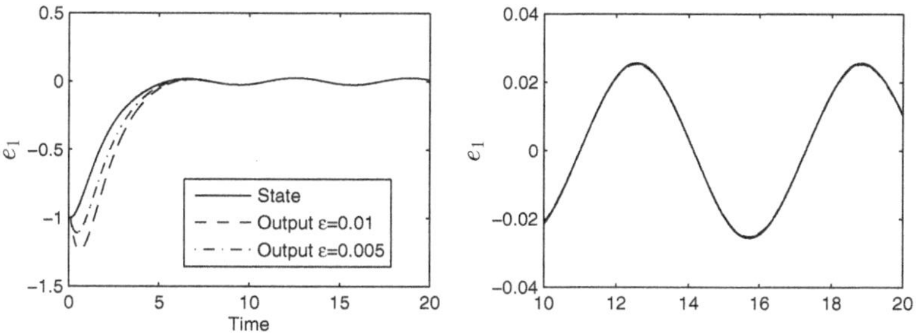

Figure 8.5. *The tracking error of Example* 8.3 *under noise-free state feedback (solid) and noisy output feedback for two values of* ε.

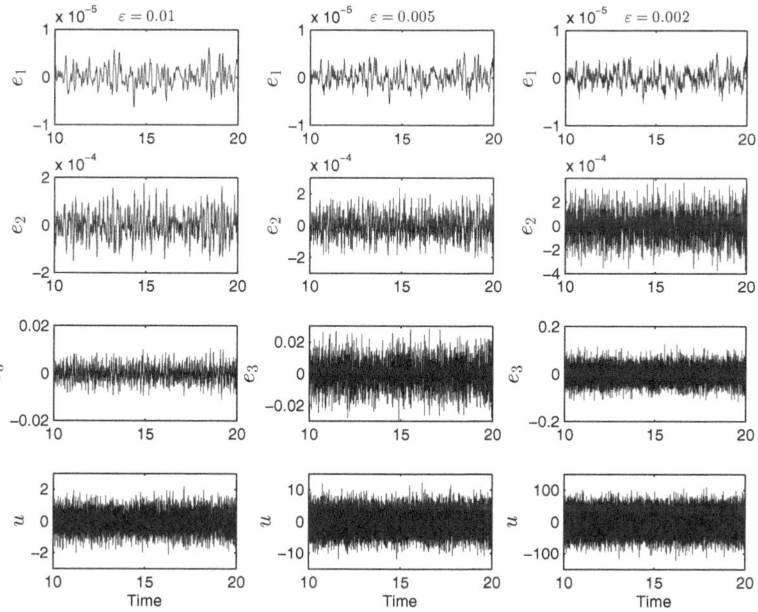

Figure 8.6. *The difference between the trajectories under noisy output feedback and noise free output feedback of Example* 8.3. *The left column is for* $\varepsilon = 0.01$, *the middle column for* $\varepsilon = 0.005$, *and the right column for* $\varepsilon = 0.002$.

different values of ε. The error is shown over the time interval $[0, 20]$ and also over the interval $[10, 20]$ to examine the transient and steady-state behavior. The simulation shows that the response under output feedback approaches the noise-free response under state feedback as ε decreases. To focus on the effect of measurement noise, Figure 8.6 shows the difference between output feedback trajectories with and without noise for $\varepsilon = 0.01$, 0.005, and 0.002. The response is shown over the time interval $[10, 20]$ to focus on the steady-state response. The figure demonstrates the ultimate bounds of Theorem 8.5. As ε decreases, e_1 and e_2 remain almost the same, e_3 grows proportional to $1/\varepsilon$, while u grows proportional to $1/\varepsilon^2$. Figure 8.7 shows the control u and the tracking error e_1 when $\varepsilon = 0.001$. In this case the control saturates beyond the peaking period of the observer. As noted in Remark 8.3, the ultimate bounds of Theorem 8.5 no longer hold, and that is clear by examining e_1. Let us now use transfer functions to interpret the results we have seen so far. Let H_1, H_2, H_3, and H_4 be the

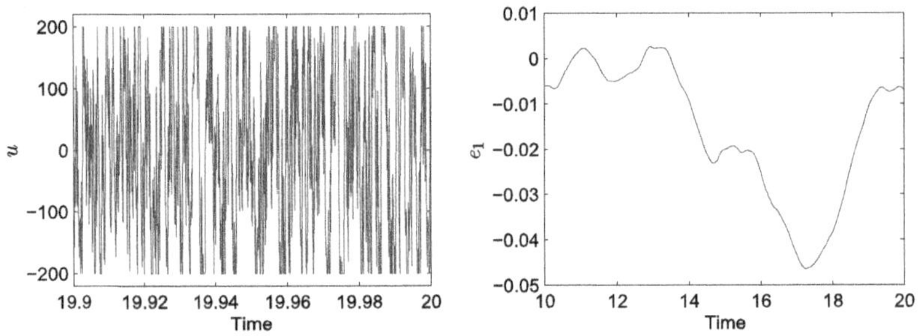

Figure 8.7. *The control signal u and the difference between tracking error e_1 under noisy output feedback and noise-free output feedback of Example* 8.3 *when $\varepsilon = 0.001$.*

transfer functions from the noise v to e_1, e_2, e_3, and u, respectively. They are given by

$$H_1 = \frac{-q(s,\varepsilon)}{\chi(s,\varepsilon)}, \quad H_2 = \frac{-sq(s,\varepsilon)}{\chi(s,\varepsilon)}, \quad H_3 = \frac{-s^2 q(s,\varepsilon)}{\chi(s,\varepsilon)}, \quad H_4 = \frac{-q(s,\varepsilon)\lambda(s)}{\chi(s,\varepsilon)},$$

where

$$\begin{aligned} \lambda(s) &= s^3 + s^2 + 1, \\ q(s,\varepsilon) &= (1 + 6\varepsilon + 3\varepsilon^2)s^2 + (2 + 3\varepsilon)s + 1, \\ \chi(s,\varepsilon) &= \mu p(\varepsilon s)\lambda(s) + q(s,\varepsilon), \\ p(\varepsilon s) &= (\varepsilon s + 1)^3. \end{aligned}$$

It can be shown that $\max_{\omega \in R} |H_i(j\omega)|$ is $O(1)$ for H_1 and H_2, $O(1/\varepsilon)$ for H_2, and $O(1/\varepsilon^2)$ for H_4. The magnitude Bode plots of these four transfer functions are shown in Figure 8.8 for $\varepsilon = 0.01$ and 0.002. It is clear that decreasing ε does not increase the peak of the magnitude plot of H_1 and H_2. However, for the smaller ε the magnitude plot does not roll off as fast as for the larger ε. Thus, while the effect of noise will not increase as a negative power of ε, more higher-frequency components of the noise will pass to e_1 and e_2. A careful examination of Figure 8.6 shows that there is an increase in the high-frequency content of e_1 and e_2 as ε decreases. For H_3 and H_4, the figure shows that the maximum of the magnitude plots of H_3 and H_4 grows by $O(1/\varepsilon)$ and $O(1/\varepsilon^2)$, respectively. Noting that the ratio of 0.01 to 0.002 is 5, $20\log(5) = 14$ dB, and $20\log(25) = 28$ dB, it can been checked from the figure that this is roughly the increase in the peaks of H_3 and H_4. △

8.4 ▪ Reducing the Effect of Measurement Noise

It is shown in the previous sections that the presence of measurement noise puts a constraint on how high the observer gain could be, which forces a trade-off between the fast convergence of the state estimates and the error due to measurement noise on the one hand, and a trade-off between the steady-state errors due to model uncertainty and measurement noise on the other. Two ideas are available to reduce the effects of measurement noise. In experimental applications of high-gain observers, it is common to use a low-pass filter to filter out the high-frequency content of the noise before feeding the measurement into the observer. In Section 8.4.1 we analyze the effect of measurement noise when a low-pass filter is used. Another idea is to adjust the observer gain

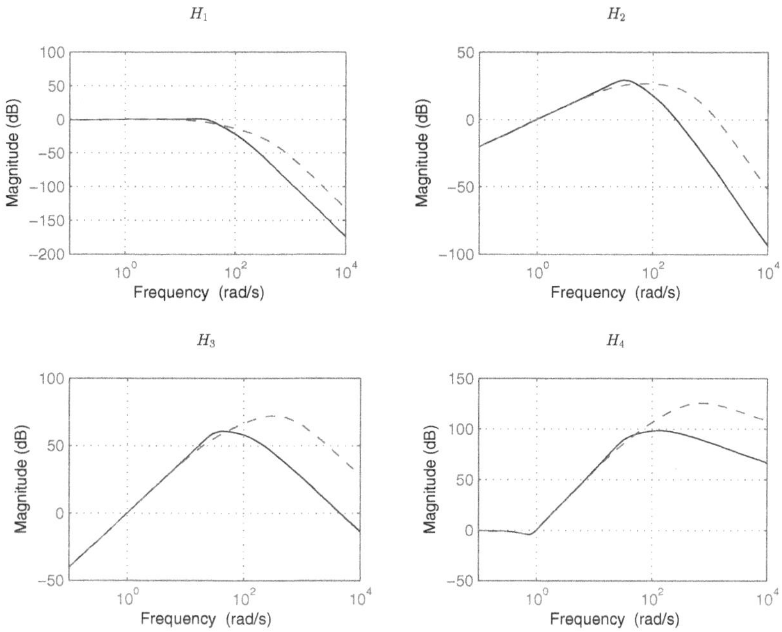

Figure 8.8. *Magnitude Bode plots of the transfer functions of Example* 8.3 *for* $\varepsilon = 0.01$ *(solid) and* $\varepsilon = 0.002$ *(dashed).*

so that a higher gain is used during the transient period to achieve fast convergence; then the gain is lowered as the estimation error approaches steady-state since the effect of measurement noise is prominent when the estimation error is small. Three techniques have been proposed in the literature to achieve this gain adjustment. The first technique switches the gain between two values [2]. The second uses an adaptive law to adjust the gain [130]. The third technique replaces the linear gain of the observer with a nonlinear one that reduces the gain when the estimation error is small [120]. The latter technique is described in Section 8.4.2.

8.4.1 ▪ Low-Pass Filters

For linear systems, the effect of inserting a low-pass filter in a feedback loop can be simplified by transfer function manipulation. Figure 8.9 shows a feedback control system with measurement noise and a low-pass filter. By transfer function manipulation, the system can be brought into the equivalent block diagram of Figure 8.10, where the effect of measurement noise appears as a filtered input to the observer. The goal of the first part of this section is to provide a nonlinear version of this manipulation.

Consider the single-output nonlinear system

$$\dot{w} = f_0(w, x, u), \tag{8.78}$$

$$\dot{x}_i = x_{i+1} \quad \text{for } 1 \le i \le \rho - 1, \tag{8.79}$$

$$\dot{x}_\rho = \phi(w, x, u), \tag{8.80}$$

$$y = x_1 + v, \tag{8.81}$$

where $w \in R^\ell$ and $x = \text{col}(x_1, x_2, \ldots, x_\rho) \in R^\rho$ form the state vector, $u \in R^m$ is the input, $y \in R$ is the measured output, and $v(t) \in R$ is the measurement noise. The

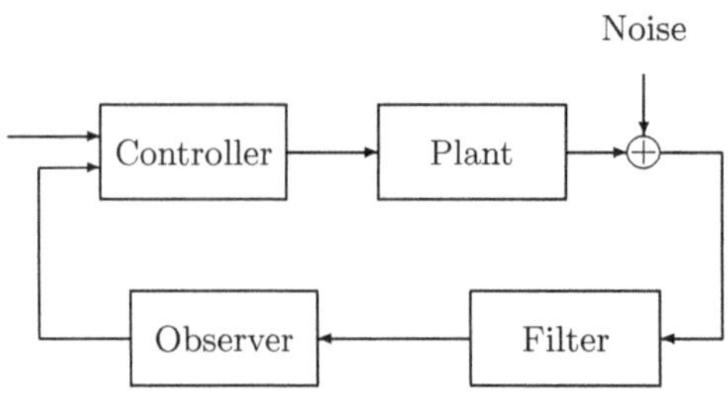

Figure 8.9. *A linear feedback control system with measurement noise and low-pass filter.*

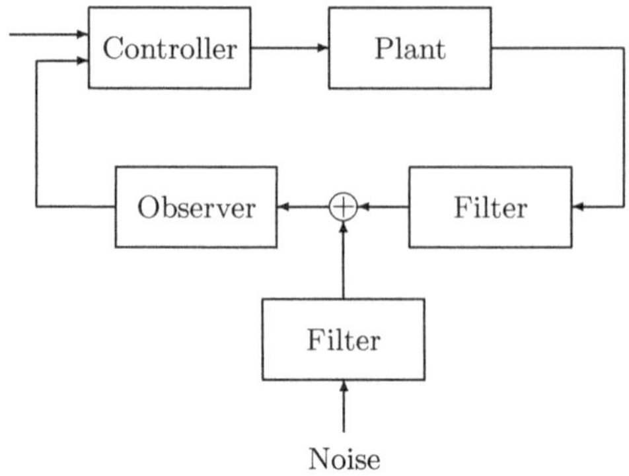

Figure 8.10. *Equivalent representation of the system of Figure 8.9.*

system (8.78)–(8.81) is a special case of the system (8.1)–(8.4), where $\psi_i = 0$ for $1 \le i \le \rho - 1$. As in Section 8.1, it is assumed that f_0 and ϕ are locally Lipschitz in their arguments, $u(t)$ and $v(t)$ are piecewise continuous functions of t, and $w(t)$, $x(t)$, $u(t)$, and $v(t)$ are bounded for all $t \ge 0$. In particular, $w(t) \in W \subset R^\ell$, $x(t) \in X \subset R^\rho$, $u(t) \in U \subset R^m$, and $|v(t)| \le N$ for all $t \ge 0$, for some compact sets W, X, and U, and a positive constant N. A high-gain observer that estimates x by $\hat{x}$ is given by

$$\dot{\hat{x}}_i = \hat{x}_{i+1} + \frac{\alpha_i}{\varepsilon^i}(y - \hat{x}_1) \quad \text{for } 1 \le i \le \rho - 1, \tag{8.82}$$

$$\dot{\hat{x}}_\rho = \phi_0(\hat{x}, u) + \frac{\alpha_\rho}{\varepsilon^\rho}(y - \hat{x}_1), \tag{8.83}$$

where ϕ_0 is a nominal model of ϕ, ε is a sufficiently small positive constant, and α_1 to α_ρ are chosen such that the polynomial

$$s^\rho + \alpha_1 s^{\rho-1} + \cdots + \alpha_{\rho-1} s + \alpha_\rho \tag{8.84}$$

is Hurwitz. It is assumed that ϕ_0 is locally Lipschitz in its arguments and globally bounded in $\hat{x}$, that is,

$$|\phi_0(\hat{x}, u)| \le M_0 \tag{8.85}$$

for all $\hat{x} \in R^\rho$ and $u \in U$. As in the proof of Theorem 8.1, the scaled estimation errors

$$\eta_i = \frac{x_i - \hat{x}_i}{\varepsilon^{\rho-i}} \quad \text{for } 1 \le i \le \rho \tag{8.86}$$

satisfy the equation

$$\varepsilon\dot{\eta} = F\eta + \varepsilon B\delta(w, x, \hat{x}, u) - (1/\varepsilon^{\rho-1})Ev, \tag{8.87}$$

where $\eta = \text{col}(\eta_1, \eta_2, \ldots, \eta_\rho)$, $B = \text{col}(0, \ldots, 0, 1)$,

$$F = \begin{bmatrix} -\alpha_1 & 1 & 0 & \cdots & 0 \\ -\alpha_2 & 0 & 1 & \cdots & 0 \\ \vdots & & \ddots & & \vdots \\ -\alpha_{\rho-1} & & & 0 & 1 \\ -\alpha_\rho & 0 & \cdots & \cdots & 0 \end{bmatrix}, \quad E = \begin{bmatrix} \alpha_1 \\ \alpha_2 \\ \vdots \\ \alpha_{\rho-1} \\ \alpha_\rho \end{bmatrix},$$

and $\delta = \phi(w,x,u) - \phi_0(\hat{x},u)$. The matrix F is Hurwitz by design. In view of (8.85) there is a positive constant M, independent of ε, such that

$$|\delta(w,x,\hat{x},u)| \le M \tag{8.88}$$

for all $w \in W$, $x \in X$, $\hat{x} \in R^\rho$, and $u \in U$. We can derive the ultimate bound on the estimation error by isolating the effect of v on η. Let θ satisfy the equation

$$\varepsilon\dot{\theta} = F\theta - Ev, \tag{8.89}$$

and let $\tilde{\eta} = \eta - (1/\varepsilon^{\rho-1})\theta$. It can been seen that $\tilde{\eta}$ satisfies the equation

$$\varepsilon\dot{\tilde{\eta}} = F\tilde{\eta} + \varepsilon B\delta. \tag{8.90}$$

Since $|v|$ is bounded by N and $|\delta|$ is bounded by M, (8.89) and (8.90) show that θ is ultimately bounded by $O(N)$ and $\tilde{\eta}$ is ultimately bounded by $O(\varepsilon M)$. Since

$$\tilde{x}_i = \varepsilon^{\rho-i}\eta_i = \varepsilon^{\rho-i}\left(\tilde{\eta}_i + \frac{1}{\varepsilon^{\rho-1}}\theta_i\right),$$

the inequality

$$|\tilde{x}_i(t)| \le \varepsilon^{\rho+1-i}cM + \frac{Nd}{\varepsilon^{i-1}} \quad \forall\, t \ge T \tag{8.91}$$

is satisfied for some positive constants c, d, and T, which is in agreement with Theorem 8.1. The point of this alternative derivation is to show that the effect of measurement noise on the steady state of $\tilde{x}_i$ is captured by the term $(1/\varepsilon^{i-1})\theta_i$.

When a low-pass filter is used, the output y is passed through a single-input–single-output system of the form

$$\tau\dot{z} = A_f z + B_f y, \quad y_f = C_f z, \tag{8.92}$$

where $\tau \ll 1$ is the filter's time constant, $z \in R^r$, A_f is a Hurwitz matrix, and the filter's dc-gain is one, that is, $-C_f A_f^{-1} B_f = 1$. Feeding y_f into the high-gain observer, its equation is given by

$$\dot{\hat{x}}_i = \hat{x}_{i+1} + \frac{\alpha_i}{\varepsilon^i}(y_f - \hat{x}_1) \quad \text{for } 1 \le i \le \rho-1, \tag{8.93}$$

$$\dot{\hat{x}}_\rho = \phi_0(\hat{x},u) + \frac{\alpha_\rho}{\varepsilon^\rho}(y_f - \hat{x}_1). \tag{8.94}$$

Let q satisfy the equation

$$\tau\dot{q} = A_f q + B_f v, \quad v_f = C_f q, \tag{8.95}$$

with $q(0) = 0$. Then $p = z - q$ satisfies the equation

$$\tau\dot{p} = A_f p + B_f x_1, \tag{8.96}$$

with $p(0) = z(0)$. Let $\sigma_1 = p + A_f^{-1}B_f x_1$, $\sigma_i = \dot{\sigma}_{i-1}$, for $2 \le i \le \rho$, and $\xi = p^{(\rho)}$, where $p^{(j)}$ is the jth derivative of p. Then σ_1 to σ_ρ and ξ satisfy the equations

$$\tau\dot{\sigma}_i = A_f\sigma_i + \tau A_f^{-1}B_f x_{i+1} \quad \text{for } 1 \le i \le \rho-1, \tag{8.97}$$

$$\tau\dot{\sigma}_\rho = A_f\sigma_\rho + \tau A_f^{-1}B_f\phi(w,x,u), \tag{8.98}$$

$$\tau\dot{\xi} = A_f\xi + B_f\phi(w,x,u). \tag{8.99}$$

By the boundedness of x and $\phi(w,x,u)$, it can be shown that ξ is ultimately bounded, uniformly in ε and τ, and σ_1 to σ_ρ are ultimately bounded by $O(\tau)$. With

$$\varphi_i = \frac{C_f\sigma_i + x_i - \hat{x}_i}{\varepsilon^{\rho-i}} \quad \text{for } 1 \le i \le \rho, \tag{8.100}$$

and using $-C_f A_f^{-1} B_f = 1$, $y_f = C_f\sigma_1 + x_1 + v_f$, and $\dot{\sigma}_\rho = \xi + A_f^{-1}B_f\dot{x}_\rho$, it can be verified that $\varphi = \text{col}(\varphi_1, \ldots, \varphi_\rho)$ satisfies the equation

$$\varepsilon\dot{\varphi} = F\varphi + \varepsilon B[C_f\xi - \phi_0(\hat{x},u)] - (1/\varepsilon^{\rho-1})Ev_f. \tag{8.101}$$

Let ψ satisfy the equation

$$\varepsilon\dot{\psi} = F\psi - Ev_f, \tag{8.102}$$

and let $\tilde{\varphi} = \varphi - (1/\varepsilon^{\rho-1})\psi$. It can been seen that $\tilde{\varphi}$ satisfies the equation

$$\varepsilon\dot{\tilde{\varphi}} = F\tilde{\varphi} + \varepsilon B[C_f\xi - \phi_0(\hat{x},u)]. \tag{8.103}$$

Since $|v|$ is bounded by N and $|C_f\xi - \phi_0|$ is bounded by some constant M_1, equations (8.102) and (8.103) show that ψ is ultimately bounded by $O(N)$ and $\tilde{\varphi}$ is ultimately bounded by $O(\varepsilon M_1)$. The estimation error $\tilde{x}_i$ is given by

$$\tilde{x}_i = -C_f\sigma_i + \varepsilon^{\rho-i}\varphi_i = -C_f\sigma_i + \varepsilon^{\rho-i}\tilde{\varphi}_i + \frac{1}{\varepsilon^{i-1}}\psi_i. \tag{8.104}$$

We have already seen that the ultimate bound on σ_i is $O(\tau)$. Thus, the effect of measurement noise on the steady state of $\tilde{x}_i$ is captured by the term $(1/\varepsilon^{i-1})\psi_i$.

To compare the effect of measurement noise on the steady-state estimation error with and without filter, we need to compare θ, which satisfies (8.89), with ψ, which satisfies (8.102). Equation (8.89) is driven by the measurement noise v, while (8.102) is driven by the filtered noise v_f. This characterization of the noise effect is the nonlinear counterpart of the block diagram of Figure 8.10.

Consider now closed-loop control of the system (8.78)–(8.81). We design a noise-free state feedback controller that stabilizes the origin, then consider the closed-loop system that arises when the controller is implemented using a high-gain observer together with a low-pass filter. The state feedback controller takes the form $u = \gamma(x)$. The closed-loop system under state feedback is represented by

$$\dot{\chi} = f_r(\chi, \gamma(x)), \tag{8.105}$$

where $\chi = \text{col}(w,x) \in R^{\ell+\rho}$ and

$$f_r(\chi,u) = \begin{bmatrix} f_0(w,x,u) \\ x_2 \\ \vdots \\ x_\rho \\ \phi(w,x,u) \end{bmatrix}.$$

The output feedback controller is taken as $u = \gamma(\hat{x})$, where $\hat{x}$ is provided by the high-gain observer (8.93)–(8.94), and y_f is the output of the low-pass filter (8.92). The closed-loop system under the output feedback controller can be written as

$$\dot{\chi} = f_r(\chi, \gamma(\hat{x})), \tag{8.106}$$

$$\dot{\hat{x}}_i = \hat{x}_{i+1} + \frac{\alpha_i}{\varepsilon^i}(C_f z - \hat{x}_1) \quad \text{for } 1 \le i \le \rho - 1, \tag{8.107}$$

$$\dot{\hat{x}}_\rho = \phi_0(\hat{x}, u) + \frac{\alpha_\rho}{\varepsilon^\rho}(C_f z - \hat{x}_1), \tag{8.108}$$

$$\tau \dot{z} = A_f z + B_f(x_1 + v). \tag{8.109}$$

Assumption 8.1.

- *f_0 is continuously differentiable, ϕ is twice continuously differentiable, $f_0(0,0,0) = 0$, and $\phi(0,0,0) = 0$;*
- *γ is twice continuously differentiable, globally bounded, and $\gamma(0) = 0$. The origin of* (8.105) *is asymptotically stable with a region of attraction $\mathscr{R}$ and locally exponentially stable;*
- *ϕ_0 is twice continuously differentiable, globally bounded, and $\phi_0(0,0) = 0$.*

Theorem 8.6. *Consider the closed-loop system* (8.106)–(8.109) *under Assumption* 8.1. *Let $\mathscr{L}$ be any compact set in the interior of $\mathscr{R}$, let $\mathscr{M}$ be any compact subset of $R^\rho \times R^r$, and suppose that $\chi(0) \in \mathscr{L}$ and $(\hat{x}(0), z(0)) \in \mathscr{M}$. The closed-loop system has the following properties:*

- *When $v = 0$, there exists $\lambda^* > 0$ such that if $\max\{\varepsilon, \tau\} < \lambda^*$, the origin ($\chi = 0$, $\hat{x} = 0, z = 0$) is an exponentially stable equilibrium point and $\mathscr{L} \times \mathscr{M}$ is a subset of its region of attraction.*
- *When $v \neq 0$ with $|v| \le N$, there exist positive constants N^* and k^* and a class $\mathscr{K}$ function ϱ such that for every $N < N^*$ and every $\mu > \varrho(N)$, there are positive constants τ^* and $\varepsilon^* > k^* N^{1/\rho}$ such that for $\tau \le \tau^*$ and $\varepsilon \in [k^* N^{1/\rho}, \varepsilon^*]$, the trajectories of the closed-loop system are bounded and*

$$\|\chi(t) - \chi_r(t)\| \le \mu \tag{8.110}$$

for all $t \ge 0$, where $\chi_r(t)$ is the solution of (8.105) *with $\chi_r(0) = \chi(0)$.* ◇

Remark 8.4. *The theorem is stated for initial states in any compact subset of the region of attraction under state feedback. When the state feedback achieves global or semiglobal stabilization, the output feedback achieves semiglobal stabilization.* ◇

Remark 8.5. *The theorem shows that exponential stability under state feedback is recovered under output feedback when ε and τ are sufficiently small. It is significant that there is no restriction on the ratio τ/ε; in other wards, it is not required that one of the parameters be much smaller than the other. This point can be easily seen in linear systems, as shown in Example* 8.4, *but it takes more work to show it in nonlinear systems. In our analysis we use a nonminimal realization of the closed-loop system by introducing the variables σ_i and ξ, whose equations are copies of the filter equation.* ◇

Remark 8.6. *The second bullet of the theorem shows that the trajectories under output feedback approach the ones under state feedback for sufficiently small ε, τ, and N. As in*

Theorem 8.4, in the presence of measurement noise, the constant μ, which measures the closeness of trajectories, cannot be made arbitrarily small because it is bounded from below by a class $\mathcal{K}$ function of N. ◇

Proof of Theorem 8.6: With the definitions of σ and ξ and the change of variables (8.100), the closed-loop system is represented by

$$\dot{\chi} = f_r(\chi, \gamma(\hat{x})), \tag{8.111}$$
$$\tau\dot{\sigma} = A_{ff}\sigma + \tau A_{ff}^{-1}B_{ff}[Ax + B\phi(\chi, \gamma(\hat{x}))], \tag{8.112}$$
$$\tau\dot{\xi} = A_f\xi + B_f\phi(\chi, \gamma(\hat{x})), \tag{8.113}$$
$$\varepsilon\dot{\varphi} = F\varphi + \varepsilon B[C_f\xi - \phi_0(\hat{x}, \gamma(\hat{x}))] - (1/\varepsilon^{\rho-1})Ev_f, \tag{8.114}$$

where

$$\begin{aligned}
\sigma &= \text{col}(\sigma_1, \dots, \sigma_\rho),\\
\hat{x} &= x + C_{ff}\sigma - \varepsilon^{\rho-1}D^{-1}(\varepsilon)\varphi,\\
A_{ff} &= \text{block diag}[A_f, \dots, A_f],\\
B_{ff} &= \text{block diag}[B_f, \dots, B_f],\\
C_{ff} &= \text{block diag}[C_f, \dots, C_f],\\
D(\varepsilon) &= \text{diag}[1, \varepsilon, \dots, \varepsilon^{\rho-2}, \varepsilon^{\rho-1}],
\end{aligned}$$

and the pair (A, B) is a controllable-canonical-form representation of a chain of ρ integrators. Consider first the case when $v = 0$. By the converse Lyapunov theorem [80, Theorem 3.9], there is a Lyapunov function $V(\chi)$ for the system $\dot{\chi} = f_r(\chi, \gamma(x))$ such that $\Omega = \{V(\chi) \le c\}$ is a compact subset of $\mathcal{R}$ for any $c > 0$. Choose c large enough to include $\mathcal{L}$ is in the interior of Ω. Because $\gamma(\hat{x})$ is globally bounded, there is time $T > 0$ such that $\chi(t) \in \Omega$ for all $t \in [0, T]$. Next it is shown that by choosing τ and ε small enough, there are time $T_1(\tau)$, with $\lim_{\tau\to 0} T_1(\tau) = 0$, and $T_2(\varepsilon)$, with $\lim_{\varepsilon\to 0} T_2(\varepsilon) = 0$, such that $\sigma = O(\tau)$ and $\varphi = O(\varepsilon)$ for all $t \in [T_1 + T_2, T]$. Over this interval, $\hat{x} = x + O(\tau) + O(\varepsilon)$, which can be used, together with the Lyapunov function V, to show that $\chi \in \Omega$ for all $t \ge 0$. Once again, using $\hat{x} = x + O(\tau) + O(\varepsilon)$, it can be shown the trajectories reach a neighborhood $\mathcal{N}$ of the origin, which can be made arbitrarily small by choosing τ and ε small enough. Exponential stability of the origin is shown by local analysis inside $\mathcal{N}$. This is done by using the composite Lyapunov function

$$V_c = bV_0(\chi) + b\tau\tilde{\xi}^T P_1\tilde{\xi} + \tau\sigma^T P_2\sigma + \varepsilon\varphi^T P_3\varphi,$$

where $V_0(\chi)$ is a Lyapunov function for $\dot{\chi} = f_r(\chi, \gamma(x))$ provided by the converse Lyapunov theorem for exponential stability [80, Theorem 3.8], P_1 is the solution of the Lyapunov equation $P_1A_f + A_f^T P_1 = -I$, $P_2 = \text{block diag}(P_1, \dots, P_1)$, P_3 is the solution of the Lyapunov equation $P_3F + F^T P_3 = -I$, b is a positive constant to be chosen, and $\tilde{\xi} = \xi + A_f^{-1}B_f\phi(\chi, \gamma(x))$. The derivative of V_c satisfies the inequality

$$\dot{V}_c \le -\mathcal{Y}^T Q\mathcal{Y},$$

where $\mathscr{Y} = \text{col}(\|\chi\|, \|\tilde{\xi}\|, \|\sigma\|, \|\varphi\|)$ and the symmetric matrix Q is given by

$$Q = \begin{bmatrix} bk_1 & -b\tau k_2 & -k_2(b+\tau) & -k_2(b+\varepsilon) \\ \star & b & -bk_2(1+\tau) & -k_2(b+b\tau+\varepsilon) \\ \star & \star & (1-2\tau k_2) & -k_2(\tau+\varepsilon) \\ \star & \star & \star & (1-2\varepsilon k_2) \end{bmatrix}$$

for some positive constants k_1 and k_2, independent of τ and ε. Setting $\tau = \varepsilon = 0$ in Q results in

$$\begin{bmatrix} bk_1 & 0 & -bk_2 & -bk_2 \\ 0 & b & -bk_2 & -bk_2 \\ \star & \star & 1 & 0 \\ \star & \star & 0 & 1 \end{bmatrix},$$

which is positive definite for sufficiently small b. Thus, by choosing b small enough, Q will be positive definite for sufficiently small τ and ε. Consider next the case $v \neq 0$. By repeating the foregoing argument it can be shown that $\chi(t) \in \Omega$ for all $t \geq 0$, provided $N/\varepsilon^{\rho-1}$ is small enough, which is achieved by requiring ε to satisfy $\varepsilon \geq k^* N^{1/\rho}$; such a choice is feasible if N is sufficiently small. Under this restriction,

$$x(t) - \hat{x}(t) = O(\tau) + O(\varepsilon) + O(N/\varepsilon^{\rho-1})$$

for all $t \geq T_1(\tau) + T_2(\varepsilon)$. Continuous dependence of the solutions on initial conditions and parameters yields the inequality (8.110). □

Example 8.4. The linear system

$$\dot{x}_1 = x_2, \quad \dot{x}_2 = u, \quad y = x_1$$

can be stabilized by the state feedback control $u = -x_1 - 2x_2$, which assigns the closed-loop eigenvalues at $-1, -1$. The output feedback controller that uses a high-gain observer is given by

$$u = -\hat{x}_1 - 2\hat{x}_2, \quad \dot{\hat{x}}_1 = \hat{x}_2 + \frac{2}{\varepsilon}(y_f - \hat{x}_1), \quad \dot{\hat{x}}_2 = \frac{1}{\varepsilon^2}(y_f - \hat{x}_1),$$

which assigns the observer eigenvalues at $-1/\varepsilon$, $-1/\varepsilon$. The observer input y_f is the output of the low-pass filter

$$\frac{1}{(\tau s + 1)^2},$$

which is driven by y. The filter poles are at $-1/\tau, -1/\tau$. In this example it is shown that, for linear systems, the exponential stability result of Theorem 8.6 can be shown by examining the closed-loop poles. The closed-loop characteristic equation is

$$s^2 \Lambda(\varepsilon s)\Lambda(\tau s) + 2(1+\varepsilon)s + 1 = 0, \tag{8.115}$$

where $\Lambda(\beta) = (\beta + 1)^2$. This equation has six roots. To study the asymptotic behavior of the roots for small ε and τ, let $\lambda = \max\{\varepsilon, \tau\}$. As $\lambda \to 0$, the characteristic equation tends to

$$s^2 + 2s + 1 = 0,$$

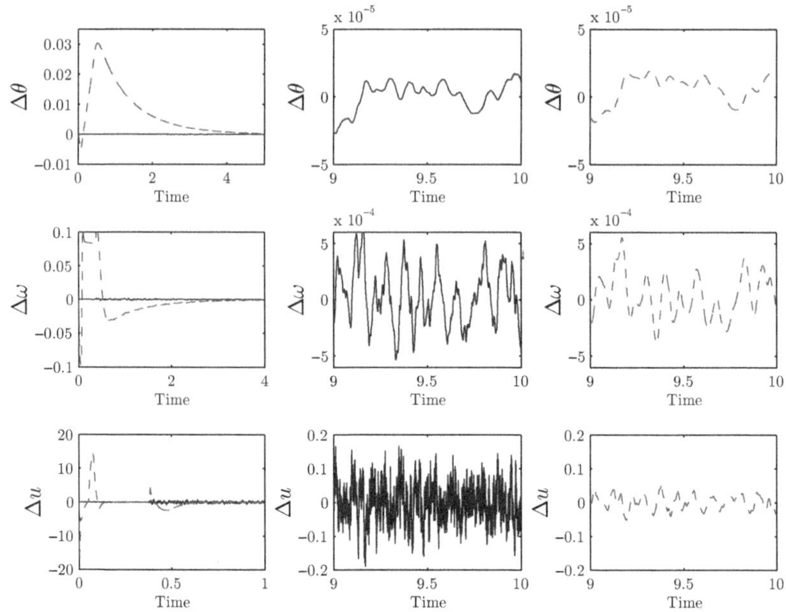

Figure 8.11. *Comparison of the trajectories of Example* 8.5 *with low-pass filter (LPF) (dashed) and without LPF (solid) when* $\varepsilon = \tau = 0.01$. *The plotted curves are the differences from the trajectories under output feedback with no measurement noise.*

which shows that two roots of (8.115) approach the eigenvalues of the closed-loop system under state feedback. Multiplying the characteristic equation by λ^2 and changing the frequency variable to $p = \lambda s$, the equation is rewritten as

$$p^2\Lambda\left(\frac{\varepsilon}{\lambda}p\right)\Lambda\left(\frac{\tau}{\lambda}p\right) + 2\lambda(1+\varepsilon)p + \lambda^2 = 0.$$

As $\lambda \to 0$, $2\lambda(1+\varepsilon)p + \lambda^2 \to 0$, and the equation reduces to

$$p^2\Lambda\left(\frac{\varepsilon}{\lambda}p\right)\Lambda\left(\frac{\tau}{\lambda}p\right) = 0.$$

Dividing by p^2 yields

$$\Lambda\left(\frac{\varepsilon}{\lambda}p\right)\Lambda\left(\frac{\tau}{\lambda}p\right) = 0,$$

whose roots are $p = -\lambda/\varepsilon,\ -\lambda/\varepsilon,\ -\lambda/\tau,\ -\lambda/\tau$. When transformed back into the frequency variable s, it is seen that four roots of (8.115) approach the observer eigenvalues $-1/\varepsilon$, $-1/\varepsilon$, and the filter poles $-1/\tau$, $-1/\tau$. The limits of the six roots of (8.115) are independent of the ratio τ/ε. △

Example 8.5. Reconsider the pendulum equation of Examples 3.4 and 8.2. The effect of adding the low-pass filter $1/(\tau s+1)^2$ is examined for the same controller, initial conditions, and measurement noise parameters as in Example 8.2. Figure 8.11 compares the trajectories of θ, ω, and u when $\varepsilon = \tau = 0.01$. Because the trajectories are fairly close to each other, the figure plots the deviation from the trajectories under no noise.

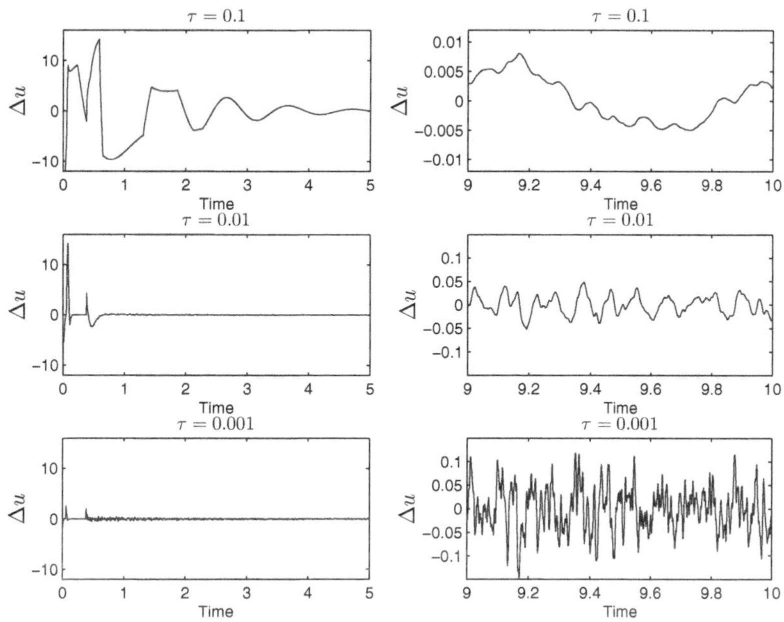

Figure 8.12. *The effect of changing τ on the control trajectory of ω of Example* 8.5.

It is seen that using the filter causes a minor degradation of the transient response while significantly reducing the steady-state error due to measurement noise, which is most obvious in the trajectories of u. To examine the effect of the filter's time constant τ, Figure 8.12 shows the Δu trajectory of Figure 8.11 for three different values of τ : 0.1, 0.01, and 0.001, while $\varepsilon = 0.01$. The large $\tau = 0.1$ smooths the trajectories at steady state, which is expected because more frequency content of the noise is filtered out, but this happens at the expense of a significant degradation of the transient response. The small $\tau = 0.001$ has almost no effect during the transient period, but there is significant increase in the effect of noise at steady state. Of the three values of τ, the best results are obtained with $\tau = 0.01$. △

Example 8.6. The effect of adding the low-pass filter $1/(\tau s + 1)^2$ is examined for the system of Example 8.3 with the same controller, initial conditions, and measurement noise parameters. Figures 8.13 and 8.14 show the trajectories of e_1, e_2, e_3, and u during the transient and steady-state periods when $\tau = \varepsilon = 0.005$. For the steady-state period, Figure 8.14 shows the deviation from the trajectories under output feedback control with no measurement noise and no filter. Adding the filter has a small effect during the transient period. At steady state, it smooths out the trajectories and makes a significant reduction in the effect of noise on the trajectories of e_3 and u. As in Example 8.3, transfer functions can be used to interpret the results. Let H_1, H_2, H_3, and H_4 be the transfer functions from the noise v to e_1, e_2, e_3, and u, respectively. They are given by

$$H_1 = \frac{-q(s,\varepsilon)}{\chi(s,\varepsilon,\tau)}, \quad H_2 = \frac{-sq(s,\varepsilon)}{\chi(s,\varepsilon,\tau)}, \quad H_3 = \frac{-s^2 q(s,\varepsilon)}{\chi(s,\varepsilon,\tau)}, \quad H_4 = \frac{-q(s,\varepsilon)\lambda(s)}{\chi(s,\varepsilon,\tau)},$$

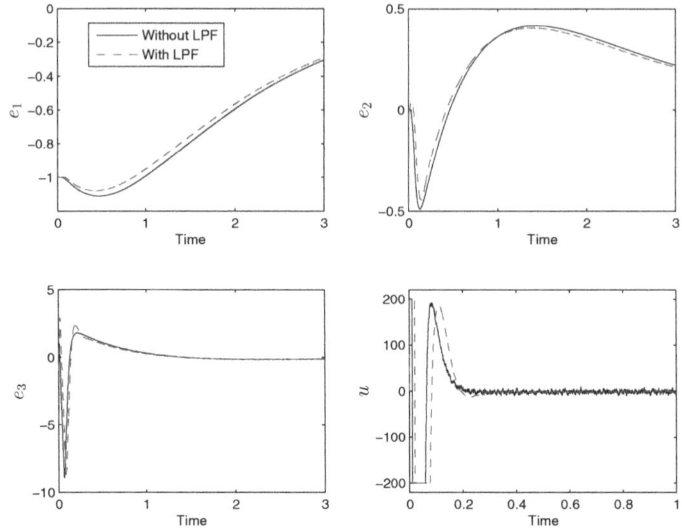

Figure 8.13. *Comparison of the trajectories of Example* 8.6 *with and without low-pass filter (LPF) when the* $\varepsilon = \tau = 0.005$.

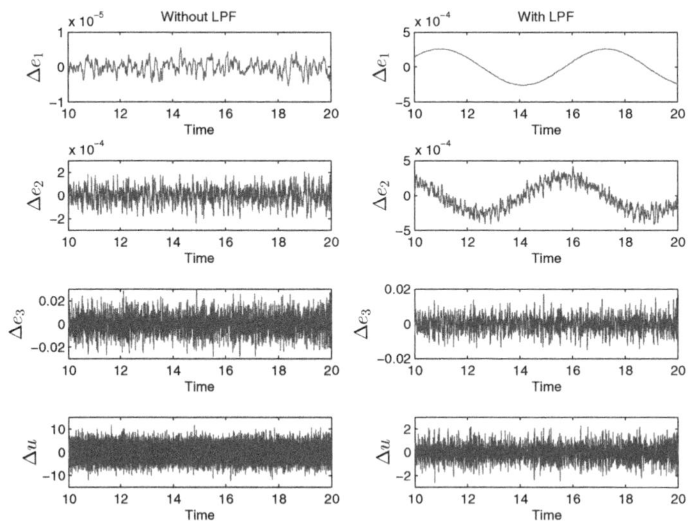

Figure 8.14. *Comparison of the trajectories of Example* 8.6 *with and without low-pass filter (LPF) when the* $\varepsilon = \tau = 0.005$. *The plotted curves are the differences from the trajectories under output feedback with no measurement noise.*

where

$$\begin{aligned}
\lambda(s) &= s^3 + s^2 + 1, \\
q(s,\varepsilon) &= (1 + 6\varepsilon + 3\varepsilon^2)s^2 + (2 + 3\varepsilon)s + 1, \\
\chi(s,\varepsilon,\tau) &= \mu p(\varepsilon s)\lambda(s)\psi(\tau s) + q(s,\varepsilon), \\
p(\varepsilon s) &= (\varepsilon s + 1)^3, \\
\psi(\tau s) &= (\tau s + 1)^2.
\end{aligned}$$

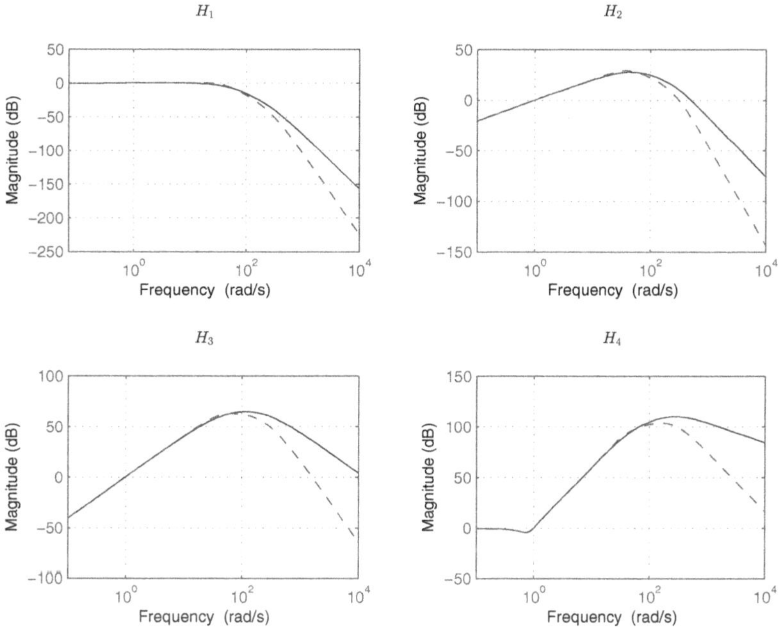

Figure 8.15. *Magnitude Bode plots of the transfer functions of Example* 8.6 *without low-pass filter (solid) and with filter (dashed) when* $\tau = \varepsilon = 0.005$.

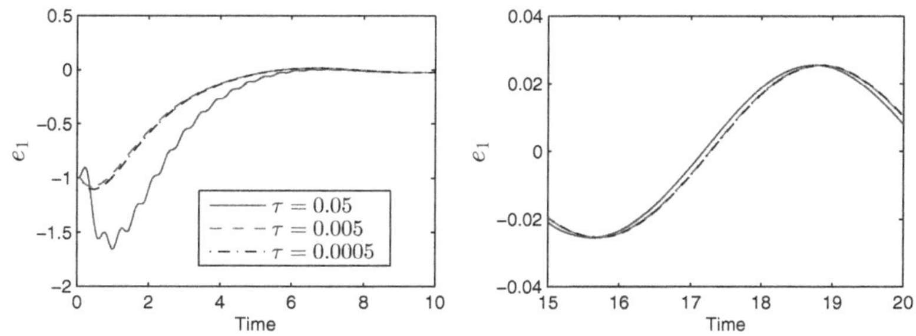

Figure 8.16. *The effect of changing* τ *on the trajectories of Example* 8.6.

The magnitude Bode plots of the four transfer functions are shown in Figure 8.15. Inserting the low-pass filter causes the frequency response to roll off at a higher rate, compared with the case without filter. This explains the attenuation of the high-frequency content of the noise. Finally, Figures 8.16 and 8.17 compare the trajectories of e_1 and u when τ is increased from 0.005 to 0.05 and decreased to 0.0005. Increasing τ deteriorates the transient response, while decreasing it deteriorates the steady-state response. △

In the previous two examples, the observer time constant ε is chosen first to achieve the desired recovery of the performance under state feedback. Then the filter time constant τ is chosen of the order of ε. A much larger τ could deteriorate the transient response of the closed-loop system, while a much smaller τ would deteriorate the steady-state response by allowing higher-frequency content of the noise to influence the response.

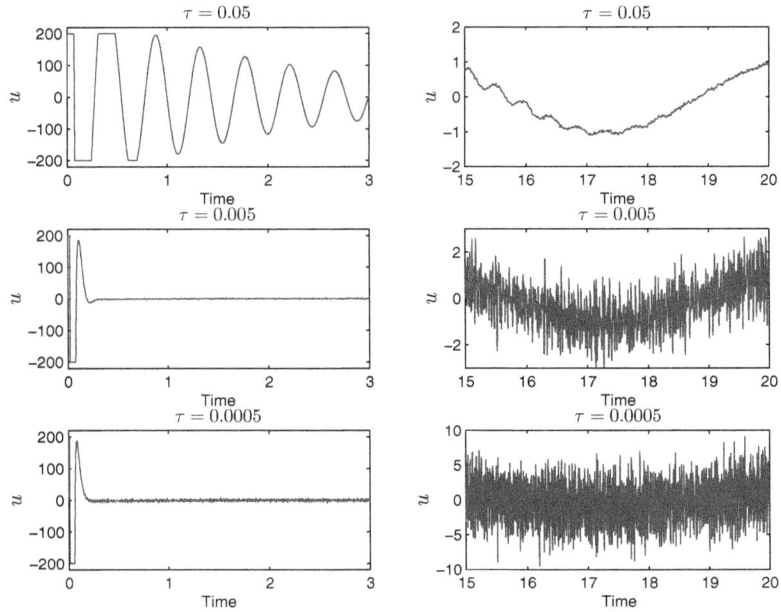

Figure 8.17. *The effect of changing τ on the trajectories of Example* 8.6.

8.4.2 ▪ Nonlinear Gain

Reconsider the system (8.78)–(8.81) under the assumptions stated in Section 8.4.1 with $|v(t)| \le N$. The choice of the parameter ε of the high-gain observer (8.82)–(8.83) is a trade-off between the speed of convergence and the steady-state error due to measurement noise. To achieve fast convergence without sacrificing the steady-state error, the divining terms

$$\frac{\alpha_i}{\varepsilon^i}(y - \hat{x}_1)$$

in (8.82)–(8.83), which have the linear gains α_i/ε^i, are replaced by the nonlinear functions $\alpha_i h_i(y - \hat{x}_1)$, where

$$h_i(y - \hat{x}_1) = \left[\frac{1}{\varepsilon_1^i}(y - \hat{x}_1) + \left(\frac{1}{\varepsilon_2^i} - \frac{1}{\varepsilon_1^i}\right) d \text{ sat}\left(\frac{y - \hat{x}_1}{d}\right)\right], \tag{8.116}$$

$\varepsilon_1 < \varepsilon_2$, and $d = kN$ with $k > 1$. A sketch of the function h_i is shown in Figure 8.18. The modified observer, which is referred to as the nonlinear-gain observer, is given by

$$\dot{\hat{x}}_i = \hat{x}_{i+1} + \alpha_i h_i(y - \hat{x}_1) \quad \text{for } 1 \le i \le \rho - 1, \tag{8.117}$$

$$\dot{\hat{x}}_\rho = \phi_0(\hat{x}, u) + \alpha_\rho h_\rho(y - \hat{x}_1). \tag{8.118}$$

Taking

$$\eta_i = \frac{x_i - \hat{x}_i}{\varepsilon_1^{\rho - i}} \quad \text{for } 1 \le i \le \rho,$$

it can be shown that

$$\varepsilon_1 \dot{\eta} = F\eta + \varepsilon_1 B\delta(w, x, \hat{x}, u) + \frac{1}{\varepsilon_1^{\rho-1}}\left[-Ev + JEd \text{ sat}\left(\frac{\varepsilon_1^{\rho-1}\eta_1 + v}{d}\right)\right], \tag{8.119}$$

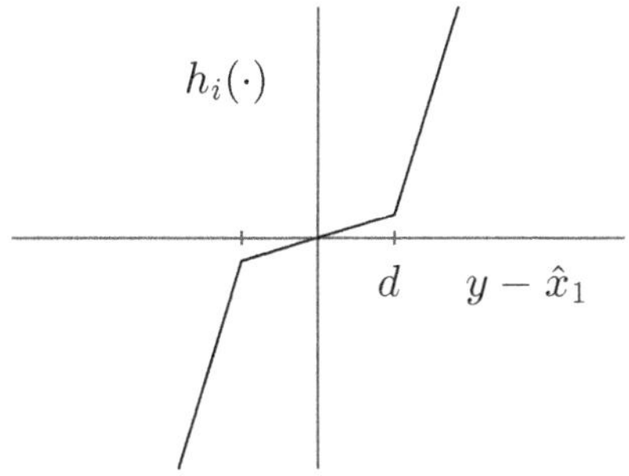

Figure 8.18. *Nonlinear gain.*

where δ, η, B, E, and F are defined after (8.87) and

$$J = \text{diag}\left[1-\left(\frac{\varepsilon_1}{\varepsilon_2}\right), 1-\left(\frac{\varepsilon_1}{\varepsilon_2}\right)^2, \dots, 1-\left(\frac{\varepsilon_1}{\varepsilon_2}\right)^\rho\right].$$

The elements J_i of J satisfy $0 < J_i < 1$. Similar to proof of Theorem 8.1 and the derivations of Section 8.4.1, it can be shown that the inequality

$$|\tilde{x}_i(t)| \leq \varepsilon^{\rho+1-i} aM + \frac{b\,N + c\,d}{\varepsilon^{i-1}} \quad \forall\, t \geq T \tag{8.120}$$

is satisfied for some positive constants a, b, c, and T. Since $d = kN$, (8.120) shows that, for sufficiently small ε_1, $\tilde{x}_1$ is ultimately bounded by $O(N)$. Knowing that the ultimate bound on $\tilde{x}_1$ is $O(N)$ does not guarantee that $y - \hat{x}_1 = \tilde{x}_1 + v$ will ultimately be confined to the zone $[-d, d]$, where the higher parameter ε_2 is effective. However, as we shall in the next two examples, that is usually the case. It is possible to choose the parameters α_1 to α_ρ to ensure that $y - \hat{x}_1$ will be confined to the zone $[-d, d]$ in finite time, but that will complicate the observer design.[74]

Example 8.7. Reconsider the pendulum equation of Examples 3.4 and 8.2. The performance of a nonlinear-gain observer with $\varepsilon_1 = 0.01$, $\varepsilon_2 = 0.1$, and $d = 0.002$ is compared with two linear-gain high-gain observers with $\varepsilon = 0.01$ and 0.1, respectively. The controller, initial conditions, and measurement noise parameters are the same as in Example 8.2. Figures 8.19 and 8.20 compare the trajectories of θ, ω, and u during the transient period and at steady state. It is seen that the nonlinear-gain observer captures the performance of the linear-gain observer with $\varepsilon = 0.01$ during the transient period and the performance of the linear-gain observer with $\varepsilon = 0.1$ at steady state.

[74]See [120] for the choice of the parameters α_1 to α_ρ to ensure that $y - \hat{x}_1$ will be confined to the zone $[-d, d]$. They are chosen such that the polynomial $s^\rho + \alpha_1 s^{\rho-1} + \cdots + \alpha_\rho$ can be factored as

$$(s^{\rho-1} + \beta_1 s^{\rho-2} + \cdots + \beta_{\rho-2} s + \beta_{\rho-1})(s + \lambda),$$

where λ is much larger than the magnitudes of the roots of $s^{\rho-1} + \beta_1 s^{\rho-2} + \cdots + \beta_{\rho-1}$. Moreover, β_1 to $\beta_{\rho-1}$ are chosen such that the transfer function

$$\frac{\beta_1 J_1 s^{\rho-2} + \cdots + \beta_{\rho-1} J_{\rho-1}}{s^{\rho-1} + \beta_1 s^{\rho-2} + \cdots + \beta_{\rho-1}}$$

has nonnegative impulse response. It is shown in [120] that such choices are always possible.

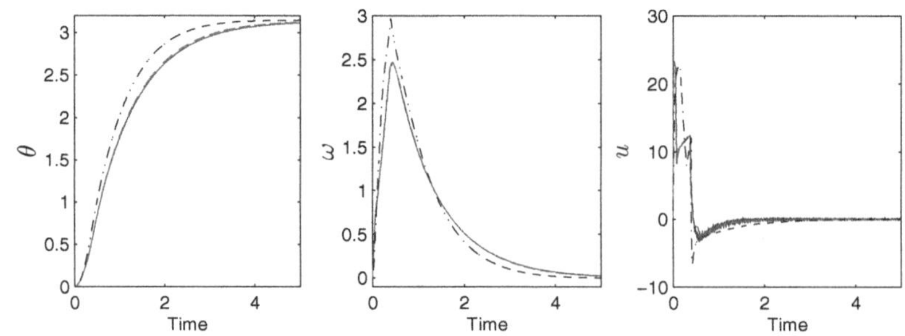

Figure 8.19. *Comparison of the trajectories of Example 8.7 for a nonlinear-gain observer with $\varepsilon_1 = 0.01$ and $\varepsilon_2 = 0.1$ (solid), a linear-gain observer with $\varepsilon = 0.01$ (dashed), and a linear-gain observer with $\varepsilon = 0.1$ (dash-dotted).*

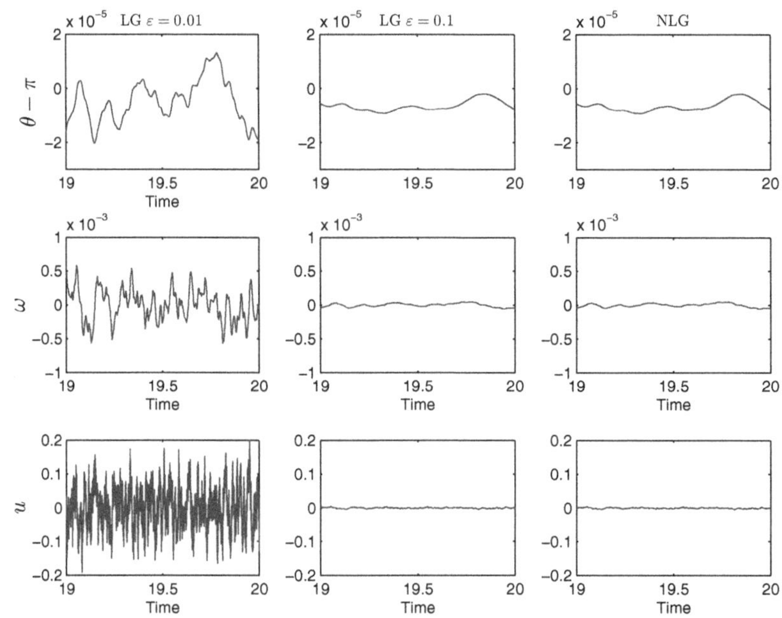

Figure 8.20. *Comparison of the steady-state trajectories of Example 8.7 for a linear-gain observer with $\varepsilon = 0.01$ (left column), a linear-gain observer with $\varepsilon = 0.1$ (middle column), and a nonlinear-gain observer with $\varepsilon_1 = 0.01$ and $\varepsilon_2 = 0.1$ (right column).*

The same is true for the estimation errors $\tilde{x}_1 = \theta - \pi - \hat{x}_1$ and $\tilde{x}_2 = \omega - \hat{x}_2$, as shown in Figures 8.21 and 8.22. Figure 8.23 shows that $y - \hat{x}_1 = \tilde{x}_1 + v$ enters the zone $[-d, d]$ in finite time. △

Example 8.8. Reconsider the tracking problem of Example 8.3. The performance of a nonlinear-gain observer with $\varepsilon_1 = 0.002$, $\varepsilon_2 = 0.01$, and $d = 0.0002$ is compared with two linear-gain high-gain observers with $\varepsilon = 0.002$ and 0.01, respectively. The controller, initial conditions, and measurement noise parameters are the same as in Example 8.3. Figures 8.24 and 8.25 compare the trajectories of e_1, e_2, and e_3 during the transient period and at steady state. It is seen that the nonlinear-gain observer captures

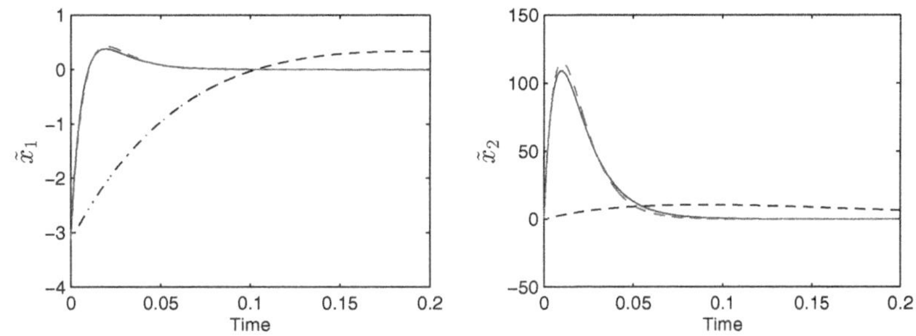

Figure 8.21. *Comparison of the estimation errors of Example* 8.7 *for a nonlinear-gain observer with* $\varepsilon_1 = 0.01$ *and* $\varepsilon_2 = 0.1$ *(solid), a linear-gain observer with* $\varepsilon = 0.01$ *(dashed), and a linear-gain observer with* $\varepsilon = 0.1$ *(dash-dotted).*

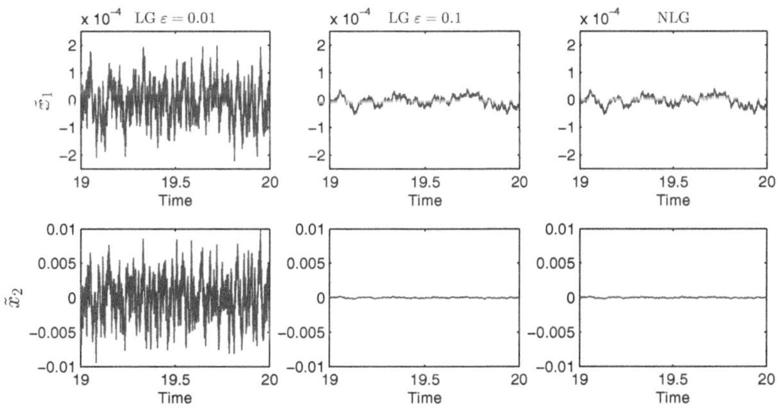

Figure 8.22. *Comparison of the steady-state estimation errors of Example* 8.7 *for a linear-gain observer with* $\varepsilon = 0.01$ *(left column), a linear-gain observer with* $\varepsilon = 0.1$ *(middle column), and a nonlinear-gain observer with* $\varepsilon_1 = 0.01$ *and* $\varepsilon_2 = 0.1$ *(right column).*

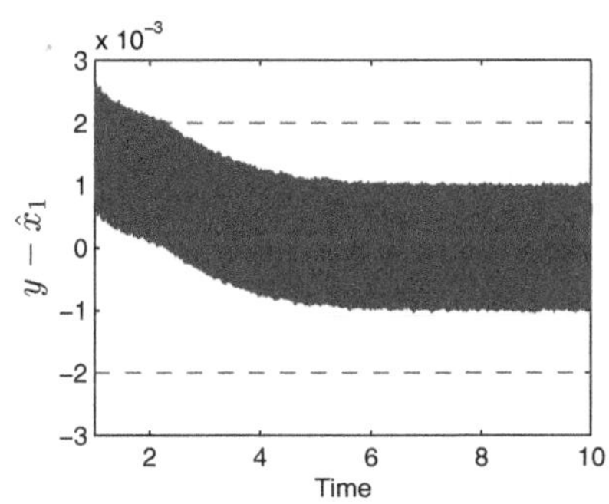

Figure 8.23. *The error term* $y - \hat{x}_1 = \tilde{x}_1 + v$ *of Example* 8.7 *for a nonlinear-gain observer with* $\varepsilon_1 = 0.01$, $\varepsilon_2 = 0.1$, *and* $d = 0.002$.

the performance of the linear-gain observer with $\varepsilon = 0.002$ during the transient period and the performance of the linear-gain observer with $\varepsilon = 0.01$ at steady state. The same is true for the estimation errors, as shown in Figures 8.26 and 8.27. Figure 8.28 shows that $y - r - \hat{e}_1 = \tilde{e}_1 + v$ enters the zone $[-d, d]$ in finite time. △

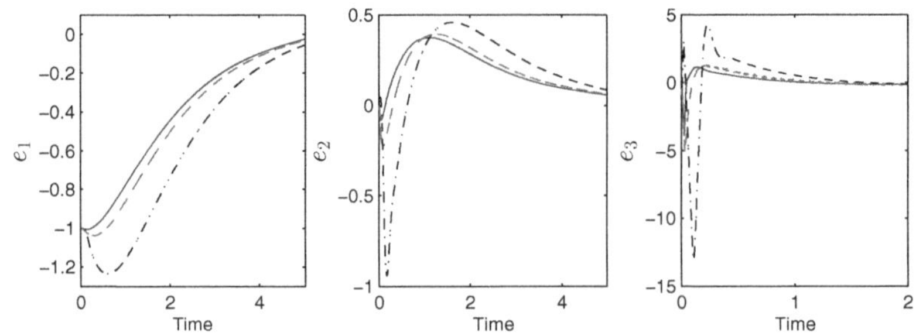

Figure 8.24. *Comparison of the trajectories of Example* 8.8 *for a nonlinear-gain observer with* $\varepsilon_1 = 0.002$ *and* $\varepsilon_2 = 0.01$ *(solid), a linear-gain observer with* $\varepsilon = 0.002$ *(dashed), and a linear-gain observer with* $\varepsilon = 0.01$ *(dash-dotted).*

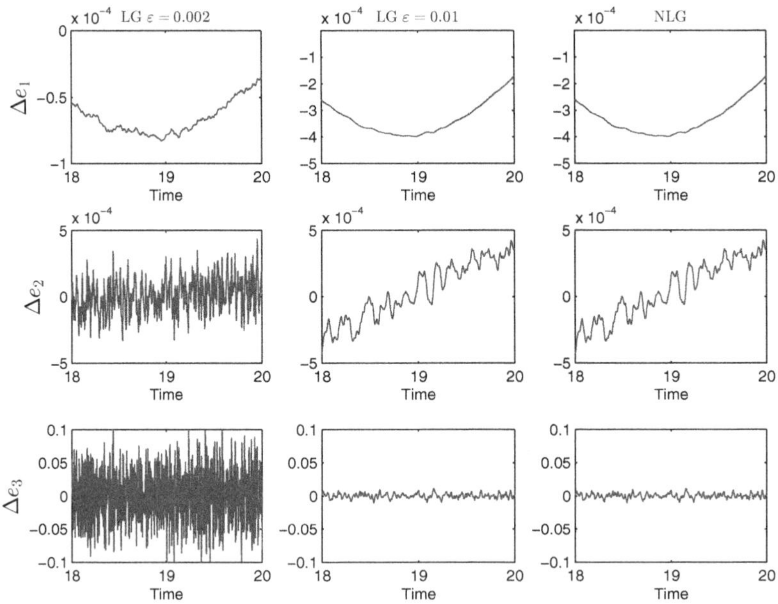

Figure 8.25. *Comparison of the steady-state trajectories of Example* 8.8 *for a linear-gain observer with* $\varepsilon = 0.002$ *(left column), a linear-gain observer with* $\varepsilon = 0.01$ *(middle column), and a nonlinear-gain observer with* $\varepsilon_1 = 0.002$ *and* $\varepsilon_2 = 0.01$ *(right column). The plots are the differences between the trajectories under noisy output feedback and the trajectories under state feedback.*

8.5 ▪ Notes and References

Theorem 8.1 is taken from [2]. A similar result with explicit numerical bounds is given in [153]. Theorem 8.2 is taken from [13]. Theorem 8.4 is taken from [2], which contains a more general result where the controller stabilizes a positively invariant compact set, not only an equilibrium point as in Theorem 8.4. Theorem 8.5 is taken from [121], which contains a result similar to Theorem 8.5 for a class of nonlinear systems of dimension up to $n = 4$. Section 8.3 is limited to deal with linear systems because of the complexity of presenting the nonlinear result. A key step in the proof of Theorem 8.5 is the change of variables (8.73), which transforms the system (8.70)–(8.71) into a block triangular form where the fast variable η is removed from the slow

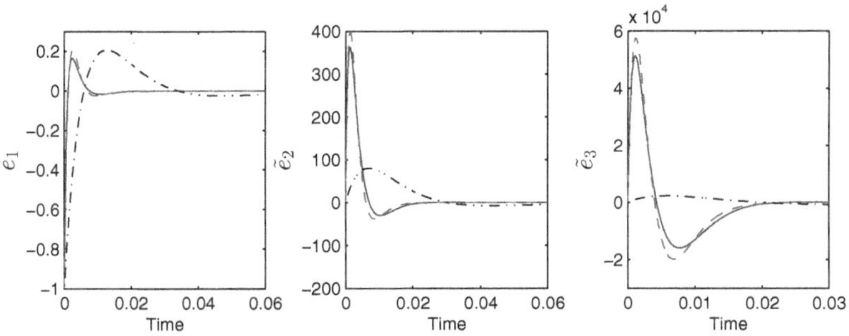

Figure 8.26. *Comparison of the estimation errors of Example 8.8 for a nonlinear-gain observer with* $\varepsilon_1 = 0.002$ *and* $\varepsilon_2 = 0.01$ *(solid), a linear-gain observer with* $\varepsilon = 0.002$ *(dashed), and a linear-gain observer with* $\varepsilon = 0.01$ *(dash-dotted).*

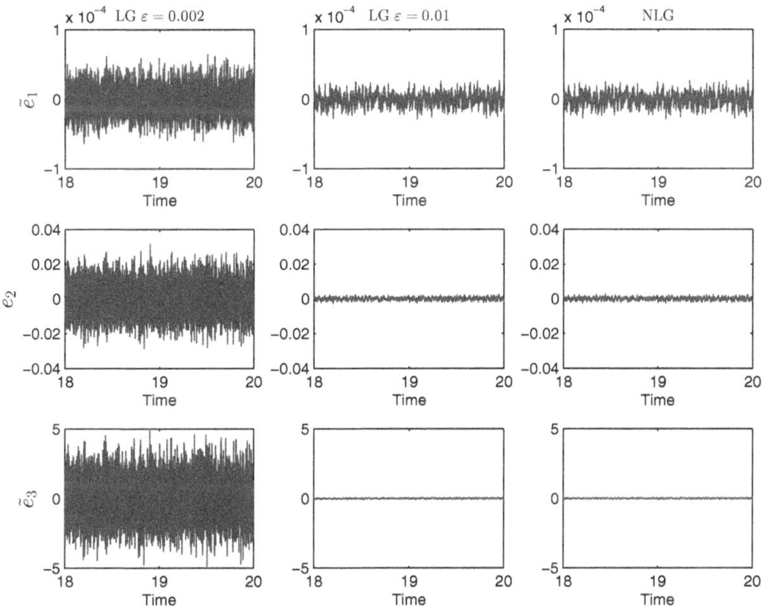

Figure 8.27. *Comparison of the steady-state estimation errors of Example 8.8 for a linear-gain observer with* $\varepsilon = 0.002$ *(left column), a linear-gain observer with* $\varepsilon = 0.01$ *(middle column), and a nonlinear-gain observer with* $\varepsilon_1 = 0.002$ *and* $\varepsilon_2 = 0.01$ *(right column).*

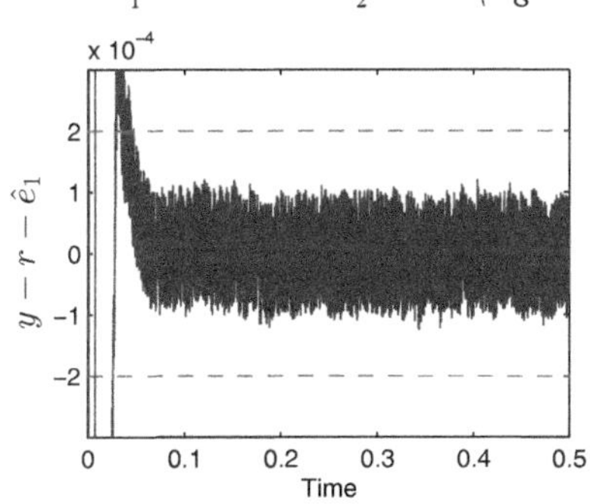

Figure 8.28. *The error term* $y - r - \hat{e}_1 = e_1 + v - \hat{e}_1$ *of Example 8.8 for a nonlinear-gain observer with* $\varepsilon_1 = 0.002$, $\varepsilon_2 = 0.01$, *and* $d = 0.0002$.

equation. This change of variables is straightforward for linear systems, as shown in [87], but it is more complicated for nonlinear systems. The nonlinear result of [121] uses a result due to [143] to achieve the desired transformation. The low-pass filter analysis of Section 8.4 is taken from [85]. The nonlinear-gain approach of the same section is based on [120].

This chapter dealt only with the standard high-gain observer. Similar results can be obtained for the reduced-order, extended, and cascade high-gain observers. It is important to note that the cascade observer of Section 7.1 has no advantage over the standard observer when it comes to measurement noise because all the saturation functions are not active beyond the peaking period. Therefore, the observer reduces to the linear cascade connection of Figure 7.1, whose transfer function is similar to the transfer function of a linear standard observer. However, the cascade observer with feedback injection of Section 7.3 has additional dynamics that act as a low-pass filter that reduces the effect of high-frequency noise [13].

Chapter 9
Digital Implementation

This chapter deals with digital implementation of high-gain observers and properties of closed loop systems under digital control. The nonlinear system under consideration is

$$\dot{w} = f_0(w, x, u), \tag{9.1}$$
$$\dot{x} = Ax + B\phi(w, x, u), \tag{9.2}$$
$$y = Cx, \tag{9.3}$$

where $u \in R^m$ is the control input, $y \in R$ is the measured output, and $w \in R^\ell$ and $x \in R^\rho$ constitute the state vector. The $\rho \times \rho$ matrix A, the $\rho \times 1$ matrix B, and the $1 \times \rho$ matrix C are given by

$$A = \begin{bmatrix} 0 & 1 & \cdots & \cdots & 0 \\ 0 & 0 & 1 & \cdots & 0 \\ \vdots & & & & \vdots \\ 0 & \cdots & \cdots & 0 & 1 \\ 0 & \cdots & \cdots & \cdots & 0 \end{bmatrix}, \quad B = \begin{bmatrix} 0 \\ 0 \\ \vdots \\ 0 \\ 1 \end{bmatrix}, \quad C = \begin{bmatrix} 1 & 0 & \cdots & \cdots & 0 \end{bmatrix}.$$

The high-gain observer for the system (9.1)–(9.3) is taken as

$$\dot{\hat{x}} = A\hat{x} + B\phi_0(\hat{x}, u) + H(y - C\hat{x}), \tag{9.4}$$

where

$$H = \begin{bmatrix} \alpha_1/\varepsilon \\ \alpha_2/\varepsilon^2 \\ \vdots \\ \alpha_{\rho-1}/\varepsilon^{\rho-1} \\ \alpha_\rho/\varepsilon^\rho \end{bmatrix}, \tag{9.5}$$

$\phi_0(x, u)$ is a nominal model of $\phi(w, x, u)$, ε is a small positive constant, and the positive constants α_i are chosen such that the polynomial

$$s^\rho + \alpha_1 s^{\rho-1} + \cdots + \alpha_{\rho-1}s + \alpha_\rho$$

is Hurwitz. The function $\phi_0(\hat{x},u)$ is locally Lipschitz in (x,u) over the domain of interest and globally bounded in x.

Section 9.1 describes the discretization of the high-gain observer for digital implementation. Both linear and nonlinear observers are discussed. The nonlinear observer is discretized using the Forward Difference method. For the linear observer, three different discretization methods are discussed. Section 9.2 deals with output feedback stabilization. Properties similar to what we have seen for continuous-time controllers are established for discrete-time controller when the observer parameter ε and the sampling period are sufficiently small. The effect of measurement noise is studied in Section 9.3, where, once again, results similar to the continuous-time case are shown. Section 9.4 presents a multirate digital control scheme in which the control sampling period is larger than the observer sampling period. This scheme is useful for computationally demanding controllers where discrete-time state feedback can be implemented with an appropriate sampling period, but such period cannot be reduced because of the computation time. Such situations arise, for example, when the controller compensates for a nonlinearity, such as hysteresis, by calculating its inverse at each time step, or in feedback linearization when the computed nonlinear terms are complicated. Another example is model predictive control where the controller solves an optimal control problem within each sampling period.[75] Such a slow sampling period might not be small enough to implement the high-gain observer, which has fast dynamics. The multirate scheme allows the observer to run with a sampling period shorter than the controller's sampling period.

9.1 ▪ Observer Discretization

The first step in discretizing the observer equation (9.4) is to scale the state estimates so as to remove the negative powers of ε in the observer gain H. Let

$$q = D\hat{x}, \tag{9.6}$$

where

$$D = \text{diag}(1, \varepsilon, \varepsilon^2, \ldots, \varepsilon^{\rho-1}).$$

The change of variables (9.6) transforms the observer equation (9.4) into

$$\varepsilon\dot{q} = Fq + Ey + \varepsilon^\rho B\phi_0(D^{-1}q, u), \tag{9.7}$$

$$\hat{x} = D^{-1}q, \tag{9.8}$$

where

$$F = \begin{bmatrix} -\alpha_1 & 1 & 0 & \cdots & 0 \\ -\alpha_2 & 0 & 1 & \cdots & 0 \\ \vdots & & \ddots & & \vdots \\ -\alpha_{\rho-1} & & & 0 & 1 \\ -\alpha_\rho & 0 & \cdots & \cdots & 0 \end{bmatrix} \quad \text{and} \quad E = \begin{bmatrix} \alpha_1 \\ \alpha_2 \\ \vdots \\ \alpha_{\rho-1} \\ \alpha_\rho \end{bmatrix}.$$

While negative powers of ε appear in (9.8), the terms Fq and Ey on the right-hand side of (9.7) have no negative powers of ε. The function ϕ_0 depends on $D^{-1}q$, but it is globally bounded. The appearance of ε multiplying $\dot{q}$ on the left-hand side of (9.7)

[75] See, for example, [29] and [124].

indicates the time scale of the observer. The change of time variable from t to $\tau = t/\varepsilon$ puts (9.7) into the form

$$\frac{dq}{d\tau} = Fq + Ey + \varepsilon^\rho B\phi_0(D^{-1}q, u). \tag{9.9}$$

If α is an appropriate sampling period for (9.9), then the sampling period for (9.7) is taken as $T = \alpha\varepsilon$.

Depending on whether or not the nominal function ϕ_0 is zero, the observer (9.7)–(9.8) could be linear or nonlinear. In the nonlinear case when $\phi_0 \neq 0$, the most practical way to discretize the observer is to use the Forward Difference method, which yields

$$\xi(k+1) = A_d\xi(k) + +B_d y(k) + \alpha\varepsilon^\rho B\phi_0(D^{-1}\xi(k), u(k)), \tag{9.10}$$
$$\hat{x}(k) = D^{-1}C_d\xi(k), \tag{9.11}$$

where $\xi = q$ and $\xi(k)$, $\hat{x}(k)$, $u(k)$, and $y(k)$ denote the signals ξ, $\hat{x}$, u, and y at the sampling points $t_k = kT$. The matrices A_d, B_d, and C_d are defined in Table 9.1. When $\phi_0 = 0$, the observer (9.7)–(9.8) reduces to the linear equation

$$\varepsilon\dot{q} = Fq + Ey, \tag{9.12}$$
$$\hat{x} = D^{-1}q \tag{9.13}$$

for which several discretization methods are available.[76] We limit our discussion to the Forward Difference, Backward Difference, and Bilinear Transformation methods, for which the discretized observer is given by

$$\xi(k+1) = A_d\xi(k) + +B_d y(k), \tag{9.14}$$
$$\hat{x}(k) = D^{-1}[C_d\xi(k) + D_d y(y)]. \tag{9.15}$$

The matrices A_d, B_d, C_d, D_d and the state ξ are given in Table 9.1 for the three methods. For the Forward Difference method, α is chosen small enough to ensure that the eigenvalues of A_d are in the interior the unit circle, while for the other two methods the eigenvalues of A_d will be in the interior the unit circle for any $\alpha > 0$ since F is Hurwitz.

An extensive simulation study was carried out to compare the performance of the digital implementation of the linear high-gain observer for different discretization methods, with and without measurement noise.[77] The best performance was obtained with the Bilinear Transformation method when both the transient and steady-state responses were taken into consideration.

The same discretization methods can be used with reduced-order observers. The previously mentioned simulation study [31] found similar results for linear reduced-order observers. In particular, the best performance was obtained with the Bilinear Transformation method. It was also found that in the presence of measurement noise the full-order high-gain observer outperforms the reduced-order one, which is not surprising in view of the simulation of Example 1.4. If the reduced-order high-gain observer

$$\dot{z} = -(\beta/\varepsilon)(z + y), \qquad \hat{x}_2 = (\beta/\varepsilon)(y + z),$$

---

[76] See, for example, [16], [45], or any digital control textbook.

[77] See [31]. The comparison included also the zero-order hold and first-order hold discretization methods.

Table 9.1. *Coefficients of the discrete-time implementation of the high-gain observer.*

	Forward Difference	*Backward Difference*	*Bilinear Transformation*
A_d	$(I+\alpha F)$	$(I-\alpha F)^{-1} \stackrel{\text{def}}{=} M_1$	$(I+\frac{\alpha}{2}F)(I-\frac{\alpha}{2}F)^{-1} \stackrel{\text{def}}{=} N_2 M_2$
B_d	αE	$\alpha M_1 E$	$\alpha M_2 E$
C_d	I	M_1	M_2
D_d	0	$\alpha M_1 E$	$\frac{\alpha}{2} M_2 E$
ξ	q	$(M_1^{-1} q - \alpha E y)$	$(M_2^{-1} q - (\alpha/2) E y)$

is used to estimate the derivative of y by $\hat{x}_2$, then Forward Difference discretization with $T = \varepsilon/\beta$ yields Euler's formula

$$\hat{x}_2(k) = \frac{1}{T}[y(k) - y(k-1)],$$

which is commonly used in engineering applications to estimate the derivative of a signal. The fact that Euler's formula is just one of the possible choices with high-gain observers shows that better results can be obtained by designing a high-gain observer rather than resorting to the simple Euler's formula.[78]

Example 9.1. Reconsider the system

$$\dot{x}_1 = x_2, \qquad \dot{x}_2 = -x_1 - 2x_2 + 0.25x_1^2 x_2 + 0.2u, \qquad y = x_1$$

from Example 1.1, where $u(t) = \sin 2t$. A continuous-time high-gain observer can be taken as

$$\dot{\hat{x}}_1 = \hat{x}_2 + \frac{2}{\varepsilon}(y - \hat{x}_1), \qquad \dot{\hat{x}}_2 = \phi_0(\hat{x}, u) + \frac{1}{\varepsilon^2}(y - \hat{x}_1).$$

We consider two choices of ϕ_0:

$$\phi_0 = -\hat{x}_1 - 2\hat{x}_2 + 0.25\hat{x}_1^2 \hat{x}_2 + 0.2u \quad \text{and} \quad \phi_0 = 0.$$

With the scaling

$$q = D\hat{x} = \begin{bmatrix} 1 & 0 \\ 0 & \varepsilon \end{bmatrix},$$

the observer is given by

$$\varepsilon \dot{q} = Fq + Ey + \varepsilon^2 B \phi_0(\hat{x}, u),$$

where

$$F = \begin{bmatrix} -2 & 1 \\ -1 & 0 \end{bmatrix}, \quad E = \begin{bmatrix} 2 \\ 1 \end{bmatrix}, \quad \text{and} \quad B = \begin{bmatrix} 0 \\ 1 \end{bmatrix}.$$

[78]This point is demonstrated experimentally in [32].

When $\phi_0 \neq 0$, the observer is discretized using the Forward Difference method to obtain

$$\xi(k+1) = A_d\xi(k) + B_d y(k) + \alpha\varepsilon^2 B\phi_0(\hat{x}(k), u(k)), \quad \hat{x}(k) = D^{-1}\xi(k),$$

where A_d and B_d are defined in Table 9.1. When $\phi_0 = 0$, the observer is discretized using the Forward Difference and the Bilinear Transformation methods. In the first case, the observer is given by

$$\xi(k+1) = A_d\xi(k) + B_d y(k), \quad \hat{x}(k) = D^{-1}\xi(k),$$

while in the second one it is given by

$$\xi(k+1) = A_d\xi(k) + B_d y(k), \quad \hat{x}(k) = D^{-1}[C_d\xi(k) + D_d y(k)],$$

where the matrices are defined in Table 9.1. In particular, for the Forward Difference method,

$$A_d = \begin{bmatrix} (1-2\alpha) & \alpha \\ -\alpha & 1 \end{bmatrix} \quad \text{and} \quad B_d = \begin{bmatrix} 2\alpha \\ \alpha \end{bmatrix},$$

and for the Bilinear Transformation method,

$$A_d = \frac{1}{1+\alpha+\alpha^2/4}\begin{bmatrix} (1-\alpha-\alpha^2/4) & \alpha \\ -\alpha & (1+\alpha-\alpha^2/4) \end{bmatrix},$$

$$B_d = \frac{\alpha}{1+\alpha+\alpha^2/4}\begin{bmatrix} 2+\alpha/2 \\ 1 \end{bmatrix}, \quad C_d = \frac{1}{1+\alpha+\alpha^2/4}\begin{bmatrix} 1 & \alpha/2 \\ -\alpha/2 & (1+\alpha) \end{bmatrix},$$

and $D_d = \frac{1}{2}B_d$. Simulation is carried out with initial conditions $x(0) = \text{col}(1, -1)$ and $\xi(0) = 0$. Figure 9.1 shows x_2 and its estimate $\hat{x}_2$ as determined by the continuous-time observer with $\phi_0 \neq 0$, the Forward Difference discrete-time observer with $\phi_0 \neq 0$, the Forward Difference discrete-time observer with $\phi_0 = 0$, and the Bilinear Transformation discrete-time observer with $\phi_0 = 0$. The observers with $\phi_0 \neq 0$ are nonlinear, while those with $\phi_0 = 0$ are linear. In all cases $\varepsilon = 0.1$ and $T = 0.01$; hence, $\alpha = T/\varepsilon = 0.1$. The performance of the four observers is comparable, with the nonlinear ones slightly better than the linear ones. Figure 9.2 shows the steady-state estimation errors of the three discrete-time observers. The errors are comparable for the two linear observers, but a smaller error is achieved with the nonlinear one. Figure 9.3 shows the impact on the estimation error $\tilde{x}_2$ when ε is reduced from 0.1 to 0.01 while keeping $\alpha = 0.1$; hence, $T = 0.01$ when $\varepsilon = 0.1$ and $T = 0.001$ when $\varepsilon = 0.01$. Similar to the continuous-time case, reducing ε causes a larger peak and faster decay. The figure shows results for the two discrete-time linear observers, but similar results are obtained for the nonlinear observer. The effect on the steady-state error is shown in Figure 9.4 for the Forward Difference linear observer, where reducing ε leads to smaller error. Finally, Figure 9.5 compares the Forward Difference and Bilinear Transformation linear observers when $\alpha = 0.5$ and $\varepsilon = 0.01$. The steady-state estimation errors are comparable, but there are differences in the transient response. The Bilinear Transformation observer has less peaking at the expense of a slightly larger settling time. △

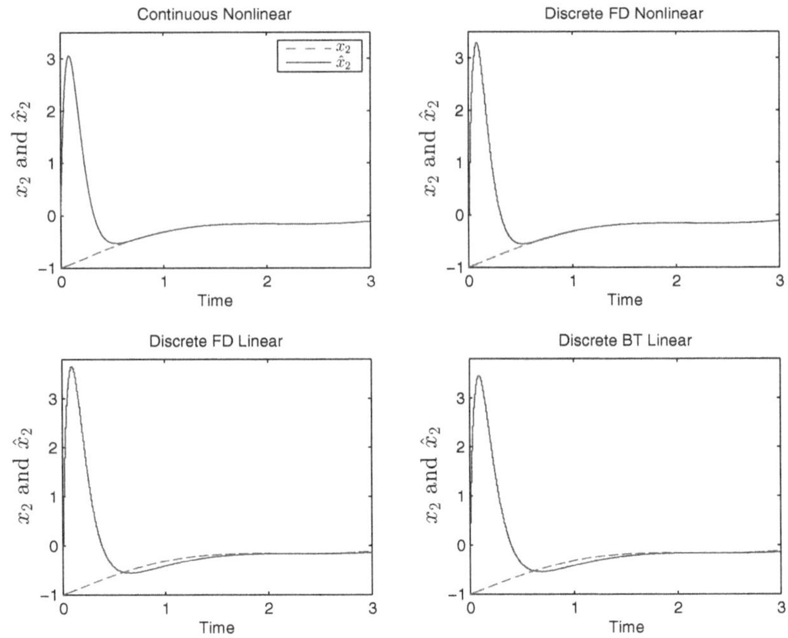

Figure 9.1. *Simulation of Example 9.1. The estimation of x_2 using four different observers: continuous nonlinear ($\phi_0 \neq 0$) observer, discrete Forward Difference nonlinear observer, discrete Forward Difference linear ($\phi_0 = 0$) observer, and discrete Bilinear Transformation linear observer.*

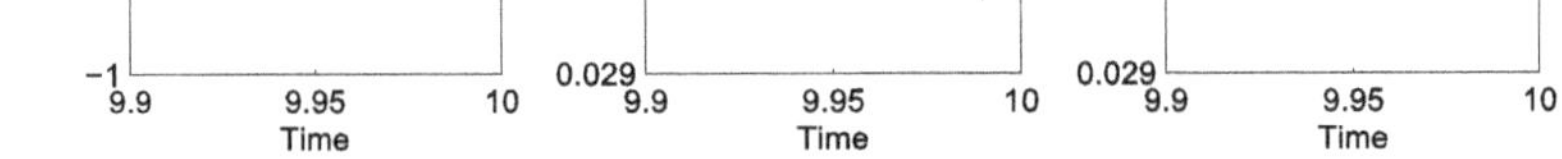

Figure 9.2. *Simulation of Example 9.1. The steady-state estimation error $\tilde{x}_2$ of the discrete Forward Difference nonlinear observer, the discrete Forward Difference linear observer, and the discrete Bilinear Transformation linear observer.*

9.2 ▪ Stabilization

Consider the single-input–single-output nonlinear system

$$\dot{w} = f_0(w, x, u), \tag{9.16}$$
$$\dot{x} = Ax + B\phi(w, x, u), \tag{9.17}$$
$$y = Cx, \tag{9.18}$$
$$z = \psi(w, x), \tag{9.19}$$

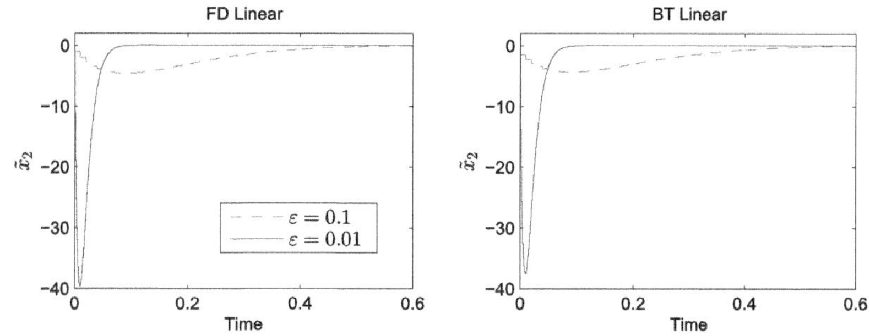

Figure 9.3. *Simulation of Example* 9.1. *The impact of reducing* ε *on the transient response of* $\tilde{x}_2$ *for the Forward Difference and Bilinear Transformation linear observers.*

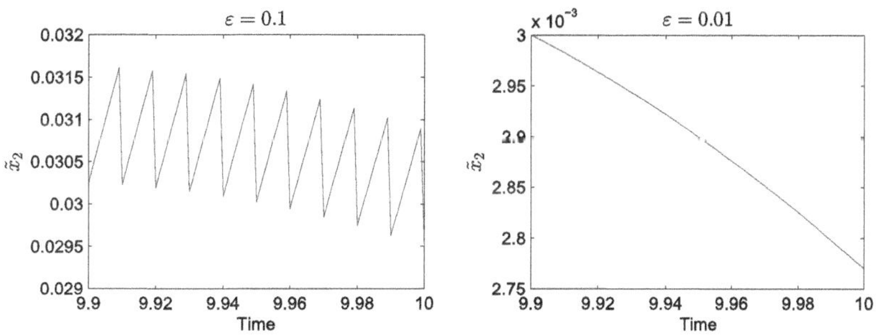

Figure 9.4. *Simulation of Example* 9.1. *The impact of reducing* ε *on the steady-state error* $\tilde{x}_2$ *for the Forward Difference linear observer.*

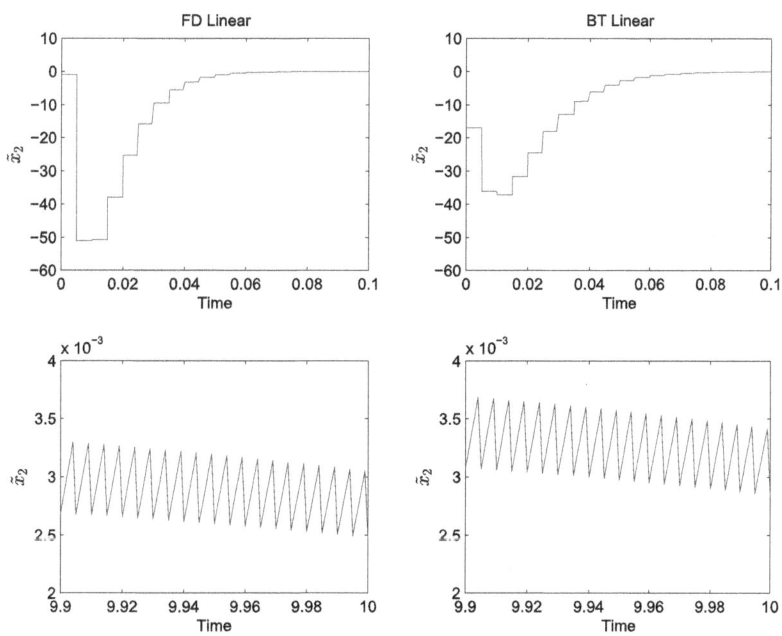

Figure 9.5. *Simulation of Example* 9.1. *Comparison of the transient and steady-state estimation error* $\tilde{x}_2$ *between the Forward Difference and Bilinear Transformation linear observers when* $\alpha = 0.5$ *and* $\varepsilon = 0.01$.

where $u \in R$ is the control input, $y \in R$ and $z \in R^s$ are measured outputs, and $w \in R^\ell$ and $x \in R^\rho$ constitute the state vector. The matrices A, B, C are defined after equation (9.3). The functions f_0, ϕ, and ψ are locally Lipschitz in their arguments for $(w, x, u) \in D_w \times D_x \times R$, where $D_w \subset R^\ell$ and $D_x \subset R^\rho$ are domains that contain their respective origins. Moreover, $f_0(0,0,0) = 0$, $\phi(0,0,0) = 0$, and $\psi(0,0) = 0$.

The goal is to design a discrete-time output feedback controller to stabilize the origin $(w = 0, x = 0)$ using only the measured outputs y and z at the sampling points. It is assumed that the digital control system uses zero-order hold (ZOH); that is, the control signal is held constant in between the sampling points. We start by designing a continuous-time partial state feedback controller that uses measurements of x and z. A high-gain observer is designed to estimate x from y. Finally, the observer and control signal are discretized to arrive at the discrete-time controller.

The continuous-time state feedback controller takes the form

$$u = \gamma(x, z), \tag{9.20}$$

where γ is locally Lipschitz in (x, z) over the domain of interest and globally bounded in x, with $\gamma(0,0) = 0$. It is designed such that the origin $\chi = 0$ of the closed-loop system

$$\dot{\chi} = f_r(\chi, \gamma(x, z)) \tag{9.21}$$

is asymptotically stable, where

$$\chi = \begin{bmatrix} w \\ x \end{bmatrix}, \qquad f_r(\chi, u) = \begin{bmatrix} f_0(w, x, u) \\ Ax + B\phi(w, x, u) \end{bmatrix}.$$

The continuous-time output feedback controller is given by

$$u = \gamma(\hat{x}, z), \tag{9.22}$$

where $\hat{x}$ is generated by the high-gain observer

$$\dot{\hat{x}} = A\hat{x} + B\phi_0(\hat{x}, z, u) + H(y - C\hat{x}), \tag{9.23}$$

where H is given by (9.5) and $\phi_0(x, z, u)$ is a nominal model of $\phi(w, x, u)$, which is locally Lipschitz in its arguments over the domain of interest and globally bounded in x, with $\phi_0(0,0,0) = 0$.

Assuming that the sampling period $T = \alpha\varepsilon$, the observer is discretized as described in the previous section. In particular, when $\phi_0 \neq 0$, the observer is discretized using the Forward Difference method to obtain the discrete-time observer

$$\begin{aligned} \xi(k+1) &= A_d\xi(k) + +B_d y(k) + \alpha\varepsilon^\rho B\phi_0(D^{-1}\xi(k), z(k), u(k)), && (9.24) \\ \hat{x}(k) &= D^{-1}C_d\xi(k), && (9.25) \end{aligned}$$

while when $\phi_0 = 0$, a linear discrete-time observer is given by

$$\begin{aligned} \xi(k+1) &= A_d\xi(k) + +B_d y(k), && (9.26) \\ \hat{x}(k) &= D^{-1}[C_d\xi(k) + D_d y(k)], && (9.27) \end{aligned}$$

where the matrices A_d, B_d, C_d, D_d and the state ξ are given in Table 9.1 for the three discretization methods. In the case of the Forward Difference method, it is assumed that α is chosen small enough such that the eigenvalues of A_d are in the interior of the unit circle. The discrete-time output feedback controller is given by

$$u(k) = \gamma(\hat{x}(k), z(k)). \tag{9.28}$$

To analyze the closed-loop system, we derive a discrete time model that describes the state variables at the sampling points. A key step in deriving the model is the representation of the system in a two-time-scale form that reflects the fact that the observer dynamics are faster than the plant dynamics. The state χ of the plant satisfies the equation

$$\dot{\chi} = f_r(\chi, u(k)) \tag{9.29}$$

over the sampling period $[kT, kT+T]$, during which the control is kept constant at $u(k)$. The solution of (9.29) is given by

$$\chi(t) = \chi(k) + (t - kT)f_r(\chi(k), u(k)) + \int_{kT}^{t} [f_r(\chi(\tau), u(k)) - f_r(\chi(k), u(k))]\, d\tau$$

for $t \in [kT, kT+T]$. Using the Lipschitz property of f_r and the Gronwall–Bellman inequality, it can be shown that

$$\|\chi(t) - \chi(k)\| \leq \frac{1}{L_1}\left[e^{(t-kT)L_1} - 1\right]\|f_r(\chi(k), u(k))\| \quad \forall\ t \in [kT, kT+T], \tag{9.30}$$

where L_1 is a Lipschitz constant of f_r with respect to χ. Therefore,

$$\chi(k+1) = \chi(k) + \varepsilon\alpha f_r(\chi(k), u(k)) + \varepsilon^2\Phi(\chi(k), u(k), \varepsilon), \tag{9.31}$$

where Φ is locally Lipschitz in (χ, u), uniformly bounded in ε, for sufficiently small ε, and $\Phi(0, 0, \varepsilon) = 0$. This equation and inequality (9.30) are sufficient to characterize the plant dynamics. However, to develop a model that describes the dynamics of the estimation error, we need a more detailed model of x that makes use of the special structure of (9.17) and the properties of the matrices A and B. The solution of (9.17) is given by

$$x(t) = e^{A(t-kT)}x(k) + \int_{kT}^{t} e^{A(t-\tau)}B\, d\tau\ \phi(w(k), x(k), u(k)) + \int_{kT}^{t} e^{A(t-\tau)}B\Delta_1(\tau)\, d\tau,$$

where $\Delta_1(\tau) = \phi(w(\tau), x(\tau), u(k)) - \phi(w(k), x(k), u(k))$. Using the fact that the matrices A and B satisfy the relations

$$A^\rho = 0, \quad \varepsilon^i DA^i = A^i D, \quad DB = \varepsilon^{\rho-1}B, \tag{9.32}$$

it can be shown that

$$De^{At}B = D\sum_{i=0}^{\rho-1}\frac{t^i}{i!}A^iB = \sum_{i=0}^{\rho-1}\frac{t^i}{\varepsilon^i i!}A^iDB = \varepsilon^{\rho-1}e^{A(t/\varepsilon)}B. \tag{9.33}$$

Hence,

$$\int_{kT}^{kT+T} e^{A(kT+T-\tau)}B\Delta_1(\tau)\, d\tau = \varepsilon^{\rho-1}D^{-1}\int_{kT}^{kT+T} e^{A(kT+T-\tau)/\varepsilon}B\Delta_1(\tau)\, d\tau. \tag{9.34}$$

Using the Lipschitz property of ϕ, (9.30) and the fact that

$$\int_{kT}^{kT+T} \left(e^{(\tau-kT)L_1}-1\right) d\tau = \frac{1}{L_1}\left(e^{L_1 T}-1\right)-T = O(T^2) = O(\varepsilon^2),$$

it can be shown that the integral on the right-hand side of (9.34) is $O(\varepsilon^2)$. Therefore, $x(k)$ can be modeled by

$$x(k+1) = e^{AT}x(k) + \int_0^T e^{A\tau}B\, d\tau\, \phi(w(k), x(k), u(k)) + \varepsilon^{\rho+1}D^{-1}R(\chi(k), u(k), \varepsilon), \tag{9.35}$$

where R is locally Lipschitz in (χ, u), uniformly bounded in ε, for sufficiently small ε, and $R(0,0,\varepsilon) = 0$.

As in the continuous-time case, define the scaled estimation errors

$$\varphi_i = \frac{x_i - \hat{x}_i}{\varepsilon^{\rho-i}} \quad \text{for } 1 \le i \le \rho,$$

which can be written as

$$\varphi = \frac{1}{\varepsilon^{\rho-1}}D(x-\hat{x}) = \frac{1}{\varepsilon^{\rho-1}}(Dx - C_d\xi - D_d y), \tag{9.36}$$

where $\varphi = \text{col}(\varphi_1, \ldots, \varphi_\rho)$. Using the relation $C = CD$, it can be shown that φ satisfies the equation

$$\begin{aligned}\varphi(k+1) = A_d\varphi(k) &+ \frac{1}{\varepsilon^{\rho-1}}MDx(k) + \varepsilon h(\chi(k), u(k), \varepsilon)\\ &- \varepsilon\alpha C_d B\phi_0(\hat{x}(k), z(k), u(k)),\end{aligned} \tag{9.37}$$

where

$$M = (I - D_dC)e^{A\alpha} - A_d(I - D_dC) - C_dB_dC$$

and

$$h = (I - D_dC)\left[\int_0^\alpha e^{A\sigma}B\, d\sigma\phi + \varepsilon R\right].$$

For the linear observer $\phi_0 = 0$. In arriving at (9.37) we used the property $C_dA_dC_d^{-1} = A_d$, which can be verified for the three discretization methods using Table 9.1. The term $(1/\varepsilon^{\rho-1})MDx(k)$ in (9.37) is eliminated by the change of variables

$$\eta = \varphi - \frac{1}{\varepsilon^{\rho-1}}LDx, \tag{9.38}$$

where L is the solution of the Sylvester equation

$$A_dL + M - Le^{A\alpha} = 0. \tag{9.39}$$

Because the eigenvalues of A_d are in the interior of the unit circle while those of $e^{A\alpha}$ are at one, (9.39) has a unique solution [156, Problem 9.6]. The new variable η satisfies the equation

$$\eta(k+1) = A_d\eta(k) + \varepsilon(h - \alpha C_dB\phi_0) - \varepsilon L\int_0^\alpha e^{A\sigma}\, d\sigma\, B\phi - \varepsilon^2 LR,$$

Table 9.2. *$D^{-1}JD$ and $D^{-1}JMD$ in terms of $X = -(1/\alpha)D^{-1}F^{-1}D$.*

	$D^{-1}JD$	$D^{-1}JMD$
Forward Difference	X	$X(e^{AT} - I - TA)$
Backward Difference	$I + X$	$X[(I - TA)e^{AT} - I]$
Bilinear Transformation	$\frac{1}{2}I + X$	$X\left[\left(I - \frac{T}{2}A\right)e^{AT} - I - \frac{T}{2}A\right]$

and the state estimate $\hat{x}$ is given by

$$\hat{x}(k) = (I - D^{-1}LD)x(k) + Q_1\eta(k),$$

where $Q_1(\varepsilon) = -\varepsilon^{\rho-1}D^{-1}$ is an analytic function of ε.

Lemma 9.1. *$D^{-1}LD = \varepsilon Q_2(\varepsilon)$, where $Q_2(\varepsilon)$ is an analytic function of ε.*

Proof: Using the property $A^\rho = 0$, it can be verified that

$$L = \sum_{i=1}^{\rho} J^i M(I - e^{A\alpha})^{i-1}, \tag{9.40}$$

where $J = (I - A_d)^{-1}$. In particular,

$$\begin{aligned}(I - A_d)L - L(I - e^{A\alpha}) &= \sum_{i=1}^{\rho} J^{i-1}M(I - e^{A\alpha})^{i-1} - \sum_{i=1}^{\rho} J^i M(I - e^{A\alpha})^i \\ &= \sum_{k=0}^{\rho-1} J^k M(I - e^{A\alpha})^k - \sum_{i=1}^{\rho} J^i M(I - e^{A\alpha})^i \\ &= M - J^\rho M(I - e^{A\alpha})^\rho = M\end{aligned}$$

since $(I - e^{A\alpha})^\rho = 0$. Using (9.40), $D^{-1}LD$ can be written as

$$D^{-1}LD = \sum_{i=1}^{\rho} (D^{-1}JD)^{i-1}(D^{-1}JMD)[D^{-1}(I - e^{A\alpha})D]^{i-1}.$$

Table 9.2 gives expressions of $D^{-1}JD$ and $D^{-1}JMD$ for the three discretization methods in terms of $X = -(1/\alpha)D^{-1}F^{-1}D$. It can be shown that

$$\begin{aligned}D^{-1}(I - e^{A\alpha})D = I - e^{AT} &= O(T) = O(\varepsilon), \\ e^{AT} - I - TA &= O(T^2) = O(\varepsilon^2), \\ (I - TA)e^{AT} - I &= O(T^2) = O(\varepsilon^2), \\ \left(I - \frac{T}{2}A\right)e^{AT} - I - \frac{T}{2}A &= O(T^2) = O(\varepsilon^2),\end{aligned}$$

and

$$D^{-1}F^{-1}D = \begin{bmatrix} 0 & 0 & \cdots & 0 & -\varepsilon^{\rho-1}/\alpha_\rho \\ 1/\varepsilon & 0 & \cdots & 0 & -\varepsilon^{\rho-2}\alpha_1/\alpha_\rho \\ & 1/\varepsilon & & & -\varepsilon^{\rho-3}\alpha_2/\alpha_\rho \\ \vdots & & & \vdots & \\ 0 & 0 & \cdots & 1/\varepsilon & -\alpha_{\rho-1}/\alpha_\rho \end{bmatrix} = O(1/\varepsilon).$$

Using Table 9.2 and the foregoing expressions we conclude that $D^{-1}JD = O(1/\varepsilon)$, $D^{-1}JMD = O(\varepsilon)$, and $D^{-1}(I - e^{A\alpha})D = O(\varepsilon)$. Hence,

$$\|D^{-1}LD\| \leq \sum_{i=1}^{\rho} \left(\frac{K_1}{\varepsilon}\right)^{i-1} K_2\varepsilon(K_3\varepsilon)^{i-1} = O(\varepsilon). \qquad \square$$

In summary, the closed-loop digital control system can be represented at the sampling points by the discrete-time model

$$\chi(k+1) = \chi(k) + \varepsilon\alpha f_r(\chi(k), u(k)) + \varepsilon^2\Phi(\chi(k), u(k), \varepsilon), \tag{9.41}$$

$$\eta(k+1) = A_d\eta(k) + \varepsilon\Gamma(\chi(k), u(k), \hat{x}(k), \varepsilon), \tag{9.42}$$

$$u(k) = \gamma(\hat{x}(k), z(k)), \tag{9.43}$$

$$\hat{x}(k) = [I - \varepsilon Q_2(\varepsilon)]x(k) + Q_1(\varepsilon)\eta(k), \tag{9.44}$$

where the functions f_r, Φ, γ, and Γ are locally Lipschitz, Φ and Γ are globally bounded in $\hat{x}$, and the matrices Q_1 and Q_2 are analytic functions of ε.

Theorem 9.1. *Consider the closed-loop system of the plant* (9.16)–(9.19) *and the output feedback digital controller* (9.28) *with the observer* (9.24)–(9.25) *or the observer* (9.26)–(9.27). *Let $\mathscr{R}$ be the region of attraction of* (9.21), *let $\mathscr{S}$ be any compact set in the interior of $\mathscr{R}$, and let $\mathscr{Q}$ be any compact subset of R^ρ. Suppose $(\chi(0), \hat{x}(0)) \in \mathscr{S} \times \mathscr{Q}$. Then*

- *there exists $\varepsilon_1^* > 0$ such that for every $\varepsilon \in (0, \varepsilon_1^*]$, $\chi(t)$ is bounded for all $t \geq 0$ and $\eta(k)$ is bounded for all $k \geq 0$;*

- *given any $\mu > 0$, there exist $\varepsilon_2^* > 0$, $T_1 > 0$, and $k^* > 0$, all dependent on μ, such that for every $\varepsilon \in (0, \varepsilon_2^*]$,*

$$\|\chi(t)\| \leq \mu \quad \forall\, t \geq T_1 \quad \textit{and} \quad \|\eta(k)\| \leq \mu \quad \forall\, k \geq k^*; \tag{9.45}$$

- *given any $\mu > 0$, there exists $\varepsilon_3^* > 0$, dependent on μ, such that for every $\varepsilon \in (0, \varepsilon_3^*]$,*

$$\|\chi(t) - \chi_r(t)\| \leq \mu \quad \forall\, t \geq 0, \tag{9.46}$$

  where χ_r is the solution of (9.21) *with $\chi_r(0) = \chi(0)$;*

- *if the origin of* (9.21) *is exponentially stable and $f_r(\chi, \gamma(x, z))$ is twice continuously differentiable in the neighborhood of the origin, then there exists $\varepsilon_4^* > 0$ such that for every $\varepsilon \in (0, \varepsilon_4^*]$, the origin of* (9.41)–(9.44) *is exponentially stable and $\mathscr{S} \times \mathscr{Q}$ is a subset of its region of attraction. Moreover, the continuous-time trajectory $\chi(t)$ decays to zero exponentially fast.* ◇

Proof: Rewrite (9.41)–(9.42) as

$$\chi(k+1) = \chi(k) + \varepsilon \mathscr{F}(\chi(k), \eta(k), \varepsilon), \tag{9.47}$$
$$\eta(k+1) = A_d \eta(k) + \varepsilon \mathscr{G}(\chi(k), \eta(k), \varepsilon), \tag{9.48}$$

where

$$\mathscr{F}(\chi, \eta, \varepsilon) = \alpha f_r(\chi, u) + \varepsilon \Phi(\chi, u, \varepsilon), \qquad \mathscr{G}(\chi(k), \eta(k), \varepsilon) = \Gamma(\chi, u, \hat{x}, \varepsilon),$$

and u and $\hat{x}$ are substituted using (9.43) and (9.44), respectively. The functions $\mathscr{F}$ and $\mathscr{G}$ are locally Lipschitz, globally bounded in η, and vanish at the origin ($\chi = 0, \eta = 0$). Since the origin of (9.21) is asymptotically stable and $\mathscr{R}$ is its region of attraction, by the converse Lyapunov theorem of asymptotic stability [78, Theorem 4.17], there is a smooth, positive definite function $V(\chi)$ and a continuous, positive definite function $U(\chi)$, both defined for all $\chi \in \mathscr{R}$, such that

$$V(\chi) \to \infty \quad \text{as } \chi \to \partial \mathscr{R},$$
$$\frac{\partial V}{\partial \chi} f_r(\chi, \gamma(x, z)) \leq -U(\chi) \quad \forall\, \chi \in \mathscr{R}$$

and for any $c > 0$, $\{V(\chi) \leq c\}$ is a compact subset of $\mathscr{R}$. Choose positive constants b and c such that $c > b > \max_{\chi \in \mathscr{S}} V(\chi)$. Then

$$\mathscr{S} \subset \Omega_b = \{V(\chi) \leq b\} \subset \Omega_c = \{V(\chi) \leq c\} \subset \mathscr{R}.$$

We use $V(\chi)$ as a Lyapunov function for the slow subsystem (9.47) and $W(\eta) = \eta^T P \eta$ as a Lyapunov function for the fast subsystem (9.48), where P is the positive definite solution of the discrete-time Lyapunov equation $A_d^T P A_d - P = -I$. It can be seen that $P = \sum_{k=0}^{\infty} (A_d^k)^T (A_d^k) > I$. Therefore, $\|P\| = \lambda_{\max}(P) > \lambda_{\min}(P) > 1$ and

$$\lambda_{\min}(P) \|\eta\|^2 \leq W(\eta) \leq \lambda_{\max}(P) \|\eta\|^2.$$

Let $\Lambda = \Omega_c \times \{W(\eta) \leq c_1 \varepsilon^2\}$, where $c_1 > 0$ will be chosen. For every $\varepsilon \in (0, \tilde{\varepsilon}]$, where $\tilde{\varepsilon} < 1$, there are positive constants K_1, K_2, K_3, and L_1, independent of ε, such that

$$\|\mathscr{F}(\chi, \eta, \varepsilon)\| \leq K_1, \quad \|\Phi(\chi, \gamma(x, z), \varepsilon)\| \leq K_2, \quad \|\mathscr{G}(\chi, \eta, \varepsilon)\| \leq K_3$$

for all $(\chi, \eta) \in \Omega \times R^\rho$ and

$$\|\mathscr{F}(\chi, \eta, \varepsilon) - \mathscr{F}(\chi, 0, \varepsilon)\| \leq L_1 \|\eta\|$$

for all $(\chi, \eta) \in \Lambda$.

We start by showing that there exist positive constants $c_1 > 0$ and $\varepsilon_1 > 0$ (dependent on c_1) such that the compact set Λ is positively invariant for $\varepsilon \in (0, \varepsilon_1]$. This is done by showing that for $\eta \in \{W(\eta) \leq c_1 \varepsilon^2\}$, Ω_c is positively invariant, and for $\chi \in \Omega_c$, $\{W(\eta) \leq c_1 \varepsilon^2\}$ is positively invariant:

$$\begin{aligned}
\Delta V(\chi(k)) = V(\chi(k+1)) - V(\chi(k)) &= \frac{\partial V}{\partial \chi}(\bar{\chi}(k))[\chi(k+1)-\chi(k)] \\
&= \frac{\partial V}{\partial \chi}(\chi(k))[\chi(k+1)-\chi(k)] \\
&\quad + \left[\frac{\partial V}{\partial \chi}(\bar{\chi}(k)) - \frac{\partial V}{\partial \chi}(\chi(k))\right][\chi(k+1)-\chi(k)] \\
&\le \frac{\partial V}{\partial \chi}(\chi(k))\varepsilon \mathscr{F}(\chi(k),\eta(k),\varepsilon) + L_2\|\chi(k+1)-\chi(k)\|^2 \\
&\le -\varepsilon\alpha U(\chi(k)) + \varepsilon \left\|\frac{\partial V}{\partial \chi}(\chi(k))\right\| \, \|\mathscr{F}(\chi(k),\eta(k),\varepsilon) - \mathscr{F}(\chi(k),0,\varepsilon)\| \\
&\quad + \varepsilon^2 \left\|\frac{\partial V}{\partial \chi}(\chi(k))\right\| \, \|\Phi(\chi(k),\gamma(x(k),z(k)),\varepsilon)\| + L_2\|\chi(k+1)-\chi(k)\|^2 \\
&\le -\varepsilon\alpha U(\chi(k)) + \varepsilon L_1 L_3 \|\eta(k)\| + \varepsilon^2 L_3 K_2 + \varepsilon^2 L_2 K_1^2,
\end{aligned}$$

where $\bar{\chi}(k)$ is a point on the line segment joining $\chi(k)$ to $\chi(k+1)$, and L_2 and L_3 are Lipschitz constant and upper bound for $[\partial V/\partial \chi]$ over Ω_c. In Λ,

$$\lambda_{\min}(P)\|\eta\|^2 \le W(\eta) \le c_1\varepsilon^2 \quad \Rightarrow \quad \|\eta\| \le \varepsilon\sqrt{\frac{c_1}{\lambda_{\min}(P)}}.$$

Thus,

$$\Delta V(\chi(k)) \le \varepsilon\alpha[-U(\chi(k)) + \varepsilon K_4], \tag{9.49}$$

where $K_4 = [L_1 L_3 \sqrt{c_1/\lambda_{\min}(P)} + L_2 K_1^2 + L_3 K_2]/\alpha$. For any positive constant $\tilde{c} < c$, trajectories starting inside $\{V(\chi) \le \tilde{c}\}$ cannot leave $\{V(\chi) \le c\}$, provided ε is sufficiently small. This is seen from

$$V(\chi(k+1)) \le V(\chi(k)) - \varepsilon\alpha U(\chi(k)) + \varepsilon^2\alpha K_4 \le \tilde{c} + \varepsilon^2\alpha K_4 < c$$

whenever $\varepsilon^2 < \varepsilon_2^2 \stackrel{\text{def}}{=} (c-\tilde{c})/(\alpha K_4)$. Trajectories starting inside $\{V(\chi) \le c\}$ but outside $\{V(\chi) \le \tilde{c}\}$ remain inside $\{V(\chi) \le c\}$, provided ε is sufficiently small. This is seen from

$$V(\chi(k+1)) \le V(\chi(k)) - \varepsilon\alpha U(\chi(k)) + \varepsilon^2\alpha K_4 \le c - \varepsilon\alpha\varrho_1 + \varepsilon^2\alpha K_4 < c$$

whenever $\varepsilon < \varepsilon_3 \stackrel{\text{def}}{=} \varrho_1/K_4$, where $\varrho_1 = \min_{\tilde{c} \le V(\chi) \le c} U(\chi)$. Thus, for $\varepsilon_1 = \min\{\varepsilon_2, \varepsilon_3\}$, the set $\{V(\chi) \le c\}$ is positively invariant for $\varepsilon < \varepsilon_1$. On the other hand,

$$\begin{aligned}
\Delta W(\eta(k)) = W(\eta(k+1)) - W(\eta(k)) &= \eta^T(k+1)P\eta(k+1) - \eta^T(k)P\eta(k) \\
&\le \eta^T(k)[A_d^T P A_d - P]\eta(k) + 2\varepsilon\|\eta(k)\| \, \|A_d\| \, \|P\| \, \|\mathscr{G}(\chi(k),\eta(k),\varepsilon)\| \\
&\quad + \varepsilon^2\|\mathscr{G}(\chi(k),\eta(k),\varepsilon)\|^2 \, \|P\| \\
&\le -\|\eta(k)\|^2 + \varepsilon(2K_3\|\eta(k)\| \, \|A_d\| \, \|P\| + \varepsilon K_3^2\|P\|).
\end{aligned}$$

Hence,

$$W(\eta(k+1)) \le \left(1 - \frac{1}{\|P\|}\right) W(\eta(k)) + \varepsilon K_3\|P\|(2\|\eta(k)\| \, \|A_d\| + \varepsilon K_3).$$

For $\eta(k) \in \{W(\eta) \le c_1\varepsilon^2\}$,

$$W(\eta(k+1)) \le \left(1 - \frac{1}{||P||}\right)c_1\varepsilon^2 + 2\varepsilon^2 K_3||P||\,||A_d||\sqrt{\frac{c_1}{\lambda_{\min}(P)}} + \varepsilon^2 K_3^2||P||.$$

Choosing c_1 large enough ensures that $W(\eta(k+1)) \le \varepsilon^2 c_1$, which shows that $\{W(\eta) \le c_1\varepsilon^2\}$ is positively invariant. This completes the proof that Λ is positively invariant.

Now consider the initial state $(\chi(0), \hat{x}(0)) \in \mathcal{S} \times \mathcal{Q}$. It can be verified that the corresponding initial state $\eta(0)$ satisfies $||\eta(0)|| \le l/\varepsilon^{\rho-1}$, where l depends on $\mathcal{S}$ and $\mathcal{Q}$. Using the fact that $\chi(0)$ is in the interior of Ω_c, it can be shown from (9.47) that

$$||\chi(k) - \chi(0)|| = \varepsilon \left\| \sum_{i=0}^{k-1} \mathcal{F}(\chi(i), \eta(i), \varepsilon) \right\| \le \varepsilon k K_1 \tag{9.50}$$

as long as $\chi(k) \in \Omega_c$. Therefore, there exists a positive constant K_5, independent of ε, such that $\chi(k) \in \Omega_c$ for all $k \le K_5/\varepsilon$. Let $\tilde{k}_0$ be the first time $\eta(k)$ enters the set $\{W(\eta) \le c_1\varepsilon^2\}$ and take $\bar{k}_0 \le \min\{\tilde{k}_0 - 1, K_5/\varepsilon\}$. During the time interval $[0, \bar{k}_0]$, $W(\eta(k)) > c_1\varepsilon^2$ and $\chi(k) \in \Omega_c$. Therefore,

$$\begin{aligned} W(\eta(k+1)) \le &\left[-\frac{1}{2||P||} W(\eta(k)) + \varepsilon K_3 ||P|| \left(2||A_d|| \sqrt{\frac{W(\eta(k))}{\lambda_{\min}(P)}} + \varepsilon K_3 \right) \right] \\ &+ \left(1 - \frac{1}{2||P||}\right) W(\eta(k)). \end{aligned}$$

Choosing c_1 large enough ensures that the bracketed term in the foregoing inequality is nonpositive. Hence,

$$W(\eta(k+1)) \le \lambda W(\eta(k)) \Rightarrow W(\eta(k)) \le \lambda^k W(\eta(0)) \Rightarrow W(\eta(k)) \le \lambda^k ||P|| \frac{l^2}{\varepsilon^{2\rho-2}},$$

where $\lambda = 1 - 1/(2||P||) \in (0,1)$. Therefore, $W(\eta(k))$ is strictly decreasing toward zero, which shows that $\eta(k)$ must enter the set $\{W(\eta) \le c_1\varepsilon^2\}$ within the finite time $[0, \ln(K_6/\varepsilon^{2\rho})/\ln(1/\lambda)]$, where $K_6 = l^2||P||/c_1$. To ensure that $\eta(k)$ enters $\{W(\eta) \le c_1\varepsilon^2\}$ before $\chi(k)$ leaves Ω_c, ε_4 is chosen small enough so that for $\varepsilon \in (0, \varepsilon_4]$,

$$\frac{\ln(K_6/\varepsilon^{2\rho})}{\ln(1/\lambda)} < \frac{K_5}{\varepsilon} - 1$$

or, equivalently,

$$\varepsilon \ln\left(\frac{1}{\lambda}\right) + \varepsilon \ln\left(\frac{K_6}{\varepsilon^{2\rho}}\right) < K_5 \ln\left(\frac{1}{\lambda}\right).$$

Such choice of ε_4 is possible because the left-hand side of the foregoing inequality tends to zero as ε tends to zero. Taking $\varepsilon_1^* = \min\{\tilde{\varepsilon}, \varepsilon_1, \varepsilon_4\}$ guarantees that for every $\varepsilon \in (0, \varepsilon_1^*]$, $(\chi(k), \eta(k))$ enters Λ during the interval $[0, \Delta_2(\varepsilon)]$ and remains inside thereafter, where

$$\Delta_2(\varepsilon) \le 1 + \frac{\ln(K_6/\varepsilon^{2\rho})}{\ln(1/\lambda)} < \frac{K_5}{\varepsilon}.$$

Thus, the trajectory is bounded for all $k \geq \Delta_2(\varepsilon)$. It is also bounded for $k \in [0, \Delta_2(\varepsilon)]$ since during this interval

$$\|\chi(k) - \chi(0)\| \leq \varepsilon k K_1 \quad \text{and} \quad W(\eta(k)) \leq \lambda^k \|P\|(l^2/\varepsilon^{2\rho-2}).$$

Next we show (9.45). We already know that $(\chi(k), \eta(k)) \in \Lambda$ for all $k \geq \tilde{k}$ for some $\tilde{k}$. We can find $\varepsilon_5 \in (0, \varepsilon_1^*]$, dependent on μ, such that for every $\varepsilon \in (0, \varepsilon_5]$, $\|\eta(k)\| \leq \mu$ for all $k \geq \tilde{k}$. The ultimate bound on χ is proved in two steps. First, it is shown that trajectories inside Λ enter the set

$$\tilde{\Lambda} = \{V(\chi) \leq c_2(\varepsilon)\} \times \{W(\eta) \leq c_1 \varepsilon^2\}$$

in finite time k^*, where $\lim_{\varepsilon \to 0} c_2(\varepsilon) = 0$. Second, it is shown that there is a positive constant $c_3(\varepsilon) > c_2(\varepsilon)$, with $\lim_{\varepsilon \to 0} c_3(\varepsilon) = 0$, such that trajectories in $\tilde{\Lambda}$ at time k^* belong to the set

$$\bar{\Lambda} = \{V(\chi) \leq c_3(\varepsilon)\} \times \{W(\eta) \leq c_1 \varepsilon^2\}$$

for all $k \geq k^* + 1$. From (9.49), it is seen that for all $(\chi, \eta) \in \Lambda$,

$$V(\chi(k+1)) \leq V(\chi(k)) - \frac{\varepsilon\alpha}{2} U(\chi(k)) - \varepsilon\alpha\left[\tfrac{1}{2} U(\chi(k)) - \varepsilon K_4\right]. \tag{9.51}$$

Since $U(\chi)$ is positive definite and continuous, the set $\{U(\chi) \leq 2\varepsilon K_4\}$ is compact for sufficiently small ε. Let $c_2(\varepsilon) = \max_{U(\chi) \leq 2\varepsilon K_4} V(\chi)$. The inequality $V(\chi(k)) > c_2(\varepsilon)$ implies that $U(\chi(k)) > 2\varepsilon K_4$ and

$$V(\chi(k+1)) \leq V(\chi(k)) - \frac{\varepsilon\alpha}{2} U(\chi(k)).$$

Starting with $V(\chi(0)) > c_2(\varepsilon)$, it can be shown by mathematical induction that, as long as $V(\chi(k)) > c_2(\varepsilon)$,

$$\begin{aligned} V(\chi(k+1)) &\leq V(\chi(0)) - \frac{\varepsilon\alpha}{2} \sum_{i=0}^{k} U(\chi(i)) \\ &< V(\chi(0)) - \frac{\varepsilon\alpha}{2} \sum_{i=0}^{k} 2\varepsilon K_4 \;=\; V(\chi(0)) - \varepsilon^2 \alpha K_4 (1+k), \end{aligned}$$

which shows that there is finite time k^* such that $V(\chi(k^*)) \leq c_2(\varepsilon)$. When $V(\chi(k)) \leq c_2(\varepsilon)$ one of two cases could happen: $U(\chi(k)) > 2\varepsilon K_4$ or $U(\chi(k)) \leq 2\varepsilon K_4$. It is seen from (9.51) that if $U(\chi(k)) > 2\varepsilon K_4$,

$$V(\chi(k+1)) \leq V(\chi(k)) - \frac{\varepsilon\alpha}{2} U(\chi(k)) \leq c_2(\varepsilon),$$

and if $U(\chi(k)) \leq 2\varepsilon K_4$

$$V(\chi(k+1)) \leq V(\chi(k)) + \varepsilon^2 \alpha K_4 \leq c_2(\varepsilon) + \varepsilon^2 \alpha K_4 \stackrel{\text{def}}{=} c_3(\varepsilon).$$

In the latter case, one of two things could happen: either $U(\chi(k+1)) > 2\varepsilon K_4$ or $U(\chi(k+1)) \leq 2\varepsilon K_4$. If $U(\chi(k+1)) \leq 2\varepsilon K_4$, then $V(\chi(k+1)) \leq c_2(\varepsilon)$. If $U(\chi(k+1)) > 2\varepsilon K_4$, then $V(\chi(k+2)) < V(\chi(k+1))$, and if $V(\chi(k+1)) > c_2(\varepsilon)$, $V(\chi)$ will decrease until eventually $V(\chi(k)) \leq c_2(\varepsilon)$ in a finite number of steps. Therefore,

$$V(\chi(k+1)) \leq c_3(\varepsilon) \qquad \forall\, k \geq k^*.$$

From (9.30) and the fact that $\lim_{\varepsilon\to 0} c_3(\varepsilon)=0$, we conclude that there is a finite time T_1 such that, for sufficiently small ε, $\|\chi(t)\|\le\mu$ for all $t\ge T_1$.

To show (9.46), note that from (9.50) and asymptotic stability of the origin of (9.21), there exists $T_2=T_2(\mu)>0$ such that

$$\|\chi(t)-\chi_r(t)\|\le\mu \qquad \forall\ t\ge T_2. \tag{9.52}$$

From (9.30) and (9.52) it is sufficient to prove that $\|\chi(k)-\chi_r(k)\|\le\mu$ for $k\in[0,K_7/\varepsilon]$ for some K_7 (independent of ε). Let $e(k)=\chi(k)-\chi_r(k)$ and divide the interval $[0,K_7/\varepsilon]$ into $[0,\Delta_2(\varepsilon)]$ and $[\Delta_2(\varepsilon),K_7/\varepsilon]$. Over $[0,\Delta_2(\varepsilon)]$ we use (9.50), a similar expression for χ_r, and $\lim_{\varepsilon\to 0}\Delta_2(\varepsilon)=0$ to show that $e(k)=O(\varepsilon\Delta_2(\varepsilon))$ for $k\in[0,\Delta_2(\varepsilon)]$. Over the interval $[\Delta_2(\varepsilon),K_7/\varepsilon]$,

$$e(k+1)=e(k)+\varepsilon\alpha[f_r(\chi(k),\gamma(x(k),z(k)))-f_r(\chi_r(k),\gamma(x_r(k),z_r(k)))]+O(\varepsilon^2).$$

Hence,

$$\|e(k+1)\|\le(1+\varepsilon\alpha L_4)\|e(k)\|+O(\varepsilon^2),$$

where L_4 is a Lipschitz constant for f_r over Ω_c. Solving the inequality results in

$$\begin{aligned}\|e(k)\|&\le(1+\varepsilon\alpha L_4)^{k-\Delta_2(\varepsilon)}O(\varepsilon\Delta_2(\varepsilon))+\sum_{\ell=\Delta_2(\varepsilon)}^{k-1}(1+\varepsilon\alpha L_4)^{k-\ell-1}O(\varepsilon^2)\\ &\le(1+\varepsilon\alpha L_4)^{k-\Delta_2(\varepsilon)}\left[O(\varepsilon\Delta_2(\varepsilon))+\sum_{r=0}^{\infty}\frac{1}{(1+\varepsilon\alpha L_4)^{r+1}}O(\varepsilon^2)\right]\\ &\le(1+\varepsilon\alpha L_4)^{K_7/\varepsilon}[O(\varepsilon\Delta_2(\varepsilon))+O(\varepsilon)].\end{aligned}$$

Using

$$\lim_{\varepsilon\to 0}\varepsilon\Delta_2(\varepsilon)=0 \quad\text{and}\quad \lim_{\varepsilon\to 0}(1+\varepsilon\alpha L_4)^{K_7/\varepsilon}=e^{\alpha L_4 K_7}$$

completes the proof of (9.46).

Finally, if the origin of (9.21) is exponentially stable, by the converse Lyapunov theorem of exponential stability,[79] there is a twice continuously differentiable Lyapunov function $V_1(\chi)$, defined over a ball $\{\|\chi\|\le r\}\subset\mathscr{R}$, and four positive constants δ_1 to δ_4 such that

$$\delta_1\|\chi\|^2\le V_1(\chi)\le\delta_2\|\chi\|^2,\quad \frac{\partial V_1}{\partial\chi}f_r(\chi,\gamma(x,z))\le-\delta_3\|\chi\|^2,\quad \left\|\frac{\partial V_1}{\partial\chi}\right\|\le\delta_4\|\chi\|$$

for all $\chi\in\{\|\chi\|\le r\}$. Using the local Lipschitz properties of $\mathscr{F}$ and $\mathscr{G}$ and the fact that $\mathscr{F}(0,0,\varepsilon)=0$ and $\mathscr{G}(0,0,\varepsilon)=0$, it can be shown that the composite Lyapunov function $V_c=V_1+W$ satisfies

$$\Delta V_c\le-\varepsilon\mathscr{Y}^T Q\mathscr{Y},$$

where

$$Q=\begin{bmatrix}(\alpha\delta_3-\varepsilon\beta_1) & -(\beta_2+\varepsilon\beta_3)\\ -(\beta_2+\varepsilon\beta_3) & ((1/\varepsilon)-\beta_4-\varepsilon\beta_5\end{bmatrix},\quad \mathscr{Y}=\begin{bmatrix}\|\chi(k)\|\\ \|\eta(k)\|\end{bmatrix}$$

[79]See [78, Theorem 4.14]. The fact that V_1 is twice continuously differentiable when f_r is twice continuously differentiable can be seen from the proof of the theorem.

for some nonnegative constants β_i. The matrix Q is positive definite for sufficiently small ε. Hence, the origin of (9.47)–(9.48) is exponentially stable, and there is a neighborhood $\mathcal{N}$ of the origin (independent of ε) such that all trajectories in $\mathcal{N}$ converge to the origin. By choosing ε small enough, $\bar{\Lambda} = \{V(\chi) \le c_3(\varepsilon)\} \times \{W(\eta) \le c_1\varepsilon^2\} \subset \mathcal{N}$. Thus, all trajectories starting in $\mathcal{S} \times \mathcal{Q}$ enter $\mathcal{N}$ in finite time and converge to zero as k tends to infinity. The fact that $\chi(t)$ decays to zero exponentially fast follows from (9.30). □

Example 9.2. Reconsider the pendulum equation

$$\dot{x}_1 = x_2, \quad \dot{x}_2 = \sin x_1 - c_1 x_2 + c_2 u, \quad y = x_1$$

from Example 3.4, where $0 \le c_1 \le 0.2$ and $0.5 \le c_2 \le 2$. It is shown there that the output feedback controller

$$\begin{aligned}
\dot{\hat{x}}_1 &= \hat{x}_2 + (2/\varepsilon)(y - \hat{x}_1),\\
\dot{\hat{x}}_2 &= \phi_0(\hat{x}, u) + (1/\varepsilon^2)(y - \hat{x}_1),\\
u &= -2\left[1.25\pi \,\text{sat}\left(\frac{|\hat{x}_1|}{1.25\pi}\right) + 2.25\pi \,\text{sat}\left(\frac{|\hat{x}_2|}{2.25\pi}\right) + 1\right] \text{sat}\left(\frac{\hat{x}_1 + \hat{x}_2}{0.1}\right)
\end{aligned}$$

stabilizes the origin ($x = 0, \hat{x} = 0$), for sufficiently small ε, and the set $\Omega = \{|x_1| \le 1.25\pi\} \times \{|x_1 + x_2| \le \pi\}$ is included in the region of attraction. Two choices of ϕ_0 are considered in Example 3.4. The choice $\phi_0 = 0$ yields a linear observer, while $\phi_0 = \sin \hat{x}_1 - 0.1\hat{x}_2 + 1.25u$ yields a nonlinear observer. The continuous-time observer is discretized using the Forward Difference method to obtain

$$\xi(k+1) = A_d\xi(k) + B_d y(k) + \alpha\varepsilon^2 B\phi_0(\hat{x}(k), u(k)), \quad \hat{x}(k) = D^{-1}\xi(k),$$

where the matrices A_d, B_d, and B are as defined in Example 9.1. Simulation results are shown in Figure 9.6 and 9.7 for the initial conditions $x_1(0) = -\pi$, $x_2(0) = \xi_1(0) = \xi_2(0) = 0$. Figure 9.6 shows the state trajectories under continuous-time state feedback and discrete-time output feedback for the linear and nonlinear observers. The parameter ε is taken as 0.05 and 0.01, with $\alpha = 0.1$ in both cases; hence, $T = 0.005$ and 0.001, respectively. The figure shows a lot of similarity with Figure 3.8 for continuous-time output feedback. In particular, as ε decreases, the trajectories under discrete-time output feedback approach the trajectories under continuous-time state feedback, as stated in Theorem 9.1. The comparison of the linear and nonlinear observers shows an advantage for the nonlinear one when $\varepsilon = 0.05$, which is more prominent in the trajectory of x_2. The choice of α plays an important role because it determines the relation between the sampling period T and the observer time constant ε. It is reasonable to have $T < \varepsilon$, and that is why $\alpha = 0.1$ is chosen. However, sometimes we might not be able to push T very small due to hardware constraints. In such cases we may have to work with α close to one. Figure 9.7 shows the trajectories of x_1 and x_2 and the estimation errors $\tilde{x}_1$ and $\tilde{x}_2$ for $\alpha = 1$, at which $T = \varepsilon = 0.01$, and compares them with the trajectories and estimation errors for $\alpha = 0.1$, at which $T = 0.001$ and $\varepsilon = 0.01$. The larger α causes a larger peak in the estimation errors, but they quickly recover the errors for the smaller α. The state trajectories, on the other hand, are almost unaffected by the change in the sampling period. This is likely because the larger sampling period is still adequate for controlling the system. The fact that the control and estimation problems may require different sampling periods is explored further in the Section 9.4, where a multirate digital control scheme is developed. △

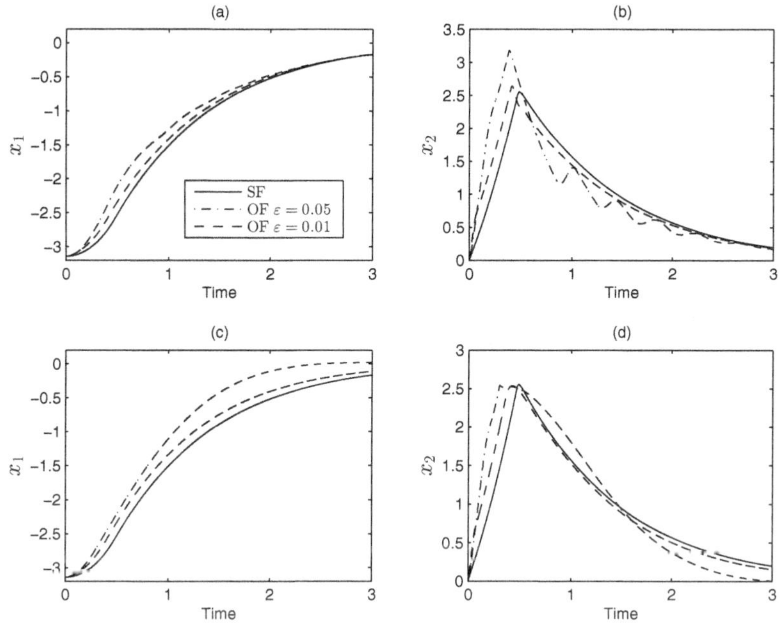

Figure 9.6. *Simulation of Example* 9.2. *Comparison of continuous-time state feedback (SF) with discrete-time output feedback (OF).* (a) *and* (b) *are for the linear observer while* (c) *and* (d) *are for the nonlinear one.*

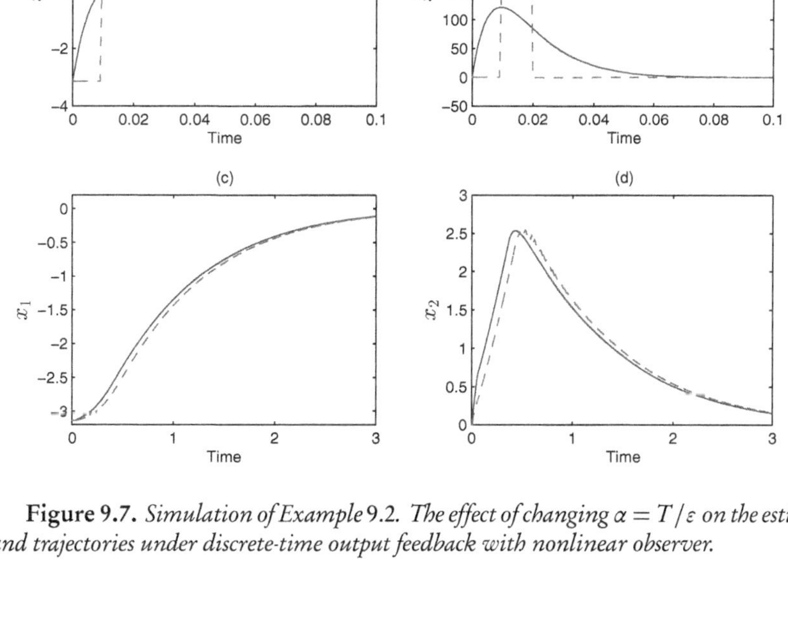

Figure 9.7. *Simulation of Example* 9.2. *The effect of changing* $\alpha = T/\varepsilon$ *on the estimation errors and trajectories under discrete-time output feedback with nonlinear observer.*

9.3 ▪ Measurement Noise

Reconsider the closed-loop system of the plant (9.16)–(9.19) and the output feedback digital controller (9.28) with the observer (9.24)–(9.25) or the observer (9.26)–(9.27) from the previous section. Suppose the measured output y is corrupted by noise v, that is, $y = Cx + v$, where $|v(k)| \leq N$. To arrive at the representation (9.41)–(9.44) of the closed-loop system, the transformations (9.36) and (9.38) were used. These transformations can be combined into the change of variables

$$\eta = \frac{1}{\varepsilon^{\rho-1}}[(I - L - D_d C)Dx - C_d \xi].$$

To include the effect of measurement noise, consider the system

$$\theta(k+1) = A_d \theta(k) + B_d v(k), \qquad \theta(0) = 0, \tag{9.53}$$
$$w(k) = C_d \theta(k) + D_d v(k) \tag{9.54}$$

and apply the change of variables

$$\eta = \frac{1}{\varepsilon^{\rho-1}}[(I - L - D_d C)Dx - C_d \xi + C_d \theta].$$

By repeating the derivations of Section 9.2, it can be shown that the closed-loop digital control system can be represented at the sampling points by the discrete-time model

$$\chi(k+1) = \chi(k) + \varepsilon \alpha f_r(\chi(k), u(k)) + \varepsilon^2 \Phi(\chi(k), u(k), \varepsilon), \tag{9.55}$$
$$\eta(k+1) = A_d \eta(k) + \varepsilon \Gamma(\chi(k), u(k), \hat{x}(k), \varepsilon), \tag{9.56}$$
$$u(k) = \gamma(\hat{x}(k), z(k)), \tag{9.57}$$
$$\hat{x}(k) = [I - \varepsilon Q_2(\varepsilon)]x(k) + Q_1(\varepsilon)\eta(k) + D^{-1}w(k). \tag{9.58}$$

From (9.58) it is seen that

$$x(k) - \hat{x}(k) = \varepsilon Q_2(\varepsilon)x(k) - Q_1(\varepsilon)\xi(k) - D^{-1}w(k).$$

Over compact sets of χ, Γ is be bounded; hence, $\eta(k)$ is ultimately bounded by an $O(\varepsilon)$ bound. The same is true for the term $\varepsilon Q_2(\varepsilon)x(k)$. On the other hand, if $|v(k)| \leq N$ for all $k \geq 0$, then using the fact that the eigenvalues of A_d are in the interior of the unit circle, we see from (9.53)–(9.54) that $w(k)$ is $O(N)$; hence, $D^{-1}w(k)$ is $O(N/\varepsilon^{\rho-1})$. Thus, we can establish an ultimate bound on $\|x(k) - \hat{x}(k)\|$ of the form $c_1\varepsilon + c_2 N/\varepsilon^{\rho-1}$ for some positive constants c_1 and c_2, independent of ε and N. Properties of this bounded are established in Lemma 8.1.

Theorem 9.2. *Consider the closed-loop system of the plant* (9.16)–(9.19) *and the output feedback digital controller* (9.28) *with the observer* (9.24)–(9.25) *or the observer* (9.26)–(9.27). *Suppose* $y(k) = Cx(k) + v(k)$, *where* $|v(k)| \leq N$ *for all* $k \geq 0$. *Let* $\mathscr{R}$ *be the region of attraction of* (9.21), *let* $\mathscr{S}$ *be any compact set in the interior of* $\mathscr{R}$, *and let* $\mathscr{Q}$ *be any compact subset of* R^ρ. *Suppose* $(\chi(0), \hat{x}(0)) \in \mathscr{S} \times \mathscr{Q}$. *Then*

- *there exist positive constants* c_a *and* N^* *such that for each* $N < N^*$ *there is a constant* $\varepsilon_a = \varepsilon_a(N) > c_a N^{1/\rho}$, *with* $\lim_{N\to 0} \varepsilon_a(N) = \varepsilon_a^* > 0$, *such that for each* $\varepsilon \in \left(c_a N^{1/\rho}, \varepsilon_a\right]$ *the solutions* $(\chi(k), \eta(k))$ *of the closed-loop system are bounded for all* $k \geq 0$, *and* $\chi(t)$ *is bounded for all* $t \geq 0$;

- *there exist* $N_1^* > 0$ *and a class* $\mathcal{K}$ *function* ϱ_1 *such that for every* $N < N_1^*$ *and every* $\mu_1 > \varrho_1(N)$, *there are constants* $T_1 = T_1(\mu_1) \geq 0$ *and* $\varepsilon_b = \varepsilon_b(N, \mu_1) > c_a N^{1/p}$, *with* $\lim_{N\to 0} \varepsilon_b(N, \mu_1) = \varepsilon_b^*(\mu_1) > 0$, *and an integer* $k^* = k^*(\mu_1)$ *such that for each* $\varepsilon \in \left(c_a N^{1/p}, \varepsilon_b\right]$ *we have*

$$\|x(k) - \hat{x}(k)\| \leq \mu_1 \qquad \forall\, k \geq k^*, \tag{9.59}$$

$$\|\chi(t)\| \leq \mu_1 \qquad \forall\, t \geq T_1; \tag{9.60}$$

- *there exist* $N_2^* > 0$ *and a class* $\mathcal{K}$ *function* ϱ_2 *such that for every* $N < N_2^*$ *and every* $\mu_2 > \varrho_2(N)$, *there is a constant* $\varepsilon_c = \varepsilon_c(N, \mu_2) > c_a N^{1/p}$, *with* $\lim_{N\to 0} \varepsilon_c(N, \mu_2) = \varepsilon_c^*(\mu_2) > 0$, *such that for each* $\varepsilon \in \left(c_a N^{1/p}, \varepsilon_c\right]$ *we have*

$$\|\chi(t) - \chi_r(t)\| \leq \mu_2 \quad \forall\, t \geq 0, \tag{9.61}$$

where $\chi_r(t)$ *is the solution of* (9.21) *with* $\chi_s(0) = \chi(0)$.

◇

The proof of Theorem 9.2 is a merging of the proofs of Theorems 8.4 and 9.1.[80] Remark 8.2 on Theorem 8.4 applies almost verbatim to Theorem 9.2.

Example 9.3. Consider the pendulum equation of Examples 3.4 and 9.2. A digital controller is derived in Example 9.2 with Forward Difference discrete-time implementation of two high-gain observers: a linear observer and a nonlinear one. In this example we consider the linear observer, and the results are representative of the nonlinear one. Since all the equations are given in Example 9.2, only simulation results are presented here. Simulation is carried out with $\alpha = 1$ and $T = \varepsilon = 0.01$. The initial conditions are $x_1(0) = -\pi$, $x_2(0) = \xi_1(0) = \xi_2(0) = 0$. The measurement noise is generated by the Simulink block "Uniform Random Number" with amplitude between ± 0.01 and sample time 0.001. The noise waveform is shown in Figure 9.8. Figure 9.9 compares the state trajectories of the system under digital control with and without measurement noise. The trajectories are fairly close, but the effect of measurement noise is seen in some jittering in x_2. Figure 9.10 focuses on that jittering part. It is typical in digital control systems to use an antialiasing filter to reduce the effect of aliasing due to signal sampling.[81] It is an analog low-pass filter with cutoff frequency equal to half the sampling frequency. An analog signal is passed through the filter before sampling. In view of our earlier discussion on low-pass filters in Chapter 8, it is reasonable to expect the antialiasing filter to reduce the effect of measurement noise. This is shown in Figures 9.11 and 9.12, which compare the state trajectories with and without filter. The antialiasing filter is a Butterworth one of order 8 and cutoff frequency 50 Hz (half the sampling frequency of 100 Hz). The state trajectories are fairly close to each other, although the filter increases the settling time a little bit, which is not surprising in view of its phase lag. Focusing on the steady state of x_2 in Figure 9.12 shows reduction in the error due to noise. △

[80]See [79] for the details of the proof.

[81]See [16], [45], or any digital control textbook.

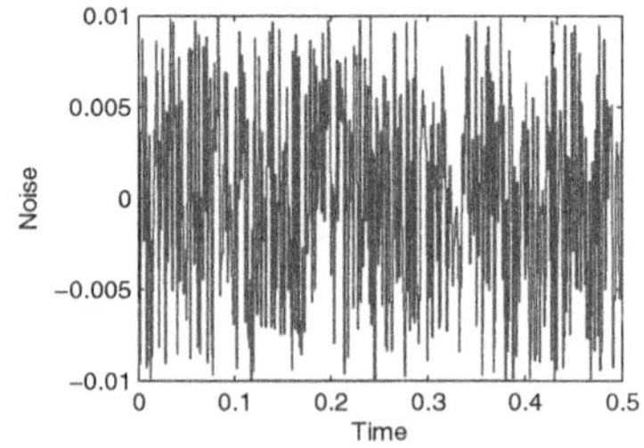

Figure 9.8. *Measurement noise of Example* 9.3.

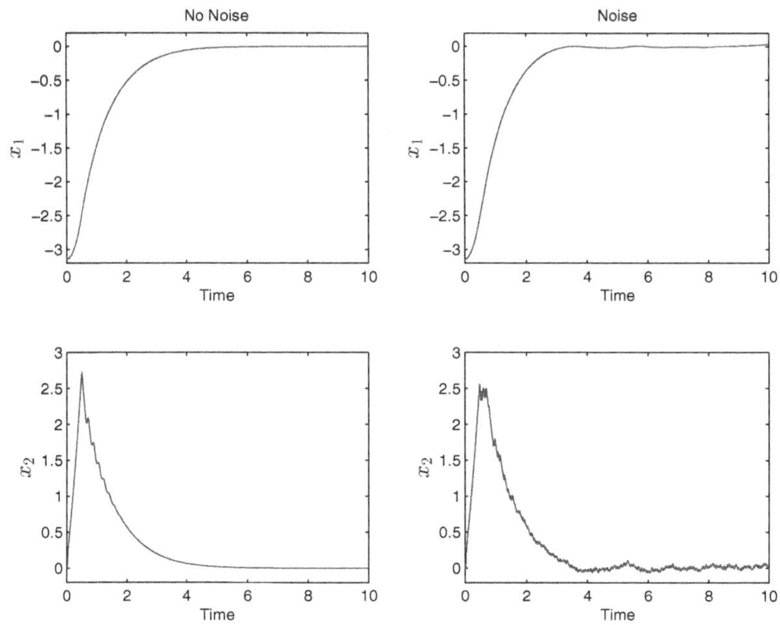

Figure 9.9. *Comparison of the trajectories of Example* 9.3 *under digital control with and without measurement noise.*

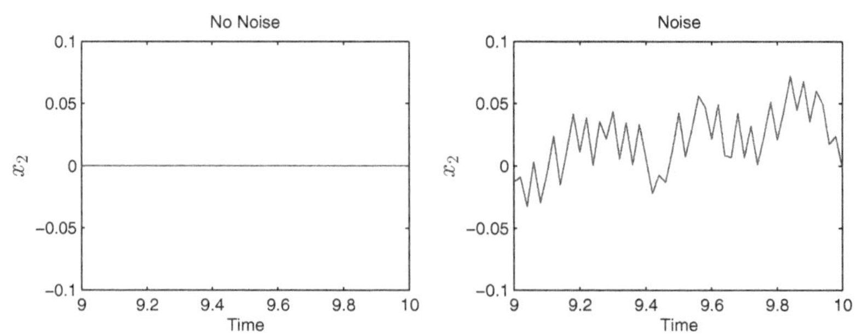

Figure 9.10. *Comparison of the* x_2 *trajectories of Example* 9.3 *under digital control with and without measurement noise.*

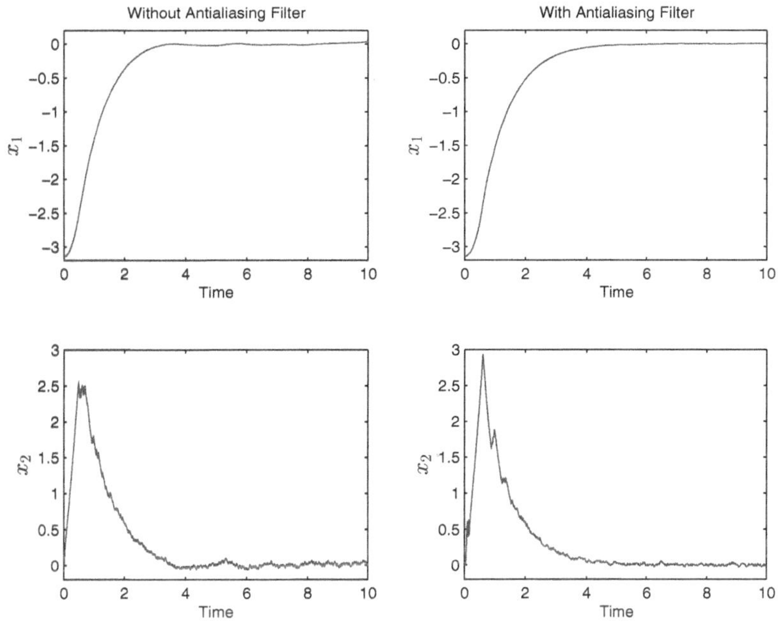

Figure 9.11. *Comparison of the trajectories of Example* 9.3 *under digital control with and without antialiasing filter.*

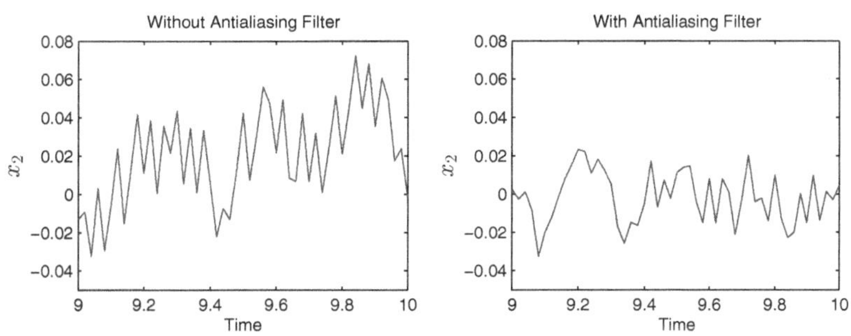

Figure 9.12. *Comparison of the* x_2 *trajectories of Example* 9.3 *under digital control with and without antialiasing filter.*

9.4 ▪ Multirate Digital Control

Reconsider the stabilization problem of Section 9.2 for the system

$$\dot{w} = f_0(w, x, u), \tag{9.62}$$

$$\dot{x} = Ax + B\phi(w, x, u), \tag{9.63}$$

$$y = Cx, \tag{9.64}$$

$$z = \psi(w, x), \tag{9.65}$$

where $u \in R$ is the control input, $y \in R$ and $z \in R^s$ are measured outputs, and $w \in R^\ell$ and $x \in R^\rho$ constitute the state vector. The matrices A, B, C are defined after equation (9.3). The functions f_0, ϕ, and ψ are locally Lipschitz in their arguments for $(w, x, u) \in D_w \times D_x \times R$, where $D_w \subset R^\ell$ and $D_x \subset R^\rho$ are domains that contain their respective origins. Moreover, $f_0(0,0,0)=0$, $\phi(0,0,0)=0$, and $\psi(0,0)=0$. Suppose the continuous-time state feedback controller

$$u=\gamma(x,z) \tag{9.66}$$

stabilizes the origin $\chi=0$ of the closed-loop system

$$\dot{\chi}=f_r(\chi,\gamma(x,z)), \tag{9.67}$$

where

$$\chi=\begin{bmatrix} w \\ x \end{bmatrix}, \qquad f_r(\chi,u)=\begin{bmatrix} f_0(w,x,u) \\ Ax+B\phi(w,x,u) \end{bmatrix},$$

and γ is locally Lipschitz in (x,z) and globally bounded in x, with $\gamma(0,0)=0$.

Assuming that with the sampling period T_s, measurements of x and z are available at the sampling points, the discrete-time state feedback controller is taken as

$$u(k)=\gamma(x(k),z(k)), \tag{9.68}$$

where $u(k)$ is constant over the period $[kT_s,(k+1)T_s)$, $x(k)=x(kT_s)$, and $z(k)=z(kT_s)$. As in Section 9.2, it can be shown that in between the sampling points,

$$\|\chi(t)-\chi(k)\|\le\frac{1}{L_1}\left[e^{(t-kT_s)L_1}-1\right]\|f_r(\chi(k),u(k))\|, \tag{9.69}$$

where L_1 is a Lipschitz constant of f_r with respect to χ. The state of the closed-loop system at the sampling points satisfies the difference equation

$$\chi(k+1)=\chi(k)+T_sf_r(\chi(k),u(k))+T_s^2\Phi_s(\chi(k),u(k),T_s), \tag{9.70}$$

where

$$\Phi_s(\chi(k),u(k),T_s)=\frac{1}{T_s^2}\int_{kT_s}^{(k+1)T_s}[f_r(\chi(\tau),u(k))-f_r(\chi(k),u(k))]\,d\tau$$

is locally Lipschitz in (χ,u) and, using (9.69),

$$\|\Phi_s(\chi(k),u(k),T_s)\|\le\frac{1}{T_s^2}\left[(e^{T_sL_1}-1)-T_s\right]\|f_r(\chi(k),u(k))\|.$$

Since

$$\lim_{T_s\to 0}\frac{1}{T_s^2}\left[\frac{1}{L_1}(e^{T_sL_1}-1)-T_s\right]=\frac{1}{2}L_1$$

there are positive constants K and T_s^* such that for all $T_s\le T_s^*$, $\|\Phi_s(\chi(k),u(k))\|\le K\|f_r(\chi(k),u(k))\|$.

The discretized high-gain observer is implemented with the fast sampling period $T_f=\alpha\varepsilon$, where $h=T_s/T_f$ is a positive integer greater than one. The observer is discretized as in Section 9.1 using the Forward Difference method to obtain

$$\xi(n+1)=A_d\xi(n)++B_dy(n)+\alpha\varepsilon^\rho B\phi_0(D^{-1}\xi(n),z(k),u(k)), \tag{9.71}$$

$$\hat{x}(n)=D^{-1}\xi(n), \tag{9.72}$$

where $A_d=A+\alpha F$, $B_d=\alpha E$, and the matrices D, E, and F are defined after equation (9.6). It is assumed that α is chosen such that the eigenvalues of A_d satisfy $|\lambda|<1$. The

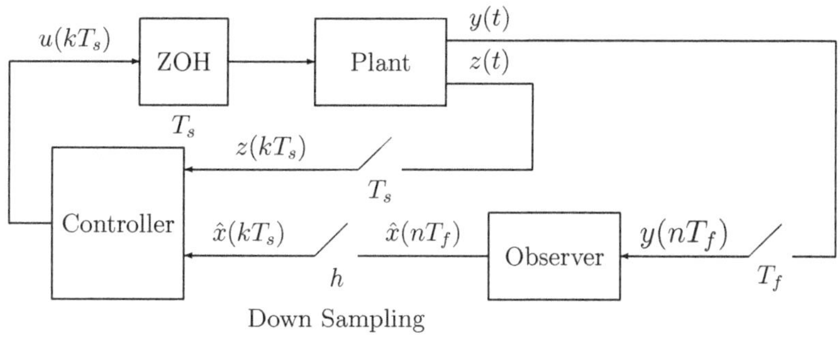

Figure 9.13. *Block diagram of the multirate digital control scheme.*

index n is used to indicate the sampling points that are equally spaced with period T_f. The index k indicates the sampling points that are equally spaced with period T_s. For each value of k, n takes h values from $n = kh$ to $n = kh + (h-1)$. The observer estimates are down-sampled for use in the output feedback controller. If $\hat{x}(n)$ is the estimate calculated by the observer for $n \subset [kh, kh + h - 1]$, then the down-sampled estimate $\hat{x}_s$ is given by $\hat{x}_s(k) = \hat{x}(hk)$ and is kept constant over the slow sampling period T_s. For clarity, $x_f(n)$ and $\chi_f(n)$ are used to denote the samples of $x(t)$ and $\chi(t)$ in the fast sampling period T_f. They are related to $x(k)$ and $\chi(k)$ by the relations $x(k) = x_f(hk)$ and $\chi(k) = \chi_f(hk)$. The output feedback controller is given by

$$u(k) = \gamma(\hat{x}_s(k), z(k)), \tag{9.73}$$

where $z(k)$ are the samples of $z(t)$ in the slow sampling period T_s. A block diagram of the multirate scheme is shown in Figure 9.13.

By repeating the derivation of Section 9.2, it can be shown that

$$\hat{x}(n) = [I - \varepsilon Q_2(\varepsilon)]x_f(n) + Q_1(\varepsilon)\eta(n), \tag{9.74}$$

where Q_1 and Q_2 are analytic functions of ε while η, which is defined by the changes of variables (9.36) and (9.38), satisfies the equation

$$\eta(n+1) = A_d\eta(n) + \varepsilon\Gamma(\chi_f(n), u(k), \hat{x}(n), \varepsilon), \tag{9.75}$$

where Γ is locally Lipschitz in $(\chi, u, \hat{x})$, globally bounded in $\hat{x}$, and uniformly bounded in ε for sufficiently small ε. The down-sampled estimate $\hat{x}_s(k)$ is given by

$$\hat{x}_s(k) = \hat{x}(hk) = [I - \varepsilon Q_2(\varepsilon)]x_f(hk) + Q_1(\varepsilon)\eta(hk) = [I - \varepsilon Q_2(\varepsilon)]x(k) + Q_1(\varepsilon)\eta_s(k),$$

where $\eta_s(k) = \eta(hk)$ is down-sampling of $\eta(n)$. Thus, the closed-loop multirate digital control system can be represented at the slow sampling points k by the discrete-time model

$$\chi(k+1) = \chi(k) + T_s f_r(\chi(k), u(k)) + T_s^2\Phi_s(\chi(k), u(k), T_s), \tag{9.76}$$
$$u(k) = \gamma(\hat{x}_s(k), z(k)), \tag{9.77}$$
$$\hat{x}_s(k) = [I - \varepsilon Q_2(\varepsilon)]x(k) + Q_1(\varepsilon)\eta_s(k). \tag{9.78}$$

Theorem 9.3. *Consider the closed-loop system of the plant* (9.62)–(9.65) *and the output feedback multirate digital controller* (9.73) *with the observer* (9.71)–(9.72). *Let* $\mathcal{R}$ *be the* *region of attraction of (9.67), let $\mathcal{S}$ be any compact set in the interior of $\mathcal{R}$, and let $\mathcal{Q}$ be any compact subset of R^ρ. Suppose $(\chi(0), \hat{x}(0)) \in \mathcal{S} \times \mathcal{Q}$. Then*

- *there exists $T_{s1}^* > 0$ and for each $T_s \in (0, T_{s1}^*]$, there is $\varepsilon_1^* > 0$ such that for every $\varepsilon \in (0, \varepsilon_1^*]$, $\chi(t)$ is bounded for all $t \geq 0$ and $\eta(n)$ is bounded for all $n \geq 0$;*
- *given any $\mu > 0$ there exists $T_{s2}^* > 0$, and for each $T_s \in (0, T_{s2}^*]$, there is $\varepsilon_2^* > 0$, $T_1 > 0$, and $n^* > 0$, all dependent on μ, such that for every $\varepsilon \in (0, \varepsilon_2^*]$,*

$$\|\chi(t)\| \leq \mu \quad \forall\, t \geq T_1 \quad \text{and} \quad \|\eta(n)\| \leq \mu \quad \forall\, n \geq n^*; \tag{9.79}$$

- *given any $\mu > 0$ there exists $T_{s3}^* > 0$, and for each $T_s \in (0, T_{s3}^*]$, there is $\varepsilon_3^* > 0$, dependent on μ, such that for every $\varepsilon \in (0, \varepsilon_3^*]$,*

$$\|\chi(t) - \chi_r(t)\| \leq \mu \quad \forall\, t \geq 0, \tag{9.80}$$

  where χ_r is the solution of (9.67) with $\chi_r(0) = \chi(0)$;
- *if the origin of (9.67) is exponentially stable and $f_r(\chi, \gamma(x,z))$ is twice continuously differentiable in the neighborhood of the origin, then there exists $T_{s4}^* > 0$, and for each $T_s \in (0, T_{s4}^*]$ there is $\varepsilon_4^* > 0$ such that for every $\varepsilon \in (0, \varepsilon_4^*]$ the discrete-time trajectories $\chi(k)$ and $\eta_s(k)$ converge to zero exponentially fast as $k \to \infty$ and the continuous-time trajectory $\chi(t)$ decays to zero exponentially fast as $t \to \infty$.* ◇

Proof: Rewrite (9.76) and (9.75) as

$$\chi(k+1) = \chi(k) + T_s \mathcal{F}(\chi(k), \eta_s(k), T_s, \varepsilon), \tag{9.81}$$

$$\eta(n+1) = A_d \eta(n) + \varepsilon \mathcal{G}(\chi_f(n), \eta(n), x(k), \eta_s(k), z(k), \varepsilon), \tag{9.82}$$

where

$$\begin{aligned}\mathcal{F}(\chi, \eta_s, T_s, \varepsilon) &= f_r(\chi, \gamma(\hat{x}_s, z)) + T_s \Phi_s(\chi, \gamma(\hat{x}_s, z), T_s)\\ &= f_r(\chi, \gamma((I - \varepsilon Q_2)x + Q_1 \eta_s, z))\\ &\quad + T_s \Phi_s(\chi, \gamma((I - \varepsilon Q_2)x + Q_1 \eta_s, z), T_s)\end{aligned}$$

and

$$\begin{aligned}\mathcal{G}(\chi_f, \eta, x, \eta_s, z, \varepsilon) &= \Gamma(\chi_f, \gamma(\hat{x}_s, z), \hat{x}, \varepsilon)\\ &= \Gamma(\chi_f, \gamma((I - \varepsilon Q_2)x + Q_1 \eta_s, z), (I - \varepsilon Q_2)x_f + Q_1 \eta, \varepsilon).\end{aligned}$$

The functions $\mathcal{F}$ and $\mathcal{G}$ are locally Lipschitz, globally bounded in η_s and η, and vanish at the origin ($\chi = 0, \eta = 0$). Since the origin of (9.67) is asymptotically stable and $\mathcal{R}$ is its region of attraction, by the converse Lyapunov theorem of asymptotic stability [78, Theorem 4.17], there is a smooth, positive definite function $V(\chi)$ and a continuous, positive definite function $U(\chi)$, both defined for all $\chi \in \mathcal{R}$, such that

$$V(\chi) \to \infty \text{ as } \chi \to \partial \mathcal{R},$$

$$\frac{\partial V}{\partial \chi} f_r(\chi, \gamma(x,z)) \leq -U(\chi) \quad \forall\, \chi \in \mathcal{R}$$

and for any $c > 0$, $\{V(\chi) \le c\}$ is a compact subset of $\mathscr{R}$. Choose positive constants b and c such that $c > b > \max_{\chi\in\mathscr{S}} V(\chi)$. Then

$$\mathscr{S} \subset \Omega_b = \{V(\chi) \le b\} \subset \Omega_c = \{V(\chi) \le c\} \subset \mathscr{R}.$$

As in the proof of Theorem 9.1, we use $V(\chi)$ as a Lyapunov function for the slow subsystem (9.81) and $W(\eta) = \eta^T P \eta$ as a Lyapunov function for the fast subsystem (9.82), where P is the positive definite solution of the discrete-time Lyapunov equation $A_d^T P A_d - P = -I$. Let $\Lambda = \Omega_c \times \{W(\eta) \le c_1 \varepsilon^2\}$, where $c_1 > 0$ will be chosen. For every $T_s \in (0, \tilde{T}_s]$ and $\varepsilon \in (0, \tilde{\varepsilon}]$, where $\tilde{T}_s$ and $\tilde{\varepsilon} < 1$ are positive constants, there are positive constants K_1 to K_6, independent of T_s and ε, such that

$$\|\mathscr{F}(\chi, \eta_s, T_s, \varepsilon)\| \le K_1, \quad \|\Phi_s(\chi, \gamma(x,z), T_s)\| \le K_2, \quad \|\mathscr{G}(\chi_f, \eta, x, \eta_s, z, \varepsilon)\| \le K_3$$

for all $(\chi, \eta) \in \Omega \times R^\rho$ and

$$\|\mathscr{F}(\chi, \eta_s, T_s, \varepsilon) - \mathscr{F}(\chi, 0, T_s, 0)\| \le K_4\|\eta_s\| + K_5\varepsilon\|x\| \le K_6\varepsilon$$

for all $(\chi, \eta) \in \Lambda$. It will now be shown that Λ is positively invariant. Since

$$\mathscr{F}(\chi, 0, T_s, 0) = f_r(\chi, \gamma(x,z)) + T_s\Phi_s(\chi, \gamma(x,z), T_s),$$

equation (9.81) can be written as a perturbation of (9.70), that is,

$$\chi(k+1) = \chi(k) + T_s f_r(\chi, \gamma(x,z)) + T_s^2\Phi_s(\chi, \gamma(x,z), T_s) + T_s\Delta_1(\cdot),$$

where $\|\Delta_1(\cdot)\| \le K_6\varepsilon$ for all $(\chi, \eta) \in \Lambda$. Using this equation in calculating $\Delta V(\chi(k)) = V(\chi(k+1)) - V(\chi(k))$ yields

$$\begin{aligned}
\Delta V(\chi(k)) &= \frac{\partial V}{\partial \chi}(\bar{\chi}(k))[\chi(k+1) - \chi(k)] \\
&= \frac{\partial V}{\partial \chi}(\chi(k))[\chi(k+1) - \chi(k)] \\
&\quad + \left[\frac{\partial V}{\partial \chi}(\bar{\chi}(k)) - \frac{\partial V}{\partial \chi}(\chi(k))\right][\chi(k+1) - \chi(k)] && (9.83) \\
&\le \frac{\partial V}{\partial \chi}(\chi(k))\big[T_s f_r(\chi, \gamma(x,z)) + T_s^2\Phi_s(\chi, \gamma(x,z), T_s) + T_s\Delta_1(\cdot)\big] && (9.84) \\
&\quad + L_2\|\chi(k+1) - \chi(k)\|^2 \\
&\le -T_s U(\chi(k)) + \left\|\frac{\partial V}{\partial \chi}(\chi(k))\right\| \left(T_s^2\|\Phi_s(\chi, \gamma(x,z), T_s)\| + T_s\|\Delta_1(\cdot)\|\right) \\
&\quad + L_2 T_s^2 K_1^2 && (9.85) \\
&\le -T_s U(\chi(k)) + T_s^2 K_7 + T_s\varepsilon K_8, && (9.86)
\end{aligned}$$

where L_2, K_7, and K_8 are positive constants independent of T_s and ε. Repeating arguments from the proof of Theorem 9.1, the foregoing inequality can be used to show that for all $\eta \in \{W(\eta) \le c_1\varepsilon^2\}$, Ω_c is positively invariant for sufficiently small T_s and ε. It can be also shown that by choosing c_1 large enough, the set $\{W(\eta) \le c_1\varepsilon^2\}$ is positively invariant for all $\chi \in \Omega_c$. Hence, $\Lambda = \Omega_c \times \{W(\eta) \le c_1\varepsilon^2\}$ is positively invariant.

With the initial state $(\chi(0), \hat{x}(0)) \in \mathscr{S} \times \mathscr{Q}$, we have $\|\eta(0)\| \leq l/\varepsilon^{\rho-1}$, where l depends on $\mathscr{S}$ and $\mathscr{Q}$. Since $\mathscr{S}$ is in the interior of Ω_c, it can be shown from (9.81) that

$$\|\chi(k) - \chi(0)\| = T_s \left\| \sum_{i=0}^{k-1} \mathscr{F}(\chi(k), \eta_s(k), T_s, \varepsilon) \right\| \leq T_s k K_1$$

as long as $\chi(k) \in \Omega_c$. Therefore, there exists a positive constant K_9, independent of T_s and ε, such that $\chi(k) \in \Omega_c$ for all $k \leq K_9/T_s$. With $\chi(k) \in \Omega_c$ and $\eta(n) \notin \{W(\eta) \leq c_1\varepsilon^2\}$, it can be shown that by choosing c_1 large enough

$$W(\eta(n)) \leq \lambda^n W(\eta(0)) \quad \Rightarrow \quad W(\eta(n)) \leq \lambda^n \|P\| \frac{l^2}{\varepsilon^{2\rho-2}},$$

where $\lambda \in (0,1)$. Hence, η enters $\{W(\eta) \leq c_1\varepsilon^2\}$ and stays therein for all

$$n \geq \bar{n}(\varepsilon) \stackrel{\text{def}}{=} \frac{\ln\left(\frac{\|P\| l^2}{c_1 \varepsilon^{2\rho}}\right)}{\ln\left(\frac{1}{\lambda}\right)}.$$

To ensure that $\eta(n)$ enters $\{W(\eta) \leq c_1\varepsilon^2\}$ before $\chi(k)$ leaves Ω_c, ε should be chosen small enough that $\bar{n}(\varepsilon) < hK_9/T_s$. Since $T_f = T_s/h = \alpha\varepsilon$, ε should be chosen small enough that $\bar{n}(\varepsilon) < K_9/(\alpha\varepsilon)$, which can be done since $\lim_{\varepsilon \to 0} \varepsilon \bar{n}(\varepsilon) = 0$.

The proof of the ultimate boundedness property (9.79) follows an argument similar to the proof of Theorem 9.1. In particular, using (9.86) it can be shown that after finite time the trajectories inside Λ will be in the set

$$\bar{\Lambda} = \{V(\chi) \leq c_3(T_s + \varepsilon)\} \times \{W(\eta) \leq c_1\varepsilon^2\},$$

where $\lim_{r \to 0} c_3(r) = 0$. Choosing T_s and ε small enough ensures that (9.79) is satisfied. The proof of (9.80) is the same as the corresponding argument in the proof of Theorem 9.1 with the error $e(k) = \chi(k) - \chi_r(k)$ satisfying an equation of the form

$$\begin{aligned} e(k+1) = e(k) + T_s[&f_r(\chi(k), \gamma(x(k), z(k))) - f_r(\chi_r(k), \gamma(x_r(k), z_r(k)))] \\ &+ O(T_s^2) + O(T_s\varepsilon). \end{aligned}$$

Finally, we prove the last bullet of the theorem. Since the origin of $\dot{\chi} = f_r(\chi, \gamma(x,z))$ is exponentially stable and $f_r(\chi, \gamma(x,z))$ is twice continuously differentiable, by the converse Lyapunov theorem for exponential stability [78, Theorem 4.14], there is a twice continuously differentiable Lyapunov function $V_1(\chi)$ that satisfies the following inequality in some neighborhood of the origin $\chi = 0$ for some positive constants β_1 to β_4:

$$\beta_1\|\chi\|^2 \leq V_1(\chi) \leq \beta_2\|\chi\|^2, \quad \frac{\partial V}{\partial \chi} f_r(\chi, \gamma(x,z)) \leq -\beta_3\|\chi\|^2, \quad \left\|\frac{\partial V}{\partial \chi}\right\| \leq \beta_4\|\chi\|.$$

Let $\mathscr{B}$ be ball centered at the origin $(\chi, \eta) = (0,0)$ such that the foregoing inequalities hold in $\mathscr{B}$. The second bullet of the theorem guarantees that, for sufficiently small T_s and ε, the trajectories will enter $\mathscr{B}$ in finite time. The forthcoming analysis is limited to the ball $\mathscr{B}$. We prove exponential stability of the origin of the discrete-time system in the slow sample time k. This requires a description of $\eta_s(k) = \eta(hk)$, which is obtained by studying $(\chi_f(n), \eta(n))$ over one slow time period $[hk, hk+1]$. In the fast sample time n, the discrete-time system is represented by

$$\chi_f(n+1) = \chi_f(n) + \alpha\varepsilon f_r(\chi_f(n), u(k)) + \alpha^2\varepsilon^2\Phi_f(\chi_f(n), u(k), \varepsilon), \tag{9.87}$$

$$\eta(n+1) = A_d\eta(n) + \varepsilon\mathscr{G}(\chi_f(n), \eta(n), x(k), \eta_s(k), z(k), \varepsilon). \tag{9.88}$$

Even though $|\lambda(A_d)| < 1$, it is not clear that performing accumulation of the solution of (9.88) over the interval $[hk, hk+1]$ will add up to $O(\varepsilon)$. This is due to the appearance of η in the function $\mathscr{G}$. We perform a change of variables to weaken the dependence of $\mathscr{G}$ on η. The system (9.87)–(9.88) has two-time-scale behavior with χ_f as the slow variable and η as the fast one. Treating χ_f as a constant input to the fast equation (9.88), its quasi-steady-state solution $\bar{\eta}(n)$ satisfies the equation

$$\bar{\eta}(n) = A_d\bar{\eta}(n) + \varepsilon\mathscr{G}(\chi_f(n), \bar{\eta}(n), x(k), \eta_s(k), z(k), \varepsilon),$$

where $x(k)$, $\eta_s(k)$, and $z(k)$ are constant during the interval $[hk, hk+1]$. We seek a solution to the foregoing equation in the form $\bar{\eta}(n) = \varepsilon\theta(\chi_f(n), x(k), \eta_s(k), z(k), \varepsilon)$, where θ is a continuously differentiable function of its arguments and $\theta(0,0,0,0,0) = 0$. Evaluating $\bar{\eta}(n) = \varepsilon\theta(\chi_f(n), x(k), \eta_s(k), z(k), \varepsilon)$ at $n+1$ and substituting for $\chi_f(n+1)$ and $\bar{\eta}(n+1)$ using (9.87) and (9.88) result in

$$A_d\theta(\chi_f(n), \ldots, \varepsilon) + \mathscr{G}(\chi_f(n), \varepsilon\theta(\cdot), \ldots, \varepsilon) = \theta(\chi_f(n) + \alpha\varepsilon f_r(\cdot) + \alpha^2\varepsilon^2\Phi_f(\cdot), \ldots, \varepsilon). \tag{9.89}$$

Setting $\varepsilon = 0$ yields

$$(I - A_d)\theta(\chi_f(n), x(k), \eta_s(k), z(k), 0) = \mathscr{G}(\chi_f(n), 0, x(k), \eta_s(k), z(k), 0).$$

Since $|\lambda(A_d)| < 1$, $(I - A_d)$ is nonsingular. Hence, the foregoing equation has a unique solution. By the implicit function theorem, there is a continuously differentiable function $\theta(\chi_f(n), x(k), \eta_s(k), z(k), \varepsilon)$ that satisfies (9.89) for sufficiently small ε. The change of variable

$$\zeta(n) = \eta(n) - \varepsilon\theta(\chi_f(n), x(k), \eta_s(k), z(k), \varepsilon) \tag{9.90}$$

transforms (9.88) into

$$\zeta(n+1) = A_d\zeta(n) + \varepsilon\tilde{\mathscr{G}}(\chi_f(n), \zeta(n), x(k), \eta_s(k), z(k), \varepsilon), \tag{9.91}$$

where $\tilde{\mathscr{G}}$ is continuously differentiable and $\tilde{\mathscr{G}}(\chi_f(n), 0, x(k), \eta_s(k), z(k), \varepsilon) = 0$. Therefore, $\zeta = 0$ is an equilibrium point of (9.91). In the ball $\mathscr{B}$, $\tilde{\mathscr{G}}$ can be represented as [80, Equation (B.4)].

$$\begin{aligned}\tilde{\mathscr{G}}(\chi_f, \zeta, x, \eta_s, z, \varepsilon) &= \int_0^1 \frac{\partial\tilde{\mathscr{G}}}{\partial\zeta}(\chi_f, \sigma\zeta, x, \eta_s, z, \varepsilon)\, d\sigma\, \zeta \\ &\stackrel{\text{def}}{=} \tilde{B}(\chi_f, \zeta, x, \eta_s, z, \varepsilon)\, \zeta,\end{aligned}$$

and (9.91) can be rewritten as the linear time-varying system

$$\zeta(n+1) = \left[A_d + \varepsilon\tilde{B}(\chi_f(n), \zeta(n), x(k), \eta_s(k), z(k), \varepsilon)\right]\zeta(n) \stackrel{\text{def}}{=} \tilde{A}_d(n)\,\zeta(n), \tag{9.92}$$

which is valid because $\chi_f(n)$ and $\eta(n)$ are bounded and belong to $\mathscr{B}$. Since $|\lambda(A_d)| \leq \lambda_1 < 1$ for some $\lambda_1 > 0$, it follows that for sufficiently small ε the transition matrix $\tilde{\Phi}$ of $\tilde{A}_d$ satisfies $\|\tilde{\Phi}(n,m)\| \leq \kappa_1 \lambda_2^{n-m}$, where $0 < \lambda_2 < 1$ and $\kappa_1 > 0$ [126, Theorem 24.7]. Because $h = T_s/T_f = T_s/(\alpha\varepsilon)$,

$$\|\tilde{\Phi}(hk+h, hk)\| \leq \kappa_1 \lambda_2^h = O(\varepsilon).$$

Therefore, $\zeta_s(k) \stackrel{\text{def}}{=} \zeta(hk)$ satisfies an equation of the form

$$\zeta_s(k+1) = \varepsilon G(\chi(k), \zeta_s(k), \varepsilon)\zeta_s(k),$$

where G is a continuous function of its arguments. Writing (9.90) at $n = hk$, we obtain

$$\zeta_s(k) = \eta_s(k) - \varepsilon\theta(\chi(k), x(k), \eta_s(k), z(k), \varepsilon).$$

At $\varepsilon = 0$, $\eta_s(k) = \zeta_s(k)$. By the implicit function theorem, it can be concluded that, for sufficiently small ε, there is a continuously differentiable function ϑ such that

$$\eta_s(k) = \zeta_s(k) + \varepsilon\vartheta(\chi(k), \zeta_s(k), \varepsilon).$$

Thus, the discrete-time system can be represented at the slow sample times k by the equations

$$\begin{aligned} \chi(k+1) = \chi(k) &+ T_s f_r(\chi(k), \gamma(x(k), z(k))) \\ &+ T_s^2 \Phi_s(\chi(k), \gamma(x(k), z(k)), T_s) + T_s \Delta_2(\cdot), \end{aligned} \tag{9.93}$$

$$\zeta_s(k+1) = \varepsilon G(\chi(k), \zeta_s(k), \varepsilon)\zeta_s(k), \tag{9.94}$$

where $\|\Delta_2(\cdot)\| \leq K_{10}(\|\zeta_s\| + \varepsilon\|\chi\|)$ and $\|G(\cdot)\| \leq K_{11}$ for some positive constants K_{10} and K_{11} independent of T_s and ε. Equation (9.93) is a perturbation of the system

$$\chi(k+1) = \chi(k) + T_s f_r(\chi(k), \gamma(x(k), z(k))) + T_s^2 \Phi_s(\chi(k), \gamma(x(k), z(k)), T_s), \tag{9.95}$$

which is the closed-loop system under state feedback. Using $V_1(\chi)$ as a Lyapunov function for this system, we obtain

$$\begin{aligned} V_1(\chi(k+1)) - V_1(\chi(k)) &\leq \frac{\partial V_1}{\partial \chi}(\chi(k))[\chi(k+1) - \chi(x)] + L_2\|\chi(k+1) - \chi(k)\|^2 \\ &\leq T_s \frac{\partial V_1}{\partial \chi}(\chi(k)) f_r(\chi(k), \gamma(x(k), z(k))) \\ &\quad + T_s^2 \frac{\partial V_1}{\partial \chi}(\chi(k)) \Phi_s(\chi(k), \gamma(x(k), z(k)), T_s) \\ &\quad + L_2 T_s^2 \|f_r(\cdot) + T_s \Phi_s(\cdot)\|^2 \\ &\leq -T_s \beta_3 \|\chi(k)\|^2 + T_s^2 K_a \|\chi(k)\|^2, \end{aligned}$$

where $K_a > 0$ is independent of T_s. Thus,

$$V_1(\chi(k+1)) - V_1(\chi(k)) \leq -\tfrac{1}{2} T_s \beta_3 \|\chi(k)\|^2 \qquad \text{for } T_s \leq \frac{\beta_3}{2K_a}.$$

Taking $V_2(\chi, \zeta_s) = V_1(\chi) + \gamma \zeta_s^T \zeta_s$, with $\gamma > 0$, as a Lyapunov function for the system (9.93)–(9.94), it can be shown that for $T_s \leq \beta_3/(2K_a)$,

$$V_2(\chi(k+1), \zeta_s(k+1)) - V_2(\chi(k), \zeta_s(k)) \leq -\mathscr{Y}^T \mathscr{Q} \mathscr{Y},$$

where

$$\mathcal{Y} = \begin{bmatrix} \chi(k) \\ \zeta_s(k) \end{bmatrix}, \qquad \mathcal{Q} = \begin{bmatrix} \frac{1}{2}T_s\beta_3 - \varepsilon T_s\delta_1 & -T_s\delta_2 \\ -T_s\delta_2 & \gamma(1-\varepsilon^2\delta_3) - \delta_4 T_s^2 \end{bmatrix},$$

and δ_1 to δ_4 are nonnegative constants independent of T_s and ε. Take ε small enough so that $\frac{1}{2}T_s\beta_3 - \varepsilon T_s\delta_1 > 0$ and $1 - \varepsilon^2\delta_3 > 0$; then choose γ large enough to make Q positive definite. Hence, the origin $(\chi, \zeta_s) = (0,0)$ of (9.93)–(9.94) is exponentially stable. Consequently, there are positive constants C_1 and $\lambda_3 < 1$ such that $\|\chi(k)\| \leq C_1\lambda_3^{k-k_0}\|\chi(k_0)\|$ for $k \geq k_0$. Set $a = (1/T_s)\ln(1/\lambda_3)$ so that $e^{-aT_s} = \lambda_3$. From inequality (9.69), there is a positive constant C_2 such that $\|\chi(t)\| \leq C_2\|\chi(kT_s)\|$ for $t \in [kT_s, (k+1)T_s)$. Hence,

$$\|\chi(t)\| \leq C_2 C_1 e^{-a(kT_s - k_0T_s)}\|\chi(k_0T_s)\| \leq C_3 e^{-a(t-k_0T_s)}\|\chi(k_0T_s)\|$$

for some $C_3 > 0$, which shows that $\chi(t)$ decays to zero exponentially fast. □

Example 9.4. Reconsider the pendulum equation

$$\dot{x}_1 = x_2, \quad \dot{x}_2 = \sin x_1 - c_1x_2 + c_2u, \quad y = x_1$$

from Example 3.4, where $0 \leq c_1 \leq 0.2$ and $0.5 \leq c_2 \leq 2$. A robust stabilizing state feedback controller is given by

$$u = -2\left[1.25\pi \text{ sat}\left(\frac{|x_1|}{1.25\pi}\right) + 2.25\pi \text{ sat}\left(\frac{|x_2|}{2.25\pi}\right) + 1\right] \text{sat}\left(\frac{x_1 + x_2}{0.1}\right).$$

Implementing this controller using zero-order hold and sampling period $T_s = 0.05$ results in adequate response, as shown in Figure 9.14. The simulation is carried out with $c_1 = 0.01$, $c_2 = 0.5$, $x_1(0) = -\pi$, and $x_2(0) = 0$. Assuming that the output $y = x_1$ is the only measured signal, a high-gain observer is used to estimate the states. Suppose we want, for some practical reason, to implement the output feedback controller such that the control is updated with the same sampling period $T_s = 0.05$. A single-rate output feedback controller will have to use the same sampling period for the control and state estimation. This corresponds to choosing $\varepsilon = 0.05$ and $\alpha = 1$ so that $\alpha\varepsilon = 0.05$. Although we see in Example 3.4 that a continuous-time output feedback controller with $\varepsilon = 0.05$ gives adequate response, in discrete-time implementation the single-rate output feedback controller fails to stabilize the system, as shown in Figure 9.15. The observer used here is the same linear observer with Forward Difference discretization that is used in Example 9.2, with $\hat{x}_1(0) = \hat{x}_2(0) = 0$. A multirate discrete-time output feedback controller is implemented with $T_s = 0.05$ and $T_f = 0.005$. The choice of T_f corresponds to $\varepsilon = 0.05$ and $\alpha = 0.1$. Simulation is carried out using the same type of observer and the same parameters and initial conditions. Figure 9.16 shows the down-sampling of the estimates $\hat{x}_1$ and $\hat{x}_2$, while Figure 9.17 shows the response of the system, which is fairly close to the discrete-time state feedback response. △

9.5 ▪ Notes and References

The observer discretization of Section 9.1 is based on [31], which considers also the zero-order hold and first-order hold discretization methods in the addition to the three

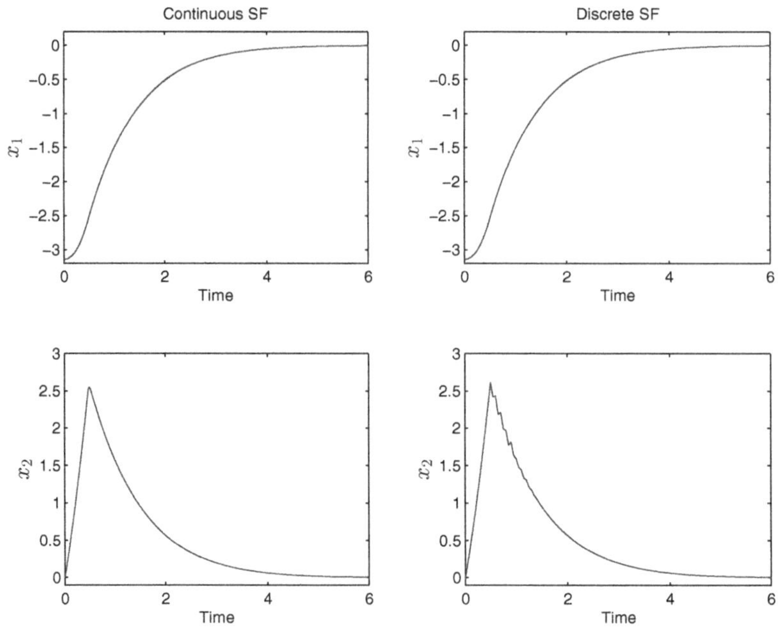

Figure 9.14. *Simulation of Example 9.4. Comparison of the closed-loop trajectories under continuous-time state feedback (SF) and discrete-time state feedback with sampling period 0.05.*

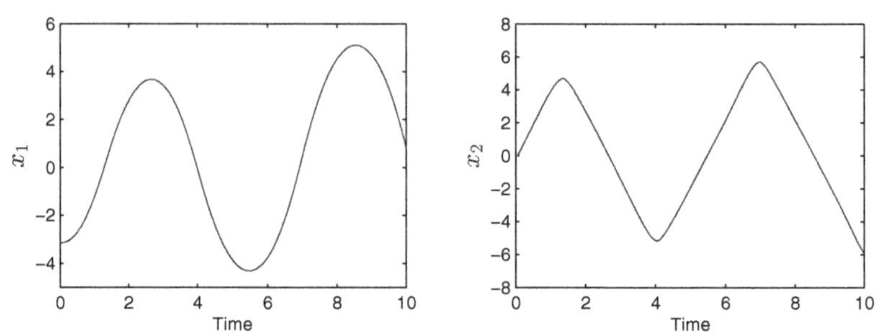

Figure 9.15. *Simulation of Example 9.4. The closed-loop trajectories under single-rate discrete-time output feedback controller with sampling period 0.05.*

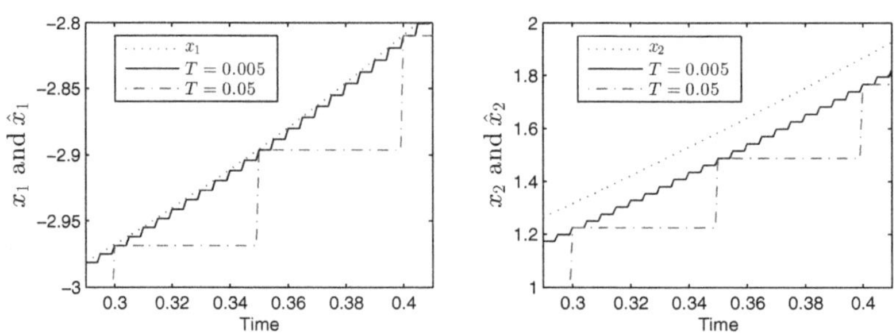

Figure 9.16. *Simulation of Example 9.4. Down sampling of the state estimates from the observer sampling period 0.005 to the control sampling period 0.05.*

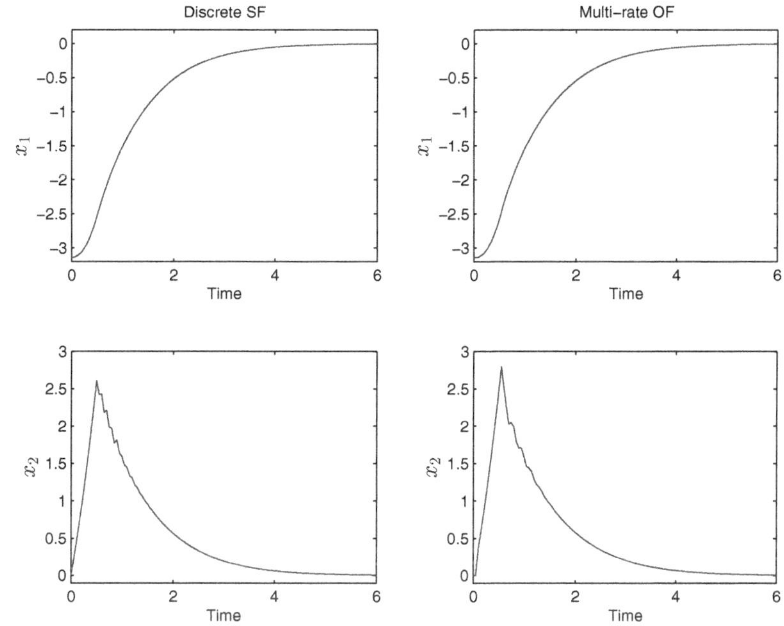

Figure 9.17. *Simulation of Example 9.4. Comparison of the closed-loop trajectories under discrete-time state feedback (SF) with sampling period 0.05 and multirate discrete-time output feedback (OF) with control sampling period 0.05 and observer sampling period 0.005.*

methods discussed here. Section 9.2 is based on [32]. This paper includes experimental testing of various discretization methods of the high-gain observer. It demonstrates the advantage of using a high-gain observer to estimate derivatives compared with the simple, yet routinely used, Euler's method. Section 9.3 is based on [79]. Section 9.4 is based on [3]. This paper deals with a more general control problem that allows for time-varying external signals and stabilization of a closed set, as opposed to stabilization of an equilibrium point. It includes also experimental application to a smart-material-actuated system in which the control computes the inverse of a hysteresis nonlinearity.

Bibliography

[1] A.H. Ahrens and H.K. Khalil. Closed-loop behavior of a class of nonlinear systems under EKF-based control. *IEEE Trans. Automat. Contr.*, 52:536–540, 2007. (Cited on p. 2)

[2] J. Ahrens and H.K. Khalil. High-gain observers in the presence of measurement noise: A switched-gain approach. *Automatica*, 41:43–54, 2009. (Cited on pp. 2, 261, 276)

[3] J.H. Ahrens, X. Tan, and H.K. Khalil. Multirate sampled-data output feedback control with application to smart material actuated systems. *IEEE Trans. Automat. Contr.*, 54:2518–2529, 2009. (Cited on p. 311)

[4] B. Aloliwi and H.K. Khalil. Robust adaptive output feedback control of nonlinear systems without persistence of excitation. *Automatica*, 33:2025–2032, 1997. (Cited on pp. 2, 106)

[5] J. Alvarez-Ramirez and R. Fermat. Robust PI stabilization of a class of chemical reactors. *Syst. Contr. Lett.*, 38:219–225, 1999. (Cited on p. 1)

[6] B.D.O. Anderson, R.R. Bitmead, C.R. Johnson, P.V. Kokotovic, R.L. Kosut, I.M.Y. Mareels, L. Praly, and B.D. Riedle. *Stability of Adaptive Systems*. MIT Press, Cambridge, MA, 1986. (Cited on p. 73)

[7] P.M. Anderson and A.A. Fouad. *Power System Control and Stability*. Iowa State University Press, 1977. (Cited on p. 207)

[8] V. Andrieu, L. Praly, and A. Astolfi. High gain observers with updated gain and homogeneous correction terms. *Automatica*, 45:422–428, 2009. (Cited on p. 1)

[9] P.J. Anstaklis and A.N. Michel. *Linear Systems Primer*. Birkhäuser, Boston, MA, 2007. (Cited on pp. 24, 131)

[10] A. Astolfi and R. Ortega. Immersion and invariance: A new tool for stabilization and adaptive control of nonlinear systems. *IEEE Trans. Automat. Contr.*, 48:590–606, 2003. (Cited on p. 73)

[11] D. Astolfi and L. Marconi. A high-gain nonlinear observer with limited gain power. *IEEE Trans. Automat. Contr.*, 3059–3064, 2015. (Cited on p. 236)

[12] D. Astolfi, L. Marconi, L. Praly, and A. Teel. Low-power peaking-free high-gain observers. *Automatica*, 2017. Under review. (Cited on p. 236)

[13] D. Astolfi, L. Marconi, L. Praly, and A. Teel. Sensitivity to high-frequency measurement noise of nonlinear high-gain observers. In 10th *IFAC Symposium on Nonlinear Control Systems*, Monterey, CA, 2016. (Cited on pp. 276, 278)

[14] D. Astolfi, L. Marconi, and A. Teel. Lower-power peaking-free high-gain observers for nonlinear systems. In *Proc. European Control Conf.*, Aablorg, Denmark, 2016. (Cited on p. 236)

[15] K.J. Astrom and B. Wittenmark. *Adaptive Control*. Addison-Wesley, Reading, MA, second edition, 1995. (Cited on p. 73)

[16] K.J. Astrom and B. Wittenmark. *Computer-Controlled Systems*. Prentice–Hall, Upper Saddle River, NJ, third edition, 1997. (Cited on pp. 121, 281, 299)

[17] A.N. Atassi. *A separation principle for the control of a class of nonlinear systems*. PhD thesis, Michigan State University, East Lansing, 1999. (Cited on p. 71)

[18] A.N. Atassi and H.K. Khalil. A separation principle for the stabilization of a class of nonlinear systems. *IEEE Trans. Automat. Contr.*, 44:1672–1687, 1999. (Cited on pp. 2, 30, 36, 70, 71)

[19] A.N. Atassi and H.K. Khalil. A separation principle for the control of a class of nonlinear systems. *IEEE Trans. Automat. Contr.*, 46:742–746, 2001. (Cited on p. 71)

[20] J.S. Baras, A. Bensoussan, and M.R. James. Dynamic observers as asymptotic limits of recursive filters: Special cases. *SIAM J. Applied Math.*, 48:1147–1158, 1988. (Cited on p. 205)

[21] G. Besançon. High-gain observation with disturbance attenuation and application to robust fault detection. *Automatica*, 39:1095–1102, 2003. (Cited on p. 1)

[22] A.M.A. Boker and H.K. Khalil. Nonlinear observers comprising high-gain observers and extended Kalman filters. *Automatica*, 49:3583–3590, 2013. (Cited on p. 210)

[23] A.M.A. Boker and H.K. Khalil. Semi-global output feedback stabilization of a class of non-minimum phase nonlinear systems. In *Proc. American Control Conf.*, Washington, DC, 2013. (Cited on p. 210)

[24] R.T. Bupp, D.S. Bernstein, and V.T. Coppola. A benchmark problem for nonlinear control design. *Int. J. Robust and Nonlinear Control*, 8:307–310, 1998. (Cited on p. 199)

[25] K. Busawon, M. Farza, and H. Hammouri. Observer design for a special class of nonlinear systems. *Int. J. Contr.*, 71:405–418, 1998. (Cited on pp. 1, 30)

[26] C.I. Byrnes, F. Celani, and A. Isidori. Omega-limit sets of a class of nonlinear systems that are semiglobally practically stabilized. *Int. J. Robust and Nonlinear Control*, 15:315–333, 2005. (Cited on p. 1)

[27] C.I. Byrnes and A. Isidori. Nonlinear internal models for output regulation. *IEEE Trans. Automat. Contr.*, 49:2244–2247, 2004. (Cited on p. 158)

[28] C.I. Byrnes, F.D. Priscoli, and A. Isidori. *Output Regulation of Uncertain Nonlinear Systems*. Birkhäuser, Boston, MA, 1997. (Cited on pp. 108, 158)

[29] E.F. Camacho and C. Bordons. *Model Predictive Control*. Springer-Verlag, London, second edition, 2007. (Cited on p. 280)

[30] Z. Chen and J. Huang. *Stabilization and Regulation of Nonlinear Systems*. Springer, London, 2015. (Cited on pp. 108, 158)

[31] A. Dabroom and H.K. Khalil. Discrete-time implementation of high-gain observers for numerical differentiation. *Int. J. Contr.*, 72:1523–1537, 1999. (Cited on pp. 281, 309)

[32] A.M. Dabroom and H.K. Khalil. Output feedback sampled-data control of nonlinear systems using high-gain observers. *IEEE Trans. Automat. Contr.*, 46, 2001. (Cited on pp. 2, 282, 311)

[33] E.J. Davison. The robust control of a servomechanism problem for linear time-invariant multivariable systems. *IEEE Trans. Automat. Contr.*, AC-21:25–34, 1976. (Cited on pp. 108, 158)

[34] C. Desoer and Y.-T. Wang. Foundation of feedback theory for nonlinear dynamical systems. *IEEE Trans. Circuits Syst.*, 27:104–123, 1980. (Cited on p. 108)

[35] F. Deza, E. Busvelle, J.P. Gauthier, and J. Rakotopara. High gain estimation for nonlinear systems. *Syst. Contr. Lett.*, 18:295–299, 1992. (Cited on p. 1)

[36] J.C. Doyle and G. Stein. Robustness with observers. *IEEE Trans. Automat. Contr.*, AC-24:607–611, 1979. (Cited on p. 1)

[37] F. Esfandiari and H.K. Khalil. Observer-based design of uncertain systems: Recovering state feedback robustness under matching conditions. In *Proc. Allerton Conf.*, pages 97–106, Monticello, IL, September 1987. (Cited on p. 1)

[38] F. Esfandiari and H.K. Khalil. Observer-based design of uncertain systems: Recovering state feedback robustness under matching conditions. In *Proc. Allerton Conf.*, pages 97–106, Monticello, IL, 1992. (Cited on p. 71)

[39] F. Esfandiari and H.K. Khalil. Output feedback stabilization of fully linearizable systems. *Int. J. Contr.*, 56:1007–1037, 1992. (Cited on pp. 1, 202)

[40] J.A. Farrell and M.M. Polycarpou. *Adaptive Approximation Based Control.* Wiley, Hoboken, NJ, 2006. (Cited on p. 101)

[41] M. Farza, M. M'Saad, M. Triki, and T. Maatougb. High gain observer for a class of non-triangular systems. *Syst. Contr. Lett.*, 60:27–35, 2011. (Cited on pp. 30, 71)

[42] M. Farza, M. Ouederab, R. Ben Abdennourb, and M. M'Saad. High gain observer with updated gain for a class of MIMO nonlinear systems. *Int. J. Contr.*, 84:270–280, 2011. (Cited on p. 30)

[43] H.A. Fertik and C.W. Ross. Direct digital control algorithm with anti-windup feature. *ISA Transactions*, 6:63–68, 1967. (Cited on p. 121)

[44] A.B. Francis and W.M. Wonham. The internal model principle for linear multivariable systems. *J. Appl. Math. Optimiz.*, 2:170–194, 1975. (Cited on pp. 108, 158)

[45] G.F. Franklin, D. Powell, and M.L. Workman. *Digital Control of Dynamic Systems.* Addison Wesley, Reading, MA, 1990. (Cited on pp. 281, 299)

[46] L.B. Freidovich and H.K. Khalil. Lyapunov-based switching control of nonlinear systems using high-gain observers. *Automatica*, 43:150–157, 2007. (Cited on p. 2)

[47] L.B. Freidovich and H.K. Khalil. Performance recovery of feedback-linearization-based designs. *IEEE Trans. Automat. Contr.*, 53:2324–2334, 2008. (Cited on p. 210)

[48] Z. Gao. Active disturbance rejection control: A paradigm shift in feedback control system design. In *Proc. American Control Conf.*, Minneapolis, MN, 2006. (Cited on p. 159)

[49] J.-P. Gauthier and I. Kupka. *Deterministic Observation Theory and Applications.* Cambridge University Press, Cambridge, UK, 2001. (Cited on pp. 1, 18, 30, 127, 205)

[50] J.P. Gauthier, H. Hammouri, and S. Othman. A simple observer for nonlinear systems application to bioreactors. *IEEE Trans. Automat. Contr.*, 37:875–880, 1992. (Cited on pp. 1, 18, 21, 30)

[51] J.P. Gauthier and I.A.K. Kupka. Observability and observers for nonlinear systems. *SIAM J. Contr. Optim.*, 32:975–994, 1994. (Cited on pp. 1, 30)

[52] S.S. Ge and J. Zhang. Neural-network control of nonaffine nonlinear system with zero dynamics by state and output feedback. *IEEE Trans. Neural Networks*, 14:900–918, 2003. (Cited on p. 106)

[53] N.H. Getz and J.K. Hedrick. An internal equilibrium manifold method of tracking for nonlinear nonminimum phase systems. In *Proc. American Control Conf.*, pages 2241–2245, Seattle, WA, 1995. (Cited on p. 43)

[54] J. Gonzalez-Trejo, J. Alvarez Ramirez, and G. Fernandez. Robust control with uncertainty estimation for feedback linearizable systems: Application to control of distillation columns. *J. Process Control*, 9:221–231, 1999. (Cited on p. 1)

[55] H.F. Grip and A. Saberi. High-gain observer design for multi-output systems: Transformation to a canonical form by dynamic output shaping. *Int. J. Robust and Nonlinear Control*, 24:1016–1042, 2014. (Cited on p. 30)

[56] H. Hammouri, G. Bornard, and K. Busawon. High gain observer for structured multi-output nonlinear systems. *IEEE Trans. Automat. Contr.*, 55:987–992, 2010. (Cited on p. 30)

[57] H. Hammouri, B. Targui, and F. Armanet. High gain observer based on a triangular structure. *Int. J. Robust and Nonlinear Control*, 12:497–518, 2002. (Cited on pp. 1, 30)

[58] J. Han. From PID to active disturbance rejection control. *IEEE Trans. Ind. Electron.*, 56:900–906, 2009. (Cited on p. 159)

[59] N. Hovakimyan, E. Lavretsky, and C. Cao. Dynamic inversion for multivariable non-affine-in-control system via time-scale separation. *Int. J. Contr.*, 81:1960–1967, 2008. (Cited on p. 183)

[60] J. Huang. *Nonlinear Output Regulation: Theory and Applications*. SIAM, Philadelphia, PA, 2004. (Cited on pp. 108, 158)

[61] J. Huang and C.F. Lin. On a robust nonlinear servomechanism problem. *IEEE Trans. Automat. Contr.*, 39:1510–1513, 1994. (Cited on p. 158)

[62] L. Imsland, R. Findeisen, E. Bullinger, F. Allgower, and B.A. Foss. A note on stability, robustness and performance of output feedback nonlinear model predictive control. *J. Process Control*, 13:633–644, 2003. (Cited on p. 1)

[63] P.A. Ioannou and J. Sun. *Robust Adaptive Control*. Prentice–Hall, Upper Saddle River, NJ, 1995. (Cited on pp. 73, 77, 80)

[64] A. Isidori. *Nonlinear Control Systems*. Springer-Verlag, Berlin, third edition, 1995. (Cited on pp. 17, 28, 32, 50, 74, 75, 109)

[65] A. Isidori. A remark on the problem of semiglobal nonlinear output regulation. *IEEE Trans. Automat. Contr.*, 42:1734–1738, 1997. (Cited on p. 1)

[66] A. Isidori. *Nonlinear Control Systems II*. Springer-Verlag, London, 1999. (Cited on pp. 162, 210)

[67] A. Isidori. A tool for semiglobal stabilization of uncertain non-minimum-phase nonlinear systems via output feedback. *IEEE Trans. Automat. Contr.*, 45:1817–1827, 2000. (Cited on pp. 1, 162, 195, 210)

[68] A. Isidori and C.I. Byrnes. Output regulation of nonlinear systems. *IEEE Trans. Automat. Contr.*, 35:131–140, 1990. (Cited on p. 158)

[69] A. Isidori, L. Marconi, and L. Praly. Robust design of nonlinear internal models without adaptation. *Automatica*, 48:2409–2419, 2012. (Cited on pp. 108, 158)

[70] A. Isidori, L. Marconi, and A. Serrani. *Robust Autonomous Guidance: An Internal Model Approach*. Springer, London, 2003. (Cited on pp. 108, 158)

[71] M. Janković. Adaptive output feedback control of nonlinear feedback linearizable systems. *Int. J. Adaptive Control and Signal Processing*, 10:1–18, 1996. (Cited on p. 1)

[72] Z.-P. Jiang, D.J. Hill, and Y. Guo. Stabilization and tracking via output feedback for the nonlinear benchmark system. *Automatica*, 34:907–915, 1998. (Cited on p. 1)

[73] H.K. Khalil. Robust servomechanism output feedback controllers for a class of feedback linearizable systems. *Automatica*, 30:1587–1599, 1994. (Cited on pp. 2, 107, 158)

[74] H.K. Khalil. Adaptive output feedback control of nonlinear systems represented by input-output models. *IEEE Trans. Automat. Contr.*, 41:177–188, 1996. (Cited on pp. 2, 106)

[75] H.K. Khalil. *Nonlinear Systems*. Prentice-Hall, Upper Saddle River, NJ, 2nd edition, 1996. (Cited on pp. 84, 152)

[76] H.K. Khalil. On the design of robust servomechanism for minimum phase nonlinear systems. *Int. J. Robust and Nonlinear Control*, 10:339–361, 2000. (Cited on pp. 2, 107, 158)

[77] H.K. Khalil. Universal integral controllers for minimum-phase nonlinear systems. *IEEE Trans. Automat. Contr.*, 45:490–494, 2000. (Cited on p. 158)

[78] H.K. Khalil. *Nonlinear Systems*. Prentice-Hall, Upper Saddle River, NJ, third edition, 2002. (Cited on pp. 22, 37, 38, 40, 53, 54, 58, 69, 70, 84, 85, 100, 115, 124, 125, 140, 147, 152, 153, 171, 185, 191, 220, 221, 222, 227, 242, 248, 250, 251, 291, 295, 304, 306)

[79] H.K. Khalil. Analysis of sampled-data high-gain observers in the presence of measurement noise. *European J. Contr.*, 15:166–176, 2009. (Cited on pp. 299, 311)

[80] H.K. Khalil. *Nonlinear Control*. Pearson, Boston, MA, 2015. (Cited on pp. 3, 4, 6, 7, 8, 17, 21, 30, 32, 42, 47, 50, 52, 53, 58, 67, 71, 100, 109, 110, 140, 162, 170, 173, 174, 180, 181, 193, 196, 197, 199, 204, 223, 239, 244, 266, 307)

[81] H.K. Khalil. Cascade high-gain observer for high-dimensional systems. *Proc. IEEE Conf. on Decision and Control*, 2016. (Cited on p. 236)

[82] H.K. Khalil. Cascade high-gain observers in output feedback control. *Automatica*, 80:110–118, 2017. (Cited on p. 236)

[83] H.K. Khalil and F. Esfandiari. Semiglobal stabilization of a class of nonlinear systems using output feedback. *IEEE Trans. Automat. Contr.*, 38(9):1412–1415, 1993. (Cited on p. 2)

[84] H.K. Khalil and L. Praly. High-gain observers in nonlinear feedback control. *Int. J. Robust and Nonlinear Control*, 24: 993–1015, 2014. (Cited on pp. 9, 10)

[85] H.K. Khalil and S. Priess. Analysis of the use of low-pass filters with high-gain observers. In *10th IFAC Symposium on Nonlinear Control Systems*, Monterey, CA, 2016. (Cited on p. 278)

[86] H.K. Khalil and A. Saberi. Adaptive stabilization of a class of nonlinear systems using high-gain feedback. *IEEE Trans. Automat. Contr.*, AC-32(11):1031–1035, 1987. (Cited on p. 1)

[87] P. Kokotovic, H.K. Khalil, and J. O'Reilly. *Singular Perturbations Methods in Control: Analysis and Design.* SIAM, Philadelphia, PA, 1999. (Cited on pp. 37, 185, 254, 256, 257, 278)

[88] A.J. Krener. The convergence of the extended Kalman filter. In A. Rantzer and C.I. Byrnes, editors, *Directions in Mathematical Systems Theory and Optimization*, pages 173–182. Springer-Verlag, Berlin, 2003. (Cited on p. 205)

[89] P. Krishnamurthy and F. Khorrami. Dynamic high-gain scaling: State and output feedback with application to systems with ISS appended dynamics driven by all states. *IEEE Trans. Automat. Contr.*, 49:2219–2239, 2004. (Cited on p. 1)

[90] M. Krstić, I. Kanellakopoulos, and P. Kokotović. *Nonlinear and Adaptive Control Design.* Wiley-Interscience, New York, 1995. (Cited on pp. 73, 162)

[91] H. Kwakernaak and R. Sivan. *Linear Optimal Control Systems.* Wiley-Interscience, New York, 1972. (Cited on pp. 22, 205)

[92] J. Lee, R. Mukherjee, and H.K. Khalil. Output feedback stabilization of inverted pendulum on a cart in the presence of uncertainty. *Automatica*, 54:146–157, 2015. (Cited on pp. 43, 44, 71, 210)

[93] J. Lee, R. Mukherjee, and H.K. Khalil. Output feedback performance recovery in the presence of uncertainties. *Syst. Contr. Lett.*, 90:31–37, 2016. (Cited on p. 210)

[94] H. Lei and W. Lin. Universal adaptive control of nonlinear systems with unknown growth rate by output feedback. *Automatica*, 42:1783–1789, 2006. (Cited on p. 1)

[95] R. Li and H.K. Khalil. Semiglobal nonlinear output regulation with adaptive conditional servocompensator. *Automatica*, 48:2550–2559, 2012. (Cited on p. 158)

[96] R. Li and H.K. Khalil. On the steady-state error of a nonlinear regulator. *Int. J. Robust and Nonlinear Control*, 23:1869–1879, 2013. (Cited on p. 158)

[97] S. Li, J. Yang, W.-H. Chen, and X. Chen. *Disturbance Observer-Based Control: Methods and Applications.* CRC Press, Boca Raton, FL, 2014. (Cited on p. 159)

[98] W. Lin and C. Qian. Semiglobal robust stabilization of nonlinear systems by partial state and output feedback. In *Proc. IEEE Conf. on Decision and Control*, pages 3105–3110, Tampa, FL, 1998. (Cited on p. 1)

[99] Z. Lin and A. Saberi. Low-and-high gain design technique for linear systems subject to input saturation—a direct method. *Int. J. Robust and Nonlinear Control*, 7:1071–1101, 1997. (Cited on p. 1)

[100] L. Liu, Z. Chen, and J. Huang. Parameter convergence and minimal internal model with an adaptive output regulation problem. *Automatica*, 45:1306–1311, 2009. (Cited on p. 148)

[101] D.G. Luenberger. *Optimization by Vector Space Methods.* Wiley, New York, 1969. (Cited on p. 257)

[102] M. Maggiore and K. Passino. Output feedback tracking: A separation principle approach. *IEEE Trans. Automat. Contr.*, 50:111–117, 2005. (Cited on p. 1)

[103] M.S. Mahmoud and M. Zribi. Output feedback stabilization of systems with nonlinear actuators. *Int. J. Contr.*, 73:187–196, 2000. (Cited on p. 1)

[104] N.A. Mahmoud and H.K. Khalil. Asymptotic regulation of minimum phase nonlinear systems using output feedback. *IEEE Trans. Automat. Contr.*, 41:1402–1412, 1996. (Cited on pp. 2, 158)

[105] N.A. Mahmoud and H.K. Khalil. Robust control for a nonlinear servomechanism problem. *Int. J. Contr.*, 66:779–802, 1997. (Cited on p. 158)

[106] R. Marino and P. Tomei. *Nonlinear Control Design: Geometric, Adaptive & Robust*. Prentice–Hall, London, 1995. (Cited on p. 73)

[107] P. Martin, S. Devasia, and B. Paden. A different look at output tracking: Control of a VTOL aircraft. *Automatica*, 32:101–107, 1996. (Cited on p. 46)

[108] A.Y. Memon and H.K. Khalil. Output regulation of nonlinear systems using conditional servocompensators. *Automatica*, 46:1119–1128, 2010. (Cited on p. 158)

[109] T. Mita. On zeros and responses of linear regulators and linear observers. *IEEE Trans. Automat. Contr.*, AC-22(3):423–428, 1977. (Cited on p. 1)

[110] K.S. Narendra and A.M. Annaswamy. *Stable Adaptive Systems*. Prentice–Hall, Englewood Cliffs, NJ, 1989. (Cited on p. 73)

[111] M.S. Nazrulla and H.K. Khalil. Robust stabilization of non-minimum phase nonlinear systems using extended high gain observers. *IEEE Trans. Automat. Contr.*, 56:802–813, 2011. (Cited on p. 210)

[112] V. Nikiforov. Adaptive nonlinear tracking with complete compensation of unknown disturbances. *European J. Contr.*, 4:132–139, 2010. (Cited on p. 132)

[113] S. Oh and H.K. Khalil. Nonlinear output feedback tracking using high-gain observer and variable structure control. *Automatica*, 33:1845–1856, 1997. (Cited on p. 2)

[114] M.A. Pai. *Power System Stability Analysis by the Direct Method of Lyapunov*. North-Holland, Amsterdam, 1981. (Cited on p. 207)

[115] A. Pavolov, N. van de Wouw, and H. Nijmeijer. *Uniform Output Regulation of Nonlinear Systems: A Convergent Dynamics Approach*. Birkhäuser, Boston, MA, 2006. (Cited on pp. 108, 158)

[116] I.R. Petersen and C.V. Holot. High-gain observers applied to problems in disturbance attenuation, H-infinity optimization and the stabilization of uncertain linear systems. In *Proc. American Control Conf.*, pages 2490–2496, Atlanta, GA, 1988. (Cited on pp. 1, 30)

[117] V.N. Polotskii. On the maximal errors of an asymptotic state identifier. *Automation and Remote Control*, 11:1116–1121, 1979. (Cited on p. 1)

[118] L. Praly. Asymptotic stabilization via output feedback for lower triangular systems with output dependent incremental rate. *IEEE Trans. Automat. Contr.*, 48:1103–1108, 2003. (Cited on p. 1)

[119] L. Praly and Z.P. Jiang. Linear output feedback with dynamic high gain for nonlinear systems. *Syst. Contr. Lett.*, 53:107–116, 2004. (Cited on p. 1)

[120] A. Prasov and H.K. Khalil. A nonlinear high-gain observer for systems with measurement noise in a feedback control framework. *IEEE Trans. Automat. Contr.*, 58:569–580, 2013. (Cited on pp. 261, 273, 278)

[121] A.A. Prasov and H.K. Khalil. Tracking performance of a high-gain observer in the presence of measurement noise. *Int. J. Adaptive Control and Signal Processing*, 30, 2016. (Cited on pp. 276, 278)

[122] F.D. Priscoli. Robust tracking for polynomial plants. In *Proc. European Control Conf.*, pages 369–373, Groningen, The Netherlands, 1993. (Cited on p. 158)

[123] C. Qian. Semi-global stabilization of a class of uncertain nonlinear systems by linear output feedback. *IEEE Trans. Circuits Syst.-II: Express Briefs*, 52, 2005. (Cited on p. 1)

[124] J.B. Rawlings and D.Q. Mayne. *Model Predictive Control: Theory and Design*. Nob Hill Publishing, Madison, WI, 2009. (Cited on p. 280)

[125] N. Rouche, P. Habets, and M. Laloy. *Stability Theory by Lyapunov's Direct Method*. Springer-Verlag, New York, 1977. (Cited on p. 140)

[126] W.J. Rugh. *Linear System Theory*. Prentice–Hall, Upper Saddle River, NJ, second edition, 1996. (Cited on pp. 24, 308)

[127] A. Saberi, B.M. Chen, and P. Sannuti. *Loop Transfer Recovery: Analysis and Design*. Springer, Berlin, 1993. (Cited on pp. 21, 30)

[128] A. Saberi, Z. Lin, and A.R. Teel. Control of linear systems with saturating actuators. *IEEE Trans. Automat. Contr.*, 41:368–378, 1996. (Cited on p. 1)

[129] A. Saberi and P. Sannuti. Observer design for loop transfer recovery and for uncertain dynamical systems. *IEEE Trans. Automat. Contr.*, 35:878–897, 1990. (Cited on pp. 1, 202)

[130] R. Sanfelice and L. Praly. On the performance of high-gain observers with gain adaptation under measurement noise. *IEEE Trans. Automat. Contr.*, 56:2165–2176, 2011. (Cited on p. 261)

[131] S. Sastry and M. Bodson. *Adaptive Control*. Prentice–Hall, Englewood Cliffs, NJ, 1989. (Cited on p. 73)

[132] L. Sciavicco and B. Siciliano. *Modeling and Control of Robot Manipulators*. Springer-Verlag, London, second edition, 2000. (Cited on p. 181)

[133] A. Serrani, A. Isidori, and L. Marconi. Semiglobal nonlinear output regulation with adaptive internal model. *IEEE Trans. Automat. Contr.*, 46:1178–1194, 2001. (Cited on pp. 1, 145, 158)

[134] S. Seshagiri and H.K. Khalil. Output feedback control of nonlinear systems using RBF neural networks. *IEEE Trans. Neural Networks*, 11:69–79, 2000. (Cited on pp. 2, 106)

[135] S. Seshagiri and H.K. Khalil. Robust output feedback regulation of minimum-phase nonlinear systems using conditional integrators. *Automatica*, 41:936–943, 2005. (Cited on pp. 2, 158)

[136] S. Seshagiri and H.K. Khalil. Robust output regulation of minimum phase nonlinear systems using conditional servocompensators. *Int. J. Robust and Nonlinear Control*, 15: 83–102, 2005. (Cited on pp. 2, 158)

[137] D. Seto and A.M. Annaswamy. Adaptive control of nonlinear systems with a triangular structure. *IEEE Trans. Automat. Contr.*, 39:1411–1428, 1994. (Cited on p. 73)

[138] H. Shim, Y.I. Son, and J.H. Seo. Semi-global observer for multi-output nonlinear systems. *Syst. Contr. Lett.*, 42:233–244, 2000. (Cited on p. 1)

[139] H. Shim and A.R. Teel. Asymptotic controllability and observability imply semiglobal practical asymptotic stabilizability by sampled-data output feedback. *Automatica*, 39:441–454, 2003. (Cited on p. 1)

[140] A. Singh and H.K. Khalil. Regulation of nonlinear systems using conditional integrators. *Int. J. Robust and Nonlinear Control*, 15:339–362, 2005. (Cited on pp. 2, 158)

[141] S.N. Singh, R. Zhang, P. Chandler, and S. Banda. Decentralized nonlinear robust control of UAVs in close formation. *Int. J. Robust and Nonlinear Control*, 13:1057–1078, 2003. (Cited on p. 1)

[142] J-J.E. Slotine and W. Li. *Applied Nonlinear Control*. Prentice–Hall, Englewood Cliffs, NJ, 1991. (Cited on p. 73)

[143] V.A. Sobolev. Integral manifolds and decomposition of singularly perturbed systems. *Syst. Contr. Lett.*, 5:169–179., 1984. (Cited on p. 278)

[144] E.D. Sontag. Smooth stabilization implies coprime factorization. *IEEE Trans. Automat. Contr.*, 34:435–443, 1989. (Cited on p. 110)

[145] E.D. Sontag and Y. Wang. On characterizations of the input-to-state stability property. *Syst. Contr. Lett.*, 24:351–359, 1995. (Cited on p. 98)

[146] M.W. Spong, S. Hutchinson, and M. Vidyasagar. *Robot Modeling and Control*. Wiley, Hoboken, NJ, 2005. (Cited on pp. 27, 76, 181)

[147] B. Srinivasan, P. Huguenin, and D. Bonvin. Global stabilization of an inverted pendulum-control theory and experimental verification. *Automatica*, 45:265–269, 2009. (Cited on p. 43)

[148] G. Tao. *Adaptive Control Design and Analysis*. Wiley-Interscience, Hoboken, NJ, 2003. (Cited on p. 73)

[149] A. Teel and L. Praly. Global stabilizability and observability imply semi-global stabilizability by output feedback. *Syst. Contr. Lett.*, 22:313–325, 1994. (Cited on pp. 1, 2, 71)

[150] A. Teel and L. Praly. Tools for semiglobal stabilization by partial state and output feedback. *SIAM J. Contr. Optim.*, 33:1443–1488, 1995. (Cited on pp. 1, 71)

[151] A. Tornambe. Use of asymptotic observers having high-gains in the state and parameter estimation. In *Proc. IEEE Conf. on Decision and Control*, pages 1792–1794, 1988. (Cited on p. 1)

[152] A. Tornambè. Output feedback stabilization of a class of non-minimum phase nonlinear systems. *Syst. Contr. Lett.*, 19:193–204, 1992. (Cited on p. 1)

[153] L.K. Vasiljevic and H.K. Khalil. Error bounds in differentiation of noisy signals by high-gain observers. *Syst. Contr. Lett.*, 57:856–862, 2008. (Cited on pp. 2, 276)

[154] F. Viel, E. Busvelle, and J.P. Gauthier. Stability of polymerization reactors using I/O linearization and a high-gain observer. *Automatica*, 31:971–984, 1995. (Cited on p. 1)

[155] L. Wang, A. Isidori, and H. Su. Output feedback stabilization of nonlinear MIMO systems having uncertain high-frequency gain. *Syst. Contr. Lett.*, 83:1–8, 2015. (Cited on pp. 179, 210)

[156] D.M. Wiberg. *Schaum's Outline of Theory and Problems of State Space and Linear Systems*. McGraw-Hill, New York, 1971. (Cited on pp. 241, 288)

[157] B. Yao and M. Tomizuka. Adaptive robust control of SISO nonlinear a semi-strict feedback form. *Automatica*, 33:893–900, 1997. (Cited on p. 73)

[158] X. Ye. Semiglobal output feedback control of uncertain nonlinear systems with unknown high frequency gain sign. *IEEE Trans. Automat. Contr.*, 45:2402–2405, 2000. (Cited on p. 1)

Index